INFLATION AND STRING THEORY

The past two decades have seen transformative advances in cosmology and string theory. Observations of the cosmic microwave background have revealed strong evidence for a period of inflationary expansion in the very early universe, while new insights about compactifications of string theory have led to a deeper understanding of inflation in a framework that unifies quantum mechanics and general relativity.

Written by two of the leading researchers in the field, this complete and accessible volume provides a modern treatment of inflationary cosmology and its connections to string theory and elementary particle theory. After an up-to-date experimental summary, the authors present the foundations of effective field theory, string theory, and string compactifications, setting the stage for a detailed examination of models of inflation in string theory. Three appendices contain background material in geometry and cosmological perturbation theory, making this a self-contained resource for graduate students and researchers in string theory, cosmology, and related fields.

DANIEL BAUMANN is Reader in Theoretical Physics at the University of Cambridge. He earned his Ph.D. from Princeton University in 2008 and was a postdoctoral researcher at Harvard University and at the Institute for Advanced Study in Princeton.

LIAM MCALLISTER is Associate Professor of Physics at Cornell University. He earned his Ph.D. from Stanford University in 2005 and was a postdoctoral researcher at Princeton University.

CAMBRIDGE MONOGRAPHS ON MATHEMATICAL PHYSICS

General Editors: P. V. Landshoff, D. R. Nelson, S. Weinberg

S. J. Aarseth *Gravitational N-Body Simulations: Tools and Algorithms*[†]
J. Ambjørn, B. Durhuus and T. Jonsson *Quantum Geometry: A Statistical Field Theory Approach*[†]
A. M. Anile *Relativistic Fluids and Magneto-fluids: With Applications in Astrophysics and Plasma Physics*
J. A. de Azcárraga and J. M. Izquierdo *Lie Groups, Lie Algebras, Cohomology and Some Applications in Physics*[†]
O. Babelon, D. Bernard and M. Talon *Introduction to Classical Integrable Systems*[†]
F. Bastianelli and P. van Nieuwenhuizen *Path Integrals and Anomalies in Curved Space*[†]
D. Baumann and L. McAllister *Inflation and String Theory*
V. Belinski and E. Verdaguer *Gravitational Solitons*[†]
J. Bernstein *Kinetic Theory in the Expanding Universe*[†]
G. F. Bertsch and R. A. Broglia *Oscillations in Finite Quantum Systems*[†]
N. D. Birrell and P. C. W. Davies *Quantum Fields in Curved Space*[†]
K. Bolejko, A. Krasiński, C. Hellaby and M-N. Célérier *Structures in the Universe by Exact Methods: Formation, Evolution, Interactions*
D. M. Brink *Semi-Classical Methods for Nucleus–Nucleus Scattering*[†]
M. Burgess *Classical Covariant Fields*[†]
E. A. Calzetta and B.-L. B. Hu *Nonequilibrium Quantum Field Theory*
S. Carlip *Quantum Gravity in 2+1 Dimensions*[†]
P. Cartier and C. DeWitt-Morette *Functional Integration: Action and Symmetries*[†]
J. C. Collins *Renormalization: An Introduction to Renormalization, the Renormalization Group and the Operator-Product Expansion*[†]
P. D. B. Collins *An Introduction to Regge Theory and High Energy Physics*[†]
M. Creutz *Quarks, Gluons and Lattices*[†]
P. D. D'Eath *Supersymmetric Quantum Cosmology*[†]
J. Dereziński and C. Gérard*Mathematics of Quantization and Quantum Fields*
F. de Felice and D. Bini *Classical Measurements in Curved Space-Times*
F. de Felice and C. J. S Clarke *Relativity on Curved Manifolds*[†]
B. DeWitt *Supermanifolds, 2nd edition*[†]
P. G. O. Freund *Introduction to Supersymmetry*[†]
F. G. Friedlander *The Wave Equation on a Curved Space-Time*[†]
J. L. Friedman and N. Stergioulas *Rotating Relativistic Stars*
Y. Frishman and J. Sonnenschein *Non-Perturbative Field Theory: From Two Dimensional Conformal Field Theory to QCD in Four Dimensions*
J. A. Fuchs *Affine Lie Algebras and Quantum Groups: An Introduction, with Applications in Conformal Field Theory*[†]
J. Fuchs and C. Schweigert *Symmetries, Lie Algebras and Representations: A Graduate Course for Physicists*[†]
Y. Fujii and K. Maeda *The Scalar-Tensor Theory of Gravitation*[†]
J. A. H. Futterman, F. A. Handler and R. A. Matzner *Scattering from Black Holes*[†]
A. S. Galperin, E. A. Ivanov, V. I. Ogievetsky and E. S. Sokatchev *Harmonic Superspace*[†]
R. Gambini and J. Pullin *Loops, Knots, Gauge Theories and Quantum Gravity*[†]
T. Gannon *Moonshine beyond the Monster: The Bridge Connecting Algebra, Modular Forms and Physics*[†]
M. Göckeler and T. Schücker *Differential Geometry, Gauge Theories, and Gravity*[†]
C. Gómez, M. Ruiz-Altaba and G. Sierra *Quantum Groups in Two-Dimensional Physics*[†]
M. B. Green, J. H. Schwarz and E. Witten *Superstring Theory Volume 1: Introduction*
M. B. Green, J. H. Schwarz and E. Witten *Superstring Theory Volume 2: Loop Amplitudes, Anomalies and Phenomenology*
V. N. Gribov *The Theory of Complex Angular Momenta: Gribov Lectures on Theoretical Physics*[†]
J. B. Griffiths and J. Podolský *Exact Space-Times in Einstein's General Relativity*[†]
S. W. Hawking and G. F. R. Ellis *The Large Scale Structure of Space-Time*[†]
F. Iachello and A. Arima *The Interacting Boson Model*[†]
F. Iachello and P. van Isacker *The Interacting Boson-Fermion Model*[†]

C. Itzykson and J. M. Drouffe *Statistical Field Theory Volume 1: From Brownian Motion to Renormalization and Lattice Gauge Theory*[†]

C. Itzykson and J. M. Drouffe *Statistical Field Theory Volume 2: Strong Coupling, Monte Carlo Methods, Conformal Field Theory and Random Systems*[†]

G. Jaroszkiewicz *Principles of Discrete Time Mechanics*

C. V. Johnson *D-Branes*[†]

P. S. Joshi *Gravitational Collapse and Spacetime Singularities*[†]

J. I. Kapusta and C. Gale *Finite-Temperature Field Theory: Principles and Applications, 2nd edition*[†]

V. E. Korepin, N. M. Bogoliubov and A. G. Izergin *Quantum Inverse Scattering Method and Correlation Functions*[†]

M. Le Bellac *Thermal Field Theory*[†]

Y. Makeenko *Methods of Contemporary Gauge Theory*[†]

N. Manton and P. Sutcliffe *Topological Solitons*[†]

N. H. March *Liquid Metals: Concepts and Theory*[†]

I. Montvay and G. Münster *Quantum Fields on a Lattice*[†]

L. O'Raifeartaigh *Group Structure of Gauge Theories*[†]

T. Ortín *Gravity and Strings, 2nd edition*

A. M. Ozorio de Almeida *Hamiltonian Systems: Chaos and Quantization*[†]

L. Parker and D. Toms *Quantum Field Theory in Curved Spacetime: Quantized Fields and Gravity*

R. Penrose and W. Rindler *Spinors and Space-Time Volume 1: Two-Spinor Calculus and Relativistic Fields*[†]

R. Penrose and W. Rindler *Spinors and Space-Time Volume 2: Spinor and Twistor Methods in Space-Time Geometry*[†]

S. Pokorski *Gauge Field Theories, 2nd edition*[†]

J. Polchinski *String Theory Volume 1: An Introduction to the Bosonic String*[†]

J. Polchinski *String Theory Volume 2: Superstring Theory and Beyond*[†]

J. C. Polkinghorne *Models of High Energy Processes*[†]

V. N. Popov *Functional Integrals and Collective Excitations*[†]

L. V. Prokhorov and S. V. Shabanov *Hamiltonian Mechanics of Gauge Systems*

A. Recknagel and V. Schomerus *Boundary Conformal Field Theory and the Worldsheet Approach to D-Branes*

R. J. Rivers *Path Integral Methods in Quantum Field Theory*[†]

R. G. Roberts *The Structure of the Proton: Deep Inelastic Scattering*[†]

C. Rovelli *Quantum Gravity*[†]

W. C. Saslaw *Gravitational Physics of Stellar and Galactic Systems*[†]

R. N. Sen *Causality, Measurement Theory and the Differentiable Structure of Space-Time*

M. Shifman and A. Yung *Supersymmetric Solitons*

H. Stephani, D. Kramer, M. MacCallum, C. Hoenselaers and E. Herlt *Exact Solutions of Einstein's Field Equations, 2nd edition*[†]

J. Stewart *Advanced General Relativity*[†]

J. C. Taylor *Gauge Theories of Weak Interactions*[†]

T. Thiemann *Modern Canonical Quantum General Relativity*[†]

D. J. Toms *The Schwinger Action Principle and Effective Action*[†]

A. Vilenkin and E. P. S. Shellard *Cosmic Strings and Other Topological Defects*[†]

R. S. Ward and R. O. Wells, Jr *Twistor Geometry and Field Theory*[†]

E. J. Weinberg *Classical Solutions in Quantum Field Theory: Solitons and Instantons in High Energy Physics*

J. R. Wilson and G. J. Mathews *Relativistic Numerical Hydrodynamics*[†]

[†] Available in paperback

Inflation and String Theory

DANIEL BAUMANN
University of Cambridge

LIAM MCALLISTER
Cornell University

CAMBRIDGE
UNIVERSITY PRESS

CAMBRIDGE
UNIVERSITY PRESS

University Printing House, Cambridge CB2 8BS, United Kingdom

Cambridge University Press is part of the University of Cambridge.

It furthers the University's mission by disseminating knowledge in the pursuit of education, learning and research at the highest international levels of excellence.

www.cambridge.org
Information on this title: www.cambridge.org/9781107089693

© D. Baumann & L. McAllister 2015

First published 2015

A catalogue record for this publication is available from the British Library

ISBN 978-1-107-08969-3 Hardback

To Anna, Fritz, and Julian

and

to Josephine, Aelwen, and Vala

Contents

Preface

The past two decades of advances in observational cosmology have brought about a revolution in our understanding of the universe. Observations of type Ia supernovae [1, 2], measurements of temperature fluctuations in the cosmic microwave background (CMB) – particularly by the *Wilkinson Microwave Anisotropy Probe* (WMAP) [3–7] and the *Planck* satellite [8–10] – and maps of the distribution of large-scale structure (LSS) [11] have established a standard model of cosmology, the ΛCDM model. This is a universe filled with 68% dark energy, 27% dark matter, and only 5% ordinary atoms [8]. There is now decisive evidence that large-scale structures were formed via gravitational instability of primordial density fluctuations, and that these initial perturbations originated from quantum fluctuations [12–17], stretched to cosmic scales during a period of inflationary expansion [18–20]. However, the microphysical origin of inflation remains a mystery, and it will require a synergy of theory and observations to unlock it.

In the standard cosmology without inflation, causal signals travel a finite distance between the time of the initial singularity and the time of formation of the first neutral atoms. However, the CMB anisotropies display vivid correlations on scales larger than this distance. This causality puzzle is known as the *horizon problem*. The horizon problem is resolved if the early universe went through an extended period of *inflationary expansion*, i.e. expansion at a nearly constant rate, with

$$|\dot{H}| \ll H^2 \,,$$

where $H \equiv \dot{a}/a$ is the Hubble parameter associated with a Friedmann–Robertson–Walker spacetime,

$$ds^2 = -dt^2 + a^2(t)d\boldsymbol{x}^2 \,.$$

Because space expands quasi-exponentially during inflation, $a(t) \propto e^{Ht}$, homogeneous initial conditions on subhorizon scales are stretched to apparently acausal superhorizon scales. Besides explaining the overall homogeneity of the universe, inflation also creates small primordial inhomogeneities, which eventually provide the seeds for large-scale structures. These perturbations are inevitable in a quantum-mechanical treatment of inflation: viewed as a quantum field, the expansion rate H experiences local zero-point fluctuations, $\delta H(t, \boldsymbol{x})$, which lead to spatial variations in the density after inflation, $\delta\rho(t, \boldsymbol{x})$. If inflation is correct, then CMB observations are probing the quantum origin of structure in the universe. By measuring the statistical properties of the CMB anisotropies we learn about the physics of inflation and about the precise mechanism that created the primordial seed fluctuations.

In this book we will describe two intertwined approaches to the physics of inflation: from the bottom up in effective field theory (EFT), and from the top down in string theory.

We speak of an *effective theory* when we do not resolve the small-scale (or high-energy) details of a more fundamental theory. Often this coarse-graining is done so automatically that it is not explicitly mentioned: for instance, we describe fluid dynamics and thermodynamics without reference to atoms, and computations of atomic spectra are in turn insensitive to the quark substructure of nucleons. Reasoning in terms of theories valid up to a critical energy scale is also how the history of particle physics developed, long before Wilson formalized the concept of effective theories. Effective theories are particularly useful when the full theory is unknown, or is specified but is not computable. In that case, one parameterizes the unknown physics associated with degrees of freedom at a high energy scale Λ by a collection of non-renormalizable interactions in the EFT, known as irrelevant interactions. At low energies, irrelevant interactions are suppressed by powers of E/Λ. In the limit $E/\Lambda \to 0$, the high-scale degrees of freedom *decouple*. However, in some contexts a low-energy observable is strongly affected by irrelevant interactions: such an observable is termed *ultraviolet (UV) sensitive*. As we shall explain, inflation is an ultraviolet-sensitive phenomenon.

The ultraviolet behavior of gravity is a foundational question for cosmology. To understand the nature of general relativity at high energies, we recall that the interactions dictated by the Einstein–Hilbert action can be encoded in Feynman rules, just as in ordinary quantum field theory (see [21] for a modern perspective). The coupling strength is set by Newton's constant G, which has negative mass dimension, so the interaction becomes stronger at higher energies. Moreover, when divergences do arise, they cannot be absorbed by renormalization of the terms in the classical Einstein–Hilbert Lagrangian; on dimensional grounds, the factors of the gravitational coupling from graviton loops must be offset by additional derivatives compared with the classical terms. General relativity is therefore non-renormalizable,[1] and for energies above the Planck scale,

$$M_P \equiv \sqrt{\frac{\hbar c}{G}} = 1.2 \times 10^{19} \, \text{GeV}/c^2 \,,$$

the theory stops making sense as a quantum theory: it violates unitarity. The conservative interpretation of this finding is that new physics has to come into play at some energy below the Planck scale, and any quantum field theory that

---

[1] Pure Einstein gravity is free of one-loop divergences, but diverges at two loops. Gravitational theories including matter fields typically diverge at one loop, except in supersymmetric cases [21].

is coupled to gravity should then be interpreted as an effective theory valid at energies below the Planck scale. This is precisely what happens in string theory: strings of characteristic size ℓ_s cut off the divergences in graviton scattering at energies of order $1/\ell_s$, where the extended nature of the string becomes important. The result is a finite quantum theory of gravity, whose long-wavelength description, at energies $E \ll 1/\ell_s$, is an effective quantum field theory that includes gravity, and whose non-renormalizable interactions include terms suppressed by the Planck scale (or the string scale). String theory therefore provides an internally consistent framework for studying quantum fields coupled to general relativity.

A striking feature of effective theories that support inflation is that they are sensitive to Planck-suppressed interactions: an otherwise successful model of inflation can be ruined by altering the spectrum and interactions of Planck-scale degrees of freedom. In *every* model of inflation, the duration of the inflationary expansion is affected by at least a small number of non-renormalizable interactions suppressed by the Planck scale. In a special class of scenarios called *large-field* models, an infinite series of interactions, of arbitrarily high dimension, affect the dynamics; this corresponds to extreme sensitivity to Planck-scale physics. The universal sensitivity of inflation to Planck-scale physics implies that a treatment in a theory of quantum gravity is required in order to address critical questions about the inflationary dynamics. This is the cardinal motivation for pursuing an understanding of inflation in string theory.

A primary subject of this book is the challenge of realizing inflationary dynamics in string theory (recommended reviews on the subject include [22–33]). Let us set the stage for our discussion by outlining the range of gains that can be expected from this undertaking.

The most conservative goal of studies of inflation in string theory is to place field-theoretic models of inflation on a firmer logical footing, giving controlled computations of quantum gravity corrections to these models. In particular, ultraviolet completion can clarify and justify symmetry assumptions made in the EFT approach. For example, realizing chaotic inflation [34] through axion monodromy in string theory [35] gives a microphysical understanding of the shift symmetry, $\phi \mapsto \phi + const.$, that ensures radiative stability of the low-energy EFT. Inflationary models relying on the shift symmetries of axions in string theory – variants of "natural inflation" [36] – have provided one of the best-controlled paths to ultraviolet-complete scenarios yielding significant gravitational waves. In favorable cases, the embedding into quantum gravity can also entail small modifications of the theory that lead to additional observational signatures. For example, in axion monodromy inflation nonperturbative corrections introduce modulations of the power spectrum [37] and the bispectrum [38]. This is an example where the structure of the ultraviolet completion could potentially be inferred from correlated signatures.

String theory is a far more constrained framework than effective field theory, and some effective theories that appear consistent at low energies do not admit ultraviolet completions in quantum gravity. Enforcing the restrictions imposed by ultraviolet completion winnows the possible models, leading to improved predictivity. For example, the Dirac–Born–Infeld (DBI) scenario [39] may be viewed as a special case of k-inflation [40]. While most versions of k-inflation are radiatively unstable, string theory makes it possible to control an infinite series of higher-derivative terms. In this case, a higher-dimensional symmetry significantly restricts the form of the four-dimensional effective action. Unlike its field theory counterpart, the observational signatures of DBI inflation are correspondingly specific [41].

String theory has also been an important source of inspiration for the development of novel effective field theories. The geometric perspective afforded by compactifications, and by D-branes moving inside them, complements the more algebraic tools used to construct effective theories in particle physics. The effective theories in D-brane inflation [42, 43], DBI inflation [39], fiber inflation [44], and axion monodromy inflation [35], for example, all exist in their own right as low-energy theories, but would likely have gone undiscovered without the approach provided by string theory. Generating effective theories from the top down in string theory also leads to modified notions of what constitutes a natural inflationary model, or a minimal one. Although we are very far from a final understanding of naturalness in string theory, one broad characteristic of existing geometric constructions is the presence of many light scalar fields, the *moduli* of the compactification. Moduli play a central role in inflation, and can affect both the background evolution and the perturbations. While theories with many "unnecessary" fields might be considered non-minimal in field-theoretic model-building, they are extremely common in string theory.

The boldest hope for the use of string theory in cosmology is that string theory will open entirely new dynamical realms that cannot be described in any effective quantum field theory with a finite number of fields, and that the resulting cosmic histories will avoid or overcome the limitations of contemporary models. While this enticing prospect has inspired work in string cosmology for more than two decades, in our opinion string theory is not yet understood at the level required for such a dramatic step. Even the low-energy effective actions governing the interactions of massless string states in non-supersymmetric vacua are not adequately characterized at present, while computing dynamics driven by the full tower of massive strings is a distant dream. Fundamental advances in the understanding of time-dependent solutions of string theory with string scale curvatures will be required if we are to move outside the aegis of the effective theory for the massless modes. In this book we will restrict our attention to conservative applications of string theory to the study of inflation: we will survey the substantial literature in which string theory underpins or informs inflationary effective theories, but does not replace them outright.

The task of making predictions in string theory is overshadowed by the problem of the *landscape*, i.e. the fact that string theory has an astronomical number of vacua (see [45] for a review). Although the dynamics that populates the landscape is poorly understood, false vacuum eternal inflation seems to be an unavoidable consequence. The cosmological constant problem, the question of pre-inflationary initial conditions, and the challenge of defining a probability measure for eternal inflation are all facets of the fundamental problem of understanding the landscape and making predictions within it. The number of vacua is too large for enumeration to be a realistic possibility [46], but it does not follow that in the landscape, "everything goes." Instead, there seem to exist strong structural constraints on the properties of the vacua in the landscape. For example, axion decay constants appear to be smaller than the Planck mass in all computationally controllable vacua [47, 48]. As we will see, this has important consequences for inflationary model-building in the context of string theory. Moreover, all four-dimensional de Sitter vacua in supersymmetric string theories are metastable, essentially because ten-dimensional Minkowski space is supersymmetric and therefore has zero energy, while a de Sitter solution has positive vacuum energy. Constructing a metastable de Sitter solution is much more difficult than finding a supersymmetric vacuum, and correspondingly, determining the prevalence of de Sitter vacua is far more subtle than counting supersymmetric solutions. In fact, de Sitter solutions appear to be exponentially sparse in comparison to unstable saddle points [49]. The formidable challenges of constructing and surveying the landscape compel us to understand dynamical selection effects in the early universe, but we have yet to see the first glimmering of a solution.

The organization of this book is as follows: in Chapter 1, we define inflation as an extended period of quasi-de Sitter evolution, and show how quantum fluctuations during this era lead to primordial density fluctuations and anisotropies in the CMB. We review the current observational evidence in favor of the inflationary hypothesis. In Chapter 2, we discuss the effective field theory approach to the physics of inflation. We explain why the effective theories supporting inflation are unusually sensitive to UV physics, and highlight the importance of symmetries for the radiative stability of inflationary models. In Chapter 3, we provide the groundwork for a discussion of inflation in string theory. We first give a brief overview of string theory, emphasizing those aspects that are particularly relevant for research in string cosmology. We examine string compactifications, discuss some leading mechanisms for moduli stabilization, and critically analyze proposals for metastable de Sitter vacua. In Chapter 4, we outline how inflation can arise in this context. In Chapter 5, we provide a more detailed discussion of several classes of inflationary models in string theory. We end, in Chapter 6, by describing some challenges and opportunities for the field.

In an effort to make this book self-contained, and accessible for a reader who is entering the field, we have included extensive background material in the appendices. In Appendix A, we collect mathematical concepts, definitions, and results that will be helpful for following the discussion in Chapters 3–5. In Appendix B, we present the effective theory of adiabatic fluctuations during inflation [50, 51]. In Appendix C, we introduce cosmological perturbation theory and derive the primordial perturbations from inflation.

We are indebted to our colleagues and collaborators for sharing their insights on the material presented in this book. Special thanks go to Peter Adshead, Nima Arkani-Hamed, Valentin Assassi, Thomas Bachlechner, Neil Barnaby, Rachel Bean, Cliff Burgess, Anthony Challinor, Xingang Chen, Miranda Cheng, David Chernoff, Michele Cicoli, Joseph Conlon, Paolo Creminelli, Sera Cremonini, Csaba Csáki, Anne Davies, Mafalda Dias, Anatoly Dymarsky, Richard Easther, Raphael Flauger, Jonathan Frazer, Daniel Green, Michael Green, Arthur Hebecker, Shamit Kachru, Renata Kallosh, Marc Kamionkowski, Igor Klebanov, Eiichiro Komatsu, Hayden Lee, Andrei Linde, Connor Long, Juan Maldacena, David Marsh, Paul McGuirk, Alberto Nicolis, Enrico Pajer, Hiranya Peiris, Maxim Perelstein, Rafael Porto, Fernando Quevedo, Sébastien Renaux-Petel, Raquel Ribeiro, David Seery, Leonardo Senatore, Paul Shellard, Eva Silverstein, Marko Simonović, David Spergel, Paul Steinhardt, Simon Su, Andrew Tolley, David Tong, Sandip Trivedi, Henry Tye, Erik Verlinde, Herman Verlinde, Filippo Vernizzi, Yi Wang, Scott Watson, Alexander Westphal, Timm Wrase, Gang Xu, and Matias Zaldarriaga.

We are grateful to Valentin Assassi, Marcus Berg, Miranda Cheng, Michele Cicoli, Joseph Conlon, Daniel Green, Hayden Lee, Emil Martinec, Enrico Pajer, Fernando Quevedo, John Stout, Amir Tajdini, and Marco Zagermann for comments on the draft, and we are particularly indebted to Alexander Westphal for extensive corrections.

Finally, we thank our editors, Vince Higgs and Katherine Law of Cambridge University Press, for their guidance and support.

D.B. gratefully acknowledges support from the European Research Council (ERC STG grant 279617), the Science and Technology Facilities Council (STFC), and the Centre for Theoretical Cosmology in Cambridge. L.M. is grateful for support provided by the National Science Foundation under grant PHY-0757868, by an NSF CAREER award, and by a Simons Fellowship.

Notation and conventions

Throughout this book, we will employ natural units with $\hbar = c \equiv 1$. Moreover, the reduced Planck mass,

$$M_{\mathrm{pl}}^{-2} \equiv 8\pi G = \left(2.4 \times 10^{18}\,\mathrm{GeV}\right)^{-2} ,$$

is often set equal to one.

Our metric signature is mostly plus, $(-+++\cdots)$. We use t for physical time and τ for conformal time. We denote ten-dimensional spacetime coordinates by X^M, four-dimensional spacetime coordinates by x^μ, three-dimensional spatial coordinates by x^i, and three-dimensional vectors by $\boldsymbol{x}$. The coordinates of extra dimensions are y^m. Worldsheet coordinates of strings and branes are σ^a. The spacetime metric in ten dimensions is G_{MN}, while for the four-dimensional counterpart we use $g_{\mu\nu}$. The metric of the three noncompact spatial directions is g_{ij}, while the metric of the six-dimensional internal space is g_{mn}. The worldsheet metric is h_{ab}. The notation $(\partial\phi)^2$ means $g^{\mu\nu}\partial_\mu\phi\partial_\nu\phi$ or $G^{MN}\partial_M\phi\partial_N\phi$, depending on the context.

The letter π stands both for $3.14159\cdots$ and for the Goldstone boson of spontaneously broken time translations. We use $\mathcal{R}$ (not ζ) for the curvature perturbation in comoving gauge. Our Fourier convention is

$$\mathcal{R}_{\boldsymbol{k}} = \int \mathrm{d}^3 x\, \mathcal{R}(\boldsymbol{x})\, e^{i\boldsymbol{k}\cdot\boldsymbol{x}} .$$

The power spectrum for a statistically homogeneous field is defined by

$$\langle \mathcal{R}_{\boldsymbol{k}}\mathcal{R}_{\boldsymbol{k}'}\rangle = (2\pi)^3 P_{\mathcal{R}}(k)\delta(\boldsymbol{k}+\boldsymbol{k}') .$$

We also use the dimensionless power spectrum

$$\Delta_{\mathcal{R}}^2(k) \equiv \frac{k^3}{2\pi^2} P_{\mathcal{R}}(k) .$$

The Hubble slow-roll parameters are

$$\varepsilon \equiv -\frac{\dot{H}}{H^2} , \qquad \tilde{\eta} \equiv \frac{\dot{\varepsilon}}{H\varepsilon} ,$$

where overdots stand for derivatives with respect to physical time t. The potential slow-roll parameters are

$$\epsilon \equiv \frac{M_{\text{pl}}^2}{2}\left(\frac{V'}{V}\right)^2 \, , \qquad \eta \equiv M_{\text{pl}}^2\frac{V''}{V} \, ,$$

where primes are derivatives with respect to the inflaton ϕ, and $V(\phi)$ is the potential energy density.

We define the string length and the string mass, respectively, as

$$\ell_{\text{s}}^2 \equiv \alpha' \, , \qquad M_{\text{s}}^2 \equiv \frac{1}{\alpha'} \, ,$$

where α' is the Regge slope. The ten-dimensional gravitational coupling is

$$2\kappa^2 = (2\pi)^7(\alpha')^4 \, .$$

Beware of factors of 2π in alternative definitions of these quantities in the literature.

1

Inflation: theory and observations

A fundamental observational fact about our universe is that on large scales it is well-described by the spatially flat Friedmann–Robertson–Walker (FRW) metric

$$ds^2 = -dt^2 + a^2(t)\,d\boldsymbol{x}^2 . \tag{1.1}$$

In Section 1.1, we first explain why the homogeneity, isotropy, and flatness of the universe encoded in (1.1) are puzzling in the standard cosmology. We then show how an early phase of quasi-de Sitter evolution drives the primordial universe towards these conditions, even if it started in an inhomogeneous, anisotropic, and curved initial state. In Section 1.2, we argue that quantum fluctuations during inflation are the origin of all structure in the universe, and we derive the power spectra of scalar and tensor fluctuations. In Section 1.3, we describe the main cosmological observables, which are used, in Section 1.4, to obtain constraints on the inflationary parameters. We then review recent experimental results. Finally, in Section 1.5, we discuss future prospects for testing the physics of inflation with cosmological observations.

1.1 Horizon problem

1.1.1 The particle horizon

To discuss the causal structure of the FRW spacetime, we write the metric (1.1) in terms of *conformal time* τ:

$$ds^2 = a^2(\tau)\left[-d\tau^2 + d\boldsymbol{x}^2\right], \tag{1.2}$$

so that the maximal *comoving distance* $|\Delta\boldsymbol{x}|$ that a particle can travel between times τ_1 and $\tau_2 = \tau_1 + \Delta\tau$ is simply $|\Delta\boldsymbol{x}| = \Delta\tau$, for any $a(\tau)$. In the standard Big Bang cosmology, the expansion at early times is driven by the energy density of radiation, and by tracing the evolution backward one finds that $a \to 0$ at sufficiently early times, and the spacetime becomes singular at this point. We choose coordinates so that the initial singularity is at $t = 0$. At some time $t > 0$, the maximal comoving distance a particle can have traversed since the initial singularity (also known as the *particle horizon*) is given by

$$\Delta\tau = \int_0^t \frac{\mathrm{d}t'}{a(t')} = \int_{-\infty}^{\ln a(t)} \frac{\mathrm{d}\ln a}{aH}, \qquad \text{where} \quad H \equiv \frac{1}{a}\frac{da}{dt}. \qquad (1.3)$$

During the standard Big Bang evolution, $\ddot{a} < 0$ and the comoving Hubble radius $(aH)^{-1} = (\dot{a})^{-1}$ grows with time. The integral in (1.3) is therefore dominated by the contributions from late times. This leads to the so-called *horizon problem*. The amount of conformal time that elapses between the singularity and the formation of the cosmic microwave background (an event known as recombination) is much smaller than the conformal time between recombination and today (see Fig. 1.1). Quantitatively, one finds that points in the CMB that are separated by more than one degree were never in causal contact, according to the standard cosmology: their past light cones do not overlap before the spacetime is terminated by the initial singularity. Yet their temperatures are observed to be the same, to one part in 10^4. Moreover, the observed temperature fluctuations are *correlated* on what seem to be superhorizon scales at recombination. Not only must we explain why the CMB is so uniform, we must also explain why its small fluctuations are correlated on apparently acausal scales.

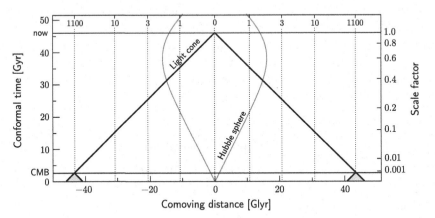

Fig. 1.1 Spacetime diagram illustrating the horizon problem in comoving coordinates (figure adapted from [52]). The dotted vertical lines correspond to the worldlines of comoving objects. We are the central worldline. The current redshifts of the comoving galaxies are labeled on each worldline. All events that we currently observe are on our past light cone. The intersection of our past light cone with the spacelike slice labeled CMB corresponds to two opposite points on the CMB surface of last-scattering. The past light cones of these points, shaded gray, do not overlap, so the points appear never to have been in causal contact.

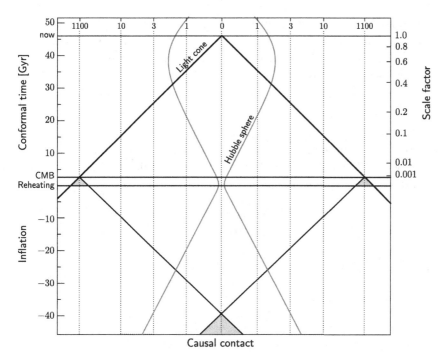

Fig. 1.2 Inflationary solution to the horizon problem. The comoving Hubble sphere shrinks during inflation and expands during the conventional Big Bang evolution (at least until dark energy takes over). Conformal time during inflation is negative. The spacelike singularity of the standard Big Bang is replaced by the reheating surface: rather than marking the beginning of time, $\tau = 0$ now corresponds to the transition from inflation to the standard Big Bang evolution. All points in the CMB have overlapping past light cones and therefore originated from a causally connected region of space.

1.1.2 Cosmic inflation

To address the horizon problem, we may postulate that the comoving Hubble radius was decreasing in the early universe, so that the integral in (1.3) is dominated by the contributions from early times. This introduces an additional span of conformal time between the singularity and recombination (see Fig. 1.2): in fact, conformal time now extends to negative values. If the period of decreasing comoving Hubble radius is sufficiently prolonged, all points in the CMB originate from a causally connected region of space. The observed correlations can therefore result from ordinary causal processes at early times.

In an expanding universe, a shrinking comoving Hubble sphere implies

$$\frac{d}{dt}(aH)^{-1} = -\frac{1}{a}\left[\frac{\dot{H}}{H^2} + 1\right] < 0 \qquad \Rightarrow \qquad \varepsilon \equiv -\frac{\dot{H}}{H^2} < 1 \,. \qquad (1.4)$$

We will take the slow evolution of the Hubble parameter, $\varepsilon < 1$, as our definition of inflation. This definition includes, but is not limited to, the dynamics of a slowly rolling scalar field (see Section 2.2.1). In the de Sitter limit, $\varepsilon \to 0$, the space grows exponentially,

$$a(t) \propto e^{Ht} , \qquad (1.5)$$

with $H \approx const.$

Inflationary expansion requires a somewhat unconventional matter content. In a spatially flat FRW universe supported by a perfect fluid, the Einstein equations lead to the Friedmann equations,

$$3M_{\mathrm{pl}}^2 H^2 = \rho , \qquad (1.6)$$

$$6M_{\mathrm{pl}}^2(\dot{H} + H^2) = -(\rho + 3P) , \qquad (1.7)$$

where ρ and P are the energy density and pressure of the fluid. Combining (1.6) and (1.7), we find

$$2M_{\mathrm{pl}}^2 \dot{H} = -(\rho + P) , \qquad (1.8)$$

and hence

$$\varepsilon = \frac{3}{2}\left(1 + \frac{P}{\rho}\right) . \qquad (1.9)$$

Inflation therefore occurs when $P < -\frac{1}{3}\rho$, corresponding to a violation of the strong energy condition (SEC), which for a perfect fluid states that $\rho + P \geq 0$ and $\rho + 3P \geq 0$. One simple energy source that can drive inflation is a positive potential energy density of a scalar field with negligible kinetic energy, but we will encounter a range of alternative mechanisms.

1.2 Primordial perturbations

With the new cosmology the universe must have been started off in some very simple way. What, then, becomes of the initial conditions required by dynamical theory? Plainly there cannot be any, or they must be trivial. We are left in a situation which would be untenable with the old mechanics. If the universe were simply the motion which follows from a given scheme of equations of motion with trivial initial conditions, it could not contain the complexity we observe. Quantum mechanics provides an escape from the difficulty. It enables us to ascribe the complexity to the quantum jumps, lying outside the scheme of equations of motion. The quantum jumps now form the uncalculable part of natural phenomena, to replace the initial conditions of the old mechanistic view.

P. A. M. Dirac [53].

Inflation not only explains the homogeneity of the universe, but also provides a mechanism to create the primordial inhomogeneities required for structure formation [12–17]. This process happens automatically when we treat the inflationary de Sitter phase quantum mechanically. Here, we briefly sketch the quantum generation of primordial fluctuations. We also present the modern view of inflation as a symmetry-breaking phenomenon [50, 51]. For more details, see Appendices B and C.

1.2.1 Goldstone action

By definition, inflation is a transient phase of accelerated expansion, corresponding approximately, but not exactly, to a de Sitter solution. In order for inflation to end, the time-translation invariance present in an eternal de Sitter spacetime must be broken. The slow evolution of the Hubble parameter $H(t)$ serves as a clock that measures the progress of inflation, breaking time-translation invariance and defining a preferred time slicing of the spacetime. The isometries of de Sitter space, $SO(4,1)$, are spontaneously broken down to just spatial rotations and translations. It is often useful to think of the time slicing as being defined by the time-dependent expectation values $\psi_m(t)$ of one or more bosonic fields ψ_m (see Fig. 1.3).

As with spontaneously broken symmetries in flat-space quantum field theory (see e.g. [54]), the broken symmetry is *nonlinearly realized* by a Goldstone boson. Focusing on symmetry breaking and on the physics of the Goldstone boson allows a model-insensitive description of fluctuations during inflation [51]. In particular, we can defer consideration of the dynamics that created the background evolution $H(t)$, though ultimately we will return to explaining the background.

The Goldstone boson associated with the spontaneous breaking of time translation invariance is introduced as a spacetime-dependent transformation along the direction of the broken symmetry, i.e. as a spacetime-dependent shift of the time coordinate [50],

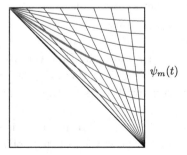

$\psi_m(t)$

Fig. 1.3 Time-dependent background fields $\psi_m(t)$ introduce a preferred time slicing of de Sitter space.

$$U(t, \boldsymbol{x}) \equiv t + \pi(t, \boldsymbol{x}) \ . \tag{1.10}$$

The Goldstone mode π parameterizes *adiabatic fluctuations* of the fields ψ_m, i.e. perturbations corresponding to a common, local shift in time of the homogeneous fields,

$$\delta\psi_m(t, \boldsymbol{x}) \equiv \psi_m\big(t + \pi(t, \boldsymbol{x})\big) - \psi_m(t) \ . \tag{1.11}$$

The Einstein equations couple the Goldstone boson π to metric fluctuations $\delta g_{\mu\nu}$. A convenient gauge for describing these fluctuations is the *spatially flat gauge*, where the spatial part of the metric is unperturbed,

$$g_{ij} = a^2(t)\,\delta_{ij} \ . \tag{1.12}$$

The remaining metric fluctuations δg_{00} and δg_{0i} are related to π by the Einstein constraint equations. The dynamics of the coupled Goldstone-metric system can therefore be described by π alone.

A second description of the same physics is sometimes convenient, especially in the cosmological context. First, we note that, for purely adiabatic fluctuations, we can perform a time reparameterization that removes all matter fluctuations, $\delta\psi_m \mapsto 0$. This takes us to *comoving gauge*, where the field π has been "eaten" by the metric $g_{\mu\nu}$. The spatial part of the metric can now be written as

$$g_{ij} = a^2(t)\,e^{2\mathcal{R}(t,x)}\,\delta_{ij} \ , \tag{1.13}$$

where $\mathcal{R}$ is called the *comoving curvature perturbation*. The other components of the metric are related to $\mathcal{R}$ by the Einstein constraint equations (see Appendix C). The relationship between π (in spatially flat gauge) and $\mathcal{R}$ (in comoving gauge) is

$$\mathcal{R} = -H\pi + \cdots , \tag{1.14}$$

where the ellipsis denotes terms that are higher order in π. This links the comoving curvature perturbation $\mathcal{R}$ with the Goldstone boson π of spontaneous symmetry breaking during inflation [55, 56].

The Goldstone mode π exists in every model of inflation. In single-field inflation, π is the unique fluctuation mode [51], while in multi-field inflation, additional light fields can contribute to $\mathcal{R}$: see Appendix B. As we will see in Chapter 4, string theory strongly motivates considering scenarios in which multiple fields are light during inflation. However, from a purely bottom-up perspective, extra light fields during inflation are not required by present observations, and in this section we will focus on the minimal case of a single light field.

One can learn a great deal about the CMB perturbations by studying the Goldstone boson fluctuations alone. The physics of the Goldstone boson is described by the low-energy effective action for π, which can be obtained by writing down the most general Lorentz-invariant action for the field $U \equiv t + \pi$:

$$S = \int d^4x \sqrt{-g} \, \mathcal{L}[U, (\partial_\mu U)^2, \Box U, \cdots] \, . \qquad (1.15)$$

The action (1.15) is manifestly invariant under spatial diffeomorphisms, but because π transforms nonlinearly under time translations, one says that time translation symmetry is nonlinearly realized in (1.15). Expanding (1.15) in powers of π and derivatives gives the effective action for the Goldstone mode. We derive the Goldstone action in detail in Appendix B, via an alternative geometric approach [50, 51], and present only the main results here. At quadratic order in π, and to leading order in derivatives, one finds (cf. Eq. (B.77))

$$S_\pi^{(2)} = \int d^4x \sqrt{-g} \, \frac{M_{\rm pl}^2 |\dot{H}|}{c_s^2} \left[\dot{\pi}^2 - \frac{c_s^2}{a^2} (\partial_i \pi)^2 + 3\varepsilon H^2 \pi^2 \right] \, , \qquad (1.16)$$

where $(\partial_i \pi)^2 \equiv \delta^{ij} \partial_i \pi \partial_j \pi$. Since Lorentz symmetry is broken by the time-dependence of the background, we have the possibility of a nontrivial *speed of sound* c_s; standard slow-roll inflation (see Section 2.2.1) is recovered for $c_s = 1$. The field π has a small mass term, which arises from the mixing between π and the metric fluctuations. Using (1.14), we can write (1.16) in terms of the curvature perturbation $\mathcal{R}$,

$$S_\mathcal{R}^{(2)} = \frac{1}{2} \int d^4x \, a^3 \, y^2(t) \left[\dot{\mathcal{R}}^2 - \frac{c_s^2}{a^2} (\partial_i \mathcal{R})^2 \right] \, , \qquad (1.17)$$

where

$$y^2 \equiv 2 M_{\rm pl}^2 \frac{\varepsilon}{c_s^2} \, . \qquad (1.18)$$

The field $\mathcal{R}$ is therefore massless, implying – as we shall see – that it is conserved on superhorizon scales [55].

For simplicity, we will assume that ε and c_s are nearly constant, so that the overall normalization of the action can be absorbed into the definition of a new, canonically normalized, field

$$v \equiv y \mathcal{R} = \int d^3k \left[v_k(t) \, a_k \, e^{i \mathbf{k} \cdot \mathbf{x}} + c.c. \right] \, . \qquad (1.19)$$

We have written v in terms of time-independent stochastic parameters a_k and time-dependent mode functions $v_k(t)$. The mode functions satisfy the *Mukhanov–Sasaki equation*,

$$\ddot{v}_k + 3H \dot{v}_k + \frac{c_s^2 k^2}{a^2} v_k = 0 \, . \qquad (1.20)$$

This is the equation of a simple harmonic oscillator with a friction term provided by the expanding background. The oscillation frequency depends on the physical momentum and is therefore time dependent:

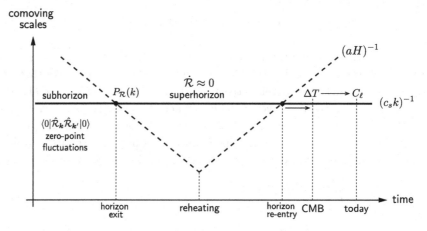

Fig. 1.4 The evolution of curvature perturbations during and after inflation: the comoving horizon $(aH)^{-1}$ shrinks during inflation and grows in the subsequent FRW evolution. This implies that comoving scales $(c_s k)^{-1}$ exit the horizon at early times and re-enter the horizon at late times. In physical coordinates, the Hubble radius H^{-1} is constant and the physical wavelength grows exponentially, $\lambda \propto a(t) \propto e^{Ht}$. For adiabatic fluctuations, the curvature perturbations $\mathcal{R}$ do not evolve outside of the horizon, so the power spectrum $P_{\mathcal{R}}(k)$ at horizon exit during inflation can be related directly to CMB observables at late times.

$$\omega_k(t) \equiv \frac{c_s k}{a(t)} \ . \tag{1.21}$$

At early times (small a), $\omega_k \gg H$ for all modes of interest. In this limit, the friction is irrelevant and the modes oscillate. However, the frequency of any given mode drops exponentially during inflation. At late times (large a), the dynamics is dominated by friction and the mode has a constant amplitude. We say that the mode "freezes" at *horizon crossing*, i.e. when $\omega_k(t_\star) = H$ or $c_s k = aH(t_\star)$. It is these constant superhorizon fluctuations that eventually become the density fluctuations that we observe in the CMB and in LSS (see Fig. 1.4).[1]

1.2.2 Vacuum fluctuations

The initial conditions for v (or $\mathcal{R}$) are computed by treating it as a quantum field in a classical inflationary background spacetime. This calculation has become textbook material [57, 58] and can also be found in many reviews (e.g. [27, 59]). We present the details in Appendix C. Here, we will restrict ourselves to a simplified, but intuitive, computation [60].

---

[1] Recall that we are assuming adiabatic initial conditions. The presence of entropy perturbations, as in multi-field models, can complicate the relation between the curvature perturbations at horizon crossing and the late-time observables – see Appendix C.

The Fourier modes of the classical field v are promoted to quantum operators

$$\hat{v}_{\boldsymbol{k}} = v_k(t)\hat{a}_{\boldsymbol{k}} + h.c. \tag{1.22}$$

At sufficiently early times, all modes of cosmological interest were deep inside the Hubble radius. In this limit, each mode behaves as an ordinary harmonic oscillator. The operators $\hat{a}_{\boldsymbol{k}}$ play the role of the annihilation operators of the quantum oscillators. The vacuum state is defined by $\hat{a}_{\boldsymbol{k}}|0\rangle = 0$. The oscillation amplitude will experience the same zero-point fluctuations as an oscillator in flat space, $\langle 0|\hat{v}_{\boldsymbol{k}}\hat{v}_{\boldsymbol{k}'}|0\rangle = (2\pi)^3|v_k|^2\delta(\boldsymbol{k}+\boldsymbol{k}')$, where

$$|v_k|^2 = \frac{1}{a^3}\frac{1}{2\omega_k} . \tag{1.23}$$

The factor of a^{-3} arises from the physical volume element in the Lagrangian (1.17) – note that the Fourier mode v_k was defined using the comoving coordinates rather than the physical coordinates. The second factor, $1/(2\omega_k)$, is the standard result for the variance of the amplitude of a harmonic oscillator in its ground state. (In inflation, this state is the *Bunch–Davies vacuum*.) As long as the physical wavelength of the mode is smaller than the Hubble radius, the ground state will evolve adiabatically. Equation (1.23) then continues to hold, and the precise time at which we define the initial condition is not important. Once a given mode gets stretched outside the Hubble radius, the adiabatic approximation breaks down and the fluctuation amplitude freezes at

$$|v_k|^2 = \frac{1}{2}\frac{1}{a_\star^3}\frac{1}{c_s k/a_\star} , \tag{1.24}$$

where $a_\star$ is the value of the scale factor at horizon crossing,

$$\frac{c_s k}{a_\star} = H . \tag{1.25}$$

Combining (1.25) and (1.24), we get

$$|v_k|^2 = \frac{1}{2}\frac{H^2}{(c_s k)^3} , \tag{1.26}$$

where from now on it is understood implicitly that the right-hand side is evaluated at horizon crossing.

1.2.3 Curvature perturbations

Using (1.19), we obtain the *power spectrum* of primordial curvature perturbations,

$$P_{\mathcal{R}}(k) \equiv |\mathcal{R}_k|^2 = \frac{1}{4}\frac{H^4}{M_{\mathrm{pl}}^2|\dot{H}|c_s}\frac{1}{k^3} . \tag{1.27}$$

The variance in real space is $\langle \mathcal{R}^2 \rangle = \int d\ln k \, \Delta_{\mathcal{R}}^2(k)$, where we have defined the dimensionless power spectrum

$$\Delta_{\mathcal{R}}^2(k) \equiv \frac{k^3}{2\pi^2} P_{\mathcal{R}}(k) = \frac{1}{8\pi^2} \frac{H^4}{M_{\mathrm{pl}}^2 |\dot{H}| c_s} \ . \tag{1.28}$$

Since the right-hand side is supposed to be evaluated at horizon crossing, $c_s k = aH$, any time dependence of H and c_s translates into a scale dependence of the power spectrum. Scale-invariant fluctuations correspond to $\Delta_{\mathcal{R}}^2(k) = const.$, and deviations from scale invariance are quantified by the *spectral tilt*

$$n_s - 1 \equiv \frac{d\ln \Delta_{\mathcal{R}}^2}{d\ln k} = -2\varepsilon - \tilde{\eta} - \kappa \ , \tag{1.29}$$

where we have defined two additional expansion parameters,

$$\tilde{\eta} \equiv \frac{\dot{\varepsilon}}{H\varepsilon} \quad \text{and} \quad \kappa \equiv \frac{\dot{c}_s}{Hc_s} \ . \tag{1.30}$$

Inflationary backgrounds typically satisfy $\{\varepsilon, |\tilde{\eta}|, |\kappa|\} \ll 1$ and hence predict $n_s \approx 1$. Inflation would not end if the slow-roll parameters vanished, so importantly, we also expect a finite deviation from perfect scale invariance, $n_s \neq 1$.

1.2.4 Gravitational waves

Arguably the cleanest prediction of inflation is a spectrum of primordial gravitational waves. These are tensor perturbations to the spatial metric,

$$g_{ij} = a^2(t)(\delta_{ij} + h_{ij}) \ , \tag{1.31}$$

where h_{ij} is transverse and traceless. Expanding the Einstein–Hilbert action leads to the quadratic action for the tensor fluctuations:

$$S_h^{(2)} = \frac{1}{2} \int d^4x \, a^3 \, y^2 \left[(\dot{h}_{ij})^2 - \frac{1}{a^2} (\partial_k h_{ij})^2 \right], \tag{1.32}$$

where

$$y^2 \equiv \frac{1}{4} M_{\mathrm{pl}}^2 \ . \tag{1.33}$$

The structure of the action is identical to that of the scalar fluctuations, Eq. (1.17), except that tensors do not have a nontrivial sound speed and the relation to the canonically normalized field does not include ε, because at linear order tensors do not feel the symmetry breaking due to the background evolution. The quantization of tensor fluctuations is therefore the same as for the scalar fluctuations. In particular, Eq. (1.26) applies to each polarization mode of the gravitational field. Adding the power spectra of the two polarization modes, one finds [61]

$$\Delta_h^2(k) \equiv \frac{k^3}{2\pi^2} P_h(k) = \frac{2}{\pi^2} \frac{H^2}{M_{\rm pl}^2} , \tag{1.34}$$

where the right-hand side is evaluated at horizon crossing, $k = aH$. While the power spectrum of scalar fluctuations, Eq. (1.28), depends on H, $\dot{H}$, and c_s, the power spectrum of tensor fluctuations is only a function of the de Sitter expansion rate H. Tensor fluctuations are therefore a direct probe of the energy scale at which inflation took place. The scale dependence of the tensor modes is determined by the time dependence of H,

$$n_t \equiv \frac{d \ln \Delta_h^2}{d \ln k} = -2\varepsilon . \tag{1.35}$$

Observational constraints on tensor modes are usually expressed in terms of the tensor-to-scalar ratio,

$$r \equiv \frac{\Delta_h^2}{\Delta_{\mathcal{R}}^2} = 16\varepsilon c_s . \tag{1.36}$$

Since the amplitude of scalar fluctuations has been measured, the tensor-to-scalar ratio quantifies the size of the tensor fluctuations. Using (1.34), we can write

$$\frac{H}{M_{\rm pl}} = \pi \Delta_{\mathcal{R}}(k_\star) \sqrt{\frac{r}{2}} , \tag{1.37}$$

which on substituting $\Delta_{\mathcal{R}}(k_\star) = 4.7 \times 10^{-5}$ becomes

$$H = 3 \times 10^{-5} \left(\frac{r}{0.1}\right)^{1/2} M_{\rm pl} . \tag{1.38}$$

Detecting inflationary tensor perturbations at the level $r \sim 0.1$ would imply that the expansion rate during inflation was about $10^{-5} M_{\rm pl}$. This is sometimes expressed in terms of the *energy scale of inflation*,

$$E_{\rm inf} \equiv (3H^2 M_{\rm pl}^2)^{1/4} = 8 \times 10^{-3} \left(\frac{r}{0.1}\right)^{1/4} M_{\rm pl} . \tag{1.39}$$

Note that reducing r by four orders of magnitude reduces $E_{\rm inf}$ by only one order of magnitude. Gravitational waves from inflation are only observable if inflation occurred near the GUT scale, $E_{\rm inf} \sim 10^{-2} M_{\rm pl} \sim 10^{16}$ GeV.

1.3 Cosmological observables

When the curvature perturbation $\mathcal{R}$ re-enters the horizon it sources fluctuations in the primordial plasma. These matter perturbations evolve into anisotropies in the cosmic microwave background (CMB) [57, 62] and inhomogeneities in large-scale structure (LSS). In this section, we describe these key cosmological observables. In the next section, we will show how these observables are used to constrain both the composition of the universe and its initial conditions.

1.3.1 CMB anisotropies

In the very early universe, photons had a small mean free path due to the high density of charged particles. At a temperature of about 0.3 eV, the formation of neutral hydrogen,

$$e + p \rightarrow H + \gamma \,, \tag{1.40}$$

termed *recombination*, became entropically favored. The free electron density dropped rapidly and Thomson scattering between electrons and photons, $e + \gamma \leftrightarrow e + \gamma$, became inefficient: the photons *decoupled*. Since the moment of *last scattering* at $t \approx 380,000$ yr, these primordial photons have been streaming freely through the universe, reaching our detectors 13.7 billion years later [63]. The observed frequency spectrum is that of an almost perfect black body with a mean temperature $\bar{T} = 2.72548 \pm 0.00057$ K [64]. Figure 1.5 shows the variation of the CMB temperature as a function of direction $\boldsymbol{n}$ on the sky,

$$\Delta T(\boldsymbol{n}) \equiv T(\boldsymbol{n}) - \bar{T} \,. \tag{1.41}$$

These anisotropies reflect inhomogeneities in the density of the primordial plasma, which can be traced back to the curvature perturbations calculated in the previous section.

For initial conditions drawn from a Gaussian probability distribution, complete information about the temperature map is contained in the correlations between the temperatures at pairs of distinct points $\boldsymbol{n}$ and $\boldsymbol{n}'$,

Fig. 1.5 CMB anisotropies as observed by the Planck satellite. The fluctuations in the blackbody temperature reflect fluctuations in the density at recombination.

$$C(\theta) \equiv \left\langle \frac{\Delta T}{\bar{T}}(\boldsymbol{n}) \frac{\Delta T}{\bar{T}}(\boldsymbol{n}') \right\rangle , \qquad (1.42)$$

where $\cos\theta \equiv \boldsymbol{n} \cdot \boldsymbol{n}'$, and the angle brackets denote an ensemble average.[2] It is convenient to describe the same information in harmonic space, by expanding the temperature field in spherical harmonics,

$$\frac{\Delta T(\boldsymbol{n})}{\bar{T}} = \sum_{\ell=0}^{\infty} \sum_{m=-\ell}^{+\ell} a_{\ell m} Y_{\ell m}(\boldsymbol{n}) , \qquad (1.43)$$

where ℓ and m are eigenvalues of differential operators on the sphere, with $\nabla^2 Y_{\ell m} = -\ell(\ell+1) Y_{\ell m}$ and $\partial_\phi Y_{\ell m} = im Y_{\ell m}$. Reality of the temperature field imposes $a_{\ell m}^* = (-1)^m a_{\ell-m}$. Statistical isotropy constrains the two-point correlation function of the multipole moments $a_{\ell m}$ to be of the form

$$\langle a_{\ell m} a_{\ell' m'}^* \rangle = C_\ell \delta_{\ell\ell'} \delta_{mm'} . \qquad (1.44)$$

The *angular power spectrum*, C_ℓ, is the Legendre transform of the two-point function (1.42),

$$C_\ell = 2\pi \int_{-1}^{1} \mathrm{d}\cos\theta \, C(\theta) \, P_\ell(\cos\theta) . \qquad (1.45)$$

Although the theory predicts ensemble-averaged quantities, we only observe a single realization of the ensemble. After extracting the multipole moments of the measured temperature map, we can construct an *estimator* for the angular power spectrum,

$$\hat{C}_\ell = \frac{1}{2\ell+1} \sum_m |a_{\ell m}|^2 . \qquad (1.46)$$

This estimator is unbiased, in that $\langle \hat{C}_\ell \rangle = C_\ell$. The variance of the estimator is called *cosmic variance*,

$$\mathrm{var}(\hat{C}_\ell) \equiv \langle \hat{C}_\ell \hat{C}_\ell \rangle - \langle \hat{C}_\ell \rangle^2 = \frac{2}{2\ell+1} C_\ell^2 . \qquad (1.47)$$

This irreducible error arises from having only $2\ell+1$ modes at each multipole moment ℓ to estimate the variance of their distribution. Figure 1.6 shows the CMB power spectrum as measured by the Planck satellite. The error bars include both cosmic variance and measurement noise, but the former dominates up to $\ell \sim 2000$.

The shape of the CMB power spectrum is well understood theoretically. Before neutral hydrogen formed, photons and baryons were strongly coupled and acted as a single fluid in which the photon pressure sustained acoustic oscillations (i.e. sound waves), driven by the gravitational force induced by the curvature

---

[2] Recall that in Section 1.2.2 we computed a quantum average. This is related to the ensemble average after decoherence turns the quantum state into a single classical state of the ensemble: see e.g. [65–69].

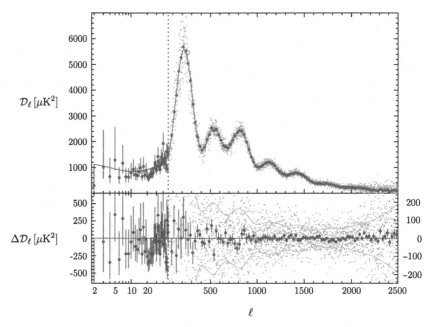

Fig. 1.6 The power spectrum of CMB temperature anisotropies measured by the Planck satellite (figure adapted from [8]). Plotted is the combination $\mathcal{D}_\ell \equiv \ell(\ell+1)C_\ell/2\pi$. Shown are both the data for individual multipoles (gray points), as well as binned averages (points with error bars). The lower plot shows the residuals with respect to the best fit ΛCDM model.

perturbations. The observed CMB fluctuations are a snapshot of these density waves. For adiabatic initial conditions, the angular power spectrum is predicted to be

$$C_\ell = \int d\ln k \, \Delta_\mathcal{R}^2(k) \, T_\ell^2(k) \,, \tag{1.48}$$

where the transfer function $T_\ell(k)$ describes both the evolution of the initial fluctuations from the moment of horizon entry to the time of recombination, as well as the projection from recombination to today [57, 62]. Since the transfer function depends only on known physics it is computable using a set of coupled Einstein–Boltzmann equations for the primordial plasma [70, 71]. The knowledge of $T_\ell(k)$ allows us to use the observed C_ℓ as a probe of the initial conditions $\Delta_\mathcal{R}(k)$. The theoretical curve in Fig. 1.6 assumes a nearly scale-invariant spectrum as predicted by inflation.

1.3.2 *CMB polarization*

Recombination was not an instantaneous process. In the time it took protons and electrons to combine into neutral hydrogen, the photons developed a quadrupole anisotropy in the local electron rest frame. Thomson scattering converted this temperature anisotropy into CMB polarization [72–74].

Linear polarization can be measured in terms of the Stokes parameters Q and U [75]. Let $\boldsymbol{n}$ be the direction of observation and $(\boldsymbol{e}_1, \boldsymbol{e}_2)$ be a basis of orthogonal unit vectors. The Stokes parameters are not invariant under a change of these coordinates: rotating the basis $(\boldsymbol{e}_1, \boldsymbol{e}_2)$ by an angle ψ leads to

$$(Q \pm iU)'(\boldsymbol{n}) = e^{\mp 2i\psi}(Q \pm iU)(\boldsymbol{n}) \ . \tag{1.49}$$

This identifies $Q \pm iU$ as a spin-2 field to be expanded in terms of spin-weighted spherical harmonics [76],

$$(Q \pm iU)(\boldsymbol{n}) = \sum_{\ell m} a_{\pm 2,\ell m} \,_{\pm 2}Y_{\ell m}(\boldsymbol{n}) \ . \tag{1.50}$$

Acting twice with a spin-lowering operator on $Q + iU$ and twice with a spin-raising operator on $Q - iU$ produces scalar (spin-0) quantities. These scalars can be collected according to their transformations under parity (the operation that takes $\boldsymbol{n}$ into $-\boldsymbol{n}$):

$$E(\boldsymbol{n}) \equiv a_{E,\ell m} Y_{\ell m}(\boldsymbol{n}) \,, \qquad a_{E,\ell m} \equiv -\frac{a_{2,\ell m} + a_{-2,\ell m}}{2} \,, \tag{1.51}$$

$$B(\boldsymbol{n}) \equiv a_{B,\ell m} Y_{\ell m}(\boldsymbol{n}) \,, \qquad a_{B,\ell m} \equiv -\frac{a_{2,\ell m} - a_{-2,\ell m}}{2i} \ . \tag{1.52}$$

The E-modes are parity-even, while the B-modes are parity-odd. Roughly, we can think of the E-mode as the gradient of a scalar and the B-mode as the curl of a vector. Typical E- and B-patterns are shown in Fig. 1.7. Given T, E, and B, we can form several types of correlation functions,

$$\langle a_{X,\ell m} a^*_{Y,\ell'm'} \rangle = C_\ell^{XY} \delta_{\ell\ell'} \delta_{mm'}, \quad X, Y \in \{T, E, B\} \ . \tag{1.53}$$

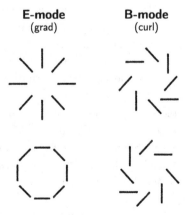

E-mode   **B-mode**
(grad)   (curl)

Fig. 1.7 Examples of E-mode and B-mode patterns of CMB polarization. While the E-mode patterns are mirror-symmetric, the B-mode patterns are interchanged under reflection about a line going through the center.

Since B is parity-odd, while T and E are parity-even, in a parity-conserving theory we expect $C_\ell^{TB} = C_\ell^{EB} = 0$.

Discussing polarization in terms of E-modes and B-modes has several distinct advantages. First of all, unlike the Stokes parameters, the parameters E and B are independent of the choice of coordinates. More importantly, symmetry forbids the generation of B-modes by scalar fluctuations [77, 78]. B-modes are therefore a crucial signature of the presence of tensor (or vector) fluctuations.

1.3.3 Large-scale structure

The density perturbations are small at recombination, but under the influence of gravity they grow [79], eventually forming the large-scale structure of the universe. A linear order, the initial conditions from inflation are related to the dark matter density contrast $\delta \equiv \delta\rho/\rho$ at redshift z via a transfer function $T_\delta(z, k)$,

$$P_\delta(z, k) = T_\delta^2(z, k) P_\mathcal{R}(k) . \tag{1.54}$$

On large scales, the transfer function is relatively easy to calculate in perturbation theory [80], while on small scales numerical N-body simulations [81] are required.

With the exception of gravitational lensing [82–84], we do not measure the dark matter density δ directly. Instead, we observe biased baryonic tracers of the dark matter field, such as galaxies, clusters, and Lyα fluctuations; see the compilation of recent measurements in Fig. 1.8. On large scales, the density contrast of these tracers, δ_g, has a linear and deterministic relationship to the underlying dark matter field,

$$\delta_g(z, \boldsymbol{x}) = b(z)\,\delta(z, \boldsymbol{x}) , \tag{1.55}$$

where $b(z)$ parameterizes the biasing. On small scales, however, the biasing can become nonlinear, nonlocal, and stochastic. This makes it challenging to relate large-scale observations to the initial conditions.

The Einstein equations couple the oscillations in the photon–baryon fluid to the dark matter density. The same oscillations that we observe in the CMB power spectrum are therefore also imprinted in the matter power spectrum. These oscillations are barely visible in Fig. 1.8 between $k = 0.01$ Mpc^{-1} and 0.1 Mpc^{-1}. Detections of these *baryon acoustic oscillations* (BAO) were first reported by the Sloan Digital Sky Survey (SDSS) in [87, 88] and more fully characterized in [89–93] (for a review of BAO see [94]). Figure 1.9 shows the measured matter two-point function in real space, $\xi(r)$. The BAO feature is clearly visible at about 100 Mpc.

Both the CMB observations and the BAO observations measure the sound horizon of the photon–baryon plasma. The observed scale in the CMB measurements depends on the angular diameter distance to recombination, $D_A(z_{\rm rec})$.

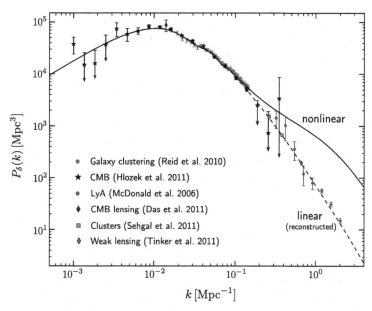

Fig. 1.8 Compilation of measurements of the matter power spectrum (figure adapted from [85]).

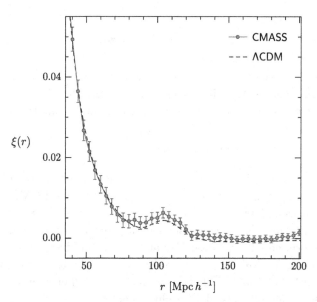

Fig. 1.9 Spherically averaged redshift-space correlation function of the CMASS Data Release 9 (DR9) sample of the Baryonic Oscillation Spectroscopic Survey (BOSS) (figure adapted from [86]). The dashed line corresponds to the best-fitting ΛCDM model.

For BAO, the observed scale depends on the (spherically averaged) distance to the effective survey redshift $\bar{z}$, which is a combination of the angular diameter distance and the Hubble parameter,

$$D_V(\bar{z}) \equiv \left[(1+\bar{z})^2 D_A^2(\bar{z}) \frac{c\bar{z}}{H(\bar{z})} \right]^{1/3} . \tag{1.56}$$

Comparing CMB and LSS measurements provides important information about the evolution of the universe after recombination and helps to break an important geometric degeneracy [95–98] that exists in the CMB-only analysis. Alternatively, the degeneracy can be broken by using the gravitational lensing of the CMB anisotropies [84].

1.4 Current tests of inflation

In March 2013, the *Planck* collaboration released its first cosmological analysis [8]. Together with the measurements of the CMB damping tail by the *Atacama Cosmology Telescope* (ACT) [99, 100] and the *South Pole Telescope* (SPT) [101, 102] this provides a beautiful picture of the first seven acoustic peaks of the CMB power spectrum. In this section, we summarize how the CMB results have tested the physics of inflation [9, 10]. Errors quoted in this section are 1σ errors (68% limits) unless otherwise specified.

1.4.1 Λ CDM model

The standard model of cosmology has six free parameters: the physical baryon density, $\omega_b \equiv \Omega_b h^2$; the physical density of cold dark matter (CDM), $\omega_c \equiv \Omega_c h^2$; the dark energy density, Ω_Λ; the optical depth τ; and the amplitude A_s and the spectral index n_s in the power law ansatz for the initial conditions,[3]

$$\Delta_{\mathcal{R}}^2(k) = A_s \left(\frac{k}{k_\star} \right)^{n_s-1} . \tag{1.57}$$

This simple model provides a superb fit to a wide range of cosmological data, from CMB to LSS. Figure 1.6 shows the power spectrum of CMB temperature fluctuations measured by Planck, as well as the best-fit curve of the ΛCDM model. Table 1.1 summarizes the best-fit parameters. The Planck data is precise enough to determine all six parameters at the percent level without recourse to external datasets. A small degeneracy between τ and A_s (and/or n_s) is broken by the addition of WMAP low-ℓ polarization data [7] (or Planck lensing data [103]).

The best-fit value for the scalar amplitude is

$$A_s = \left(2.196^{+0.051}_{-0.060} \right) \times 10^{-9} . \tag{1.58}$$

---

[3] In the Planck analysis, A_s and n_s are defined at the pivot scale $k_\star = 0.05$ Mpc^{-1}.

Table 1.1 *Parameters of the ΛCDM baseline model (with 2σ errors). The first four parameters describe the composition of the universe, the last two its initial conditions. The BAO data improves the constraint on Ω_Λ. The small-scale CMB data hardly affects the constraints but helps with a characterization of foregrounds, which becomes essential when going beyond the ΛCDM model.*

Parameter	Planck	$\cdots$ + WMAP + ACT	CMB + BAO
$\Omega_b h^2$	0.02207 ± 0.00067	0.02207 ± 0.00054	0.02214 ± 0.00048
$\Omega_c h^2$	0.1196 ± 0.0061	0.1198 ± 0.0052	0.1187 ± 0.0034
Ω_Λ	0.683 ± 0.040	0.685 ± 0.033	0.692 ± 0.021
τ	0.097 ± 0.080	0.091 ± 0.027	0.092 ± 0.026
$10^9 A_s$	2.23 ± 0.32	2.20 ± 0.11	2.20 ± 0.11
n_s	0.962 ± 0.019	0.959 ± 0.014	0.961 ± 0.011

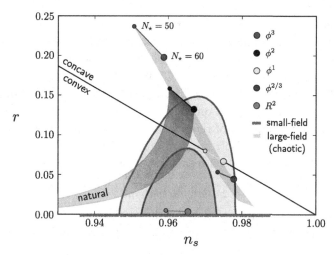

Fig. 1.10 Planck+WMAP+BAO constraints on n_s and r (figure adapted from [9]).

A scale-invariant primordial power spectrum is now excluded at almost 6σ significance,

$$n_s = 0.9603 \pm 0.0073 \,. \tag{1.59}$$

This result assumes that tensor fluctuations make a negligible contribution to the temperature fluctuations. Allowing for tensors introduces a new parameter, the tensor-to-scalar ratio r, cf. (1.36). With earlier datasets, including r in the fit weakened the evidence for $n_s < 1$, but with Planck this result is now robust: see Fig. 1.10 and Table 1.3.

Table 1.2 *Constraints on the geometry of the universe (with 2σ errors).*
The inclusion of BAO data plays an important role.

Parameter	Planck	$\cdots$ + WMAP + ACT	CMB + BAO
Ω_K	-0.072 ± 0.081	-0.037 ± 0.049	-0.0005 ± 0.0066

1.4.2 Probing inflation

The Planck collaboration has tested for deviations from the standard assumptions for the initial conditions, such as deviations from Gaussianity, adiabaticity, power law scaling, and flatness. Here we summarize their findings [9, 10].

Geometry

Inflation very effectively solves the flatness problem [18]. The baseline analysis of Planck has therefore fixed the curvature parameter to be vanishing, $\Omega_K = 0$. On the other hand, including Ω_K in the fit allows a test of this key prediction of inflation. Table 1.2 shows the constraints on the parameter Ω_K, after marginalizing over the other parameters of the ΛCDM model. Here, the BAO data plays a crucial role in breaking the geometric degeneracy between Ω_m and H_0 and reducing the error on Ω_K by an order of magnitude. Even at this new level of precision the observable patch of the universe is consistent with spatial flatness. Planck has also tested the isotropy assumption [104]. Except perhaps on the largest scales, the universe indeed seems to be statistically isotropic.

Scalar fluctuations

The observations of the primordial scalar fluctuations are in striking agreement with the predictions of inflation, both qualitatively and quantitatively.

Coherent phases

A telling feature of the CMB anisotropies is that they span *superhorizon* scales at recombination (corresponding to $\ell < 200$) and have *coherent phases*. This fact is observed unambiguously through the low-ℓ peak in the cross-correlation between temperature fluctuations and E-mode polarization (see Fig. 1.11). In the absence of phase coherence, this peak would disappear [105, 106]. It is easy to see why the inflationary mechanism for generating fluctuations leads to phase coherence. Modes freeze when their physical wavelengths become larger than the Hubble radius and they only start evolving again when they re-enter the horizon. All modes with the same wavenumber k, but possibly distinct wavevectors $\boldsymbol{k}$, therefore start their evolution at the same time. This phase coherence allows for constructive interference of the modes and yields acoustic oscillations in the CMB. Alternative mechanisms for structure formation involving topological defects (e.g. cosmic strings, see Section 4.5.2) source perturbations with

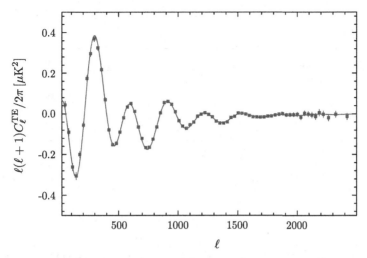

Fig. 1.11 The cross-correlation of CMB temperature anisotropies and E-mode polarization (figure adapted from [8]). The curve is not a fit, but a prediction! The low-ℓ peak is a signature of phase coherence of the initial conditions.

incoherent phases, smearing out the peaks [107], and are therefore ruled out by the CMB observations. Isocurvature fluctuations also destroy some of the phase coherence[4] and are hence significantly constrained by the data (see below).

Power law spectrum
We have seen above that slow-roll inflation predicts a power law spectrum with a percent-level deviation from perfect scale-invariance, which Planck has detected at high significance. At second order in the slow-roll expansion, inflation predicts a small correction to the power law spectrum,

$$\Delta_{\mathcal{R}}^2(k) = A_s \left(\frac{k}{k_\star} \right)^{n_s - 1 + \frac{1}{2}\alpha_s \ln(k/k_\star)} . \tag{1.60}$$

The data is not yet precise enough to detect the expected *running* of the spectrum, $\alpha_s \sim (n_s - 1)^2$, and a detection of running at a level accessible to Planck would in fact be in conflict with the expectation from slow-roll inflation. It is nevertheless interesting to include α_s as a free parameter in the fit. Table 1.3 summarizes the latest constraints on α_s, which depend on whether the tensor-to-scalar ratio r is included as a parameter or is set to zero. At present, there are no clear indications for a departure from the inflationary power law spectrum.

---

[4] In contrived scenarios, causal evolution inside the horizon yields isocurvature perturbations that lead to acoustic peaks [108] – see the review [109].

Table 1.3 *Constraints on tensor modes and on deviations from the power law spectrum (with 2σ errors).*

Parameter	Planck	$\cdots$ + WMAP + ACT	CMB + BAO
n_s	0.963 ± 0.019	0.960 ± 0.014	0.962 ± 0.011
r	< 0.115	< 0.117	< 0.119
n_s	0.974 ± 0.030	0.955 ± 0.015	0.960 ± 0.012
α_s	-0.034 ± 0.035	-0.015 ± 0.017	-0.013 ± 0.018
n_s	0.976 ± 0.030	0.957 ± 0.015	0.959 ± 0.011
r	< 0.228	< 0.230	< 0.235
α_s	-0.041 ± 0.037	-0.022 ± 0.021	-0.022 ± 0.022

Tensor fluctuations

Tensor modes contribute to the CMB temperature power spectrum in a specific way and are therefore constrained by the Planck analysis. Figure 1.10 shows the current constraints on the parameters n_s and r. Marginalizing over n_s gives an upper limit on the tensor-to-scalar ratio [9]

$$r < 0.12 \quad (95\% \text{ limit}) . \tag{1.61}$$

This constraint is at the limit of what can be achieved with CMB temperature data alone [110]. To probe smaller values of r requires measurements of CMB polarization; as we explained in Section 1.3.2, B-modes are a unique signature of inflationary tensor modes.

In March 2014, the *BICEP2* collaboration announced the first detection of B-mode polarization on degree angular scales [137, 144] (see Fig. 1.12). At the time of writing, it is not known whether the signal seen by BICEP2 is partially cosmological in origin [144], or instead was created by polarized emission from dust grains in our own galaxy [142, 143]. A detection of primordial B-modes is such a significant result that it is essential to address this question with further observations. Fortunately, CMB polarization measurements are rapidly improving. B-mode measurements by Planck will soon weigh in on the issue. Furthermore, there are many experiments looking for B-modes on degree scales, including KeckArray [144], EBEX [145], SPIDER [146], ABS [147], and CLASS [148]; and on arcminute scales, including POLARBEAR [149], SPTpol [150], and ACTpol [151].

Non-Gaussianity

The CMB power spectrum in Fig. 1.6 reduces the Planck data from about 50 million pixels to 10^3 multipole moments. This enormous data compression is justified if the primordial perturbations are isotropic and Gaussian. On the other hand, a wealth of information may be contained in deviations from a perfectly Gaussian distribution [111–113]. Among the major accomplishments of the Planck mission

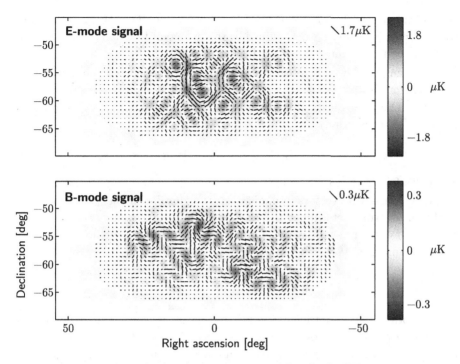

Fig. 1.12 E-mode and B-mode maps measured by the BICEP2 experiment (figure adapted from [137]). An excess over the lensing B-mode is detected with high signal-to-noise.

are the significant upper bounds placed on higher-order CMB correlations, or *non-Gaussianity* [10]. (For previous results from WMAP see [114, 115].) This has allowed the study of primordial quantum fields to move beyond the free field limit and start to place meaningful constraints on interactions.

In Section 1.2, we computed the two-point function (or power spectrum) of primordial curvature perturbations,

$$\langle 0 | \hat{\mathcal{R}}_{\boldsymbol{k}_1} \hat{\mathcal{R}}_{\boldsymbol{k}_2} | 0 \rangle = (2\pi)^3 \, P_{\mathcal{R}}(k_1) \, \delta(\boldsymbol{k}_1 + \boldsymbol{k}_2) \, , \qquad (1.62)$$

where $|0\rangle$ denotes the vacuum state and $\hat{\mathcal{R}}$ is the quantum operator associated with the field $\mathcal{R}$. In principle, there is more information in the vacuum expectation values of higher-order n-point functions. Schematically, we can write these as the path integral

$$\langle \Omega | \hat{\mathcal{R}}_{\boldsymbol{k}_1} \cdots \hat{\mathcal{R}}_{\boldsymbol{k}_n} | \Omega \rangle \propto \int [\mathcal{DR}] \, \mathcal{R}_{\boldsymbol{k}_1} \cdots \mathcal{R}_{\boldsymbol{k}_n} \, e^{iS[\mathcal{R}]} \, , \qquad (1.63)$$

where S is the inflationary action and $|\Omega\rangle$ is the vacuum of the interacting theory. For a *free* field theory, the action is a quadratic functional $S_{(2)}$, cf. Eq. (1.17), and the e^{iS} weighting of the path integral is a Gaussian (after Wick rotating to

Euclidean time). All correlation functions with n odd then vanish, while those with n even are completely determined by the two-point function (1.62). However, including nontrivial interactions in the action, $S_{\text{int}} = S_{(3)} + S_{(4)} + \cdots$, makes the e^{iS} weighting of the path integral *non-Gaussian*. This allows non-zero n-point functions for all n.

The primary diagnostic for primordial non-Gaussianity is the three-point function (or bispectrum),

$$\langle \Omega | \hat{\mathcal{R}}_{\boldsymbol{k}_1} \hat{\mathcal{R}}_{\boldsymbol{k}_2} \hat{\mathcal{R}}_{\boldsymbol{k}_3} | \Omega \rangle = (2\pi)^3 \, B_{\mathcal{R}}(k_1, k_2, k_3) \, \delta(\boldsymbol{k}_1 + \boldsymbol{k}_2 + \boldsymbol{k}_3) \; . \tag{1.64}$$

The delta-function is a consequence of statistical homogeneity; it enforces that the three momentum vectors form a closed triangle. The momentum dependence of the bispectrum determines the amount of non-Gaussianity associated with triangles of different shapes. A useful measure of the size of the non-Gaussianity is the parameter

$$f_{\text{NL}} \equiv \frac{5}{18} \frac{B_{\mathcal{R}}(k, k, k)}{P_{\mathcal{R}}^2(k)} \; , \tag{1.65}$$

i.e. the normalized amplitude of the bispectrum in the equilateral configuration, $k_1 = k_2 = k_3 \equiv k$. The momentum dependence of the bispectrum $B_{\mathcal{R}}(k_1, k_2, k_3)$ potentially contains substantial information about the physics that generated the primordial perturbations. The Planck analysis [10] has tested for shapes of non-Gaussianity parameterized by the following templates:

$$B_{\text{local}} \equiv \frac{6}{5} \left(P_1 P_2 + \text{perms.} \right) , \tag{1.66}$$

$$B_{\text{equil}} \equiv \frac{3}{5} \left(6 \left(P_1^3 P_2^2 P_3 \right)^{1/3} - 3 P_1 P_2 - 2 \left(P_1 P_2 P_3 \right)^{2/3} + \text{perms.} \right) , \tag{1.67}$$

$$B_{\text{ortho}} \equiv \frac{3}{5} \left(18 \left(P_1^3 P_2^2 P_3 \right)^{1/3} - 9 P_1 P_2 - 8 \left(P_1 P_2 P_3 \right)^{2/3} + \text{perms.} \right) , \tag{1.68}$$

where $P_i \equiv P_{\mathcal{R}}(k_i)$. We comment briefly on the physical motivations for these choices of bispectrum shapes.

Local non-Gaussianity

The shape (1.66) arises from the following ansatz in real space [116–118]:

$$\mathcal{R}(\boldsymbol{x}) \equiv \mathcal{R}_g(\boldsymbol{x}) + \frac{3}{5} f_{\text{NL}}^{\text{local}} \left[\mathcal{R}_g^2(\boldsymbol{x}) - \langle \mathcal{R}_g^2 \rangle \right] , \tag{1.69}$$

where $\mathcal{R}_g(\boldsymbol{x})$ is a Gaussian random field. In momentum space, the signal peaks for *squeezed* triangles, e.g. $k_1 \ll k_2 \sim k_3$ (see Fig. 1.13). This shape of non-Gaussianity arises in models of *multi-field inflation* – see Appendix C. On the other hand, in single-field inflation (i.e. in models in which only the adiabatic mode π is excited) the signal vanishes in the squeezed limit. This important theorem is known as the *single-field consistency relation* [119, 120]. Under mild

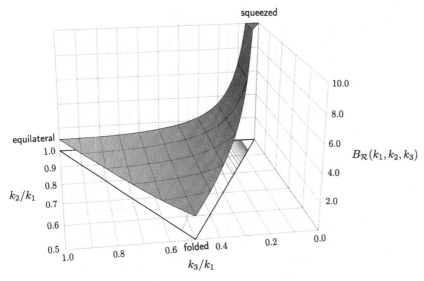

Fig. 1.13 Bispectrum of the local ansatz. The signal is peaked for squeezed triangles.

assumptions about the inflationary action and the initial state, it is possible to show that the bispectrum in single-field inflation satisfies[5]

$$\lim_{k_1 \to 0} \frac{B_{\mathcal{R}}(k_1, k_2, k_3)}{P_{\mathcal{R}}(k_1) P_{\mathcal{R}}(k_2)} = (1 - n_s) \ll 1 . \tag{1.70}$$

In terms of the shapes (1.66)–(1.68), this implies that $f_{\mathrm{NL}}^{\mathrm{local}} \ll 1$, as only the local shape peaks in the squeezed limit. Observing a signal in the squeezed limit ($f_{\mathrm{NL}}^{\mathrm{local}} \gtrsim 1$) would rule out *all* models of single-field inflation, not just slow-roll models. Planck has now severely constrained this possibility (see below).

Equilateral non-Gaussianity

Large non-Gaussianity in single-field inflation can nevertheless arise from higher-derivative interactions [124, 125]. This leads to signals that peak in equilateral triangle configurations, i.e. $k_1 \sim k_2 \sim k_3$. To characterize this type of non-Gaussianity, we return to the Goldstone action. At cubic order and to lowest order in derivatives, we get [51] (see Appendix B for the derivation)

$$S_\pi^{(3)} = \int d^4 x \sqrt{-g} \, \frac{M_{\mathrm{pl}}^2 \dot{H}}{c_s^2} (1 - c_s^2) \left(\frac{\dot{\pi}(\partial_i \pi)^2}{a^2} + \frac{A}{c_s^2} \dot{\pi}^3 \right) . \tag{1.71}$$

---

[5] This theorem can be interpreted as a Ward identity associated with the nonlinearly realized dilatation symmetry of the background [56, 121–123], and is the analogue of the Adler zero in pion physics.

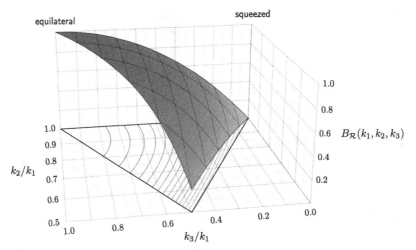

Fig. 1.14 Bispectrum of the interaction $\dot{\pi}(\partial_i \pi)^2$. The signal is peaked for equilateral triangles.

We have two cubic operators, $\dot{\pi}(\partial_i \pi)^2$ and $\dot{\pi}^3$, but only one new parameter, A. This is a consequence of the nonlinearly-realized time translation symmetry, which relates the amplitude of the operator $\dot{\pi}(\partial_i \pi)^2$ to the sound speed. In DBI inflation (see Section 5.3) one has $A = -1$ [41], while more generally, naturalness arguments suggest $A \sim \mathcal{O}(1)$ [115]. Both $\dot{\pi}(\partial_i \pi)^2$ and $\dot{\pi}^3$ produce bispectra that are well approximated by the equilateral template (1.67) (see Fig. 1.14).

Orthogonal non-Gaussianity

The two equilateral bispectra are not identical, so one can find a linear combination of the two operators $\dot{\pi}(\partial_i \pi)^2$ and $\dot{\pi}^3$ that is orthogonal in a well-defined sense [126] to the shape (1.67), and also to the local shape (1.66). This is the *orthogonal* template (1.68) [115]. In terms of the parameters of the Lagrangian (1.71), the total bispectrum is mostly of the orthogonal shape – specifically, with greater than 70% correlation with the orthogonal template – for $3.1 \lesssim A \lesssim 4.2$.

The Planck collaboration has reported the following constraints on the amplitudes of the templates (1.66), (1.67), and (1.68) [10]:

$$f_{\rm NL}^{\rm local} = 2.7 \pm 5.8 \,, \tag{1.72}$$

$$f_{\rm NL}^{\rm equil} = -42 \pm 75 \,, \tag{1.73}$$

$$f_{\rm NL}^{\rm ortho} = -25 \pm 39 \,. \tag{1.74}$$

Equation (1.72) is a very strong constraint on multi-field inflation. The limits (1.73) and (1.74) are strong, but they do not make a future detection

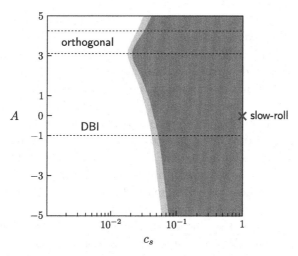

Fig. 1.15 Planck constraints on non-Gaussianity in single-field inflation (figure adapted from [127]). Shown are the 68% and 95% constraints on the sound speed c_s and interaction coefficient A; cf. Eq. (1.71).

inconceivable (see Section 1.5). Observational constraints on the parameters in the Goldstone action (1.71) are shown in Fig. 1.15.

Non-adiabaticity

As we have seen, single-field inflation predicts initial fluctuations that are *adiabatic*. Adiabatic perturbations have the property that the local state of matter (determined, for example, by the energy density ρ) at some spacetime point $(t, \boldsymbol{x})$ of the perturbed universe is the same as in the background universe at some slightly different time $t + \pi(t, \boldsymbol{x})$. That is, some parts of the universe are "ahead" and others "behind" in the evolution. At recombination, the universe consists of a mixture of several fluids: photons (γ), baryons (b), dark matter (c), and neutrinos (ν). For adiabatic initial conditions, the density perturbations in each species "I" are proportional to the Goldstone boson of broken time translations,

$$\delta_I(t, \boldsymbol{x}) \equiv \frac{\rho_I(t + \pi(t, \boldsymbol{x})) - \rho_I(t)}{\rho_I(t)} \approx \frac{\dot{\rho}_I}{\rho_I} \pi(t, \boldsymbol{x}) \,. \tag{1.75}$$

All matter perturbations therefore have the same density contrast (e.g. $\delta_b = \delta_c$) and are proportional to the radiation perturbations (e.g. $\delta_c = \frac{3}{4}\delta_\gamma$). For adiabatic initial conditions, all species fluctuate synchronously and lead to the curvature perturbation $\mathcal{R}$.

In multi-field inflation, it is possible to generate so-called *isocurvature* perturbations, where an overdensity in one species compensates for an underdensity in another, resulting in no net curvature perturbation. For example, we can define the following isocurvature perturbation for dark matter and photons:

$$\mathcal{S} \equiv \delta_c - \tfrac{3}{4}\delta_\gamma \,. \tag{1.76}$$

If this field were significantly different from zero it would lead to a measurable effect in the CMB power spectrum.

We digress briefly to describe two classic distinctions between isocurvature perturbations and curvature perturbations, at the level of the acoustic peaks [128, 129]. The first distinction involves the angular positions of successive peaks. Adiabatic perturbations from single-field inflation have fixed amplitude outside the horizon, and begin to evolve upon entering the horizon. The resulting evolution may be thought of as a cosine mode. The curvature perturbations sourced by cosmic defects, in contrast, have negligible amplitude as they enter the horizon, and grow subsequently through causal processes. The result is typically a sine mode. These two cases make different predictions for the angular positions of subsequent peaks, which are in the ratio 1 : 2 : 3 in the cosine case, and 1 : 3 : 5 in the sine case. The relative heights of even and odd peaks provide another means of testing adiabaticity. Acoustic peaks corresponding to compression waves – namely, the odd peaks – are enhanced compared to even peaks in the adiabatic case, but suppressed compared to even peaks in the isocurvature case.

Definitive evidence against isocurvature models involving causal evolution inside the horizon, without an inflationary phase, comes from measurements of CMB polarization. A characteristic signature of these models is that the temperature and E-mode polarization perturbations are positively correlated on large angular scales [128], while in inflation these perturbations are anti-correlated. The measurement of TE anti-correlation on superhorizon scales [130] shows that superhorizon adiabatic perturbations were present when the CMB decoupled.

Although purely isocurvature perturbations are now ruled out, it is possible that the observed anisotropies originate from a combination of adiabatic and isocurvature perturbations. To quantify the isocurvature contribution, it is conventional to define the relative amplitude of the power spectra of the isocurvature field and the curvature perturbation,

$$\alpha \equiv \frac{P_{\mathcal{S}}}{P_{\mathcal{R}}} \, . \tag{1.77}$$

Assuming that $\mathcal{S}$ and $\mathcal{R}$ are uncorrelated (motivated by axion isocurvature models [131–133]), Planck has constrained this ratio [9],

$$\alpha_0 < 0.036 \, . \tag{1.78}$$

The constraint strengthens if $\mathcal{S}$ and $\mathcal{R}$ are perfectly correlated (as in curvaton isocurvature models [134, 135]),

$$\alpha_{+1} < 0.0025 \, . \tag{1.79}$$

Observing an isocurvature contribution to the primordial fluctuations is another way to rule out single-field inflation, since only the presence of additional light fields can give rise to non-adiabaticity. Unfortunately, the amplitude

of the signal depends on the post-inflationary evolution: the primordial perturbations become adiabatic if the particles produced after inflation reach a suitable
thermal equilibrium [136]. Correspondingly, observable isocurvature is possible
only when one or more particle species has an abundance determined by physics
beyond thermal equilibrium.

1.5 Future tests of inflation

Cosmological observations show no signs of slowing down. CMB observations
continue to provide important measurements of the primordial fluctuations,
especially on small angular scales [99–102]. A large number of ground-based and
balloon-borne experiments are targeting high-precision measurements of CMB
polarization. Current and future large-scale structure surveys will provide additional information (see Table 1.4). In this section, we discuss what one can hope
to learn from measuring the primordial perturbations with increased precision
and over a wider range of scales.

1.5.1 Tensor tilt

In inflation, the tilt of the tensor spectrum is related to the time evolution of the
Hubble parameter,

$$n_t = 2 \frac{\dot{H}}{H^2} \ . \tag{1.80}$$

Finding a nearly scale-invariant spectrum, $|n_t| \ll 1$, would confirm that
$|\dot{H}| \ll H^2$ in the early universe. Since $|\dot{H}| \ll H^2$ was our definition of inflation in Section 1.1.2, this would be as much of a "proof" of inflation as we could
ever hope to get.

The sign of n_t is also informative. Consider a spatially flat FRW universe filled
with a perfect fluid with pressure P and energy density ρ. From (1.8) we see that
$\dot{H} > 0$ is only possible if $\rho + P < 0$, corresponding to a violation of the null
energy condition (NEC).[6] Thus, in all theories for which n_t is given by (1.80)
and the NEC holds, we predict $n_t < 0$.

Finally, in single-field slow-roll inflation, a consistency relation links the tensor
tilt to the tensor-to-scalar-ratio,

$$n_t = -\frac{r}{8} \ . \tag{1.81}$$

Even for a large tensor signal, $r \sim \mathcal{O}(0.1)$, the tensor tilt is small, $n_t \sim \mathcal{O}(0.0125)$.
Measuring n_t at this level will be very challenging, but does not seem impossible.
Testing the consistency relation will be one of the main targets of future CMB
polarization experiments. Forecasts of experimental sensitivities can be found
in [110, 152–154].

---

[6] The NEC states that the stress tensor satisfies $T_{\mu\nu}n^\mu n^\nu \geq 0$, for all null vectors n^μ. For a
perfect fluid, the NEC reduces to $\rho + P \geq 0$.

1.5.2 Scalar tilt and running

Models of inflation make specific predictions for the parameters n_s and r. Improving the measurement of either of these parameters will therefore play a vital role in narrowing down the number of viable models [155]. Future galaxy surveys [156] may reduce the error on n_s by a factor of 5. At the same time, future CMB polarization experiments [26, 157–160] have the potential to reduce the error on r to the percent level.

A test of the slow-roll paradigm may come from measurements of the running of the scalar spectrum. At second order in the Hubble slow-roll parameters, and for $c_s = 1$, the running of the scalar spectrum is

$$\alpha_s = -2\varepsilon\tilde{\eta} - \tilde{\eta}\chi\,, \tag{1.82}$$

where $\chi \equiv \dot{\tilde{\eta}}/(H\tilde{\eta})$. Measuring α_s would test the consistency of the slow-roll expansion. However, because the running is second order in slow-roll, we expect it to be small, $\alpha_s \sim (n_s - 1)^2$. Current bounds on α_s are still two orders of magnitude larger than this target, but future galaxy surveys may allow such a measurement [161] (see also [162]). Any detection of a larger level of running would be a challenge for slow-roll inflation.

1.5.3 Non-Gaussianity

The constraints on primordial non-Gaussianity from the CMB have almost reached their limit. Silk damping of the small-scale anisotropies prohibits use of multipoles larger than $\ell_{\max} \sim 2000$ to extract information about initial conditions. This limits the number of modes available in the CMB to

$$\mathcal{N}^{\mathrm{CMB}} \sim (\ell_{\max}/\ell_{\min})^2 \sim 10^6\,, \tag{1.83}$$

which is nearly saturated by the recent Planck measurements.

More modes are in principle accessible through large-scale structure measurements. This is because galaxy surveys probe the three-dimensional cosmic density field, while the CMB is only a two-dimensional projection. Hence, while $\mathcal{N}^{\mathrm{CMB}} \propto \ell_{\max}^2$ for the CMB, we have $\mathcal{N}^{\mathrm{LSS}} \propto k_{\max}^3$ for LSS (but see [163, 164]), where $k_{\max}$ is associated with the smallest scale that is both measurable and under theoretical control. Pushing to smaller scales (larger $k_{\max}$) therefore increases rather dramatically the amount of information that can be extracted from the data. The total number of (quasi-)linear modes in LSS is estimated to be

$$\mathcal{N}^{\mathrm{LSS}}_{\mathrm{linear}} \sim (k_{\max}/k_{\min})^3 \sim 10^9\,, \tag{1.84}$$

where we have taken $k_{\max} \sim 0.1\,\mathrm{Mpc}^{-1}$ and $k_{\min} \sim 10^{-4}\,\mathrm{Mpc}^{-1}$. Although this shows the great potential of LSS observations, it assumes that we measure the entire volume at low redshift. More realistically, we have $k_{\min} \sim 10^{-3}\,\mathrm{Mpc}^{-1}$ (e.g. for the Euclid mission), and hence [165]

$$\mathcal{N}^{\mathrm{Euclid}}_{\mathrm{linear}} \sim (k_{\max}/k_{\min})^3 \sim 10^6\,, \tag{1.85}$$

Table 1.4 *Compilation of current and future LSS surveys. Here, z_{max} refers to the maximal redshift of the survey, V is the survey volume, and n_g is the mean comoving number density of objects. The projected errors on $f_{\mathrm{NL}}^{\mathrm{local}}$ are from measurements of the galaxy bispectrum. (Data collected by Donghui Jeong.)*

Name	z_{max}	$V\,[(\mathrm{Gpc/h})^3]$	$n_g\,[(\mathrm{Mpc/h})^{-3}]$	$k_{\mathrm{max}}\,[\mathrm{h/Mpc}]$	$\Delta f_{\mathrm{NL}}^{\mathrm{local}}$
SDSS LRG	0.315	1.48	1.36×10^{-3}	0.1	5.62
BOSS	0.35	5.66	0.27×10^{-3}	0.1	3.34
BigBOSS	0.5	13.1	0.30×10^{-3}	0.1	2.27
HETDEX	2.7	2.96	0.27×10^{-3}	0.2	3.65
CIP	2.25	6.54	0.50×10^{-3}	0.2	1.03
EUCLID	1.0	102.9	0.16×10^{-3}	0.1	0.92
WFIRST	1.5	107.3	0.94×10^{-3}	0.1	1.11

which is comparable to the result (1.83) for the CMB. However, while ℓ_{max} for the CMB cannot be extended, for LSS, k_{max} might be pushed to larger values through a better understanding of nonlinearities in the dark matter evolution, the biasing, and the redshift space distortions. This is one of the objectives of the recently developed "effective theory of large-scale structure" [166, 167] (see also [168–173]).

Even from the CMB alone, the limit on the amplitude of local non-Gaussianity, cf. (1.72), is approaching an interesting threshold for multi-field inflation. The conversion of hidden sector non-Gaussianity during reheating [174–176] or after inflation [177, 178] typically leads to

$$|f_{\mathrm{NL}}^{\mathrm{local}}| \gtrsim \mathcal{O}(1)\,. \tag{1.86}$$

This possibility is now highly constrained, and further data from Planck and LSS surveys (see Table 1.4) has the potential to rule out the natural parameter space of a large class of multi-field models.

As we explain in Appendix B (see Section B.3.2), a similar threshold exists for equilateral non-Gaussianity [179]:

$$f_{\mathrm{NL}}^{\mathrm{equil}} \sim \mathcal{O}(1)\,. \tag{1.87}$$

Not seeing a signal at the level of (1.87) would allow us to conclude that the UV completion of the effective theory is slow-roll inflation, up to perturbative higher-derivative corrections. Conversely, a detection of equilateral non-Gaussianity with $f_{\mathrm{NL}}^{\mathrm{equil}} > \mathcal{O}(1)$ would imply that the theory has to be UV-completed by something other than slow-roll inflation, such as DBI inflation. The threshold (1.87) therefore provides an important observational distinction between UV completions of inflation corresponding to weakly coupled backgrounds and those that involve strongly coupled backgrounds. Unfortunately, $f_{\mathrm{NL}}^{\mathrm{equil}} \sim \mathcal{O}(1)$

is almost two orders of magnitude smaller than the CMB bound (1.73). Future CMB data may improve the bound by a factor of a few, but not by enough to reach the threshold (1.87). However, optimistic estimates of galaxy lensing tomography suggest that this may not be completely out of reach for future LSS observations [180].

2

Inflation in effective field theory

Inflation is a well-understood phenomenon in quantum field theory coupled to gravity, and many field theories that support inflationary phases have been proposed. Nevertheless, deriving the inflationary action from a more fundamental principle, or in the context of a well-motivated parent theory, remains a central problem.

There are two approaches or perspectives that can be used to obtain a quantum field theory suitable for inflation. In the "top-down" approach, one begins with a complete theory in the ultraviolet (UV), such as string theory, and tries to derive inflation as one of its low-energy consequences. This undertaking is discussed at length in Chapters 4 and 5. The more conservative "bottom-up" approach starts from the low-energy (IR) degrees of freedom and parameterizes our ignorance about the UV theory. Both approaches arrive at an *effective field theory* (EFT) that is valid at inflationary energies, but they do so from opposite directions. The two approaches are complementary and can inform each other.

The outline of this chapter is as follows: in Section 2.1, we provide a general overview of the essential principles of effective field theory.[1] We apply these concepts to inflation in Section 2.2, highlighting the sensitivity of inflation to Planck-scale physics in Section 2.3.

2.1 Principles of effective field theory

Natural phenomena occur across a vast range of length scales. Fortunately, in many cases one can analyze physical processes involving distinct scales by examining one relevant scale at a time. Figure 2.1 illustrates this logic using a few famous examples from the history of particle physics. For instance, at low energies, Fermi theory describes neutron–proton interactions by a four-fermion contact interaction with coupling constant $G_F = (293.6\,\text{GeV})^{-2}$. This

---

[1] More details on effective field theory can be found in [181–184].

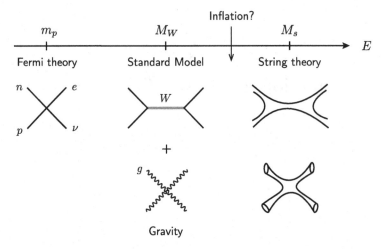

Fig. 2.1 Effective field theories in particle physics. Both Fermi theory and general relativity are non-renormalizable and are interpreted as effective theories.

theory is incomplete and breaks down (violates perturbative unitarity) at about 100 GeV. What actually happens close to 100 GeV is that we start to resolve the W-boson exchange interaction and Fermi theory is replaced by the electroweak theory of the Standard Model. Similarly, interactions of the Standard Model fields with gravitational degrees of freedom are determined by Newton's constant $G_N = (1.2 \times 10^{19} \text{ GeV})^{-2}$. Just like the Fermi theory, this theory breaks down at high energies, this time at the Planck scale,

$$M_{\text{pl}} \equiv \frac{1}{\sqrt{8\pi G_N}} = 2.4 \times 10^{18} \text{ GeV}. \qquad (2.1)$$

The Standard Model plus general relativity should therefore also be viewed as an effective theory, to be replaced by a more fundamental theory at some energy at or below the Planck scale. In much of this work we will assume that this more fundamental theory is string theory.

2.1.1 Effective action

The first step in constructing effective field theories is identifying the degrees of freedom that are relevant for the measurements of interest. For instance, in particle physics we distinguish light and heavy degrees of freedom on the basis of whether the corresponding particles can be produced on-shell at the energies available to the experiment. Formally, we introduce a cutoff scale Λ to define the regime of validity of the EFT (see Fig. 2.2). *Light* particles ϕ, of mass $m < \Lambda$, are included in the effective theory, while *heavy* particles Ψ, of mass $M > \Lambda$, are "integrated out," in a sense that we will make precise.

Fig. 2.2 Effective field theories describe the physics of light degrees of freedom below a cutoff scale Λ. We arrive at these theories either by integrating out the heavy fields (if the complete UV theory is known) or by parameterizing their effects (if the UV theory is not known or is not computable). In the latter case, symmetries inform the choice of allowed interactions.

Top down: integrating out

Imagine that we knew the full Lagrangian of the UV theory,

$$\mathcal{L}[\phi, \Psi] = \mathcal{L}_l[\phi] + \mathcal{L}_h[\Psi] + \mathcal{L}_{lh}[\phi, \Psi] , \tag{2.2}$$

where $\mathcal{L}_l$ ($\mathcal{L}_h$) describes the part of $\mathcal{L}$ involving only the light (heavy) fields, and $\mathcal{L}_{lh}$ includes all interactions involving both sets of fields. The *Wilsonian effective action* S_{eff} is defined via a path integral over the heavy modes (and over the high-frequency contributions of the light fields):

$$e^{iS_{\text{eff}}[\phi]} = \int [\mathcal{D}\Psi] \, e^{iS[\phi, \Psi]} . \tag{2.3}$$

In practice, the effective action is rarely found by performing the path integral. Instead a so-called matching calculation, order by order in perturbation theory, is usually more practical [181–184].

In the classical approximation, performing the path integral over the heavy modes corresponds to using the equations of motion to eliminate the heavy fields Ψ. In the language of Feynman diagrams, this is the tree-level approximation. The complete path integral, however, also includes loops of the heavy fields. These loops describe how the heavy degrees of freedom are eliminated from the quantum theory. The result is usually nonlocal, meaning that it contains terms such as $\phi(-\Box + M^2)^{-1}\phi$. However, at low energies, $E \ll \Lambda \leq M$, these terms can be expanded in derivatives – for example,

$$\phi(-\Box + M^2)^{-1}\phi = \frac{\phi}{M^2}\left(1 + \frac{\Box}{M^2} + \cdots\right)\phi , \tag{2.4}$$

and the EFT becomes approximately local. In other words, the effective action admits a systematic expansion in powers of the ratio (E/M),

$$\mathcal{L}_{\text{eff}}[\phi] = \mathcal{L}_l[\phi] + \sum_i c_i(g) \frac{\mathcal{O}_i[\phi]}{M^{\delta_i - 4}} , \tag{2.5}$$

where c_i are dimensionless constants that depend on the couplings g of the UV theory, and $\mathcal{O}_i$ are local operators of dimension δ_i. This procedure typically generates all operators $\mathcal{O}_i$ consistent with the symmetries of the UV theory. The absence from the effective theory of an operator allowed by the symmetries of the

UV theory, or an anomalously small coefficient for such an operator, is described as a fine-tuning.[2]

In (2.5) we have split the effective action into a renormalizable part $\mathcal{L}_l$ and a sum of non-renormalizable corrections. Note that non-renormalizable terms arise in the EFT even if the UV theory is renormalizable. Operators of dimension less than four (in four spacetime dimensions) are called *relevant operators*. They dominate in the IR and become small in the UV. Unsurprisingly, operators of dimension greater than four are called *irrelevant operators*.[3] These operators dominate in the UV but become small in the IR: the contribution of an operator $\mathcal{O}_i$ of dimension δ_i to low-energy observables is proportional to $(E/M)^{\delta_i - 4}$. As a result, although the sum in (2.5) includes operators of arbitrarily large dimension δ_i, only a finite number of terms are required to predict the results of experiments to a given accuracy. On the other hand, by studying the low-energy effects of irrelevant operators, we can learn about the structure of the UV theory. Measuring or constraining irrelevant operators can therefore be very informative.

A toy model

Let us illustrate this procedure with a simple toy example: we take the Lagrangian of the UV theory to be[4]

$$\mathcal{L}[\phi, \Psi] = -\frac{1}{2}(\partial\phi)^2 - \frac{1}{2}m^2\phi^2 - \frac{1}{4!}\lambda\phi^4 - \frac{1}{2}(\partial\Psi)^2 - \frac{1}{2}M^2\Psi^2 - \frac{1}{4}g\phi^2\Psi^2 \ . \quad (2.6)$$

Note that this Lagrangian respects the $\mathbb{Z}_2$ symmetry $\phi \to -\phi$. The effective Lagrangian for the light field ϕ takes the form

$$\mathcal{L}_{\text{eff}}[\phi] = -\frac{1}{2}(\partial\phi)^2 - \frac{1}{2}m_R^2\phi^2 - \frac{1}{4!}\lambda_R\phi^4$$

$$- \sum_{i=1}^{\infty} \left(\frac{c_i(g)}{M^{2i}}\phi^{4+2i} + \frac{d_i(g)}{M^{2i}}(\partial\phi)^2\phi^{2i} + \cdots \right) \ . \quad (2.7)$$

The parameters in (2.7) can be determined in an expansion in the couplings of the UV theory (here, λ and g). For example, the bare values of the mass m and the quartic coupling λ receive loop corrections with the heavy particle Ψ running in the loop,[5]

---

[2] An important exception is an *accidental symmetry*: if all operators in the UV theory violating a symmetry $\mathcal{S}$ are irrelevant in the sense of the renormalization group (RG), then $\mathcal{S}$ is an approximate symmetry of the low-energy theory. However, in this case the smallness of the $\mathcal{S}$-violating terms is not mysterious: it is simply a consequence of RG flow.

[3] Operators of dimension equal to four are called *marginal operators*. Quantum corrections decide whether a marginal operator is relevant or irrelevant in the IR.

[4] This is a simplified version of the example studied in [184].

[5] Of course, there are similar diagrams with the light particle ϕ running in the loop. For simplicity, we hide those terms in the ellipses of Eqs. (2.8) and (2.9).

$$m_R^2 = \underline{\quad\quad} + \underline{\quad\bigcirc\quad} + \cdots , \qquad (2.8)$$

$$\lambda_R = \times + \times\hspace{-0.3em}\bigcirc\hspace{-0.3em}\times + \cdots . \qquad (2.9)$$

The loop contributions diverge in the UV and have to be regularized. Cutting off the (Euclidean) momentum integrals at the scale Λ, we find [181],

$$m_R^2 = m^2 + \frac{g}{32\pi^2}(\Lambda^2 - M^2 L) + \cdots , \qquad (2.10)$$

$$\lambda_R = \lambda - \frac{3g^2}{32\pi^2}L + \cdots , \qquad (2.11)$$

where $L \equiv \ln(\Lambda^2/\mu^2)$, with μ being an arbitrary renormalization scale. Dimensional regularization would give the same result, except that we would not find the Λ^2 term in (2.10), and we would replace L by

$$L \to \frac{1}{\epsilon} + \gamma - \ln(4\pi) , \qquad (2.12)$$

where $\epsilon \equiv 4 - d$ and $\gamma \equiv 0.577\cdots$. We see that the quadratic divergence in (2.10) is scheme dependent and hence not physical. However, notice that the unphysical term comes with the same coupling g as the physical contribution to the renormalized mass proportional to M^2. It is therefore common to use the dependence on the cutoff Λ as a proxy for the physical dependence on the mass M of the heavy particles. However, see [185] for examples where this logic fails. Finally, the Wilson coefficients in (2.7) can also be computed in a loop expansion; e.g. the coupling of the operator ϕ^6 is

$$c_1 = \times\hspace{-0.3em}\bigcirc\hspace{-0.3em}\times + \cdots \sim \mathcal{O}(g^3) + \cdots . \qquad (2.13)$$

The trouble with light scalars
From (2.10) we see that the effective mass m_R receives a large contribution from the mass of the heavy field M. Having a light scalar field in the EFT is therefore *unnatural* [186] in the sense that large quantum corrections, $\mathcal{O}(M)$, must be canceled by a large bare mass m with opposite sign to achieve $m_R \ll M$ (see Section 2.1.2 for a detailed discussion). This is a real problem, since for natural values of m_R the field ϕ is not even part of the low-energy EFT! Notice that because the loop correction is proportional to the coupling g, a hierarchy $m_R \ll M$ can be natural if we have reason to believe that $g \ll 1$ (and $m \ll M$). Thus, the strength of the coupling between the light and heavy fields is a critical parameter, and symmetry structures in the UV theory that achieve $g \ll 1$

play an important role in discussions of light scalar fields (see Section 2.1.3). The apparent need to fine-tune the Higgs boson mass is the famous electroweak hierarchy problem of the Standard Model. As we will see in Section 2.3, a qualitatively similar, but quantitatively less dramatic, hierarchy problem exists in inflationary models driven by scalar fields.

Decoupling
All the divergences in (2.10) and (2.11) can be absorbed into a renormalization of the parameters of the Lagrangian. Moreover, in the limit $M \to \infty$ the effects of the heavy particles disappear completely. This *decoupling* of UV physics [187] ensures that the physical effects of massive particles are suppressed for large M. Decoupling is one reason that the Standard Model of particle physics was constructed by focusing only on renormalizable theories, although today we view it as an effective theory.

Bottom up: parameterizing ignorance
Often we do not know the complete UV theory, so that we cannot construct the EFT explicitly by integrating out the heavy modes. Instead we parameterize our ignorance about the UV physics, by making assumptions about the symmetries of the UV theory, and writing down the most general effective action consistent with these symmetries:

$$\mathcal{L}_{\text{eff}}[\phi] = \mathcal{L}_l[\phi] + \sum_i c_i \frac{\mathcal{O}_i[\phi]}{\Lambda^{\delta_i - 4}} , \qquad (2.14)$$

where the sum runs over all operators $\mathcal{O}_i[\phi]$, of dimension δ_i, allowed by the symmetries of the UV theory. The size of the higher-dimension operators is estimated in terms of the cutoff scale Λ, while the prefactors c_i are dimensionless *Wilson coefficients*. Equation (2.14) will be our starting point for discussing inflation in EFT.

Several comments are important at this stage. First, making assumptions about the symmetries of the UV theory can be nontrivial: not all low-energy symmetries admit ultraviolet completions. In Section 2.1.4, we discuss this issue in the context of string theory. Second, in writing (2.14) we have introduced an energy scale Λ and a collection of dimensionless coefficients c_i.[6] It is clearly important to understand how to assign values for the scale Λ and the coefficients c_i. The guiding principle for this undertaking, and more generally for the construction and interpretation of effective field theories, is *naturalness*.

---

[6] At the level of (2.14), the joint rescaling $\Lambda \to \kappa\Lambda$ and $c_i \to \kappa^{\delta_i - 4} c_i$ leaves the theory unchanged. Formalizing this leads to the renormalization group and to the running of the renormalized couplings with energy.

2.1.2 Naturalness

Naturalness arguments work in two directions, from the top down and from the bottom up. We will discuss these two aspects of naturalness in turn.

Top-down naturalness

The top-down version of naturalness asserts that the Wilson coefficients in an EFT will be of order unity if the cutoff Λ is chosen to be of order the characteristic mass scale M of the UV theory.[7]

We emphasize that the coefficients in question are those of operators $\mathcal{O}_i$ *allowed* by the symmetries of the UV theory; approximate or exact symmetries of the ultraviolet theory can lead to small or vanishing coefficients for the corresponding operators in the effective theory. This top-down version of naturalness is merely a formalization of the expectation of genericity. When the ultraviolet theory is computable, naturalness is hardly needed, as the effective theory can be constructed directly by integrating out the heavy modes, and the Wilson coefficients can be obtained explicitly in terms of the parameters of the UV theory. However, this favorable circumstance is very rare, and in particular it does not arise in presently studied compactifications of string theory. We often have only partial information about the UV theory. For instance, we might know the relevant scales, but not the couplings to all light fields. Top-down naturalness is then widely used as a systematic framework for guessing how the calculation of the effective theory would turn out if we were strong enough to perform it.

Bottom-up naturalness

The bottom-up version of naturalness allows one to infer, based on properties of a given low-energy effective theory, the plausible scale M of new physics. Here, new physics refers to the scale of the lightest degrees of freedom that are part of the ultraviolet theory but not of the effective theory. Because we generally learn about nature beginning at low energies and proceeding to higher energies, the bottom-up version of naturalness can be a predictive tool for the results of experiments.

The logic of bottom-up naturalness arguments is the following: suppose we are presented with partial information about an EFT like (2.14). Imagine that we know all the renormalizable couplings of the light fields, but the UV cutoff Λ and the higher-dimension contributions are unknown – this is, for example, the situation for the Standard Model of particle physics. We would like to make an educated guess about the size of Λ. In the low-energy theory, we can calculate loop corrections to the parameters of the renormalizable Lagrangian as functions of an unknown cutoff scale Λ. The parameters are said to be bottom-up

---

[7] The dimensionless couplings g of the light fields to the heavy fields are assumed to be of order unity for this purpose. Systematically weaker couplings should be incorporated by defining a higher effective mass scale, $\tilde{M} \sim M/\sqrt{g}$.

natural[8] as long as their measured values are larger than the loop corrections. As we extrapolate the EFT to higher energies and increase Λ, some parameters may become unnatural, and insisting on natural parameters therefore defines a maximal scale for the effective theory, $\Lambda = \Lambda_{\max}$. Bottom-up naturalness asserts that "new physics" should appear at some scale $\Lambda \leq \Lambda_{\max}$ and modify the effective theory, thereby explaining the smallness of the measured parameter values. The unnatural alternative would be that multiple loop and/or bare contributions cancel against each other for reasons beyond the purview of the effective theory. Bottom-up naturalness is a formalization of the expectation – or more properly, the hope – that this is not the case. In the quintessential example of the Higgs boson, bottom-up naturalness predicts that new physics should appear at $\mathcal{O}(10^2 - 10^3)$ GeV to cut off the quadratic divergence of the Higgs mass m_H and explain why $m_H = 125$ GeV.

The naturalness criterion has been profoundly influential in motivating physics beyond the Standard Model and it often plays an important role in inflationary model-building. However, there are reasons to use it with care. We digress briefly to discuss a few instructive examples illustrating both successes and failures of the naturalness principle.[9]

Successes of naturalness

Let us first look at examples where insisting on natural parameter values has led (or could have led) to the correct physics.

Positron In classical electromagnetism the mass of the electron is unnatural. The electric field around an electron carries energy $\Delta E = \alpha/r_e$, where $\alpha \approx 1/137$ is the fine structure constant and r_e is the size of the electron, which is introduced to regulate the divergence. This Coulomb self-energy of the electron contributes to its mass:

$$\Delta m_e = \alpha \Lambda \,, \tag{2.15}$$

where $\Lambda \equiv r_e^{-1}$. In order for the observed mass of the electron ($m_e \approx 0.511$ MeV) to be natural we require that $\Lambda < 70$ MeV. Indeed, in quantum field theory, new physics in the form of the positron comes to the rescue. In [190], Weisskopf showed that virtual positrons surrounding the electron precisely cancel the linear divergence in (2.15), leaving only a logarithmic dependence on the cutoff,

$$\Delta m_e = \alpha \, m_e \ln\left(\Lambda/m_e\right) \,. \tag{2.16}$$

In the new effective theory, containing both the electron and the positron, the small electron mass is natural, even for large Λ.

---

[8] We should stress that what we call "bottom-up naturalness" here is universally referred to as "naturalness," but we find the distinction useful for the present exposition.

[9] These comments are based on [188, 189] and on private communications with Nima Arkani-Hamed.

Rho meson The mass difference between the charged pions and the neutral pion receives a quantum correction from photon loops,

$$m_{\pi^+}^2 - m_{\pi^0}^2 = \frac{3\alpha}{4\pi} \Lambda^2 , \qquad (2.17)$$

where Λ is the UV cutoff of the effective theory of pions. In order for (2.17) not to exceed the measured mass splitting, $m_{\pi^+}^2 - m_{\pi^0}^2 = (33.5\,\text{MeV})^2$, we require that $\Lambda < 850$ MeV. New physics in the form of the rho meson with $m_\rho = 770$ MeV comes in at exactly the scale suggested by naturalness. The charged pion mass is natural in the new EFT that includes the rho meson.

Charm quark Historically, one of the most interesting applications of the naturalness principle is K^0–$\bar{K}^0$ mixing. In an effective theory valid below the kaon mass scale the mass splitting between the K_L^0 and K_S^0 states takes the form

$$\frac{m_{K_L^0} - m_{K_S^0}}{m_{K_L^0}} = \frac{G_F^2 f_K^2}{6\pi^2} \sin^2 \theta_c \Lambda^2 , \qquad (2.18)$$

in terms of the cutoff Λ, the Cabibbo angle $\sin\theta_c \approx 0.22$, and the kaon decay constant $f_K = 114$ MeV. For (2.18) to be compatible with the measured splitting $(m_{K_L^0} - m_{K_S^0})/m_{K_L^0} = 7\times10^{-15}$, we require that $\Lambda < 2$ GeV. In fact, new physics in the form of the charm quark, with $m_c \approx 1.3$ GeV, modifies the UV behavior of the theory. Gaillard and Lee used this naturalness argument in a successful prediction of the charm quark mass [191].

Failures of naturalness?
Naturalness arguments are not always applicable, and need to be used with care.

Phase transitions Condensed matter systems near critical points are described by effective theories with fine-tuned parameters, and correspondingly large correlation lengths. This "unnatural" situation is simply a consequence of explicit fine-tunings performed by the experimentalist.

Nuclear physics An example of fine-tuning in nature occurs in nuclear physics [192] (for reviews see [182, 193]). In [194–196], Weinberg observed that the scattering lengths measured in low-energy nucleon-nucleon scattering are larger than would be expected from chiral perturbation theory. The fundamental scale of the EFT is the Compton wavelength of the pion, $m_\pi^{-1} = (140\,\text{MeV})^{-1}$, but the scattering lengths in the spin singlet state, $a_s \approx (8\,\text{MeV})^{-1}$, and in the spin triplet state, $a_t \approx (36\,\text{MeV})^{-1}$, are much larger than m_π^{-1}. Correspondingly, two neutrons fail to form a bound state by only 60 keV, and the deuteron binding energy is just 2 MeV, even though the natural expectation would involve energies of order $m_\pi = 140$ MeV. These results can be attributed to approximate cancellations of the kinetic and potential energies of the nucleons, but the underlying reason for these cancellations is poorly understood.

Electroweak scale At the time of writing, experiments at the Large Hadron Collider (LHC) have discovered the Higgs boson, but have not yet revealed whether the physics determining the hierarchy between the electroweak scale and the Planck scale is natural. In the Standard Model, the dominant quantum correction to the Higgs mass comes from the coupling to the top quark,

$$\Delta m_H^2 \sim \frac{y_t^2}{(4\pi)^2} \Lambda^2 , \qquad (2.19)$$

where $y_t \sim 1$ is the top Yukawa coupling. For the observed value of the Higgs mass ($m_H = 125$ GeV) to be natural, we require $\Lambda < 1.5$ TeV. This argument, which suggests that physics beyond the Standard Model should appear at or below the TeV scale, has had far-reaching impact on decades of work in theoretical and experimental particle physics. Many physicists have anticipated detecting evidence for a natural mechanism stabilizing the electroweak hierarchy, e.g. supersymmetric partners of known particles, both in earlier experiments and in the first stages of the LHC. No such direct evidence has yet materialized, and we must continue to wait for guidance from experiment. We should emphasize that properly defining and characterizing the predictions of natural mechanisms, such as supersymmetry, is a major area of research, and it is far too soon to conclude that all such mechanisms are excluded. Because of the close analogy between the electroweak hierarchy problem and the problem of naturalness of the inflaton mass (see Section 2.3), we may hope that the ultimate resolution of the former will suggest a particular approach to the latter.

Dark energy No discussion of naturalness is complete without addressing the cosmological constant problem. The vacuum of a quantum field theory with local Lorentz invariance corresponds to a stress–energy tensor of the form

$$\langle T_{\mu\nu} \rangle = -\rho_{\text{vac}} \, g_{\mu\nu} . \qquad (2.20)$$

Quantum contributions to the vacuum energy scale as

$$\Delta \rho_{\text{vac}} \sim \Lambda^4 . \qquad (2.21)$$

Naturalness of the observed vacuum energy, $\rho_{\text{vac}} \sim (10^{-3} \, \text{eV})^4$, therefore suggests new physics at $\Lambda \lesssim 10^{-3}$ eV. Indeed, if the world were supersymmetric down to 10^{-3} eV, the small value of the cosmological constant would be natural. But the world is not supersymmetric at low energies, and we have also not seen any other new physics at 10^{-3} eV that could account for the smallness of the vacuum energy.[10] In the absence of a mechanism explaining the small value of the cosmological constant, we have to entertain the possibility that it is simply a fine-tuned parameter. Moreover, in the string theory landscape it is conceivable that

---

[10] One or two neutrino masses may have the correct scale, but this has not led to a solution to the cosmological constant problem.

the observed value is environmentally selected [197], consistent with anthropic arguments [198].

It seems clear from these examples that naturalness can at best serve as a tentative guide towards new physics, rather than as a law of nature.

2.1.3 Symmetries

The interplay between symmetry structures in ultraviolet theories and light scalars in effective theories is crucial for understanding inflation in effective field theory and string theory, so we now discuss these issues in more depth. A pivotal question for inflationary model-building in effective field theory is whether light scalars with $m \ll H$ can be natural (see Section 2.3). As we have just explained, whether a given effective theory can be considered natural depends on the properties of the ultraviolet theory, whether known or assumed. Symmetries often play a central role in the radiative stability of the low-energy theory. In this section, we explain this fact for theories in flat space. In the next section, we will discuss some subtle aspects that arise in the generalization to ultraviolet completions that include gravity.

SUSY in flat space

We have seen that, in the absence of symmetries, scalar masses receive loop corrections of the form

$$\Delta m^2 \propto \Lambda^2 . \tag{2.22}$$

There are only a few known ways to protect scalars from these effects. One elegant possibility is unbroken supersymmetry (SUSY), which obliges boson and fermion loops to cancel, so that the scalar mass is not renormalized. However, SUSY is necessarily broken during inflation, generating a mass of the order of the Hubble scale H. Although $m \sim H$ can be significantly smaller than Λ, it still inhibits successful inflation. Below we will have more to say about this.

Global symmetries in flat space

Another possibility for stabilizing light scalars is a global internal symmetry. As a concrete example, suppose that the renormalizable part of the EFT, $\mathcal{L}_l[\phi]$, respects the *shift symmetry*

$$\phi \mapsto \phi + const. \tag{2.23}$$

This may arise, for example, if ϕ is the Goldstone boson of a spontaneously broken $U(1)$ symmetry (corresponding to the angular flat direction in the familiar Mexican hat potential). If (2.23) is exact, it forbids the mass term, or any potential terms for that matter. To get nontrivial dynamics, we are usually interested in the case where the shift symmetry (2.23) is only *approximate*. Concretely, we imagine that the symmetry is broken by a small mass term, $\Delta V = \frac{1}{2}m^2\phi^2$, with $m \ll \Lambda$. Loop corrections to the tree-level mass must then scale with the symmetry-breaking parameter (which is m), so that

$$\Delta m^2 \propto m^2 \ . \tag{2.24}$$

At most, the correction can now scale logarithmically with the cutoff Λ. Moreover, in the limit $m \to 0$, the exact symmetry (2.23) is restored and ϕ becomes massless. A small mass for ϕ is therefore *technically natural* [199]: the smallness of the symmetry-breaking parameter controls the renormalization. At the level of model building, one is free to set the mass at any desired level without risking destabilization through quantum effects. On the other hand, it still makes sense to ask whether the fact that the symmetry is weakly broken in the first place is dictated by some mechanism and is natural in the top-down sense. In principle, a symmetry can be broken explicitly by an operator whose coefficient is small purely because of fine-tuning, and the resulting small parameters are technically natural but not top-down natural.

Ultraviolet completion

Exact or approximate symmetries of the UV theory can control the sizes of the Wilson coefficients in the non-renormalizable part of the effective Lagrangian. If the symmetry is weakly broken by the heavy degrees of freedom, or if the light fields couple only weakly to the symmetry-breaking terms, then the EFT enjoys an approximate symmetry, and the Wilson coefficients of all symmetry-breaking operators will be naturally small. This can be seen in our toy model (2.6): the coupling $g\phi^2\Psi^2$ breaks the shift symmetry in the UV. In the EFT, this breaking shows up through symmetry-breaking irrelevant operators. Since the symmetry is restored in the limit $g \to 0$, the Wilson coefficients of all symmetry-breaking operators in (2.7) must satisfy

$$\lim_{g \to 0} c_i(g) = 0 \ . \tag{2.25}$$

For finite g, the c_i are proportional to positive powers of the symmetry-breaking parameter g. An approximate symmetry in the UV would explain $g \ll 1$ and hence $c_i \ll 1$ in the EFT. We emphasize that assuming $c_i \ll 1$ in the effective theory amounts to assuming something about the couplings to the degrees of freedom at the cutoff scale Λ. Whether a given low-energy theory is deemed natural can hinge on which symmetries are thought to be permissible in the UV completion. Consulting a UV-complete theory like string theory can be valuable when general reasoning about what ought to be typical does not give a sharp answer.

2.1.4 Gravity

Gravity plays a fundamental role in any description of cosmology, so our effective theory must include gravitational degrees of freedom. Moreover, semiclassical gravity itself has a limited range of validity. At or below the Planck scale, graviton–graviton scattering violates perturbative unitarity, and we expect new degrees of freedom to become relevant. In this section, we discuss to what extent the UV completion of gravity can affect the matter interactions in the low-energy effective theory.

Gravity as an effective theory

The low-energy degree of freedom of gravity is the spacetime metric $g_{\mu\nu}$, whose leading interactions are determined by the Einstein–Hilbert action,

$$S_{\rm EH} = \frac{M_{\rm pl}^2}{2} \int {\rm d}^4x \sqrt{-g}\, R \,. \tag{2.26}$$

This theory is non-renormalizable and should be understood as an effective theory [181, 200]. To see this, let us expand the metric in terms of small perturbations around flat space, $g_{\mu\nu} \equiv \eta_{\mu\nu} + \frac{1}{M_{\rm pl}} h_{\mu\nu}$. Schematically, the Einstein–Hilbert action then becomes

$$S_{\rm EH} = \int {\rm d}^4x \left[(\partial h)^2 + \frac{1}{M_{\rm pl}} h(\partial h)^2 + \frac{1}{M_{\rm pl}^2} h^2 (\partial h)^2 + \cdots \right]. \tag{2.27}$$

This weak-field expansion of the Einstein–Hilbert action looks similar to the action of Yang–Mills theory,

$$S_{\rm YM} = \int {\rm d}^4x \left[(\partial A)^2 + g A^2 \partial A + g^2 A^4 \right]. \tag{2.28}$$

However, while the Yang–Mills action terminates at a finite order, the expansion of the Einstein–Hilbert action contains an infinite number of terms, coming from the expansion of $\sqrt{-g}$ and $g^{\mu\nu}$. Gravity is therefore interpreted as an effective quantum field theory with cutoff $\Lambda = M_{\rm pl}$. The quantum perturbation theory of gravitons is organized in terms of the dimensionless ratio $(E/M_{\rm pl})^2$, where E is the energy of the process, and this theory breaks down at the Planck scale. At this point either new degrees of freedom become important (like the massive excitations in string theory; see Section 3.1) or a nonperturbative miracle happens (as in asymptotic safety [201]). In the absence of detailed information about the UV completion, the simplest assumption is that the low-energy effective action contains all terms that are consistent with general coordinate invariance. We can organize this as a derivative expansion,[11]

$$S_g = \int {\rm d}^4x \sqrt{-g} \left[M_\Lambda^4 + \frac{M_{\rm pl}^2}{2} R + c_1 R^2 + c_2 R_{\mu\nu} R^{\mu\nu} \right.$$

$$\left. + \frac{1}{M^2}(d_1 R^3 + \cdots) + \cdots \right], \tag{2.29}$$

---

[11] There is no $R_{\mu\nu\sigma\rho} R^{\mu\nu\sigma\rho}$ term, since by the generalized Gauss–Bonnet theorem $R^2 - 4R_{\mu\nu} R^{\mu\nu} + R_{\mu\nu\sigma\rho} R^{\mu\nu\sigma\rho}$ is topological, and equals the Euler class of the spacetime.

where c_i and d_i are dimensionless numbers that may be expected to be of order unity. For pure gravity, the scale M should also be the Planck scale $M_{\rm pl}$. However, couplings to matter fields might lead to a hierarchy between M and $M_{\rm pl}$. Note also that the renormalized value of the cosmological constant M_Λ deduced from cosmological experiments is extremely far from its natural value $M_{\rm pl}$. This is, of course, the famous cosmological constant problem.

In many string theories, the higher-curvature terms in (2.29) can be computed order by order in the α' and $g_{\rm s}$ expansions detailed in Section 3.1. An important example is type IIB string theory in ten dimensions, where one finds[12] [202]

$$S_g = \int {\rm d}^{10}X\sqrt{-G}\left[\frac{M_{10}^2}{2}R + \frac{\zeta(3)}{3\cdot 2^5}\frac{1}{M^6}\mathcal{R}^4 + \cdots\right] , \qquad (2.30)$$

where M_{10} is the ten-dimensional Planck mass; ζ denotes the Riemann zeta function, with $\zeta(3) \approx 1.202$; $\mathcal{R}^4$ is a particular quartic invariant constructed from the Riemann tensor; and the omitted terms are subleading in $1/M$ and/or in the string coupling $g_{\rm s}$. The mass M appearing in (2.30) corresponds to the mass of the first excited level of the type II superstring, given by

$$M^2 = \frac{4}{\alpha'} , \qquad (2.31)$$

where α' is the inverse string tension defined in Section 3.1. This is the proper physical cutoff scale because the higher-derivative term in (2.30) arises upon integrating out the massive excitations of the string, which have the mass spectrum $m^2 = 4N/\alpha'$, $N \in \mathbb{Z}$. In the regime of weak coupling and weak curvature where the corrections to the Einstein–Hilbert action are small, the string scale M is small compared to the Planck scale $M_{\rm pl}$, so the higher-derivative contributions shown above are more significant in string theory than general reasoning about quantum gravity would suggest. On the other hand, the coefficient $\zeta(3)/(3\cdot 2^5) \approx 10^{-2}$ is rather small, illustrating that the general expectation of order-unity Wilson coefficients should not be viewed as a precise and immutable law.

Global symmetries in quantum gravity

A number of "folk theorems" state that exact continuous global symmetries are impossible in a theory of quantum gravity. Instead, any continuous global symmetry must be merely an accidental symmetry of the low-energy effective theory, broken by irrelevant operators at a scale not parametrically larger than $M_{\rm pl}$. We will briefly recall some of the arguments; see [203] for a modern discussion of related issues.

---

[12] The ten-dimensional cosmological constant is set to zero in (2.30) because we are displaying the supersymmetric effective action, but the four-dimensional cosmological constant arising upon compactification and supersymmetry breaking is subject to the usual cosmological constant problem.

The simplest argument against global symmetries in quantum gravity is that Hawking evaporation of a black hole can destroy global charges. For example, imagine tossing a substantial clump of baryons into a macroscopic black hole, assumed to be large enough so that protons and neutrons make up an arbitrarily small fraction of the Hawking quanta. The black hole will lose most of its mass to light quanta that carry zero baryon number, and by the time the Hawking temperature is high enough for baryons to be emitted, the black hole mass will be smaller than the mass of the initial clump of baryons. Unless the theory contains states with an arbitrarily high ratio of baryon number to mass, a sufficiently large black hole will be unable to radiate away all of its initial baryon number in Hawking quanta, or deposit this baryon number in a highly charged remnant. Baryon number is therefore violated in the black hole evaporation process and cannot be an exact symmetry. The same applies to any continuous global internal symmetry with a well-defined conserved charge.[13] Another class of arguments appeals to the destruction of global charges by wormholes [204–206].

We should be clear that although these arguments show that global symmetries are broken by Planck-scale effects, they do not show that the breaking is necessarily of order unity. Indeed, it is conceivable that the Wilson coefficients for the symmetry-breaking higher-dimension operators might be suppressed for some reason. We know too little about the degrees of freedom at the Planck scale to make definitive statements about the strength of the symmetry breaking, although in concrete examples in string theory it is often possible to compute the symmetry-breaking effects.[14]

Global symmetries in string theory

The absence of exact continuous global internal[15] symmetries is also a theorem in perturbative string theory [208]. Suppose that there is an exactly conserved global symmetry of the conformal field theory on the string worldsheet, so that by Noether's theorem there is a corresponding conserved current on the worldsheet. This current can be used to construct a vertex operator that corresponds to the emission of a massless excitation of the string, which turns out to be nothing other than a gauge boson associated with the symmetry [208]. Thus, the postulated symmetry must be a gauge symmetry in the target spacetime.

We conclude that general arguments in quantum gravity, and specific findings in string theory, limit the sorts of global symmetries that are allowed in an ultraviolet completion. Asserting a symmetry structure for the UV theory

---

[13] Shift symmetries are an important example where the absence of a conserved charge requires a refinement of the black hole arguments.

[14] For example, in Section 5.4, we will study axions in string theory. An axion ϕ with infinite periodicity $(f/M_{\mathrm{pl}} \to \infty)$ enjoys the exact shift symmetry $\phi \mapsto \phi + const.$, so the above arguments suggest that such axions are not possible in quantum gravity. And indeed, direct searches for axions with $f \gtrsim M_{\mathrm{pl}}$ in parametrically controlled string compactifications have been unsuccessful [47, 48]. A different argument against axions with $f \gg M_{\mathrm{pl}}$ appears in [207].

[15] Exact Lorentz symmetry is possible in string theory, cf. [208].

and taking natural coefficients for the operators in the resulting effective action may lead to an effective theory that is consistent at low energies but cannot be embedded in a theory of quantum gravity. Because constraints from quantum gravity can play a critical role in determining the effective action, we view it as prudent to examine any postulated symmetry structure in a theory of quantum gravity.

Coupling quantum field theory to gravity

Thus far we have discussed flat-space quantum field theories, as well as purely gravitational theories, but the theories of interest in cosmology are quantum field theories coupled to gravity. Let us illustrate this by coupling the toy model of (2.7) to a gravitational theory with higher-curvature corrections. The resulting effective theory takes the form

$$S_{\text{eff}}[\phi, g] = S_g + S_{\text{eff}}[\phi] + S_{g,\phi} , \tag{2.32}$$

where S_g is given in (2.29), $S_{\text{eff}}[\phi]$ is the action corresponding to the Lagrangian density (2.7), and

$$S_{g,\phi} = \int \mathrm{d}^4x \sqrt{-g} \left[\sum_i c_i \frac{\mathcal{O}_i[g, \phi]}{\Lambda^{\delta_i - 4}} \right] . \tag{2.33}$$

Here, $\mathcal{O}_i[g, \phi]$ are operators constructed from curvature invariants and from ϕ and its derivatives. In spacetimes where the curvature is small in units of the cutoff Λ, the only important coupling in $S_{g,\phi}$ is

$$S_{g,\phi}^{(4)} = \int \mathrm{d}^4x \sqrt{-g} \, \xi \, \phi^2 R , \tag{2.34}$$

where ξ is a dimensionless coefficient. One can perform a Weyl rescaling of the metric,

$$g_{\mu\nu} \mapsto \bar{g}_{\mu\nu} \equiv e^{2\omega(\phi)} g_{\mu\nu} , \tag{2.35}$$

so that by a suitable choice of the function $2\omega(\phi)$, one arrives at $\xi = 0$, known as *minimal coupling*. However, the rescaling (2.35) also changes any other terms in the full action that are not conformally invariant.

2.1.5 Time dependence

To complete our survey of the basic principles of effective field theory, we need to discuss if and how effective field theory applies to time-dependent settings such as those arising in cosmology. (For further discussion, see [181, 209, 210].)

An immediate concern might be that we have classified heavy and light states relative to a cutoff energy, but energy conservation is inapplicable in time-dependent backgrounds, and when the background evolution is rapid, high-energy modes can be produced out of low-energy modes. Fortunately, in many cosmological applications the time dependence of the background is sufficiently slow to be treated *adiabatically*. We can then define an adiabatic notion of energy at a given time, and define the split into light and heavy fields relative to a slowly evolving cutoff $\Lambda(t)$. When the background evolution is sufficiently rapid to allow the production of heavy states, $|\dot{\Lambda}|/\Lambda^2 \gg 1$, then the system may not admit a description in terms of an effective theory containing only the light fields; the solutions to the equations of motion of the EFT will contain only the adiabatic solutions of the full theory. In inflation, the adiabatic approximation is justified as long as (i) we start in the *Bunch–Davies vacuum* (also called the adiabatic vacuum) and (ii) the subsequent evolution is adiabatic. This situation applies to a very broad range of inflationary models, but there are interesting exceptions; for a recent discussion of this issue see [211, 212].

Even for slowly evolving backgrounds, there may be *level crossings*: the slow evolution of the cutoff, $\Lambda(t)$, may cause some light fields to leave the EFT, and/or may draw in light fields that were previously heavy enough to integrate out. Thus, one effective theory evolves into another over time. We will encounter this possibility in large-field inflation (see Section 2.3).

In summary, the methods of effective field theory are applicable in backgrounds whose time evolution is sufficiently adiabatic, provided also that the initial state is the Bunch–Davies vacuum. In this setting we can focus on the evolution of low-energy states, without having to worry about the production of high-energy states. In the rest of this work, we will mostly consider adiabatic evolution, and any violations of adiabaticity will be noted.

2.2 Effective theories of inflation

In Chapter 1, we defined inflation as an extended period of quasi-de Sitter evolution, with $-\dot{H} \ll H^2$, but we did not specify the physical origin of the inflationary background $H(t)$. In this section, we will show that the dynamics of a slowly rolling scalar field leads to inflation. However, at this level, models of slow-roll inflation are toy models that lack a clear connection to the rest of physics. To make the models more realistic, we will embed them into low-energy effective theories, allowing us to discuss high-scale corrections to the slow-roll actions. A striking feature is that the inflationary dynamics is sensitive even to Planck-suppressed corrections, as we will explain in Section 2.3.

2.2.1 Slow-roll: dynamics and perturbations

One of the earliest and most influential models of inflation uses a single scalar field, the *inflaton* ϕ, minimally coupled to gravity [19, 20],

$$S = \int d^4x \sqrt{-g} \left[\frac{M_{\rm pl}^2}{2} R - \frac{1}{2}(\partial\phi)^2 - V(\phi) \right], \qquad (2.36)$$

where we have allowed for an arbitrary inflaton potential $V(\phi)$ (see Fig. 2.3).

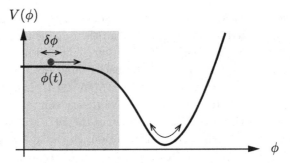

Fig. 2.3 Example of a slow-roll potential. Inflation occurs in the shaded part of the potential. In addition to the homogeneous evolution $\phi(t)$, the inflaton experiences spatially varying quantum fluctuations $\delta\phi(t, \boldsymbol{x})$.

Classical dynamics

The Friedmann equation and the Klein–Gordon equation for the homogeneous background field $\phi(t)$ are

$$3M_{\mathrm{pl}}^2 H^2 = \frac{1}{2}\dot{\phi}^2 + V \qquad \text{and} \qquad \ddot{\phi} + 3H\dot{\phi} = -V' \,, \qquad (2.37)$$

where $V' \equiv \partial_\phi V$. These equations can be combined into

$$\varepsilon \equiv -\frac{\dot{H}}{H^2} = \frac{\frac{1}{2}\dot{\phi}^2}{M_{\mathrm{pl}}^2 H^2} \,. \qquad (2.38)$$

Inflation ($\varepsilon < 1$) therefore occurs when the potential energy of the field dominates over the kinetic energy, $V \gg \frac{1}{2}\dot{\phi}^2$. The kinetic energy stays small and slow-roll persists if the acceleration of the field is small, $|\ddot{\phi}| \ll 3H|\dot{\phi}|$.

The conditions for prolonged slow-roll inflation can be expressed as conditions on the shape of the potential [213]:

$$\epsilon \equiv \frac{M_{\mathrm{pl}}^2}{2}\left(\frac{V'}{V}\right)^2 \ll 1 \,, \qquad |\eta| \equiv M_{\mathrm{pl}}^2 \frac{|V''|}{V} \ll 1 \,. \qquad (2.39)$$

During a slow-roll period, the "potential slow-roll parameters" ϵ and η are related to the "Hubble slow-roll parameters" ε and $\tilde{\eta}$ (see Section 1.2) via $\epsilon \approx \varepsilon$ and $\eta \approx 2\varepsilon - \tilde{\eta}/2$. We will see that realizing the slow-roll conditions (2.39) in a theory of fundamental physics is a nontrivial task. In particular, $|\eta| \ll 1$ requires a small hierarchy between the inflaton mass and the Hubble scale, $m^2 = V'' \ll 3H^2 \approx V/M_{\mathrm{pl}}^2$. Explaining the small inflaton mass is one of the key challenges for any microscopic theory of inflation.

Quantum fluctuations

As we explained in Chapter 1, light fields experience quantum fluctuations during inflation. As a result of fluctuations in the inflaton, $\delta\phi(t, \boldsymbol{x})$, some regions of space remain potential dominated longer than others, and different parts of the universe undergo slightly different evolution. After inflation, these differences in the evolution induce curvature perturbations $\mathcal{R}(t, \boldsymbol{x})$, which lead to density perturbations $\delta\rho(t, \boldsymbol{x})$.

We note that the inflaton fluctuation $\delta\phi$ plays the role of the Goldstone boson π of broken time translations (see Section 1.2). In spatially flat gauge the two are simply proportional,

$$\pi = \frac{\delta\phi}{\dot{\phi}} \ . \tag{2.40}$$

Using (1.14), we then find

$$\mathcal{R}(t, \boldsymbol{x}) = -H\pi(t, \boldsymbol{x}) = -\frac{H}{\dot{\phi}}\delta\phi(t, \boldsymbol{x}) \ . \tag{2.41}$$

During the slow-roll period, the sound speed of the Goldstone boson is equal to the speed of light, $c_s = 1$ (see Appendix B). The analysis of Section 1.2 then implies the following results for the spectra of scalar and tensor fluctuations:

$$\Delta_{\mathcal{R}}^2 = \frac{1}{24\pi^2}\frac{1}{\epsilon}\frac{V}{M_{\mathrm{pl}}^4} \ , \qquad \Delta_h^2 = \frac{2}{3\pi^2}\frac{V}{M_{\mathrm{pl}}^4} \ . \tag{2.42}$$

The scalar spectral index and the tensor-to-scalar ratio are

$$n_s - 1 = 2\eta - 6\epsilon \ , \tag{2.43}$$

$$r = 16\epsilon \ . \tag{2.44}$$

These observables should be evaluated at the time when the *pivot scale*[16] – a representative scale among the scales probed by the CMB – exited the horizon. This moment corresponds to a specific point in field space, $\phi_\star$, at which the number of *e*-folds of inflation remaining is (for $\phi_\star > \phi_{\mathrm{end}}$)

$$N_\star = \int_{\phi_{\mathrm{end}}}^{\phi_\star} \frac{d\phi}{M_{\mathrm{pl}}}\frac{1}{\sqrt{2\epsilon}} \ . \tag{2.45}$$

The value of $N_\star$ depends on the inflationary model and on the details of reheating. Typically, one finds $40 \lesssim N_\star \lesssim 60$. In Section 2.2.2, we give a few examples of specific slow-roll models and their observational predictions.

2.2.2 *Slow-roll: selected models*

You know how sometimes you meet somebody and they're really nice, so you invite them over to your house and you keep talking with them and they keep telling you more and more cool stuff? But then at some point you're like, maybe we should call it a day, but they just won't leave and they keep talking and as more stuff comes up it becomes more and more disturbing and you're like, just stop already? That's kind of what happened with inflation.

Max Tegmark [214].

---

[16] In the WMAP analysis the pivot scale was chosen to be $k_\star = 0.002$ Mpc^{-1}, while for Planck $k_\star = 0.05$ Mpc^{-1}.

We will not provide a comprehensive account of the vast landscape of slow-roll models, but instead give a brief sketch of some of the most important classes of models (see Fig. 2.4). For more details on slow-roll model-building we refer the reader to [215–218].

Chaotic inflation with a monomial potential

An important class of inflationary models arises when the potential is a simple monomial,

$$V(\phi) = \mu^{4-p}\phi^p \ , \tag{2.46}$$

where $p > 0$, and μ is a parameter with the dimensions of mass.[17] The slow-roll parameters for the potential (2.46) are

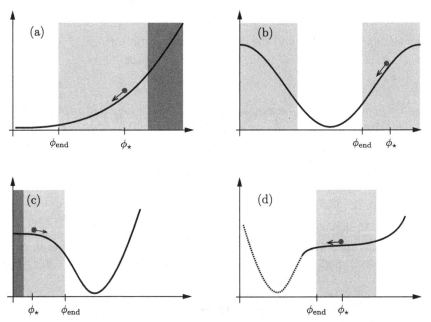

Fig. 2.4 Examples of different classes of slow-roll potentials: (a) chaotic inflation, (b) natural inflation, (c) hilltop inflation, and (d) inflection point inflation. The light gray regions indicate the parts of the potential where slow-roll inflation occurs. The dark gray regions denote regions of *eternal inflation*. The figures are *not* drawn to scale: (a)+(b) correspond to large-field models ($\Delta\phi > M_{\mathrm{pl}}$), while (c)+(d) are small-field models ($\Delta\phi < M_{\mathrm{pl}}$).

---

[17] In much of the literature, *chaotic inflation* [34] is taken to mean inflation with a potential of the monomial form (2.46). In fact, the idea of chaotic inflation proposed by Linde in [34] is far more general – see e.g. [219, 220] for explanations and historical background. A more accurate term for (2.46) is "chaotic inflation with a monomial potential."

$$\epsilon = \frac{p^2}{2}\left(\frac{M_{\rm pl}}{\phi}\right)^2 \; , \quad \eta = p(p-1)\left(\frac{M_{\rm pl}}{\phi}\right)^2 \; . \tag{2.47}$$

Notice that ϵ and η do not depend on the scale μ. Using (2.45), the number of e-folds occurring in the region $\phi \le \phi_\star$ is found to be

$$N_\star \approx \frac{1}{2p}\left(\frac{\phi_\star}{M_{\rm pl}}\right)^2 \; , \tag{2.48}$$

implying that in these models, prolonged inflationary expansion requires a super-Planckian displacement, $\phi_\star \gg M_{\rm pl}$. At the pivot scale, the spectral index and the tensor-to-scalar ratio are

$$n_s - 1 = -\frac{(2+p)}{2N_\star} \; , \quad r = \frac{4p}{N_\star} \; . \tag{2.49}$$

Let us illustrate these results in a few simple cases (see Fig. 2.5), setting $N_\star = 60$ for definiteness:

$$p = 1: \quad n_s \approx 0.975 \; , \quad r \approx 0.07 \; , \quad \phi_\star \approx 11 M_{\rm pl} \; , \tag{2.50}$$

$$p = 2: \quad n_s \approx 0.967 \; , \quad r \approx 0.13 \; , \quad \phi_\star \approx 15 M_{\rm pl} \; , \tag{2.51}$$

$$p = 3: \quad n_s \approx 0.958 \; , \quad r \approx 0.20 \; , \quad \phi_\star \approx 19 M_{\rm pl} \; , \tag{2.52}$$

$$p = 4: \quad n_s \approx 0.950 \; , \quad r \approx 0.27 \; , \quad \phi_\star \approx 22 M_{\rm pl} \; . \tag{2.53}$$

An approximate shift symmetry can make a model of chaotic inflation bottom-up natural (see Section 2.3.3), but to establish top-down naturalness a realization in string theory is necessary. Attempts to embed chaotic inflation in string theory are described in Section 5.4.

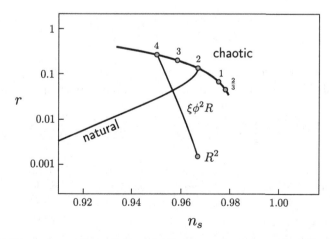

Fig. 2.5 Some slow-roll predictions in the n_s-r plane, assuming 60 e-folds of inflation.

Natural inflation

An influential idea in inflationary model-building is that the inflaton could be a pseudoscalar *axion*. At the perturbative level, an axion enjoys a continuous shift symmetry, but this is broken nonperturbatively to a discrete symmetry, leading to a potential of the form

$$V(\phi) = \frac{V_0}{2}\left[1 - \cos\left(\frac{\phi}{f}\right)\right],\tag{2.54}$$

where f is the axion decay constant. For $f \gtrsim 4M_{\rm pl}$, the potential (2.54) supports *natural inflation* [36]. Because of the shift symmetry, the model is bottom-up natural. Establishing top-down naturalness requires finding an axion in string theory whose effective decay constant can be larger than the Planck scale – see Section 5.4. At the pivot scale, one finds the following expressions for the scalar tilt and the tensor-to-scalar ratio [216]:

$$n_s - 1 = -\alpha\,\frac{e^{N_\star\alpha}+1}{e^{N_\star\alpha}-1}\ \xrightarrow{\ \alpha\ll1\ }\ -\frac{2}{N_\star},\tag{2.55}$$

$$r = 8\alpha\,\frac{1}{e^{N_\star\alpha}-1}\ \xrightarrow{\ \alpha\ll1\ }\ +\frac{8}{N_\star},\tag{2.56}$$

where we have defined $\alpha \equiv M_{\rm pl}^2/f^2$. As expected, the predictions for natural inflation reduce to those of $m^2\phi^2$ chaotic inflation for $f \gg M_{\rm pl}$: cf. Eq. (2.49) with $p = 2$.

Hilltop inflation

Consider the situation where inflation occurs near the fixed point of a symmetry, that is at a point in field space with $V' = 0$. Expanding the potential around this point gives

$$V(\phi) = V_0 + \frac{1}{2}m^2\phi^2 + \cdots.\tag{2.57}$$

For positive m^2, the symmetry is intact, while for negative m^2 the symmetry is spontaneously broken. Consider the latter case and write the potential as

$$V(\phi) = V_0\left[1 + \frac{1}{2}\eta_0\frac{\phi^2}{M_{\rm pl}^2} + \cdots\right],\qquad \text{where}\quad \eta \approx \eta_0 < 0\,.\tag{2.58}$$

For small η_0, *hilltop inflation* [221, 222] occurs (see Fig. 2.4c). The higher-order terms in (2.58) become important for large values of ϕ. They define the precise value $\phi_{\rm end}$ at which inflation ends, and determine the value of the cosmological constant in the global vacuum after inflation. We will assume that $\phi_{\rm end} \lesssim M_{\rm pl}$. If the higher-order terms in (2.58) are irrelevant when the pivot scale exits the horizon, then the spectral tilt and the tensor-to-scalar ratio are [222, 223]

$$n_s - 1 = 2\eta_0 \, , \tag{2.59}$$

$$r = 2(1 - n_s)^2 e^{-N_\star(1-n_s)} \left(\frac{\phi_{\text{end}}}{M_{\text{pl}}}\right)^2 \approx 10^{-3} \left(\frac{\phi_{\text{end}}}{M_{\text{pl}}}\right)^2 . \tag{2.60}$$

The model has two free parameters: the curvature of the hilltop, η_0, and the field value at the end of inflation, ϕ_{end}.

Inflection point inflation

Away from any symmetry points, a generic potential has the expansion, around $\phi = 0$,

$$V(\phi) = V_0 \left[1 + \lambda_0 \frac{\phi}{M_{\text{pl}}} + \frac{1}{2}\eta_0 \frac{\phi^2}{M_{\text{pl}}^2} + \frac{1}{3!}\mu_0 \frac{\phi^3}{M_{\text{pl}}^3} + \cdots \right] . \tag{2.61}$$

Again, higher-order terms may become important towards the end of inflation, but are assumed to be irrelevant when the pivot scale exits the horizon. To get enough e-folds of inflation, we require $|\eta_0| \ll 1$. The special case $\eta_0 = 0$ corresponds to *inflection point inflation* ($V'' = 0$, see Fig. 2.4d):

$$V(\phi) \approx V_0 \left[1 + \lambda_0 \frac{\phi}{M_{\text{pl}}} + \frac{1}{3!}\mu_0 \frac{\phi^3}{M_{\text{pl}}^3} + \cdots \right] . \tag{2.62}$$

This type of potential arises in D-brane inflation [224, 225] (see Section 5.1) as well as in field-theoretic constructions in the Minimal Supersymmetric Standard Model [226]. The spectral tilt derived from the potential (2.62) is [225, 227]

$$n_s - 1 = -4\sqrt{\frac{\lambda_0\mu_0}{2}} \cot\left(N_\star \sqrt{\frac{\lambda_0\mu_0}{2}}\right) , \tag{2.63}$$

and is uncorrelated with the value of the tensor-to-scalar ratio,

$$r = 16\lambda_0^2 . \tag{2.64}$$

Constraints on the total number of e-folds and the scalar amplitude typically force λ_0 to be very small and the tensor signal to be unobservable, $r \ll 0.01$.

Hybrid inflation

Inflationary models with small-field potentials, such as (c) and (d) in Fig. 2.4, often end through an instability induced by coupling the inflaton field ϕ to an additional "waterfall" field Ψ. The combination of a slow-roll potential and a waterfall instability is called *hybrid inflation* [228]. As a simple example, consider the two-field potential (see Fig. 2.6),

$$V(\phi, \Psi) = V(\phi) + V(\Psi) + \frac{1}{2}g\phi^2\Psi^2 , \tag{2.65}$$

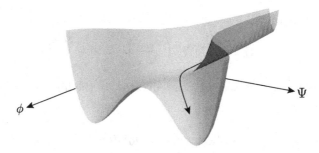

Fig. 2.6 A typical potential of hybrid inflation. A tachyonic instability ends inflation while the slow-roll parameter ϵ is still small.

where $V(\phi)$ is the slow-roll potential and $V(\Psi)$ is a potential of symmetry-breaking type,

$$V(\Psi) \equiv \frac{1}{4\lambda}\left(M^2 - \lambda\Psi^2\right)^2 . \tag{2.66}$$

We assume that $V(\phi) \ll M^4/4\lambda$, so that the dominant contribution to the inflationary energy density comes from the false vacuum energy of the symmetry-breaking potential. The coupling between ϕ and Ψ induces an effective mass for the waterfall field that depends on the value of the inflaton,

$$M_\Psi^2(\phi) = -M^2 + g\phi^2 . \tag{2.67}$$

This vanishes at the special point $\phi = \phi_c \equiv M/\sqrt{g}$. For $\phi > \phi_c$, the field Ψ is stabilized at $\Psi = 0$, and can be integrated out, so that the theory reduces to that of single-field slow-roll inflation with $V_{\rm eff}(\phi) \approx M^4/4\lambda + V(\phi)$. As ϕ approaches ϕ_c from above, Ψ becomes light and the effective description involves both fields. Finally, for $\phi < \phi_c$, the field Ψ becomes tachyonic and ends inflation. Notice that hybrid inflation requires a hierarchy between the masses of the two fields, $V_{,\phi\phi} \ll M^2$. This issue is discussed, and technically natural examples are constructed, in [229, 230].

Starobinsky's R^2 inflation

One of the earliest models of inflation was written down by Starobinsky in 1980 [231].[18] Motivated by [237], Starobinsky considered one-loop corrections to the Einstein–Hilbert action. These lead to an effective action of the form of (2.29). Starobinsky's model considers only the R^2 correction,

$$S = \frac{M_{\rm pl}^2}{2} \int {\rm d}^4x\sqrt{-g}\left(R + \frac{\alpha}{2M_{\rm pl}^2}R^2\right) . \tag{2.68}$$

---

[18] The Starobinsky model has recently received renewed attention – see [232] for a superconformal generalization and [233–236] for supergravity constructions.

For sufficiently large α, this action leads to inflationary dynamics. The easiest way to see this is to perform a conformal transformation, $g_{\mu\nu} \mapsto \tilde{g}_{\mu\nu} = \Omega^2 g_{\mu\nu}$, with $\Omega^2 \equiv 1 + \alpha R/M_{\mathrm{pl}}^2$, to arrive at the action of a minimally coupled scalar field, $\phi \equiv M_{\mathrm{pl}}\sqrt{\frac{2}{3}}\ln(1 + \alpha R/M_{\mathrm{pl}}^2)$,

$$S = \int \mathrm{d}^4 x \sqrt{-\tilde{g}}\left(\frac{M_{\mathrm{pl}}^2}{2}\tilde{R} - \frac{1}{2}(\partial\phi)^2 - V(\phi) \right) , \tag{2.69}$$

with potential

$$V(\phi) = \frac{M_{\mathrm{pl}}^4}{4\alpha}\left(1 - \exp\left[-\sqrt{\frac{2}{3}}\frac{\phi}{M_{\mathrm{pl}}} \right] \right)^2 . \tag{2.70}$$

The slow-roll parameters associated with the potential (2.70) are

$$\eta = -\frac{4}{3}e^{-\sqrt{2/3}\,\phi/M_{\mathrm{pl}}} , \quad \epsilon = \frac{3}{4}\eta^2 . \tag{2.71}$$

Inflation occurs for $\phi > M_{\mathrm{pl}}$. The normalization of the scalar amplitude requires that

$$\alpha = 2.2 \times 10^8 . \tag{2.72}$$

Such a large parameter seems unnatural from a top-down perspective. Its bottom-up naturalness is discussed in [238, 239]. The scalar spectral tilt and the tensor-to-scalar ratio are

$$n_s - 1 \approx -\frac{2}{N_\star} , \quad r \approx \frac{12}{N_\star^2} . \tag{2.73}$$

Non-minimally coupled inflation

It is also interesting to consider a scalar field φ with a non-minimal coupling to gravity [240–245]. The simplest such coupling is the operator $\varphi^2 R$.[19] Adding this to the action (2.36), we get

$$S = \int \mathrm{d}^4 x \sqrt{-g}\left[\frac{M_{\mathrm{pl}}^2}{2}\left(1 + \xi\frac{\varphi^2}{M_{\mathrm{pl}}^2} \right)R - \frac{1}{2}(\partial\varphi)^2 - \frac{\lambda}{4}\varphi^4 \right] , \tag{2.74}$$

where, for concreteness, we have chosen a quartic term for the inflaton potential. The parameter ξ determines the strength of the non-minimal coupling to gravity. Again, it is convenient to go to Einstein frame by performing a conformal rescaling, $\tilde{g}_{\mu\nu} = \Omega^2 g_{\mu\nu}$, with $\Omega^2 \equiv 1 + \xi\varphi^2/M_{\mathrm{pl}}^2$. The action then takes the form

$$S = \int \mathrm{d}^4 x \sqrt{-\tilde{g}}\left[\frac{M_{\mathrm{pl}}^2}{2}\tilde{R} - \frac{1}{2}k(\varphi)(\partial\varphi)^2 - V(\varphi) \right] , \tag{2.75}$$

---

[19] This interaction plays a fundamental role in Higgs inflation [246].

where

$$k(\varphi) = \frac{1 + (6\xi + 1)\psi^2}{(1 + \psi^2)^2} \,, \tag{2.76}$$

$$V(\varphi) = \frac{\lambda M_{\rm pl}^4}{4\xi^2} \frac{\psi^4}{(1 + \psi^2)^2} \,, \qquad \psi^2 \equiv \frac{\xi\varphi^2}{M_{\rm pl}^2} \,. \tag{2.77}$$

The canonically normalized field, $\phi = \int \sqrt{k(\varphi)} \, d\varphi$, is

$$\frac{\phi}{M_{\rm pl}} = \sqrt{\frac{6\xi + 1}{\xi}} \sinh^{-1} \left(\sqrt{6\xi + 1} \, \psi \right) - \sqrt{6} \sinh^{-1} \left(\sqrt{6\xi} \frac{\psi}{\sqrt{1 + \psi^2}} \right) \,. \tag{2.78}$$

For $\xi \gg 1$, this can be approximated as

$$\frac{\phi}{M_{\rm pl}} \approx \sqrt{\frac{3}{2}} \ln(1 + \psi^2) \,, \tag{2.79}$$

and the potential becomes

$$V(\phi) = \frac{\lambda M_{\rm pl}^4}{4\xi^2} \left(1 - \exp\left[-\sqrt{\frac{2}{3}} \frac{\phi}{M_{\rm pl}} \right] \right)^2 \,. \tag{2.80}$$

This is identical to the potential (2.70) in the Starobinsky model. In the limit $\xi \gg 1$, the model (2.74) therefore has the same phenomenology as (2.68). The constraint (2.72) translates into

$$\xi = 47000\sqrt{\lambda} \,, \tag{2.81}$$

and the predictions for n_s and r are those of (2.73). The predictions for general ξ were derived in [244, 245]:

$$n_s - 1 = -\frac{32\xi}{16\xi N_\star - 1} \xrightarrow{\xi \gg 1} -\frac{2}{N_\star} \,, \tag{2.82}$$

$$r = +\frac{12}{N_\star^2} \frac{6\xi + 1}{6\xi} \xrightarrow{\xi \gg 1} +\frac{12}{N_\star^2} \,. \tag{2.83}$$

We see that the model interpolates between ϕ^4 chaotic inflation (for $\xi = 0$) and the Starobinsky model (for $\xi \gg 1$) – see Fig. 2.5.

2.2.3 Non-slow-roll: k-inflation

So far, we have only considered slow-roll models with canonical kinetic terms. An alternative class of models – known as *k-inflation* [40, 247] or $P(X)$ *theories* [124] – considers the possibility that inflation was driven by nontrivial kinetic effects rather than by a flat potential. An efficient way to model these effects is through the action

$$S = \int d^4x \sqrt{-g} \left[\frac{M_{\rm pl}^2}{2} R + P(X, \phi) \right] \,, \tag{2.84}$$

where $P(X, \phi)$ is (so far) an arbitrary function of the inflaton field ϕ and of its kinetic energy $X \equiv -\frac{1}{2}(\partial \phi)^2$. The stress–energy tensor arising from (2.84) corresponds to a perfect fluid with pressure P and energy density $\rho = 2X P_{,X} - P$, where $P_{,X}$ denotes a derivative with respect to X. The Friedmann equation and the Klein–Gordon equation are

$$3M_{\text{pl}}^2 H^2 = 2P_{,X} X - P \qquad \text{and} \qquad \frac{d}{dt}\left(a^3 P_{,X} \dot{\phi}\right) = a^3 P_{,\phi} , \qquad (2.85)$$

so the inflationary parameter (1.4) becomes

$$\varepsilon = -\frac{\dot{H}}{H^2} = \frac{3X P_{,X}}{2X P_{,X} - P} . \qquad (2.86)$$

The condition for inflation is still $\varepsilon \ll 1$, but it is now a condition on the functional form of $P(X)$. The fluctuations in $P(X)$ theories propagate with a nontrivial speed of sound (see Appendix B),

$$c_s^2 = \frac{dP}{d\rho} = \frac{P_{,X}}{P_{,X} + 2X P_{,XX}} . \qquad (2.87)$$

The predictions for n_s and r are the same as in Section 1.2:

$$n_s - 1 = -2\varepsilon - \tilde{\eta} - \kappa , \qquad (2.88)$$

$$r = 16\varepsilon c_s , \qquad (2.89)$$

where $\tilde{\eta}$ and κ were defined in (1.30). We saw in Section 1.4.2 that a small sound speed leads to observable equilateral-type non-Gaussianity. We will discuss this further in Section 2.3.4, where we also emphasize the need to UV-complete theories such as (2.84). In Section 5.3, we present *DBI inflation* [39, 41] as a specific example in string theory.

2.2.4 Inflation in effective field theory

The models that we have presented so far are toy models; they are decoupled from the rest of physics and lack ultraviolet completions. The most conservative way to address these deficiencies is to work in effective field theory. In the remainder of this chapter, we will discuss the embedding of slow-roll inflation in the framework of effective field theory.

The starting point is the EFT Lagrangian (2.14) minimally coupled to gravity,

$$S_{\text{eff}}[\phi] = \int d^4 x \sqrt{-g} \left[\frac{M_{\text{pl}}^2}{2} R + \mathcal{L}_l[\phi] + \sum_i c_i \frac{\mathcal{O}_i[\phi]}{\Lambda^{\delta_i - 4}} \right] , \qquad (2.90)$$

where $\mathcal{L}_l[\phi]$ includes the canonical kinetic term $-\frac{1}{2}(\partial \phi)^2$ as well as any renormalizable interactions. As we explained at length above, the sum over non-renormalizable terms parameterizes the effects of massive fields on the EFT of the light fields. When the UV theory is unknown, one can at best make assumptions about the symmetry structure of the UV theory, and then include all higher-dimension operators $\mathcal{O}_i$ consistent with these symmetries. Following the remarks in Section 2.1.4, the maximal cutoff of the EFT is the Planck scale, $\Lambda \lesssim M_{\rm pl}$. In order for the effective theory (2.90) to remain valid during the freeze-out of cosmological perturbations, i.e. when $\omega = H$, the minimal cutoff is the inflationary Hubble scale, $\Lambda \gtrsim H$. Thus, all fields with masses $m \lesssim H$ are part of the EFT. We will begin by discussing the case where the only light degrees of freedom are the graviton and a single real inflaton scalar; models with multiple light scalars are discussed in subsequent sections and in Appendix C.

In most particle physics applications of effective field theory, higher-dimension operators only contribute small ("irrelevant") corrections to the leading dynamics. As the cutoff is pushed to the Planck scale, these contributions typically become negligible. (One notable exception is gravity-mediated supersymmetry breaking.) It is a special feature of the effective theory of inflation (2.90) that some irrelevant operators play a crucial role at low energies, not just for precision observables, but even for the zeroth-order dynamics. Slow-roll inflation is sensitive even to Planck-suppressed operators. The next section is devoted to a careful discussion of this important fact.

2.3 Ultraviolet sensitivity

We will highlight four aspects of the UV sensitivity of inflation. The first two (eta problems I and II) are universal and apply to any slow-roll[20] model of inflation. The last two (super-Planckian displacements and non-Gaussianity) only apply to specific classes of inflationary theories.

2.3.1 Eta problem I: radiative corrections

As we have explained in Section 2.1, the unknown heavy physics above the cutoff scale has two effects: (i) it renormalizes the couplings of the light fields, and (ii) it introduces new non-renormalizable interactions. Both effects have to be addressed in a complete discussion of the inflationary dynamics.

We have seen that quantum corrections tend to drive scalar masses to the cutoff scale, unless the fields are protected by symmetries. In the case of inflation, this implies the following quantum correction to the inflaton mass:

$$\Delta m^2 \sim \Lambda^2 \,. \tag{2.91}$$

Since consistency of the EFT treatment requires that $\Lambda > H$, we find a large renormalization of the inflationary eta parameter (2.39),

---

[20] Variations of these problems arise in most non-slow-roll models as well. For example, when nontrivial kinetic terms make a rapidly varying potential innocuous (cf. Section 2.2.3), one must still ensure that the necessary kinetic terms are not affected by Planck-suppressed contributions. The general problem is to arrange that the *action*, not just the potential, changes slowly during inflation.

$$\Delta\eta \sim \frac{\Lambda^2}{H^2} \gtrsim 1 \,, \tag{2.92}$$

and sustained slow-roll inflation appears to be unnatural. This difficulty is known as the *eta problem*. The eta problem in the context of supergravity was emphasized long ago in [248]. However, the issue is actually far more general, afflicting any construction of slow-roll inflation in effective field theory.

Two strategies are available for addressing the eta problem: fine-tuning the potential, or appealing to symmetries. The problem is a resilient one because approaches based on symmetries face serious limitations, and have only occasionally been successful. The symmetry options are the same ones discussed in Section 2.1.3: supersymmetry and/or global internal symmetries. We will discuss these in turn, but it is worth stating the upshot in advance: supersymmetry ameliorates but cannot completely solve the problem, while global symmetry arguments require precise control of Planck-suppressed operators breaking the symmetry, motivating a treatment in quantum gravity.

Supersymmetry

Even if the inflaton is part of a supersymmetric action, the inflationary background solution spontaneously breaks SUSY, because the energy density is necessarily positive. Nevertheless, SUSY still limits the size of radiative corrections, because sufficiently high frequency modes are insensitive to the effects of the spacetime curvature during inflation. The cancellation between boson and fermion loops therefore still applies in the high-energy regime, just as in flat space. On the other hand, modes with frequencies below the Hubble scale, $\omega \lesssim H$, do experience nontrivial effects from the expanding background. Boson and fermion propagators are then modified by the coupling to the spacetime curvature, with mass splittings within supermultiplets that are typically of order H, and the corresponding loops no longer cancel. Radiative corrections to the inflaton mass are therefore naturally of order of the Hubble scale,

$$\Delta m^2 \sim H^2 \,. \tag{2.93}$$

This is smaller than the correction in (2.91), but not small enough to evade the eta problem:

$$\Delta\eta \sim 1 \,. \tag{2.94}$$

This qualitative argument is confirmed in detail by investigations of inflation in supergravity [248] and in string theory (see Section 4.2). Hence, although SUSY ameliorates the eta problem, it does not solve it; successful inflation still requires fine-tuning of the mass term [249, 250], although much less than in an EFT without SUSY.

The degree of fine-tuning implied by (2.94) depends to some extent on the underlying model. In small-field models with $\epsilon \ll \eta$, the value of η at horizon crossing is related to the scalar spectral index, $\eta \approx \frac{1}{2}(n_s - 1)$. For the Planck best-fit, $n_s \approx 0.96$, this implies $\eta \approx 0.02$, so the required fine-tuning is at the percent level.

Global symmetries

We have discussed global symmetries extensively in Section 2.1.3. As we explained there, a small scalar mass is "bottom-up natural" if the renormalizable part of the Lagrangian (the IR theory) respects an approximate shift symmetry,

$$\phi \mapsto \phi + const. \tag{2.95}$$

In other words, the theory contains no relevant or marginal operators that violate (2.95). Then, loops of the light fields do not drive the scalar mass up to the cutoff; quantum corrections to the scalar mass are suppressed by the parameter measuring the weak breaking of the symmetry. Whether inflationary models based on (2.95) are "top-down natural" is an important question for a theory of quantum gravity.

2.3.2 Eta problem II: higher-dimension operators

We saw in Section 2.1.3 that not all desirable symmetries of the IR theory can be realized in a consistent UV theory; in particular, we recalled the common lore that quantum gravity breaks all continuous global symmetries. Correspondingly, although a low-energy theory with light scalars that respects (2.95) is radiatively stable, such a theory is not necessarily "top-down natural": *irrelevant operators* may spoil the desired symmetry. Whether the symmetry survives is a question for the ultraviolet completion, and cannot be addressed by studying the renormalizable Lagrangian.

As an example, consider the dimension-six operator

$$\mathcal{O}_6 = cV_l(\phi)\frac{\phi^2}{\Lambda^2} \,, \tag{2.96}$$

where c is a constant, and $V_l(\phi)$ consists of the renormalizable terms in the potential, cf. (2.90). Even if $V_l(\phi)$ respects an approximate shift symmetry, this is broken by $\mathcal{O}_6$. Provided that the inflaton vev is smaller than the cutoff, $\phi < \Lambda$, the operator $\mathcal{O}_6$ makes only a small correction to the inflationary potential, $\Delta V \ll V(\phi)$. Nevertheless, its effect on the inflaton mass is significant:

$$\Delta\eta \approx 2c \left(\frac{M_{\mathrm{pl}}}{\Lambda}\right)^2 \,. \tag{2.97}$$

For $c \sim \mathcal{O}(1)$ and $\Lambda < M_{\mathrm{pl}}$, the theory again suffers from the eta problem. (Notice that the overall scale of the potential cancels in (2.97).) Even if the operator in (2.96) is Planck-suppressed, $\Lambda \to M_{\mathrm{pl}}$, it cannot be ignored in discussions of the inflationary dynamics.

In a theory with a single real scalar ϕ, it is difficult to give a convincing argument for the absence of couplings of the form (2.96). Note in particular that if one forbids ϕ^2 via a global symmetry under which ϕ transforms linearly, one would simultaneously exclude the kinetic term $-\frac{1}{2}(\partial\phi)^2$. An influential approach is therefore to take ϕ to transform nonlinearly under a global symmetry (e.g. by taking ϕ to be an axion – see Section 5.4) and/or to be the phase of a complex scalar (cf. e.g. [251]).

Although operators of the simple form (2.96) arise in many ultraviolet completions (see Section 4.2 and Chapter 5), more general non-renormalizable interactions are also common, and can give comparable (or larger) effects. Consider an operator of the form

$$\mathcal{O}_\delta = c\langle V\rangle \left(\frac{\phi}{\Lambda}\right)^{\delta-4} , \tag{2.98}$$

where $\langle V\rangle$ is the vacuum energy at some stage of inflation. The correction to η is

$$\Delta\eta \approx c(\delta-4)(\delta-5)\left(\frac{M_{\rm pl}}{\Lambda}\right)^2\left(\frac{\phi}{\Lambda}\right)^{\delta-6} . \tag{2.99}$$

If $\Lambda = M_{\rm pl}$ and $\phi < \Lambda$, operators with $\delta \gg 6$ can be neglected. For this reason, addressing the eta problem in small-field inflation requires, at a *minimum*, characterizing Planck-suppressed interactions up to dimension six. However, operators with δ slightly larger than six are not strictly negligible unless $\phi \ll \Lambda$, while taking $\Lambda < M_{\rm pl}$ increases $\Delta\eta$ in (2.99).[21] The threshold beyond which non-renormalizable interactions can be neglected therefore varies from model to model, and depends on $\Lambda/M_{\rm pl}$ and $\phi/M_{\rm pl}$.

Explaining the absence of operators like (2.98) – including the special case (2.96) – requires an understanding of the leading high-energy corrections to the inflationary Lagrangian. When a symmetry is assumed in the EFT, one must demonstrate, in the context of an ultraviolet completion, that the symmetry survives non-renormalizable corrections such as (2.98). This sensitivity to UV physics is the key challenge for realizing inflation in a theory of fundamental physics. At the same time, the fact that Planck-scale effects do not decouple from inflation presents a striking opportunity: one can hope to use cosmological observations as a laboratory for physics at the highest energy scales.

2.3.3 Gravity waves and super-Planckian fields

Inflationary models that predict a detectably large primordial gravitational wave signal are extraordinarily sensitive to ultraviolet physics. To see this, we will

---

[21] Notice that for $4 < \delta < 6$, $\delta \neq 5$, the correction $\Delta\eta$ increases for small ϕ/Λ. Irrelevant operators with non-integer dimensions $\delta < 6$ can therefore dominate the dynamics in small-field inflation. For an example, see Section 5.1.

derive the *Lyth bound* [252], which relates observable tensor modes to super-Planckian displacements of the inflaton, $\Delta\phi \gtrsim M_{\rm pl}$. We will begin with a derivation of the Lyth bound in single-field slow-roll inflation, and then present extensions to more general scenarios.

The Lyth bound

Substituting (2.38) into $r = 16\varepsilon$, we can relate the tensor-to-scalar ratio r to the evolution of the inflaton field,

$$r = 8 \left(\frac{1}{M_{\rm pl}} \frac{{\rm d}\phi}{{\rm d}N} \right)^2 , \qquad \text{where} \quad {\rm d}N \equiv H {\rm d}t . \tag{2.100}$$

Integrating (2.100) from the time $N_\star$ when modes that are observable in the CMB exited the horizon, until the end of inflation at $N_{\rm end} \equiv 0$ (see Fig. 2.7), we get [252]

$$\frac{\Delta\phi}{M_{\rm pl}} = \int_0^{N_\star} {\rm d}N \sqrt{\frac{r(N)}{8}} . \tag{2.101}$$

To evaluate the integral in (2.101), it is useful to define

$$N_{\rm eff} \equiv \int_0^{N_\star} {\rm d}N \sqrt{\frac{r(N)}{r_\star}} , \tag{2.102}$$

where $r_\star$ is the tensor-to-scalar ratio measured in the CMB, so that

$$\frac{\Delta\phi}{M_{\rm pl}} = N_{\rm eff} \sqrt{\frac{r_\star}{8}} . \tag{2.103}$$

In slow-roll inflation, the relation

$$\frac{{\rm d}\ln r}{{\rm d}N} = - \left[n_s - 1 + \frac{r}{8} \right] , \tag{2.104}$$

combined with the observational constraints on $n_s - 1$ and r described in Chapter 1, imply that $N_{\rm eff} \sim N_\star$ (see e.g. [253, 254]). Taking $N_{\rm eff} \gtrsim 60$, we conclude that[22]

$$\frac{\Delta\phi}{M_{\rm pl}} \gtrsim 2 \times \left(\frac{r}{0.01} \right)^{1/2} . \tag{2.105}$$

To arrive at a maximally conservative bound in single-field slow-roll inflation, one can assume that slow-roll is valid only while the *observed* multipoles of the CMB exit the horizon, corresponding to $N_{\rm eff} \approx 7$. This leads to (cf. [252], which used a smaller $N_{\rm eff}$ because fewer multipoles had been observed in 1996)

$$\frac{\Delta\phi}{M_{\rm pl}} \gtrsim 0.25 \times \left(\frac{r}{0.01} \right)^{1/2} . \tag{2.106}$$

---

[22] One should not assume that simple models will approximately saturate (2.105); for example, chaotic inflation scenarios involve displacements roughly twice as large as required by the bound.

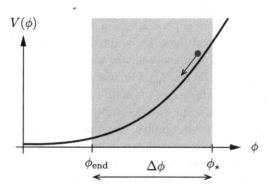

Fig. 2.7 Evolution of the inflaton field from the time when modes that are observable in the CMB exited the horizon, $\phi_\star$, to the end of inflation, ϕ_{end}. The total field displacement $\Delta\phi$ is related to the tensor-to-scalar ratio r by (2.105).

It is quite a remarkable coincidence that the level of tensors that is experimentally accessible (roughly $r \gtrsim 0.01$) is tied to the fundamental scale of quantum gravity, M_{pl}.

We strongly caution against viewing $\Delta\phi = M_{\text{pl}}$ as an absolute dividing line: the theoretical challenges of models with $\Delta\phi > M_{\text{pl}}$ are shared by models with slightly smaller displacements. In particular, although gravity itself becomes strongly coupled around the scale M_{pl}, parametrically controlled ultraviolet completions of gravity generally involve additional scales $\Lambda < M_{\text{pl}}$. For instance, the string scale and the Kaluza–Klein scale (see Section 4.1) are typically well below the Planck scale. Field excursions that are large compared to those scales raise concerns similar to the super-Planckian issues we describe below.

Finally, let us emphasize that the Lyth bound (2.105) is a purely kinematic statement, relating r to the distance in field space over which the inflaton moves. Although the bound has profound consequences in the context of effective field theory reasoning about natural Planck-suppressed interactions (see below), the derivation of (2.105) relied in no way on notions of naturalness, or on a Taylor expansion of the potential.

Super-Planckian fields in effective field theory

The simplest scenarios for large-field inflation involve a scalar field minimally coupled to gravity, with a monomial (or sinusoidal) potential that varies slowly over super-Planckian distances in field space. To understand the theoretical status of these models, it is instructive to examine them first from a purely bottom-up perspective, in quantum field theory coupled to general relativity, *without* accounting for the necessity of an ultraviolet completion of gravity. The only degrees of freedom are then the graviton and a single scalar inflaton. From this perspective, there are two issues that appear dangerous at first glance, but are in fact not at all problematic [219] (see [255] for a recent discussion).

First, one might worry that super-Planckian displacements of the inflaton will lead to super-Planckian energy densities, and correspondingly large gravitational backreaction. This concern is misplaced: the normalization (2.42) of the scalar fluctuations requires that $V \ll M_{\mathrm{pl}}^4$. For instance, in $m^2\phi^2$ chaotic inflation, (2.42) implies that the inflaton mass is small, $m \sim 10^{-5}M_{\mathrm{pl}}$, so that the energy density never becomes significant even for super-Planckian fields.

A second concern is radiative stability: do quantum corrections, from graviton loops and/or ϕ loops, destabilize the classical potential $V(\phi)$? No: the small value of the inflaton mass[23] m is technically natural, because the theory enjoys a shift symmetry in the limit $m \to 0$. Quantum corrections therefore do not destabilize the potential. In particular, the one-loop correction from graviton loops is [256]

$$\frac{\Delta V}{V} = c_1 \frac{V''}{M_{\mathrm{pl}}^2} + c_2 \frac{V}{M_{\mathrm{pl}}^4} \,, \tag{2.107}$$

where c_1 and c_2 are order-one numbers. Because $m \ll M_{\mathrm{pl}}$ and $V \ll M_{\mathrm{pl}}^4$, this is a small correction.

To summarize, in the low-energy theory of the inflaton and the graviton, potentials supporting large-field inflation can be radiatively stable, and in particular free of significant corrections from inflaton–graviton interactions. Thus, from the bottom-up perspective, large-field inflation is not problematic.

The essence of the problem of large-field inflation is that gravity requires an ultraviolet completion, and couplings of the inflaton to the degrees of freedom that provide this ultraviolet completion do not *necessarily* respect the symmetry structures needed to protect the inflaton potential in the low-energy theory. The effects of classical gravity (i.e. backreaction) and of semiclassical gravity (i.e. graviton loops) are not problematic, but full *quantum gravity* effects – corresponding to integrating out fields with Planck-scale or string-scale masses – are subtle, and have the potential to be ruinous.[24]

From our general discussion of effective actions, we know that integrating out fields of mass Λ with order-unity couplings to the inflaton ϕ will lead to an effective theory of the form

$$\mathcal{L}_{\mathrm{eff}}[\phi] = \mathcal{L}_l[\phi] + \sum_{i=1}^{\infty} \left(\frac{c_i}{\Lambda^{2i}} \phi^{4+2i} + \frac{d_i}{\Lambda^{2i}} (\partial\phi)^2 \phi^{2i} + \frac{e_i}{\Lambda^{4i}} (\partial\phi)^{2(i+1)} + \cdots \right) \,, \tag{2.108}$$

---

[23] A parallel argument applies to chaotic inflation with a non-quadratic monomial potential.

[24] Some authors have argued that quantum gravity effects are necessarily small when all energy densities are sub-Planckian. As a general statement about an arbitrary quantum gravity theory, this is false: the fact that the eta problem appears in string theory, cf. Section 4.2, is one simple counterexample, and the diverse failure modes of large-field models in string theory discussed in Section 5.4 provide many more. Ignoring quantum gravity effects purely because all energy densities are small in Planck units amounts to attributing to the quantum gravity theory underlying our universe a property that is not seen in string theory. This is a logically consistent position but is very far from being agnostic about quantum gravity.

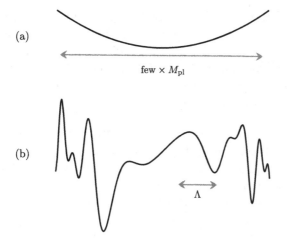

Fig. 2.8 (a) Observable tensor modes require a smooth inflaton potential over a super-Planckian range. (b) In the absence of symmetries, effective field theory predicts that generic potentials have structure on sub-Planckian scales, $\Lambda < M_{\rm pl}$.

where the omitted terms involve additional derivatives acting on ϕ (or on the metric), and c_i, d_i, e_i are dimensionless Wilson coefficients that are typically of order unity.

To begin, we discuss contributions to the potential, i.e. the terms involving c_i in (2.108). For $c_i \sim \mathcal{O}(1)$, we expect that the dominant functional form of the potential will change when the field moves a distance of order Λ: there is "structure" in the potential on scales of order Λ (see Fig. 2.8). Even under the optimistic assumption that $\Lambda = M_{\rm pl}$, the potential (2.108) will not support large-field inflation unless one effectively fine-tunes the infinite set of Wilson coefficients c_i. One might object that the expansion (2.108) is not a useful description over distances $\gtrsim \Lambda$: as a practical matter one would not compute an infinite number of corrections. This is true, but (2.108) nevertheless serves to show how badly an expansion in low-dimension operators can fail in large-field inflation; the challenge is then to show that a more sensible potential arises in some controlled setting.

As should be clear from the effective Lagrangian (2.108), the problem of controlling super-Planckian displacements is not simply a matter of protecting the potential: other terms receive equally dramatic corrections from higher-dimension operators. The terms involving d_i in (2.108) correspond to modifications of the two-derivative kinetic term for ϕ, i.e. corrections to the metric on moduli space. For $d_i \sim \mathcal{O}(1)$, these corrections are large over distances of order Λ.

The leading idea for implementing large-field inflation is to use a symmetry to suppress the dangerous higher-dimension contributions in (2.108). For example, an unbroken shift symmetry

$$\phi \mapsto \phi + const. \tag{2.109}$$

forbids all non-derivative operators in (2.108), including the desirable parts of the inflaton potential, while a suitable weakly broken shift symmetry[25] can give rise to a radiatively stable model of large-field inflation. Whether such a shift symmetry can be UV-completed is a subtle and important question for a Planck-scale theory like string theory. We will return to the problem of super-Planckian fields in Chapter 4 (see also Section 5.4).

Evading the Lyth bound?

Because the Lyth bound raises the specter of catastrophic quantum gravity corrections in all models producing detectable primordial gravitational waves, it is natural to pursue models that evade the bound by violating one or more of the assumptions that entered the derivation. As we now review, the bound is quite robust.

Let us be precise about what it means to "evade the Lyth bound." In a model involving a single inflaton with canonical kinetic term, beginning in the Bunch–Davies vacuum, and with slow-roll unbroken throughout inflation, the bound (2.105) generally applies. If slow-roll holds only during the $N_{\text{eff}} \approx 7$ directly observed e-folds, and the subsequent evolution is arbitrary, the more conservative bound (2.106) remains applicable. Particularly because $M_{\text{pl}} = 2.4 \times 10^{18}$ GeV is not a precise and absolute marker of the realm where quantum gravity corrections are large, one should be wary of the claim that a marginal violation of (2.105), or even of (2.106), diminishes the problem of ultraviolet sensitivity in large-field inflation.[26] For this reason, to truly evade the physical content of the Lyth bound, one should alter an assumption entering the derivation in such a way as to *parametrically* violate the conservative bound (2.106). In other words, the task is to avoid the conclusion that "detectable gravitational waves imply displacements of order M_{pl}."

With this in mind, we comment on a few ideas for evasion of the Lyth bound.

Nontrivial evolution

A number of authors have proposed scenarios in which nontrivial evolution after the horizon exit of the CMB fluctuation – for example, a steep drop in the potential – renders (2.105) inapplicable: see e.g. [258–261]. However, these models do still satisfy (2.106), and correspondingly involve displacements of order the Planck scale for $r \gtrsim 0.01$. For a recent discussion see [262].

---

[25] Note that to realize an approximate shift symmetry in the low-energy theory, it would suffice for the inflaton to have weak couplings $g \ll 1$ to all the degrees of freedom of the UV completion; the Wilson coefficients in (2.108) would then be suppressed by powers of g. Equivalently, the effective cutoff scale would become $M_{\text{pl}}/g \gg M_{\text{pl}}$. The coupling of the inflaton to any additional degrees of freedom would be weaker than gravitational [257].

[26] The Lyth bound is sometimes misunderstood as the statement that "detectable gravitational waves imply $\Delta\phi > 1.0 \times M_{\text{pl}}$," or equivalently – upon imposing a legalistic definition of small-field inflation as inflation with $\Delta\phi < 1.0 \times M_{\text{pl}}$ – that detectable gravitational waves are impossible in "small-field" inflation. Neither statement is true, so exhibiting counterexamples has limited utility.

Non-canonical kinetic terms

It is natural to ask whether the Lyth bound can be evaded if the inflationary phase is supported by kinetic energy. As an explicit example, consider the $P(X)$ theories of Section 2.2.3. The naive bound for the excursion of ϕ becomes [254]

$$\frac{\Delta\phi}{M_{\mathrm{pl}}} = (c_s P_{,X})^{-1/2} \sqrt{\frac{r}{8}} \Delta N . \tag{2.110}$$

This seems to suggest that the Lyth bound could be evaded by choosing $P_{,X} \gg 1$ for fixed c_s. However, when $P_{,X} \gg 1$ we are far from a canonical kinetic term for ϕ, and must worry about corrections to the entire $P(X)$ action, not just to the potential. In particular, one should inquire about Planck-suppressed corrections of the form

$$\Delta\mathcal{L} = P\left(X - V(\phi)\frac{\phi^2}{M_{\mathrm{pl}}^2}\right) = P(X) - P_{,X}V(\phi)\frac{\phi^2}{M_{\mathrm{pl}}^2} + \cdots . \tag{2.111}$$

For $P_{,X} \gg 1$, the corrections (2.111) are enhanced over potential corrections. Thus, even though taking $P_{,X} \gg 1$ does technically lead to models violating (2.106), the problem of Planck-suppressed corrections to the effective Lagrangian is undiminished, and merely moved from one class of terms to another. A generalization of the Lyth bound to a totally general single-field Lagrangian [51] was derived in [263] – see Appendix B for further details.

Multiple fields: arc length versus geodesic distance

The distance $\Delta\phi$ that enters (2.105) is the arc length along the inflaton trajectory, not the geodesic distance between the start and end points of the trajectory. The importance of this distinction is that large displacements appear unnatural when the inflaton travels outside the radius of convergence of a Taylor expansion of the low-energy potential. If one can arrange that the inflaton trajectory winds or meanders to achieve a large arc length, while remaining within the radius of convergence, then the problem of ultraviolet sensitivity is much diminished, even though the resulting models do obey (2.105). This point was stressed in [264], where monodromy in a two-axion system leads to a winding trajectory (see Section 5.4.2).

Multiple fields: modified scalar perturbations

Because the scalar amplitude (1.28) entering the derivation of (2.105) applies only to a single inflaton scalar, one can ask whether contributions to the scalar perturbations by other light fields can lead to a weaker bound. Let us first consider a multi-field inflation model in which as the observed CMB multipoles exit the horizon, slow-roll is applicable, and moreover the field trajectory does not bend sharply.[27] Then, as explained in Appendix C, the fluctuations of

---

[27] Specifically, we require that the usual slow-roll parameters are small, and that the parameter $\eta_\perp$ defined in (C.98) obeys $\eta_\perp \ll 1$.

fields transverse to the inflationary trajectory make strictly *positive* contributions to $\Delta_\mathcal{R}^2$; see (C.117). As such, the contributions of additional fields actually
strengthen the Lyth bound, increasing the displacement $\Delta\phi$ required to produce
a given observed value of r.

The bound can be (rather weakly) violated if the slow-roll, slow-turn approximations assumed above are invalid; fluctuations of additional fields can then
contribute negatively to $\Delta_\mathcal{R}^2$, increasing the effective value of r – see [265]
for explicit examples. However, we are not aware of a plausible construction
in which this effect is large enough to induce a meaningful weakening of the
bound (2.106).

A more dramatic example of the effect of multiple fields arises if the inflaton contribution to the scalar perturbations is negligible in comparison to the
perturbations arising from a curvaton [134], or through modulated reheating
[176, 266]. In typical scenarios the inflaton does still fluctuate during inflation,
but the modulated contributions imprinted later are much larger, substantially
increasing the power in scalar perturbations and correspondingly strengthening
the Lyth bound.

Modifications of the initial state
The tensor amplitude (1.34) is applicable when the initial state in which the
two-point function is computed is the Bunch–Davies vacuum. A significant modification of the initial state may allow violations of (2.106) [267, 268], and the
resulting tensor spectrum can be expected to display significant scale dependence
[268, 269].

Other sources of gravitational waves
An alternative mechanism for generating gravitational waves during inflation, as
in e.g. [270–272], can readily violate (2.106), as the bound incorporates only the
primordial gravitational waves from quantum fluctuations of the gravitational
field. A zeroth-order challenge in such approaches is to ensure that the dynamics
producing gravitational waves does not make the scalar spectrum unacceptably
non-Gaussian [273].

2.3.4 Non-Gaussianity

Single-field slow-roll inflation has an approximate shift symmetry[28] (2.95)
that constrains inflaton self-interactions in the potential and prevents large
non-Gaussianity: $f_{\mathrm{NL}} \sim \mathcal{O}(\epsilon, \eta) \ll 1$ [119]. To generate observable levels of non-
Gaussianity requires either higher-derivative interactions or couplings to extra
fields. Both options can be ultraviolet sensitive.

---

[28] This symmetry does not have to be fundamental, but may be the result of fine-tuning. Its
presence is motivated by the observed scale invariance of the primordial fluctuations.

Non-Gaussianity from higher derivatives

When higher-derivative interactions are important, the dynamics deviates significantly from slow-roll. In Section 2.2.3, we presented $P(X)$ theories as a specific example. We mentioned that fluctuations propagate with a nontrivial speed of sound,

$$c_s^2 = \frac{P_{,X}}{P_{,X} + 2X P_{,XX}} . \qquad (2.112)$$

However, the effective theories corresponding to $c_s \ll 1$ cry out for ultraviolet completion. In an EFT, one thinks of the function $P(X)$ in a derivative expansion, cf. Eq. (2.108),

$$P = X + \frac{1}{2}\frac{X^2}{\Lambda^4} + \cdots , \qquad (2.113)$$

which truncates to a finite number of terms if $X \ll \Lambda^4$. However, the condition $X \ll \Lambda^4$ also implies that the deviation from the slow-roll action, $P_{\text{s.r.}} \equiv X - V(\phi)$, is a perturbative correction, and the non-Gaussianity is correspondingly small [124, 125, 274]:

$$\left| f_{\text{NL}}^{\text{equil}} \right| \sim \frac{1}{c_s^2} - 1 \approx \frac{X}{\Lambda^4} + \cdots \ll 1 . \qquad (2.114)$$

On the other hand, $|f_{\text{NL}}^{\text{equil}}| > 1$ can only arise for $X \gtrsim \Lambda^4$, in which case it is inconsistent to truncate the expansion in (2.113). Instead, an infinite number of higher-derivative terms – those proportional to e_i in (2.108) – become relevant. Observably large non-Gaussianity in single-field inflation is therefore UV sensitive.[29] Special symmetries, such as the higher-dimensional boost symmetry of DBI inflation [39] (see Section 5.3), are required to make sense of the UV completion of (2.113).[30]

Non-Gaussianity from hidden sectors

As we will see in Chapters 4 and 5, ultraviolet completions of inflation invariably involve extra fields coupled to the inflaton. We will collectively denote these fields by ψ. If these fields are sufficiently heavy $(m_\psi \gg H)$, they can be integrated out and only affect the couplings of the single-field EFT. Light hidden sector fields

---

[29] This issue is also visible in the effective theory of fluctuations [51] (see Appendix B). In the limit of observable non-Gaussianity, the theory of the fluctuations becomes strongly coupled *below* the symmetry-breaking scale $\dot{\phi}$ [275], and must therefore be UV-completed below $\dot{\phi}$. This is in contrast to the slow-roll limit, where questions about the UV completion are deferred to scales above $\dot{\phi}$.

[30] Another class of ghost-free, radiatively stable higher-derivative models is *Galileon inflation* [276]. In these models the renormalization of the action is protected by the *Galilean symmetry* $\phi \mapsto \phi + b_\mu x^\mu + c$, which is a combination of the shift symmetry (2.95) and a spacetime translation. No candidate for an ultraviolet completion of a Galileon model in string theory has been proposed, and whether one exists is an open question.

$(m_\psi < H)$, on the other hand, can affect the inflationary fluctuations and may therefore leave imprints in cosmological observables.

Although the approximate shift symmetry (2.95) sharply limits the non-Gaussianity that can arise from self-interactions of the inflaton, the couplings of hidden sector fields are much less constrained, and hidden-sector self interactions can lead to visible non-Gaussianity, as we now explain. Suppose that the shift symmetry of the inflaton is preserved by the coupling to a hidden sector field ψ. Then the leading interaction between the hidden sector and the visible sector is the dimension-five operator [277]

$$\mathcal{O}_5 = \frac{\psi X}{\Lambda} . \tag{2.115}$$

This coupling converts any non-Gaussianity in the hidden sector into observable non-Gaussianity in the inflaton sector.

Under rather natural circumstances, the fluctuations in the hidden sector can be highly non-Gaussian. For example, suppose that supersymmetry is spontaneously broken during inflation. A generic hidden sector scalar field ψ that is not sequestered from the inflationary supersymmetry breaking will acquire a soft mass $m_\psi \sim H$ and cubic coupling (or A-term) $A\psi^3$, with $A \sim H$ [277, 278], by coupling to the inflationary vacuum energy. Unless ψ has a large supersymmetric mass, it can fluctuate during inflation, and because $A \sim H$, the correlations of ψ are order-one non-Gaussian. Via the operator (2.115), this gets communicated to the visible sector.[31] The signal can be large while keeping the effective theory under perturbative control, with $X < \Lambda^4$.

Through the coupling (2.115), the Planck limits (1.72)–(1.74) become precision constraints on light hidden sector scalars [277, 279]. For scalars with cubic couplings $\sim H\psi^3$, one finds the bound [277]

$$\Lambda \gtrsim 10^5 H . \tag{2.116}$$

This is a constraint on physics many orders of magnitude above the inflationary Hubble scale. Using (1.38), one can write the bound (2.116) in terms of the Planck scale,

$$\Lambda \gtrsim \left(\frac{r}{0.01}\right)^{1/2} M_{\mathrm{pl}} . \tag{2.117}$$

It is a striking coincidence that a detection of primordial tensors, $r > 0.01$, would push the lower bound on Λ to the Planck scale. The bispectrum results of Planck would then imply constraints on Planck-suppressed couplings to hidden sectors. Specifically, we would learn that all hidden sector scalars are either massive $(m_\psi \gg H)$, sequestered from inflationary supersymmetry breaking $(A \ll H)$, or sequestered from the inflaton itself $(\Lambda > M_{\mathrm{pl}})$.

---

[31] Order-one non-Gaussianity in the observed curvature perturbations would correspond to $f_{\mathrm{NL}} \Delta_{\mathcal{R}} \sim 1$, not $f_{\mathrm{NL}} \sim 1$.

To understand the strength of a constraint of the form $\Lambda \gtrsim M_{\text{pl}}$, one should recognize that in parametrically controlled ultraviolet completions of gravity, the actual cutoff scale of an inflationary EFT is generally far below the Planck mass. Thus, an unambiguous detection of primordial tensors would exclude a broad range of constructions involving light hidden sector fields, providing a powerful selection principle for ultraviolet completions of inflation.

3

Elements of string theory

String theory is the subject of a vast literature.[1] Our aim in this section is to assemble the results that are most relevant for the study of string inflation (the subject of Chapters 4 and 5), making no pretense of completeness. We will particularly focus on the four-dimensional effective actions arising in cosmologically realistic solutions of string theory. Careful attention is paid to the problem of moduli stabilization, and de Sitter solutions are critically analyzed.

3.1 Fundamentals

3.1.1 From worldsheet to spacetime

An elementary starting point for string theory is the worldsheet action for a string, which defines a (1+1)-dimensional quantum field theory. We will begin by describing bosonic string theory, and then turn to string theories whose worldsheet theories include fermionic fields.

Bosonic string theory

The *Polyakov action* for a bosonic string propagating in D-dimensional Minkowski space [294, 295] is

$$S_{\mathrm{P}} = -\frac{1}{4\pi\alpha'} \int \mathrm{d}^2\sigma \sqrt{-h}\, h^{ab} \partial_a X^M(\sigma) \partial_b X^N(\sigma) \eta_{MN} \, , \qquad (3.1)$$

where X^M, with $M = 0, \ldots, D-1$, are the coordinates in the target spacetime; σ^a, with $a = 0, 1$, are the coordinates on the string worldsheet; h^{ab} is an independent metric on the worldsheet; and $2\pi\alpha'$ is the inverse of the string tension. The action (3.1) describes a two-dimensional field theory with D scalar fields.

---

[1] The fundamentals of the theory can be found in classic textbooks [280–283], as well as lecture notes [284–290]. More recent advances are described in [291–293].

At the classical level, this theory is invariant under two-dimensional diffeomorphisms and under the Weyl symmetry $h_{ab} \mapsto e^{2\omega(\sigma)} h_{ab}$. Famously, these classical symmetries are non-anomalous if and only if $D = 26$ [296]. The symmetries can be used to set $h_{ab} \mapsto \eta_{ab}$,[2] known as *conformal gauge*, so that the action takes the more convenient form

$$S_{\mathrm{P}} = -\frac{1}{4\pi\alpha'} \int \mathrm{d}^2\sigma\, \partial^a X^M \partial_a X_M \,, \tag{3.2}$$

in which the X^M are recognized as D free fields that respect a global $SO(D-1,1)$ symmetry.

Upon quantizing the string, one finds that the massless spectrum consists of a graviton G_{MN}, an antisymmetric tensor B_{MN}, and a scalar Φ known as the dilaton. In addition, the spectrum contains massive excitations with scale set by $M_{\mathrm{s}} \equiv (\alpha')^{-1/2}$. The Polyakov action (3.1) can be extended to a nonlinear σ-model action describing strings propagating in a target spacetime involving background profiles for the massless excitations:

$$S_\sigma = -\frac{1}{4\pi\alpha'} \int \mathrm{d}^2\sigma \sqrt{-h} \left(\left[h^{ab} G_{MN}(X) + \epsilon^{ab} B_{MN}(X) \right] \partial_a X^M \partial_b X^N \right.$$

$$\left. + \alpha' \Phi(X) R(h) \right), \tag{3.3}$$

where $R(h)$ is the Ricci scalar constructed from h_{ab}. Expanding the background fields around a given point, $X^M = X_{(0)}^M + \delta X^M$, one finds interaction terms such as $h^{ab} \partial_P G_{MN}(X_{(0)}) \delta X^P \partial_a \delta X^M \partial_b \delta X^N$. The nonlinear σ-model defined by (3.3) therefore describes an interacting quantum field theory. When the gradients of the background fields are small in units of α' – and in particular, when all curvatures are small in string units – these interactions can be treated perturbatively. The corresponding expansion is known as the σ-model expansion or the α' expansion. Absence of anomalies in the quantum field theory defined by (3.3) requires that the background fields in the target spacetime obey certain differential equations that can be obtained order by order in the α' expansion. Consistency of string theory at the quantum level on the worldsheet therefore imposes equations of motion in the target spacetime [297]. Remarkably, the equations of motion for $G_{MN}(X)$ at leading order in α' are the Einstein equations!

The equations of motion for the background fields can also be shown to follow from a D-dimensional spacetime action that parameterizes the interactions of the massless excitations of the bosonic string. The idea is to construct an effective action in the sense described in Chapter 2; one imagines performing the path integral by first integrating out massive excitations of the string, leaving an

---

[2] This assumes that there is no topological obstruction to the existence of a metric that is flat everywhere.

effective action for the massless modes. The theory that emerges at energies below the string scale M_s takes the form (see [282] for details)

$$S_B = \frac{1}{2\kappa_D^2} \int d^D X \sqrt{-G} \, e^{-2\Phi} \left(R + 4(\partial\Phi)^2 - \frac{1}{2}|H_3|^2 - \frac{2(D-26)}{3\alpha'} + \mathcal{O}(\alpha') \right),$$

(3.4)

where κ_D is a coupling constant, and $H_3 = dB_2$ is the field strength of the antisymmetric tensor B_{MN}, or equivalently of the two-form B_2. Although the effective action (3.4) lacks the good ultraviolet behavior of the full string theory (it violates perturbative unitarity at $E \sim M_s$), it is nevertheless a convenient way to organize the interactions at energies below the cutoff, $E \ll M_s$. The omitted terms of higher order in α' correspond to higher-dimension operators, including invariants constructed from the Riemann curvature of the target space.

In practice, the effective action (3.4) is obtained by computing scattering amplitudes for strings via a path integral over worldsheets connecting initial and final states. The path integral involves a sum over surfaces connecting the initial and final configurations, and the genus of the surface is a loop counting parameter: worldsheets of Euler number χ appear in the path integral with weight

$$e^{-\Phi\chi} = e^{-\Phi(2-2g)} \equiv g_s^{2g-2},$$

(3.5)

where g is the genus of the worldsheet and $g_s \equiv e^\Phi$ is the string coupling. Amplitudes are then defined order by order in the genus expansion, although except in special cases only one-loop results are available. One can then ask which effective action in D-dimensional spacetime results in the same scattering amplitudes. The amplitudes obtained at tree level in the genus expansion can be shown to follow from the effective action (3.4), the very theory whose equations of motion enforce the absence of anomalies in the worldsheet theory (3.3).

In summary, the full D-dimensional action can be expressed in a double expansion, in g_s and in α'. The genus expansion corresponds to the $\hbar$ expansion in the effective theory, while the α' expansion controls the appearance of certain higher-dimension operators. These expansions are controlled by vevs of dynamical fields, rather than by fundamental dimensionless parameters: the coupling "constant" in the genus expansion, $g_s(\Phi)$, is the expectation value of the dilaton, while the expansion parameter of the σ-model is the curvature of the target spacetime in units of α'.

Superstring theories

The bosonic string theory defined by (3.1) is unsuitable as a description of nature: the spacetime spectrum is devoid of fermions, and the theory suffers from a tachyonic instability [282]. Supersymmetric string theories are far more promising, and differ in important details. Most fundamentally, the worldsheet actions involve additional fermionic terms: in the simplest case, known as $\mathcal{N} = (1,1)$ worldsheet supersymmetry, the total action in conformal gauge takes the form [280]

$$S = S_{\mathrm{P}} + S_{\mathrm{F}} = -\frac{1}{4\pi\alpha'} \int \mathrm{d}^2\sigma \left(\partial^a X^M \partial_a X_M - i\bar\psi^M \rho^a \partial_a \psi_M \right) . \tag{3.6}$$

Here, ρ^a are two-dimensional Dirac matrices obeying the Dirac (or Clifford) algebra

$$\{\rho^a, \rho^b\} = -2\eta^{ab} , \tag{3.7}$$

and ψ^M is a Dirac spinor on the worldsheet that transforms as a vector under Lorentz transformations in the target space (which correspond to global symmetry transformations of the worldsheet theory). In terms of the two independent components of ψ^M,

$$\psi^M \equiv \begin{pmatrix} \psi_-^M \\ \psi_+^M \end{pmatrix} , \tag{3.8}$$

the fermion action takes the form

$$S_{\mathrm{F}} = \frac{i}{2\pi\alpha'} \int \mathrm{d}^2\sigma \left(\psi_-^M \partial_+ \psi_-^N + \psi_+^M \partial_- \psi_+^N \right) \eta_{MN} , \tag{3.9}$$

where $\partial_\pm \equiv \frac{1}{2}(\partial_\tau \pm \partial_\sigma)$, with $\tau \equiv \sigma^0$ and $\sigma \equiv \sigma^1$. The worldsheet fermions therefore separate into left-moving and right-moving modes. The fermions $\psi_\pm^M$ contribute to the central charge of the worldsheet field theory, so that the theory defined by $S = S_{\mathrm{P}} + S_{\mathrm{F}}$, with S_{P} given in (3.2), has the critical dimension $D = 10$.

The action (3.9) for the worldsheet fermions does not completely determine the spacetime spectrum of the theory; one must also specify the periodicity of the fermions under transport around the closed string worldsheet. Periodic fermions obeying $\psi_\pm^M(\sigma + \pi) = +\psi_\pm^M(\sigma)$ are said to be in the *Ramond* sector, while antiperiodic fermions with $\psi_\pm^M(\sigma + \pi) = -\psi_\pm^M(\sigma)$ are said to be in the *Neveu–Schwarz* sector. This choice can be made separately for the left-moving and right-moving fermions, so that there are four possible sectors: NS-NS, R-R, R-NS, and NS-R. The ten-dimensional effective actions describing the interactions of massless states of the superstring are supergravity theories involving additional fermionic and bosonic fields in comparison to (3.4). Bosonic fields in the target spacetime arise from string states in the NS-NS and R-R sectors, while the R-NS and NS-R sectors give rise to spacetime fermions.

To construct a consistent closed string theory with spacetime fermions, it turns out to be necessary to impose a particular projection, the *GSO projection*, on the spectrum. This entails one further choice: one can perform identical GSO projections in the R-NS and NS-R sectors, or opposite projections. The former choice leads to *type IIB string theory*, which has a chiral spectrum in spacetime – in particular, the two gravitinos have the same chirality. The latter choice produces *type IIA string theory*, which has a non-chiral spectrum.

Three other consistent superstring theories are known. To arrive at *type I string theory*, we consider the worldsheet parity operation Ω, which reverses the orientation of the string worldsheet, and hence relates left-moving and right-moving modes. In type IIB string theory, the R-NS and NS-R sectors have the same spectra, so that worldsheet parity is a symmetry of the theory, and it is consistent to project the spectrum onto states with $\Omega = +1$. This operation, which corresponds to gauging the discrete symmetry of worldsheet parity, leads to a theory of *unoriented* strings, because for any given string its orientation-reversed image under Ω is also retained. The projection removes one of the two gravitinos from the spectrum, yielding a theory with $\mathcal{N} = 1$ supersymmetry in ten dimensions, the type I string.

The two remaining theories also have ten-dimensional $\mathcal{N} = 1$ supersymmetry, but have a different structure on the worldsheet. While above we have discussed theories with left-moving and right-moving bosons, and left-moving and right-moving fermions, it is also consistent to take the left-moving sector to be that of the bosonic string, and the right-moving sector to be that of the superstring. Two supersymmetric *heterotic string theories* arise from this construction: the $SO(32)$ heterotic string, and the $E_8 \times E_8$ heterotic string.

The five superstring theories described above are interrelated by a number of dualities (see Fig. 3.1), and correspond to different limits of an underlying theory called *M-theory*.

Supergravity limit

The low-energy limit of each of the consistent superstring theories is a ten-dimensional supergravity theory. We will now describe the corresponding effective actions for type IIA and type IIB string theory, focusing on the bosonic fields, which are directly relevant for obtaining classical solutions. The actions for type I string theory and the heterotic string theories may be found in e.g. [283].

The NS-NS sector of type II supergravity in ten dimensions contains the metric G_{MN}, the dilaton Φ, and the two-form B_2. The action for these fields is

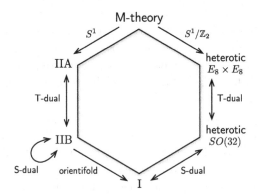

Fig. 3.1 Dualities relating the supersymmetric string theories and M-theory. S-duality exchanges strong coupling and weak coupling, while in compactification on a circle, T-duality exchanges momentum and winding.

$$S_{\mathrm{NS}} = \frac{1}{2\kappa^2} \int \mathrm{d}^{10}X\sqrt{-G}\,e^{-2\Phi}\left(R + 4(\partial\Phi)^2 - \frac{1}{2}|H_3|^2\right) , \qquad (3.10)$$

where $H_3 = \mathrm{d}B_2$. The coupling constant κ^2, corresponding with the Newton constant in ten dimensions, can be related to the string tension by comparing the worldsheet and supergravity actions; one finds [282]

$$2\kappa^2 = (2\pi)^7(\alpha')^4 . \qquad (3.11)$$

In addition, type IIA supergravity has an R-R one-form C_1 and a three-form C_3. The complete action then takes the form

$$S_{\mathrm{IIA}} = S_{\mathrm{NS}} + S_{\mathrm{R}}^{(\mathrm{IIA})} + S_{\mathrm{CS}}^{(\mathrm{IIA})} , \qquad (3.12)$$

where

$$S_{\mathrm{R}}^{(\mathrm{IIA})} = -\frac{1}{4\kappa^2} \int \mathrm{d}^{10}X\sqrt{-G}\left(|F_2|^2 + |\tilde{F}_4|^2\right) , \qquad (3.13)$$

$$S_{\mathrm{CS}}^{(\mathrm{IIA})} = -\frac{1}{4\kappa^2} \int B_2 \wedge F_4 \wedge F_4 , \qquad (3.14)$$

with $F_p = \mathrm{d}C_{p-1}$ and $\tilde{F}_4 = F_4 + C_1 \wedge H_3$. The R-R fields in type IIB supergravity are a zero-form (scalar) C_0, a two-form C_2, and a four-form C_4 with self-dual field strength. The complete action is

$$S_{\mathrm{IIB}} = S_{\mathrm{NS}} + S_{\mathrm{R}}^{(\mathrm{IIB})} + S_{\mathrm{CS}}^{(\mathrm{IIB})} , \qquad (3.15)$$

where

$$S_{\mathrm{R}}^{(\mathrm{IIB})} = -\frac{1}{4\kappa^2} \int \mathrm{d}^{10}X\sqrt{-G}\left(|F_1|^2 + |\tilde{F}_3|^2 + \frac{1}{2}|\tilde{F}_5|^2\right) , \qquad (3.16)$$

$$S_{\mathrm{CS}}^{(\mathrm{IIB})} = -\frac{1}{4\kappa^2} \int C_4 \wedge H_3 \wedge F_3 , \qquad (3.17)$$

with $F_p = \mathrm{d}C_{p-1}$, $\tilde{F}_3 = F_3 - C_0 \wedge H_3$, and $\tilde{F}_5 = F_5 - \frac{1}{2}C_2 \wedge H_3 + \frac{1}{2}B_2 \wedge F_3$. In addition, one must impose the self-duality constraint

$$\tilde{F}_5 = \star\tilde{F}_5 . \qquad (3.18)$$

We have written the NS-NS sector (3.10) of the actions (3.12) and (3.15) in *string frame*, meaning that the Ricci scalar R appears with the dilaton-dependent prefactor $e^{-2\Phi}$. This frame is convenient for comparing to the results of string perturbation theory. However, for many questions involving gravity, it is more practical to work in *Einstein frame*, in which the dilaton prefactor is absent. The action can be written in Einstein frame by performing the Weyl rescaling

$$G_{E,MN} \equiv e^{-\Phi/2}G_{MN} . \qquad (3.19)$$

In type IIB string theory, it is convenient to define the combinations

$$G_3 \equiv F_3 - \tau H_3 , \qquad (3.20)$$

$$\tau \equiv C_0 + ie^{-\Phi} , \qquad (3.21)$$

in terms of which the action (3.15), written in Einstein frame, takes the form

$$S_{\text{IIB}} = \frac{1}{2\kappa^2} \int d^{10}X \sqrt{-G_E} \left[R_E - \frac{|\partial\tau|^2}{2(\text{Im}(\tau))^2} - \frac{|G_3|^2}{2\,\text{Im}(\tau)} - \frac{|\tilde{F}_5|^2}{4} \right]$$
$$- \frac{i}{8\kappa^2} \int \frac{C_4 \wedge G_3 \wedge \bar{G}_3}{\text{Im}(\tau)} \,. \tag{3.22}$$

The action (3.22) is the starting point for our discussion of type IIB flux compactifications in Section 3.3.1.

3.1.2 D-branes

In addition to fundamental strings, string theory contains solitonic objects. Most famous are D-branes, which are charged under the gauge symmetry of the R-R fields. A Dp-brane is an object with p spatial dimensions that is charged under C_{p+1} via the electric coupling[3]

$$S_{\text{CS}} = \mu_p \int_{\Sigma_{p+1}} C_{p+1} \,, \tag{3.23}$$

where Σ_{p+1} is the Dp-brane worldvolume and μ_p is the brane charge. The *Chern–Simons* action (3.23) is simply a higher-dimensional generalization of the coupling of a charged point particle to a gauge potential, $\int dx^\mu A_\mu \equiv \int A_1$.

A defining characteristic of D-branes is that they are surfaces on which strings can end. The D in D-brane stands for Dirichlet, referring to the fact that open strings ending on a D-brane have Dirichlet boundary conditions in the directions transverse to the brane, i.e. the open string endpoints cannot leave the D-brane. Open strings have Neumann boundary conditions in the directions along the spatial extent of a Dp-brane with $p > 0$: the endpoints are free to slide along the D-brane.

Quantization of the open strings residing on a D-brane yields a spectrum of bosonic and fermionic fields living on the worldvolume. At the massless level, one finds scalar fields parameterizing fluctuations of the D-brane position, a world volume gauge field A_a with field strength F_{ab}, and their superpartners. The effective action for these fields is an important object, because it encapsulates the low-energy dynamics of the D-brane. Just as the low-energy effective action for the massless modes of a closed string could be determined by computing closed string scattering amplitudes in perturbation theory, the low-energy

---

[3] Type IIA string theory contains stable Dp-branes with p *even*, while type IIB string theory contains stable Dp-branes with p *odd*. Type I string theory has stable Dp-branes with $p = 1, 5, 9$. Dp-branes and D$(6 - p)$-branes are charged under the R-R potentials C_{p+1} and C_{7-p}, whose field strengths F_{p+2} and F_{8-p} are dual to each other, $\star F_{8-p} = F_{p+2}$. Thus, D$(6 - p)$-branes carry magnetic charge under C_{p+1}. In string theories with an NS-NS two-form B_2, fundamental strings carry charge under B_2 and are stable. Moreover, there is an additional soliton, the NS5-brane, which is magnetically charged under B_2.

effective action that governs the massless fields on a D-brane can be derived by computing scattering amplitudes involving open strings ending on the D-brane. Moreover, by computing amplitudes in which open strings on the D-brane interact with closed strings, one can determine the couplings of the D-brane to a closed string background.

A general background solution of type II or type I[4] string theory will involve profiles for all the massless bosonic fields. We would like to understand the effective action for the light fields on a Dp-brane in such a background. For simplicity, we will restrict attention to the bosonic sector.

D-brane action

An uncharged p-dimensional membrane moving in a curved spacetime with metric G_{MN} can be described by the *Dirac* action, which is simply a higher-dimensional generalization of the Polyakov action (3.1):

$$S_{\mathrm{D}} = -T_p \int \mathrm{d}^{p+1}\sigma \sqrt{-\det(G_{ab})} \,, \tag{3.24}$$

where

$$G_{ab} \equiv \frac{\partial X^M}{\partial \sigma^a} \frac{\partial X^N}{\partial \sigma^b} G_{MN} \,. \tag{3.25}$$

Here, G_{ab} is the pullback of the metric of the target spacetime, and T_p is the tension of the membrane.

Next, we recall *Born–Infeld* theory, a nonlinear generalization of Maxwell's electromagnetism. The Born–Infeld action in $(p+1)$ flat spacetime dimensions, for an Abelian gauge field A_a with field strength strength F_{ab}, is

$$\begin{aligned} S_{\mathrm{BI}} &= -Q_p \int \mathrm{d}^{p+1}\sigma \sqrt{-\det(\eta_{ab} + 2\pi\alpha' F_{ab})} \\ &= -Q_p \int \mathrm{d}^{p+1}\sigma \left(1 + \frac{(2\pi\alpha')^2}{4} F_{ab}F^{ab} + \cdots\right) \,, \end{aligned} \tag{3.26}$$

where Q_p is a constant with the dimensions of a p-brane tension.

By computing open string amplitudes, and open+closed amplitudes, one finds that the action for a Dp-brane in a general closed string background involves a combination of the Dirac and Born–Infeld actions, the *Dirac–Born–Infeld* action

$$S_{\mathrm{DBI}} = -g_{\mathrm{s}} T_p \int \mathrm{d}^{p+1}\sigma \, e^{-\Phi} \sqrt{-\det(G_{ab} + \mathcal{F}_{ab})} \,, \tag{3.27}$$

where $\mathcal{F}_{ab}$ is the gauge-invariant field strength

$$\mathcal{F}_{ab} \equiv B_{ab} + 2\pi\alpha' F_{ab} \,, \tag{3.28}$$

---

[4] The heterotic string theories contain no R-R fields, and correspondingly lack D-branes.

and B_{ab} is the pullback of B_{MN} onto the D-brane worldvolume. From the string amplitude computations one infers that the Dp-brane tension is[5]

$$T_p \equiv \frac{1}{(2\pi)^p g_s (\alpha')^{(p+1)/2}} ,$$ (3.29)

leading to the important result that Dp-branes are heavy at weak string coupling, $g_s \ll 1$. Next, the D-brane tension T_p can be related to the charge μ_p appearing in (3.23). The stable D-branes of type I and type II string theories are *BPS objects*, and preserve half of the spacetime supersymmetries. A BPS Dp-brane ($p > 0$) corresponds to a higher-dimensional generalization of an extremal black hole, with tension equal to its charge when expressed in appropriate units. In our conventions, one finds $\mu_p = g_s T_p$.

The Chern–Simons action in the presence of background fields in the target space and on the D-brane worldvolume takes the form

$$S_{\mathrm{CS}} = i\,\mu_p \int_{\Sigma_{p+1}} \sum_n C_n \wedge e^{\mathcal{F}} ,$$ (3.30)

where the sum runs over the R-R n-forms of the theory in question, and only ($p+1$)-forms contribute to the integral in (3.30). The complete bosonic action for D-branes in a supergravity background is then the sum of the Dirac–Born–Infeld action (3.27) and the Chern–Simons action (3.30),

$$S_{\mathrm{D}p} = S_{\mathrm{DBI}} + S_{\mathrm{CS}} .$$ (3.31)

When N Dp-branes coincide, the worldvolume gauge theory becomes non-Abelian, and the action becomes much more complicated, with a potential involving commutators of the worldvolume scalars – see [298].

D-branes as sources

The coupling of D-branes to the background fields has important consequences: in addition to responding to the closed string solution in which it is embedded, a D-brane contributes to the profiles of the massless closed string fields. Specifically, a D-brane provides a localized source of stress-energy and R-R charge, causing it to source curvature and R-R fluxes in proportion to its tension and charge. Incorporating this "backreaction" is sometimes difficult, as explained in Section 4.1.4.

The supergravity solution sourced by one or more Dp-branes corresponds to a spatially extended extremal black hole, or *extremal p-brane*. For N coincident Dp-branes, the characteristic radius of curvature r_+ is given by [299]

$$r_+^{7-p} = d_p\, g_s N (\alpha')^{\frac{1}{2}(7-p)} ,$$ (3.32)

---

[5] In a background with constant dilaton Φ, one has $g_s e^{-\Phi} = 1$.

where $d_p \equiv (4\pi)^{\frac{1}{2}(5-p)} \Gamma\left(\frac{1}{2}(7-p)\right)$. The dilaton profile in the radial (r) direction takes the form

$$e^{\Phi} = g_s \left(1 + \left(\frac{r_+}{r}\right)^{7-p}\right)^{\frac{1}{4}(3-p)} . \tag{3.33}$$

A classical supergravity description is valid if the curvatures are small in string units and moreover $g_s \ll 1$. For $p < 7$, we find from (3.32) that the curvature is small at large $g_s N$ – see Section 4.1.4 for further discussion of this point. Moreover, for the important special case $p = 3$, the dilaton is constant, and can be small everywhere: D3-branes decouple from the dilaton. Thus, if $p = 3$ and

$$1 \ll g_s N \ll N , \tag{3.34}$$

the α' and g_s corrections to leading-order classical supergravity can be neglected everywhere. On the other hand, for $p \neq 3$ the nontrivial dilaton profile (3.33) presents an obstacle to extending the classical supergravity solution over the entire spacetime. See [300] for a more extensive review of extremal p-brane solutions.

3.2 Compactification

For our purposes, a *solution* of string theory is a configuration of the massless fields that solves the equations of motion of the effective theory, and hence leads to a worldsheet theory without anomalies. In supersymmetric string theories, a geometric solution involves a ten-dimensional spacetime $\mathcal{M}_{10}$, and the solutions that are most relevant for cosmology include a spacetime $\mathcal{M}_4$ with four large dimensions. We therefore consider geometries of the form

$$\mathcal{M}_{10} = \mathcal{M}_4 \times X_6 , \tag{3.35}$$

where X_6 is a compact six-manifold. This is referred to as a *compactification* of string theory on X_6.

Vacuum compactifications

We will distinguish vacuum configurations, i.e. solutions of the ten-dimensional vacuum Einstein equations[6] without sources of stress-energy, from solutions involving sources, and begin by considering vacuum solutions. A suitable ansatz for vacuum configurations is

$$G_{MN}\, \mathrm{d}X^M \mathrm{d}X^N = \eta_{\mu\nu}\mathrm{d}x^{\mu}\mathrm{d}x^{\nu} + g_{mn}\mathrm{d}y^m\mathrm{d}y^n , \tag{3.36}$$

where y^m, $m = 1, \ldots, 6$, are coordinates on X_6, and g_{mn} is a metric on X_6. Such a geometry is an allowed vacuum configuration if and only if G_{MN} solves the

---

[6] The Einstein equations receive corrections in the α' expansion, which can be important at large curvatures.

ten-dimensional vacuum Einstein equations, i.e. if and only if $R_{\mu\nu} = R_{mn} = 0$. Thus, in vacuum solutions the compactification manifold must have vanishing Ricci tensor. The best-understood nontrivial vacuum configurations take the form (3.35) with X_6 a Calabi–Yau threefold.

Warped compactifications

Vacuum configurations of the form (3.35), solving the vacuum Einstein equations, are a simple and well-understood starting point. However, the ten-dimensional effective action involves fields beyond just the metric, and these fields can contribute stress-energy. Furthermore, the extended objects described above (fundamental strings, D-branes, and NS5-branes) are local sources of stress-energy and of charge. In non-vacuum solutions containing these sources – or carrying the corresponding charges without any local sources – the compactification manifold is generally not Ricci-flat. For non-vacuum configurations with maximal symmetry in the noncompact spacetime, the product ansatz (3.36) is generalized to a "warped" product:

$$G_{MN}\mathrm{d}X^M\mathrm{d}X^N = e^{2A(y)}g_{\mu\nu}\mathrm{d}x^\mu\mathrm{d}x^\nu + e^{-2A(y)}g_{mn}\mathrm{d}y^m\mathrm{d}y^n \,, \tag{3.37}$$

where now $g_{\mu\nu}$ is the metric of a maximally symmetric spacetime, the *warp factor* $A(y)$ is a function on X_6, and the internal metric g_{mn} is not necessarily Ricci-flat.

Supersymmetric compactifications

The supergravity actions described in Section 3.1.1 enjoy $\mathcal{N} = 1$ or $\mathcal{N} = 2$ supersymmetry in ten dimensions, but the solutions of the equations of motion need not preserve supersymmetry. Nevertheless, the best-understood solutions of string theory are supersymmetric,[7] for reasons that we briefly explain. The first reason is geometric: the Ricci-flatness condition required in vacuum solutions is closely tied to reduced holonomy (see Appendix A for details). Suitably reduced holonomy leads to the existence of invariant spinors and hence to unbroken supersymmetry in four dimensions. Most notably, Calabi–Yau threefolds have holonomy $SU(3)$ and correspondingly preserve one quarter of the ten-dimensional supersymmetry; a Calabi–Yau compactification of type II string theory has $\mathcal{N} = 2$ supersymmetry in four dimensions. The second reason is that unbroken supersymmetry provides unrivaled theoretical control, by guaranteeing stability and by imposing intricate relations among the couplings in the effective theory. A third reason is the historical and continuing interest in finding solutions of string theory with $\mathcal{N} = 1$ supersymmetry broken near the electroweak scale, in order to address the hierarchy problem.

---

[7] Compactifications that break supersymmetry at the Kaluza–Klein scale (for constructions with stabilized moduli, see e.g. [301–303]) present an important alternative to supersymmetric compactifications.

Inflationary configurations necessarily break supersymmetry, but a fruitful strategy is to study solutions with minimal ($\mathcal{N} = 1$) supersymmetry in four dimensions, and use these as the foundation for determining the effective action in $\mathcal{N} = 0$ solutions. Before examining supersymmetric compactifications, we will first briefly describe the structure of the corresponding supergravity theories.

$\mathcal{N} = 1$ supergravity in four dimensions

The bosonic fields of a general four-dimensional $\mathcal{N} = 1$ supergravity theory are the metric $g_{\mu\nu}$, gauge potentials A_μ^a, and complex scalar fields ϕ^i. The low-energy interactions of the scalars are encoded by the *superpotential* $W(\phi^i)$, which is a holomorphic function of the ϕ^i, and by the *Kähler potential* $K(\phi^i, \bar{\phi}^{\bar{\imath}})$, which is a real analytic function of the fields. In the absence of gauge interactions, the Lagrangian for the scalar fields is

$$\mathcal{L}_\Phi = -K_{i\bar{\jmath}} \, \partial^\mu \phi^i \partial_\mu \bar{\phi}^{\bar{\jmath}} - V_F \,, \tag{3.38}$$

where $K_{i\bar{\jmath}} \equiv \partial_i \partial_{\bar{\jmath}} K$ is the Kähler metric. The F-term potential V_F appearing in (3.38) is

$$V_F(\phi^i, \bar{\phi}^{\bar{\imath}}) = e^{K/M_{\mathrm{pl}}^2} \left[K^{i\bar{\jmath}} D_i W \overline{D_j W} - \frac{3}{M_{\mathrm{pl}}^2} |W|^2 \right] \,, \tag{3.39}$$

where $K^{i\bar{\jmath}}$ is the inverse Kähler metric and $D_i W \equiv \partial_i W + M_{\mathrm{pl}}^{-2}(\partial_i K) W$.

A primary task in studying a string compactification with $\mathcal{N} = 1$ supersymmetry is to compute the superpotential and Kähler potential in terms of geometric data. Through (3.38) and (3.39) these data determine the four-dimensional effective theory, to leading order in the low-energy (derivative) expansion.

3.2.1 Dimensional reduction

To compute the four-dimensional effective action of a string compactification, one begins with the appropriate ten-dimensional action and performs a Kaluza–Klein reduction. In order to develop intuition for this process, we will begin with a simple example.

Consider the ten-dimensional geometry

$$G_{MN} \mathrm{d}X^M \mathrm{d}X^N = e^{-6u(x)} g_{\mu\nu} \mathrm{d}x^\mu \mathrm{d}x^\nu + e^{2u(x)} \hat{g}_{mn} \mathrm{d}y^m \mathrm{d}y^n \,, \tag{3.40}$$

where $\hat{g}_{mn}$ is a reference metric with fixed volume,

$$\int_{X_6} \mathrm{d}^6 y \sqrt{\hat{g}} \equiv \mathcal{V} \,, \tag{3.41}$$

while $e^{u(x)}$ is a "breathing mode" that represents the variations in size of the internal space X_6 as a function of the four-dimensional coordinate x^μ. The factor of $e^{-6u(x)}$ in the first term is a convenient choice for which the gravitational action in four dimensions will appear in Einstein frame. We now examine the dimensional reduction of the Einstein–Hilbert term,

$$S_{\text{EH}}^{(10)} = \frac{1}{2\kappa^2} \int d^{10}X \sqrt{-G} \, e^{-2\Phi} R_{10} \,, \tag{3.42}$$

where R_{10} is the Ricci scalar constructed from G_{MN}. We would like to express R_{10} in terms of R_4 and $\hat{R}_6$, the Ricci scalars constructed from $g_{\mu\nu}$ and $\hat{g}_{mn}$, respectively. For this purpose we note that if two D-dimensional metrics g_{MN} and $\bar{g}_{MN}$ are related by the conformal rescaling

$$\bar{g}_{MN} = e^{2\omega(x)} g_{MN} \,, \tag{3.43}$$

then the corresponding Ricci scalars are related by

$$e^{2\omega} \bar{R} = R - 2(D-1)\nabla^2\omega - (D-2)(D-1)g^{MN}\nabla_M\omega\nabla_N\omega \,. \tag{3.44}$$

Similarly, the Laplacians constructed from g_{MN} and $\bar{g}_{MN}$ are related by

$$e^{2\omega}\bar{\nabla}^2 = \nabla^2 + (D-2)g^{MN}\nabla_M\omega\nabla_N \,. \tag{3.45}$$

Using these results, we find

$$S_{\text{EH}}^{(10)} = \frac{1}{2\kappa^2} \int d^4x \sqrt{-g} \int_{X_6} d^6y \sqrt{\hat{g}} \, e^{-2\Phi} \left(R_4 + e^{-8u}\hat{R}_6 + 12\partial_\mu u \partial^\mu u \right) \,. \tag{3.46}$$

If the string coupling $g_{\text{s}} \equiv e^{\Phi}$ is constant over the internal space, then the four-dimensional Einstein–Hilbert term can be written

$$S_{\text{EH}}^{(4)} = \frac{M_{\text{pl}}^2}{2} \int d^4x \sqrt{-g} \, R_4 \,, \tag{3.47}$$

with the four-dimensional Planck mass defined as

$$M_{\text{pl}}^2 \equiv \frac{\mathcal{V}}{g_{\text{s}}^2 \kappa^2} \,. \tag{3.48}$$

We recognize the combination of derivatives of $u(x)$ appearing in (3.46) as the kinetic term for a four-dimensional scalar field $u(x)$. This field is a *modulus* corresponding to a spacetime-dependent deformation of the ten-dimensional solution. As we will see below, in Calabi–Yau compactifications the breathing mode u corresponds to one of the Kähler moduli; the kinetic term for u in (3.46) follows from the Kähler potential

$$K = -3\ln\left(T + \bar{T}\right) \,, \tag{3.49}$$

where we have set $M_{\text{pl}} = 1$, and T is a complex scalar field with $\text{Re}(T) = e^{4u}$. (The imaginary part of T comes from the dimensional reduction of the four-form potential: see Section 3.2.3.)

Notice that the Ricci scalar $\hat{R}_6$ yields a potential term for the scalar u: positive internal curvature ($\hat{R}_6 > 0$) contributes a negative potential term $V \propto -e^{-8u}$ in four dimensions, driving the compactification toward small volume, while negative internal curvature contributes a positive potential term $V \propto +e^{-8u}$, leading to a decompactification instability. In Ricci-flat compactifications, the internal curvature term is absent and u has vanishing potential in the classical theory.

More general Kaluza–Klein reductions involve both more complicated ten-dimensional actions, for example involving p-form fields, as well as geometric deformations that generalize the very simple breathing mode described above. However, the principles underlying the general analysis are captured by the above example.

3.2.2 Moduli

In the simple Kaluza–Klein reduction described above, the breathing mode corresponding to an overall dilation of the internal space gave rise to a four-dimensional scalar field $u(x)$ parameterizing spacetime-dependent changes in the compactification volume. We will now describe the analogous moduli fields that arise in Calabi–Yau compactifications. To simplify the presentation, we primarily discuss four-dimensional scalars, i.e. moduli, leaving the actions for vector and tensor fields to the references.[8]

Calabi–Yau compactifications with $\mathcal{N} = 2$ supersymmetry

We begin by summarizing the effective theory that results from Kaluza–Klein reduction in Calabi–Yau compactifications of type II string theory. Consider the ten-dimensional geometry (3.36), with g_{mn} the Ricci-flat metric on a Calabi–Yau threefold X_6. Compactification of type II string theory on this background leads to a four-dimensional theory with $\mathcal{N} = 2$ supersymmetry.[9] The geometric moduli of this compactification are scalar fields corresponding to deformations of the metric g_{mn} that preserve the Calabi–Yau condition: the *Kähler moduli* are deformations of the Kähler form

$$J \equiv i\, g_{i\bar{j}}\, \mathrm{d}z^i \wedge \mathrm{d}\bar{z}^{\bar{j}} \,, \tag{3.50}$$

where z^i, $\bar{z}^{\bar{j}}$, with $i, \bar{j} = 1, 2, 3$, are complex coordinates on X_6, while *complex structure moduli* are deformations of the complex structure on X_6. See Appendix A for mathematical background and further details.

To parameterize the moduli, we introduce a set of harmonic (1,1)-forms ω^I, $I = 1, \ldots, h^{1,1}$, comprising a basis for the Dolbeault cohomology group $H_{\bar{\partial}}^{1,1}(X_6, \mathbb{C})$, as well as a set of harmonic (2,1)-forms χ^A, $A = 1, \ldots, h^{2,1}$, that form a basis for $H_{\bar{\partial}}^{2,1}(X_6, \mathbb{C})$. In terms of this basis, the Kähler form is

---

[8] A complete treatment can be found in [304], which we follow in this section. See e.g. [292] for background on Calabi–Yau geometry.

[9] Calabi–Yau compactifications of type I string theory, or of the heterotic string, yield $\mathcal{N} = 1$ supersymmetry. We will primarily discuss type II compactifications.

$$J = t^I(x)\,\omega_I \,, \tag{3.51}$$

where $t^I(x)$ are $h^{1,1}$ four-dimensional real scalar fields, the Kähler moduli. Similarly, complex structure deformations $\delta g_{\bar{i}\bar{j}}$ correspond to harmonic (2,1)-forms via

$$\delta g^A_{\bar{i}\bar{j}} = c\,\zeta^A(x)(\chi_A)_{kl\bar{i}}\,\bar{\Omega}^{kl}{}_{\bar{j}} \,, \tag{3.52}$$

where Ω is the holomorphic (3,0)-form of X_6, whose normalization determines the numerical constant c. The $h^{2,1}$ four-dimensional complex scalar fields $\zeta^A(x)$ are the complex structure moduli.

Additional scalar fields arise from expanding the NS-NS and R-R potentials in the bases of harmonic forms. We henceforth specialize to type IIB string theory, where the relevant forms are B_2, C_2, C_4, with the expansions

$$B_2 = B_2(x) + b^I(x)\omega_I \,, \tag{3.53}$$

$$C_2 = C_2(x) + c^I(x)\omega_I \,, \tag{3.54}$$

$$C_4 = \vartheta^I(x)\tilde{\omega}_I \,. \tag{3.55}$$

Here, $B_2(x)$ denotes the four-dimensional two-form $B_{\mu\nu}\mathrm{d}x^\mu \wedge \mathrm{d}x^\nu$, to be distinguished from the ten-dimensional two-form B_2, and similarly for C_2. In (3.55), we have suppressed vector field contributions to the final equality (see [304] for the complete expression), and have defined $\tilde{\omega}_I$, $I = 1,\dots,h^{1,1}$, as a basis for $H^{2,2}$. Finally, the dilaton Φ and the R-R zero-form C_0 give rise to two more real scalars.

The scalar fields just described appear in multiplets of four-dimensional $\mathcal{N} = 2$ supersymmetry. The $4h^{1,1}$ scalars $t^I, \vartheta^I, b^I, c^I$ furnish the bosonic content of $h^{1,1}$ hypermultiplets, while the $h^{1,2}$ real scalars ζ^A appear in $\mathcal{N} = 2$ vector multiplets (in combination with the vector fields $V^\mu_A \sim C^\mu_{ij\bar{k}}(\chi_A)^{ij\bar{k}}$ from the dimensional reduction of C_4, which we have suppressed above). Finally, Φ, C_0, B_2, C_2 form the "universal hypermultiplet," after Hodge dualizing the two-forms to scalars in four dimensions – see Appendix A.

Calabi–Yau orientifolds with $\mathcal{N} = 1$ supersymmetry

Type II Calabi–Yau compactifications with unbroken $\mathcal{N} = 2$ supersymmetry do not yield realistic models of nature; in particular, $\mathcal{N} = 2$ supersymmetry does not allow fermions in chiral representations of gauge groups. More promising are type II compactifications that include local sources, such as D-branes, in addition to p-form fluxes. The resulting gauge theories can be rich enough to include the Standard Model, and spontaneous breaking of supersymmetry in a metastable vacuum is plausibly achievable.

A fundamental consistency requirement for flux compactifications with D-branes is cancellation of all tadpoles associated with the charge and tension of the sources. Most dramatically, the gravitational tadpole associated with the positive tension of a D-brane requires the presence of a negative-tension source

[305]. The best-understood negative-tension objects are *orientifold planes*, which are non-dynamical extended objects that appear at the fixed point loci of an involution $\mathcal{O}$ that reverses the orientation of the string worldsheet.

We will describe the essential aspects of orientifolds here, referring the reader to [282] for a complete treatment. An orientifold action $\mathcal{O}$ is a symmetry that includes the worldsheet orientation reversal Ω_{ws}. The orientifold actions of primary interest here take the form

$$\mathcal{O} = (-1)^{F_L} \Omega_{ws} \sigma \,, \tag{3.56}$$

where $(-1)^{F_L}$ is the worldsheet fermion number in the left-moving sector – cf. the decomposition implied by (3.9) – and the geometric involution σ reverses the sign of the holomorphic $(3,0)$ form Ω of X_6, but leaves the metric and complex structure invariant. The fixed point loci of an orientifold action of the form (3.56) are points or four-cycles in X_6. Because the geometric action on the noncompact dimensions is trivial, the resulting orientifold planes have three or seven spatial dimensions, and are known as O3-planes and O7-planes, respectively.

Under the action (3.56), the cohomology group $H^{1,1}$ can be decomposed as

$$H^{1,1} = H^{1,1}_+ \oplus H^{1,1}_- \,, \tag{3.57}$$

with the subscript denoting the parity of the corresponding two-forms under the orientifold action. Correspondingly, the basis ω^I, $I = 1, \ldots, h^{1,1}$, for $H^{1,1}$ decomposes into a basis for the even eigenspace, ω^i, $i = 1, \ldots, h^{1,1}_+$, and a basis for the odd eigenspace ω^α, $\alpha = 1, \ldots, h^{1,1}_-$.

To understand the effect of orientifolding on the four-dimensional fields, we note that t^I, ϑ^I, Φ, C_0 are even under the orientifold action, while ζ^A, b^I, c^I, $B_2(x)$, $C_2(x)$ are odd. Invariant four-dimensional fields arise from even ten-dimensional fields expanded in terms of even forms, or from odd ten-dimensional fields expanded in terms of odd forms. The Kähler form can be written

$$J = t^i(x)\omega_i \,, \tag{3.58}$$

so that the orientifold-invariant Kähler moduli are the $h^{1,1}_+$ real scalars t^i, which measure the volumes of two-cycles that are even under the involution. Similarly, noting that the orientifold action projects out the four-dimensional two-forms $B_2(x)$ and $C_2(x)$, we have the invariant fields (again omitting vector contributions)

$$B_2 = b^\alpha(x)\omega_\alpha \,, \tag{3.59}$$
$$C_2 = c^\alpha(x)\omega_\alpha \,, \tag{3.60}$$
$$C_4 = \vartheta^i(x)\tilde{\omega}_i \,. \tag{3.61}$$

Likewise, the invariant complex structure moduli are ζ^a, for $a = 1, \ldots, h^{1,2}_-$. Finally, Φ and C_0 are automatically invariant.

It is important to assemble the invariant scalars into the bosonic components of chiral multiplets of four-dimensional $\mathcal{N} = 1$ supersymmetry, i.e. to determine the proper Kähler coordinates on the moduli space. First of all, the axion C_0 and dilaton Φ combine to form the complex axiodilaton,

$$\tau = C_0 + ie^{-\Phi} . \tag{3.62}$$

The complex structure moduli ζ^a are automatically good Kähler coordinates. The "two-form scalars" b_α and c_α form the complex combination

$$G_\alpha \equiv c_\alpha - \tau\, b_\alpha . \tag{3.63}$$

To go further, we note that the compactification volume $\mathcal{V}$ can be written in terms of the Kähler form J as follows:

$$\mathcal{V} = \frac{1}{6} \int_{X_6} J \wedge J \wedge J = \frac{1}{6} c_{ijk} t^i t^j t^k , \tag{3.64}$$

where c_{ijk} are the triple intersection numbers of X_6. Then, the Kähler coordinates describing complexified four-cycle volumes are [304]

$$T_i \equiv \frac{1}{2} c_{ijk} t^j t^k + i\vartheta_i + \frac{1}{4} e^\Phi c_{i\alpha\beta} G^\alpha (G - \bar{G})^\beta . \tag{3.65}$$

The expression (3.65) is not supposed to be obvious, but we can provide some intuition by dropping the contribution of G^α, so that

$$T_i = \frac{1}{2} c_{ijk} t^j t^k + i\vartheta_i . \tag{3.66}$$

Now, we use the fact that the two-cycle volumes t^i are related to the four-cycle volumes τ_i by

$$\tau_i = \frac{\partial \mathcal{V}}{\partial t^i} = \frac{1}{2} c_{ijk} t^j t^k , \tag{3.67}$$

so that (3.66) can be recognized as

$$T_i = \tau_i + i\vartheta_i . \tag{3.68}$$

This is the familiar complexification of four-cycle volumes τ_i by ϑ_i, i.e. by the integral of C_4 over the corresponding four-cycle. The more involved expression (3.65) shows that the corresponding proper Kähler coordinate depends on the vev of the two-form G^α.[10]

In summary, the Kähler coordinates on the moduli space are the $h^{1,1}_+$ complexified four-cycle volumes T_i (3.65), the $h^{1,1}_-$ two-form scalars G^α (3.63), the axiodilaton τ (3.62), and the $h^{1,2}_-$ complex structure moduli ζ^a. All told, a compactification of type IIB string theory on an O3/O7 orientifold of a Calabi–Yau

---

[10] This fact might seem to be an irrelevant technicality, but we will see in Section 5.4.2 that the mixing (3.65) is the fatal flaw in one otherwise compelling scenario for inflation in string theory.

manifold leads to $h^{1,1}_+ + h^{1,1}_- + h^{1,2}_- + 1 = h^{1,1} + h^{1,2}_- + 1$ complex moduli scalars in the four-dimensional theory. Further scalar fields can arise from the open string sector.

3.2.3 Axions

One class of fields deserves special discussion: these are *axions*, i.e. pseudoscalar fields enjoying Peccei–Quinn (PQ) shift symmetries of the form

$$a \mapsto a + const. \tag{3.69}$$

The QCD axion is the original and most famous example of an axion, and some authors reserve the word "axion" for this field alone, but we stress that the axionic fields discussed here need not couple to QCD.

Axions from p-forms

Axions arise in string compactifications from the integration of p-form gauge potentials over p-cycles of the compact space. For example, in type IIB string theory, there are axions associated with the NS-NS two-form B_2, the R-R two-form C_2, and the R-R four-form C_4, integrated over suitable two-cycles Σ^I_2 and four-cycles Σ^I_4:

$$b_I = \frac{1}{\alpha'} \int_{\Sigma^I_2} B_2 \,, \qquad c_I = \frac{1}{\alpha'} \int_{\Sigma^I_2} C_2 \,, \qquad \vartheta_I = \frac{1}{(\alpha')^2} \int_{\Sigma^I_4} C_4 \,, \tag{3.70}$$

where we have chosen the following normalizations for the forms in (3.55):

$$\int_{\Sigma^I_2} \omega^J = \alpha' \delta_I{}^J \,, \qquad \int_{\Sigma^I_4} \tilde{\omega}^J = (\alpha')^2 \delta_I{}^J \,. \tag{3.71}$$

Finally, there are three universal contributions: the R-R axion C_0, and two axions, b and c, from dualizing $B_2(x)$ and $C_2(x)$, respectively. In sum, a hypermultiplet arising in $\mathcal{N} = 2$ Calabi–Yau compactifications of type IIB string theory contains three axions: for the $h^{1,1}$ "non-universal" hypermultiplets, the axions descend from B_2, C_2, and C_4, while the axions in the universal hypermultiplet are C_0, b_U, and c_U.[11] We will collectively call these axions $a \equiv \{b_I, c_I, \vartheta_I, C_0, b_U, c_U\}$. Orientifolding by an involution (3.56) with O3/O7 fixed planes projects out some of the axions. Those that remain are C_0; b_α and c_α, for $\alpha = 1, \ldots, h^{1,1}_-$; and ϑ_i, for $i = 1, \ldots, h^{1,1}_+$.

Axionic shift symmetries

At the classical level, each axion inherits a *continuous* shift symmetry, $a \mapsto a + const.$, from the corresponding p-form gauge invariances of the ten-dimensional theory. Specifically, in a background with vanishing fluxes, the type IIB action

---

[11] The structure of shift symmetries arising in the universal hypermultiplet is described in [306–308].

(3.22) is independent of C_0, C_2, C_4, B_2, and involves only the associated field strengths. The continuous shift symmetry holds to all orders in perturbation theory, but is broken nonperturbatively, by instanton effects. What remains is a *discrete* symmetry, $a \mapsto a + (2\pi)^2$.

We now explain this important point in the concrete example of the b axion, following the classic arguments by Dine, Seiberg, Wen, and Witten [309–312] that established the shift symmetry to all orders in the g_s and α' expansions. The extension to axions from other p-forms is straightforward. We start with (3.3), the worldsheet coupling of the two-form B_2,

$$S_\sigma \supset -\frac{1}{4\pi\alpha'} \int_{\Sigma_2} d^2\sigma \, \epsilon^{ab} \partial_a X^M \partial_b X^N B_{MN}(X) \,, \tag{3.72}$$

or, equivalently,

$$S_\sigma \supset -\frac{1}{2\pi\alpha'} \int_{\Sigma_2} B_2 \equiv -\frac{b}{2\pi} \,, \tag{3.73}$$

where the integral is taken over the string worldsheet. We recognize (3.73) as a topological coupling. Expanding $B_{MN}(X)$ around a fiducial point $X_{(0)} \equiv 0$ yields

$$B_{MN}(X) = B_{MN}(X_{(0)}) + X^P \partial_P B_{MN}(X_{(0)}) + \cdots \,. \tag{3.74}$$

The constant term $B_{MN}(X_{(0)})$ gives rise in (3.72) to a *worldsheet total derivative*,

$$-\frac{1}{4\pi\alpha'} \int_{\Sigma_2} d^2\sigma \, \partial_a \left(\epsilon^{ab} X^M \partial_b X^N B_{MN}(X_{(0)}) \right) \,, \tag{3.75}$$

which vanishes unless the worldsheet either wraps a topologically nontrivial cycle, or has a boundary. The remaining terms in (3.74) involving spacetime derivatives of B_{MN} are nonvanishing in general, but correspond to finite-momentum couplings (i.e. derivative interactions involving only $\partial_\mu b$ in the effective theory). As derivative interactions do not break the shift symmetry, it suffices, for the purpose of ascertaining the symmetry structure, to consider the zero-momentum coupling arising from $B_{MN}(X_{(0)})$.

We conclude that the shift symmetry $b \mapsto b + const.$ can only be broken if the string worldsheet wraps a nontrivial cycle in the target spacetime, or has a boundary. Both sources of symmetry breaking play significant roles in model-building, and we will discuss them in turn. At any order in σ-model perturbation theory, the string worldsheet wraps a topologically trivial cycle, but the fundamental nonperturbative contribution in the σ-model is a *worldsheet instanton*, i.e. a worldsheet wrapping a nontrivial cycle Σ_2. The corresponding spontaneous breaking of the shift symmetry is nonperturbative in α', and is measured by the Euclidean action

$$S_{\text{inst}} = \exp\left(-\frac{1}{2\pi\alpha'} \int_{\Sigma_2} (J + iB_2) \right) \propto \exp\left(-i\frac{b}{2\pi} \right) \,, \tag{3.76}$$

where J is the Kähler form. The result is a periodic potential for b, with periodicity $(2\pi)^2$.[12]

Next, we consider the string loop expansion. The preceding arguments made no assumption about the genus of the worldsheet, and so must hold to any order in the string loop expansion. However, nonperturbatively in g_s a new possibility arises: the closed string worldsheet can break open on a soliton (i.e. a D-brane) and hence acquire a boundary. Correspondingly, the shift symmetry can be broken by the presence of spacetime-filling D-branes.

Finally, certain types of Euclidean D-branes can break the shift symmetry, because B_2 appears in the Euclidean D-brane action. Just as for worldsheet instantons, the resulting contribution to the potential is periodic, with scale

$$S_{\mathrm{ED}p} = \exp\left(-T_p\,\mathrm{Vol}(\Sigma_p)\right), \tag{3.77}$$

for a Euclidean Dp-brane wrapping a cycle Σ_p.

We conclude that the axion field b in the four-dimensional effective theory enjoys a continuous shift symmetry $b \mapsto b + const.$ that is spontaneously broken by worldsheet and/or D-brane instantons to a discrete shift symmetry $b \mapsto b + (2\pi)^2$, and may be explicitly broken if D-branes are present in the compactification.

Axion decay constants

The discrete shift symmetry $a \mapsto a + (2\pi)^2$ constrains the axion Lagrangian to take the form

$$\mathcal{L}(a) = -\frac{1}{2}f^2(\partial a)^2 - \Lambda^4\left[1 - \cos(a/2\pi)\right] + \cdots, \tag{3.78}$$

where Λ is a dynamically generated scale; f is a constant with dimensions of mass, known as the *axion decay constant*; and the omitted terms contain higher-derivative interactions and multi-instanton contributions. In terms of the canonically normalized field $\phi \equiv af$, the axion periodicity is $(2\pi)^2 f$.

After dimensional reduction, the decay constants can be deduced from the effective Kähler potential. On the other hand, it is also instructive to compute them directly. We again take the b axions as an example, in an O3/O7 orientifold. The two-form B_2 can be expanded in terms of the four-dimensional fields $b_\alpha(x)$ and the (1,1)-forms ω^α, $\alpha = 1, \ldots, h_-^{1,1}$:

$$B_2 = b_\alpha(x)\omega^\alpha. \tag{3.79}$$

To determine the axion kinetic terms, and hence the decay constants, we dimensionally reduce the ten-dimensional action for the two-form,

$$\frac{1}{2(2\pi)^7 g_s^2(\alpha')^4}\int \mathrm{d}^{10}X\,|\mathrm{d}B_2|^2 \supset \frac{1}{2}\int \mathrm{d}^4x\sqrt{-g}\,\gamma^{\alpha\beta}(\partial^\mu b_\alpha \partial_\mu b_\beta), \tag{3.80}$$

---

[12] In this section we follow the conventions of [35].

where

$$\gamma^{\alpha\beta} \equiv \frac{1}{6\,(2\pi)^7 g_{\rm s}^2 (\alpha')^4} \int_{X_6} \omega^\alpha \wedge \star_6 \omega^\beta \;. \tag{3.81}$$

Performing the integral in (3.81) and diagonalizing the result[13] (i.e. $\gamma_{\alpha\beta} \mapsto f_\alpha^2 \delta_{\alpha\beta}$), one can extract the axion decay constants f_α. For purposes of illustration, we consider an isotropic compactification with characteristic length L and volume $\mathcal{V} = L^6/\alpha'^3$. Using (3.48) to relate the compactification volume to the four-dimensional Planck mass, we find

$$\frac{f^2}{M_{\rm pl}^2} \approx \frac{1}{6} \frac{\alpha'^2}{L^4} \;. \tag{3.82}$$

Since computational control requires $L \gg \sqrt{\alpha'}$, we infer that $f \ll M_{\rm pl}$. Qualitatively similar upper bounds on the decay constants occur in all computable limits of string theory that have been explored to date [47, 48].

3.3 Moduli stabilization

Generic Calabi–Yau compactifications come with many *moduli*,[14] i.e. zero-energy deformations arising from the plethora of topologically distinct cycles in typical Calabi–Yau manifolds. Understanding the dynamics of moduli is crucial for describing cosmological evolution. During inflation, the positive vacuum energy tends to induce instabilities of massless scalar fields, along directions that reduce the energy and swiftly end inflation. Moreover, quantum fluctuations of moduli during inflation contribute to the primordial perturbations. Furthermore, the impact of moduli on cosmology after the time of inflation is profound and complex: moduli can affect Big Bang nucleosynthesis, overclose the universe, comprise some of the dark matter, decay to dark radiation, or mediate long-range interactions. However, a modulus that acquires a mass $m \gtrsim 30$ TeV decays before nucleosynthesis, eliminating nearly all[15] late-time effects. A full treatment of the cosmological moduli problem is beyond the scope of this book, and we will content ourselves with describing the effects of moduli during inflation.

A principal challenge in the search for cosmological models in string theory is the task of controlling instabilities associated with the moduli, i.e. finding vacua

---

[13] See [313] for an explanation of the preferred basis implied by (3.78).

[14] The word "moduli" actually has several different meanings in different contexts, so a clarification is appropriate. The geometric notion is that moduli parameterize continuous families of solutions, for example families of Ricci-flat metrics. In physics, a modulus is a scalar field with gravitational-strength couplings that has vanishing potential at some level of approximation. Some moduli have exactly vanishing potential before supersymmetry breaking, while others have vanishing classical potential but obtain a mass from quantum effects. In some contexts, "moduli" refers exclusively to parity-even real scalar fields, as distinguished from pseudoscalar axions, but we will generally refer to complex moduli.

[15] Moduli that decay early, but to fields that themselves linger and affect late-time observables, are an interesting exception: see e.g. [314–320].

in which all the moduli have positive masses-squared; this is known as *moduli stabilization*. As we will explain in Chapter 4, giving non-zero masses to all moduli does not suffice to dispel the moduli problem – for this purpose, the masses must be large compared to the scales accessed during inflation. Even so, identifying the leading contributions to the moduli potential is an essential first step toward constructing realistic models. We now turn to a characterization of the moduli potential in the example of flux compactifications of type IIB string theory.

3.3.1 Classical solutions

In this section, we will review the essential features of type IIB flux compactifications on Calabi–Yau orientifolds, following the pioneering work by Giddings, Kachru, and Polchinski (GKP) [305]. Space limitations prevent us from detailing the many advances generalizing and extending the analysis of [305], most notably to time-dependent backgrounds and to solutions with strong warping (see [321–324]). The literature on flux compactifications beyond type IIB orientifolds is so extensive that we will not attempt to summarize it; more complete discussions of flux compactifications, where the original references can be found, include [45, 325, 326].

Type IIB supergravity

At leading order in α' and g_{s}, the ten-dimensional action for the bosonic fields in Einstein frame is given by (3.22). In addition, there may be local sources, such as D-branes and orientifold planes, with corresponding action S_{loc}. We search for warped solutions with the ansatz[16] (3.37), but now taking $g_{\mu\nu} = \eta_{\mu\nu}$:

$$\mathrm{d}s^2 = e^{2A(y)}\eta_{\mu\nu}\mathrm{d}x^\mu\mathrm{d}x^\nu + e^{-2A(y)}g_{mn}\mathrm{d}y^m\mathrm{d}y^n \ . \tag{3.83}$$

Four-dimensional Poincaré invariance requires that the three-form flux G_3 has no nonvanishing components in the noncompact spacetime, while the self-dual five-form flux takes the form

$$\tilde{F}_5 = (1 + \star_{10})\,\mathrm{d}\alpha(y) \wedge \mathrm{d}x^0 \wedge \mathrm{d}x^1 \wedge \mathrm{d}x^2 \wedge \mathrm{d}x^3 \ , \tag{3.84}$$

where $\star_{10}$ is the ten-dimensional Hodge star and $\alpha(y)$ is a scalar function on X_6.

Equations of motion

The trace of the ten-dimensional Einstein equation yields

$$\nabla^2 e^{4A} = \frac{e^{8A}}{2\,\mathrm{Im}(\tau)}|G_3|^2 + e^{-4A}\left(|\partial\alpha|^2 + |\partial e^{4A}|^2\right) + 2\kappa^2 e^{2A}\mathcal{J}_{\mathrm{loc}} \ , \tag{3.85}$$

where ∇^2 is the Laplacian on X_6, and the effects of local sources are parameterized as

---

[16] For time-dependent solutions, we would require a more general ansatz [321].

$$\mathcal{J}_{\text{loc}} \equiv \frac{1}{4} \left(\sum_{M=4}^{9} T^M{}_M - \sum_{M=0}^{3} T^M{}_M \right)_{\text{loc}} , \qquad (3.86)$$

with T_{MN} the stress-energy tensor derived from S_{loc}. In the absence of local sources, i.e. for $\mathcal{J}_{\text{loc}} = 0$, the solution is trivial, with constant A, constant α, and vanishing G_3. (To see this, note that the left-hand side of (3.85) integrates to zero on X_6, while the first three terms on the right-hand side are all non-negative.) A nontrivial warped compactification requires one or more sources with $\mathcal{J}_{\text{loc}} < 0$ [327], for example orientifold planes.

Next, the Bianchi identity for the five-form flux is

$$d\tilde{F}_5 = H_3 \wedge F_3 + 2\kappa^2 T_3 \rho_3^{\text{loc}} , \qquad (3.87)$$

where ρ_3^{loc} is the D3-brane charge density due to the local sources. Because $\tilde{F}_5$ is self-dual, (3.87) may also be thought of as an equation of motion. Integrating (3.87) over X_6 leads to a tadpole-cancellation condition (i.e. Gauss's law constraint),

$$\frac{1}{2\kappa^2 T_3} \int_{X_6} H_3 \wedge F_3 + Q_3^{\text{loc}} = 0 , \qquad (3.88)$$

where Q_3^{loc} is the total charge associated with ρ_3^{loc}. Substituting (3.84) into (3.87) and combining with (3.85), we get[17]

$$\nabla^2 \left(e^{4A} - \alpha \right) = \frac{e^{8A}}{24 \text{Im}(\tau)} |iG_3 - \star_6 G_3|^2 + e^{-4A} |\partial(e^{4A} - \alpha)|^2$$
$$+ 2\kappa^2 e^{2A} \left(\mathcal{J}_{\text{loc}} - \mathcal{Q}_{\text{loc}} \right) , \qquad (3.89)$$

where $\star_6$ is the six-dimensional Hodge star and $\mathcal{Q}_{\text{loc}} \equiv T_3 \rho_3^{\text{loc}}$. The left-hand side of (3.89) integrates to zero on X_6, while the non-localized sources on the right-hand side are non-negative. As for the localized contribution $\mathcal{J}_{\text{loc}} - \mathcal{Q}_{\text{loc}}$, many well-understood localized sources satisfy the BPS-like condition

$$\mathcal{J}_{\text{loc}} \geq \mathcal{Q}_{\text{loc}} . \qquad (3.90)$$

The condition (3.90) is saturated by D3-branes and O3-planes, and by D7-branes wrapping four-cycles (in such a way as to respect the $\mathcal{N} = 1$ supersymmetry preserved by D3-branes). It is satisfied, but not saturated, by anti-D3-branes and by D5-branes wrapped on collapsed two-cycles. However, $\overline{\text{O3}}$-planes and O5-planes violate (3.90).

Consider a compactification in which all sources satisfy (3.90). Integrating (3.89) reveals that we must in fact demand that all sources saturate (3.90) – i.e. only D3-branes, O3-planes, and D7-branes are allowed – and that the three-form flux is imaginary self-dual (ISD),

$$\star_6 G_3 = iG_3 , \qquad (3.91)$$

---

[17] This corrects the numerical factor appearing in [305], cf. [328].

while the warp factor is equal to the four-form potential

$$e^{4A} = \alpha .$$
(3.92)

A configuration meeting these criteria is called an *ISD solution*.

To recapitulate, the Einstein equation and five-form Bianchi identity can be combined to give, at leading order in α' and g_s, the key relations (3.85) and (3.89). These expressions are parallel in form: the left-hand side expressions integrate to zero, while the right-hand side in each case involves a sum of non-localized ("bulk") terms that are everywhere non-negative, as well as a localized contribution. If the localized contribution is non-negative, it must in fact be zero, and then the bulk terms must be identically zero. In the case of the Einstein equation (3.85), this implies that *in the absence of negative tension sources, only unwarped solutions (without positive tension sources) are allowed*. From the Einstein equation minus Bianchi identity (3.89), we learn that *in the absence of sources violating* (3.90), *only ISD solutions are allowed*. Because well-understood supersymmetric configurations of O3-planes and O7-planes (as well as D3-branes and D7-branes) yield negative tension without violating (3.90), it is straightforward to exhibit ISD warped solutions. Non-ISD solutions are much less studied at present, because of the difficulty of controlling the comparatively exotic orientifold planes that violate (3.90).

A significant property of compactifications with three-form flux, including ISD solutions, is that the complex structure moduli ζ^a and the axiodilaton τ experience a potential. To see this, we note that the ten-dimensional type IIB action (3.22) contains the term

$$V_{\text{flux}} = \frac{1}{2\kappa^2} \int d^{10}X \sqrt{-G_E} \left[-\frac{|G_3|^2}{2\text{Im}(\tau)} \right] ,$$
(3.93)

which involves the complex structure moduli via the metric contraction, and the axiodilaton both through the denominator and through the definition (3.20) of G_3. As a result, for a generic choice of quantized fluxes, τ and all of the ζ^a receive masses at the classical level, at leading order in α'.

Effective supergravity

The data of the four-dimensional effective theory of an ISD compactification can be usefully repackaged in terms of a Kähler potential and superpotential of $\mathcal{N} = 1$ supergravity. At leading order in the α' and string loop expansions, the Kähler potential is

$$K_0 = -2\ln(\mathcal{V}) - \ln\left(-i(\tau - \bar{\tau})\right) - \ln\left(-i \int \Omega \wedge \bar{\Omega}\right) .$$
(3.94)

Here, the volume $\mathcal{V}$ and the holomorphic three-form Ω depend implicitly on the Kähler moduli T_i and the complex structure moduli ζ^a, respectively.

The ISD condition (3.91) can be derived from the Gukov–Vafa–Witten flux superpotential [329]

$$W_0 = \frac{c}{\alpha'} \int G_3 \wedge \Omega \, , \tag{3.95}$$

where c is a constant (see [330]). Since G_3 depends on the dilaton and Ω involves the complex structure moduli, the superpotential (3.95) leads to a nontrivial potential for these moduli. The scalar potential associated with K_0 and W_0 is

$$V_F = e^{K_0} \left[K_0^{I\bar{J}} D_I W_0 \overline{D_J W_0} - 3|W_0|^2 \right] \, , \tag{3.96}$$

where I, J run over all the moduli (T_i, G_α, ζ^a and τ). Supersymmetry is preserved if all F-terms vanish,[18] i.e. if

$$D_I W_0 \equiv \partial_I W_0 + (\partial_I K) W_0 = 0 \, , \tag{3.97}$$

for all I.

No-scale structure

The Kähler potential (3.94) is of a specific form that satisfies

$$\sum_{I,J=T_i,G_\alpha} K_0^{I\bar{J}} \partial_I K_0 \partial_{\bar{J}} K_0 = 3 \, . \tag{3.98}$$

Since the superpotential (3.95) is independent of the Kähler moduli, the scalar potential (3.96) is of the *no-scale* type, i.e. it is independent of the F-terms of the Kähler moduli and two-form scalars,

$$V_F = e^{K_0} \sum_{I,J \neq T_i,G_\alpha} K_0^{I\bar{J}} D_I W_0 \overline{D_J W_0} \, . \tag{3.99}$$

This potential is positive semi-definite, and $V_F = 0$ when $D_{I \neq T_i,G_\alpha} W_0 = 0$. The minimum is not necessarily supersymmetric, as in general we may have $D_{T_i,G_\alpha} W_0 \neq 0$.

No-scale structure and D3-branes

Thus far we have discussed the effective action for massless closed string fields, but the positions of D-branes provide an important additional class of *open string moduli*. Consider a D3-brane that fills spacetime and sits at a point in a flux compactification on a Calabi–Yau manifold. Evaluating the DBI+CS action (3.31) in an ISD background, one finds that the potential energy for D3-brane motion vanishes identically: the complex scalars z_α, $\alpha = 1, 2, 3$, that parameterize the D3-brane position are massless moduli. The four-dimensional action derived from the dimensional reduction of (3.31) can be expressed in $\mathcal{N} = 1$ supergravity

---

[18] When gauge multiplets are present in the effective theory, D-term contributions are an important alternative source of supersymmetry breaking, but our present discussion is confined to the moduli sector.

via the *DeWolfe–Giddings Kähler potential*, which for a compactification with a single Kähler modulus T takes the form [331]

$$K(T, \bar{T}, z_\alpha, \bar{z}_\alpha) = -3 \ln \left[T + \bar{T} - \gamma k(z_\alpha, \bar{z}_\alpha) \right], \qquad (3.100)$$

where γ is a constant, and $k(z_\alpha, \bar{z}_\alpha)$ is the Kähler potential for the metric on the Calabi–Yau manifold. The Kähler potential (3.100) is of no-scale type; if the superpotential W is independent of T and of the z_α, then the F-terms of these fields do not appear in the F-term potential. The mixing between the Kähler modulus T and the D3-brane position moduli implied by (3.100) has significant ramifications for inflationary model building with D3-branes: see Section 5.1.

In summary, in a "no-scale" compactification with imaginary self-dual fluxes, one finds, at leading order in α' and g_s, that the vacuum energy vanishes,[19] the complex structure moduli and axiodilaton are stabilized, and the Kähler moduli, two-form moduli, and D3-brane position moduli have vanishing potential.

3.3.2 Quantum effects

Perturbative and nonperturbative corrections to the effective action are known to break the no-scale structure, lifting or destabilizing the flat directions and altering the vacuum energy. We will begin by discussing perturbative corrections to the Kähler potential, in both the α' and g_s expansions, and then discuss nonperturbative corrections to the superpotential.

Perturbative corrections

The most famous perturbative correction to the Kähler potential descends from an $(\alpha')^3$ curvature correction in ten dimensions, namely the quartic invariant $\mathcal{R}^4$ appearing in (2.30). This term is part of the classical, higher-curvature ten-dimensional supergravity theory; it arises via a four-loop correction to the β-function of the worldsheet σ-model [202], rather than from a loop in spacetime. In the four-dimensional effective theory, the result takes the form [332]

$$K = -2 \ln \left[\mathcal{V} + \frac{\xi}{2g_s^{3/2}} \right], \qquad \xi \equiv -\frac{\chi(X_6)\zeta(3)}{2(2\pi)^3}, \qquad (3.101)$$

where $\chi(X_6)$ is the Euler characteristic of X_6, and $\zeta(3) \approx 1.202$ is Apéry's constant. The Kähler potential (3.101) does not satisfy the no-scale condition (3.98) (unless $\chi = 0$).

Perturbative corrections from loop effects in spacetime, i.e. from higher-genus string worldsheets, will also generically spoil the no-scale structure (3.98). The only explicit results available are for $\mathcal{N} = 1$ and $\mathcal{N} = 2$ compactifications on

---

[19] Having a non-supersymmetric vacuum with vanishing vacuum energy seems too good to be true, and it is; no-scale structure on its own is not a solution to the cosmological constant problem, because it does not survive quantum corrections.

certain toroidal orientifolds, such as $T^6/(\mathbb{Z}_2 \times \mathbb{Z}_2)$ [333, 334]. To give a concrete picture of string loop corrections, we now sketch this specific result. The correction to the Kähler potential takes the form

$$\delta K_{(g_s)} = \delta K^{\text{KK}}_{(g_s)} + \delta K^{\text{W}}_{(g_s)} , \qquad (3.102)$$

where the term $\delta K^{\text{KK}}_{(g_s)}$ comes from the exchange of closed strings with Kaluza–Klein (KK) momentum between D7-branes and D3-branes, while $\delta K^{\text{W}}_{(g_s)}$ comes from the exchange of closed strings with nonvanishing winding (W). The former is given by

$$\delta K^{\text{KK}}_{(g_s)} = -\frac{1}{128\pi^2} \sum_{i=1}^{3} \frac{\mathcal{E}_i^{\text{KK}}(\zeta, \bar{\zeta})}{\text{Re}(\tau)\,\tau_i} , \qquad (3.103)$$

where τ_i stands for the Kähler modulus associated with the four-cycle wrapped by the ith D7-brane. The second term in (3.102) takes the form

$$\delta K^{\text{W}}_{(g_s)} = -\frac{1}{128\pi^2} \sum_{i=1}^{3} \frac{\mathcal{E}_i^{\text{W}}(\zeta, \bar{\zeta})}{\tau_j \tau_k}\bigg|_{j \neq k \neq i} . \qquad (3.104)$$

These results have a complicated dependence on the complex structure moduli ζ (encoded by the functions $\mathcal{E}_i^{\text{KK}}(\zeta, \bar{\zeta})$ and $\mathcal{E}_i^{\text{W}}(\zeta, \bar{\zeta})$ given in [333, 334]), but have a simple scaling with the Kähler moduli τ_i. A conjectural generalization of the results of [333, 334] to general Calabi–Yau threefolds appears in [335] (see also [336] for related earlier work), but giving an explicit characterization of this leading string loop correction remains an open problem.

Even though the perturbative corrections (3.103) and (3.104) manifestly violate no-scale structure, the corresponding contributions to the scalar potential cancel to some extent; see the discussion in Section 5.5.2.

Nonperturbative effects

Although the Kähler potential for the Kähler moduli receives perturbative corrections in the α' and g_s expansions, the superpotential receives no corrections in either expansion, to any order in perturbation theory, as we now explain.

The fact that the superpotential of a supersymmetric field theory receives no perturbative corrections in the ordinary $\hbar$ expansion – corresponding to the g_s expansion in string theory – was originally established directly [337]. Elegant non-renormalization theorems in string theory [309–312] arrived at the same end by combining holomorphy and shift symmetry arguments. In the heterotic string setting emphasized in [309–312], the argument for non-renormalization in g_s is more straightforward than in type IIB flux compactifications, because the classical superpotential in the heterotic string is independent of the dilaton, whereas the classical GVW flux superpotential (3.95) involves the dilaton through the definition (3.20) of G_3. A careful demonstration of the absence of string loop corrections to (3.95) appears in [338].

Next, to address α' corrections, we recall that the axionic imaginary parts of the Kähler moduli (3.68) are protected by shift symmetries, $\vartheta_i \mapsto \vartheta_i + const.$, which hold to all orders in perturbation theory, as explained in Section 3.2.3. (These shift symmetries in no way rely on supersymmetry.) Holomorphy dictates that the superpotential can only depend on T_i, rather than on $T_i + \bar{T}_i$, but no nontrivial polynomial in T_i is invariant under the shift of the axion. Thus, the superpotential can depend on T_i only nonperturbatively. Because corrections in the α' expansion must change in magnitude as the T_i are varied, but the superpotential is independent of the T_i to all orders, it follows that W receives no perturbative α' corrections.

Let us now discuss nonperturbative contributions to the superpotential. Consider a compactification in which a stack of N_c D7-branes wraps a four-cycle Σ_4. The worldvolume theory of the D7-branes includes a Yang–Mills action for four-dimensional gauge fields A_μ, of the form

$$S = \frac{1}{2g_7^2} \int_{\Sigma_4} \mathrm{d}^4\sigma \sqrt{g_{\mathrm{ind}}}\, e^{-4A(y)} \cdot \int \mathrm{d}^4 x \sqrt{-g}\, \mathrm{Tr}\left[F_{\mu\nu} F^{\mu\nu}\right] \,, \qquad (3.105)$$

where the indices are raised with the unwarped metric $g_{\mu\nu}$, and g_7 is the gauge coupling of the (7+1)-dimensional Yang–Mills theory,

$$g_7^2 = 2(2\pi)^5 (\alpha')^2 \,. \qquad (3.106)$$

The gauge coupling of the four-dimensional Yang–Mills theory is

$$\frac{1}{g^2} = \frac{T_3 \mathcal{V}_4}{8\pi^2} \,, \qquad (3.107)$$

where we have defined the volume of Σ_4 as

$$\mathcal{V}_4 \equiv \int_{\Sigma_4} \mathrm{d}^4\sigma \sqrt{g_{\mathrm{ind}}}\, e^{-4A(y)} \,, \qquad (3.108)$$

and g_{ind} is the induced metric on the D7-brane. Because of the appearance of $e^{-4A(y)}$, $\mathcal{V}_4$ as defined in (3.108) is sometimes called the "warped volume."

Given certain topological conditions on Σ_4, discussed further below – heuristically, one asks that Σ_4 should have no deformations that could correspond to charged matter fields – the four-dimensional gauge theory arising upon dimensional reduction is *pure glue* $\mathcal{N} = 1$ super Yang–Mills theory. At low energies, this field theory generates a nonperturbative superpotential from gaugino condensation [339–344] (cf. [345]):

$$|W_{\lambda\lambda}| = 16\pi^2 M_{\mathrm{UV}}^3 \exp\left(-\frac{1}{N_c} \frac{8\pi^2}{g^2}\right) \propto \exp\left(-\frac{T_3 \mathcal{V}_4}{N_c}\right) \,. \qquad (3.109)$$

The volume $\mathcal{V}_4$ is proportional to the real part[20] of a corresponding Kähler modulus T, so the gaugino condensate superpotential may be written as

$$W_{\lambda\lambda} = \mathcal{A}\,e^{-aT}, \tag{3.110}$$

where $a = \frac{2\pi}{N_c}$ and the prefactor $\mathcal{A}$ is independent of all the Kähler moduli, but generally depends on the complex structure moduli, the axiodilaton, and the positions of any D-branes. One might suspect from (3.109) that $\mathcal{A} \propto M_{\mathrm{KK}}^3$, but because $M_{\mathrm{KK}}/M_{\mathrm{pl}}$ depends on $T + \bar{T}$, such a dependence would not be holomorphic. Instead, for typical complex structure moduli vevs and D-brane positions, one has $\mathcal{A} \sim M_{\mathrm{pl}}^3$: see [35, 330].

A very similar superpotential contribution arises if Σ_4 is wrapped not by spacetime-filling D7-branes, but by *Euclidean D3-branes*, also known as *D3-brane instantons* [348] (see [349] for a review). A Euclidean Dp-brane is an instantonic contribution to the path integral whose Euclidean action has a real part that is proportional to the volume of the $(p + 1)$-cycle wrapped by the Euclidean brane, while the imaginary part is determined by the corresponding Chern–Simons action. For a Euclidean D3-brane wrapping Σ_4, the resulting superpotential term is

$$W_{\mathrm{ED3}} = \mathcal{A}\,e^{-aT}, \tag{3.111}$$

where $a = 2\pi$, and as in (3.110) the prefactor $\mathcal{A}$ can depend on the complex structure moduli, axiodilaton, and D-brane positions, but is independent of the Kähler moduli.

We now turn to the necessary and sufficient topological conditions for the generation of a nonperturbative superpotential, focusing on the case of Euclidean D3-branes. These conditions can be expressed most simply in terms of an auxiliary eight-dimensional geometry Y, in which the axiodilaton τ parameterizes an elliptic fibration over the six-dimensional manifold X on which the type IIB theory is compactified.[21] This construction is known as *F-theory* [351]: one says that F-theory has been compactified on Y, which is an elliptically fibered fourfold over the base X.

Witten observed in [348] that a necessary condition for a non-vanishing Euclidean D3-brane superpotential term associated with a four-cycle $\Sigma_4 \subset X$ is that Σ_4 is the projection of a six-cycle $D \subset Y$ obeying

---

[20] Supersymmetry requires that the superpotential be a holomorphic function of the moduli, but verifying that $\mathcal{V}_4$ is the real part of a holomorphic function is highly nontrivial [225, 346]. When D3-branes are present, their backreaction on the volume $\mathcal{V}_4$ must be incorporated in order to maintain holomorphy [346]. This effect was first understood in the open string channel, as a threshold correction to the gauge coupling g [347].

[21] An elliptic fibration is a fibration in which almost all fibers are non-singular and have the topology of two-tori, but a finite number of singular fibers can appear. The possible singular fibers have been classified by Kodaira in [350].

$$\chi(\mathcal{O}_D) \equiv \sum_{i=0}^{3} (-1)^i h^{0,i}(D) = 1 \, , \tag{3.112}$$

where $\mathcal{O}_D$ denotes the trivial line bundle defined on D (see e.g. [352] for the relevant mathematical background). The number $\chi(\mathcal{O}_D)$ is known as the holomorphic Euler characteristic of D [352], or the arithmetic genus of D.[22] Next, a sufficient condition for a non-vanishing Euclidean D3-brane superpotential is [348]

$$h^{0,1}(D) = h^{0,2}(D) = h^{0,3}(D) = 0 \, . \tag{3.113}$$

The Hodge numbers in (3.113) count the independent deformations of D, so a six-cycle D obeying (3.113) is said to be *rigid*.

The sufficient condition (3.113) is unmodified by the presence of flux, but in flux backgrounds the necessary condition (3.112) is modified and becomes less restrictive [345, 354–359]. Couplings to flux can give mass to (some of) the deformations of Euclidean D3-branes, and of D7-branes, counted by $h^{0,2}$. Generalizations of (3.112) to backgrounds with flux, and further consistency conditions, are described in [345, 354–363] and reviewed in [349, 364].

3.3.3 Volume stabilization

Having assembled the known perturbative and nonperturbative corrections to the potential for the Kähler moduli in type IIB flux compactifications, we are in a position to ask whether the quantum-corrected theory has cosmologically interesting metastable vacua, even though the classical theory has unstabilized Kähler moduli.

There is a very general problem [365] underlying any search for string compactifications in which flat directions are stabilized by perturbative or non-perturbative corrections. The *Dine–Seiberg problem* [365] can be summarized as follows: when corrections are important, they are not computable, and when they are computable, they are not important [364]. To understand the observation of [365] in more detail, let ρ be a modulus that controls a weak coupling expansion, such that $\rho \to \infty$ is the free limit. Concretely, ρ could be the Kähler modulus that measures the compactification volume, $\rho = T + \bar{T}$, so that $\rho \to \infty$ corresponds to decompactification to ten dimensions; or, for the string loop expansion, $\rho = g_s^{-1} = e^{-\Phi}$. We now ask whether perturbative or nonperturbative corrections generate a potential for ρ that has a minimum at finite ρ. Because the leading-order classical action is valid for $\rho \to \infty$, the potential $V(\rho)$ generated

---

[22] In the mathematics literature, some authors define the arithmetic genus $p_a(D)$ so that $p_a(D) = 1 - \chi(\mathcal{O}_D)$, for D a six-manifold [352]. Here, as in most of the string theory literature, the arithmetic genus and the holomorphic Euler characteristic are both equal to $\chi(\mathcal{O}_D)$, cf. [348, 353]. In (3.112), the notation $\chi(\mathcal{O}_D)$ for the holomorphic Euler characteristic is used instead of $\chi(D)$, because the latter can be confused with the more familiar topological Euler characteristic.

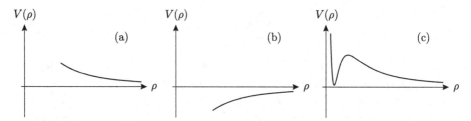

Fig. 3.2 The Dine–Seiberg problem [365] for a modulus ρ. In case (a), there is a runaway to $\rho = \infty$, where the theory is free. In case (b), the leading correction drives the theory toward small ρ, where it is strongly coupled. The existence of the minimum in case (c) requires competition among at least three terms.

by perturbative and nonperturbative corrections must vanish for $\rho \to \infty$. In particular, $V(\rho)$ must approach zero from above or from below as $\rho \to \infty$ (see Fig. 3.2). If $V(\rho)$ is positive for large ρ, then the leading correction term in $V(\rho)$, which dominates for $\rho \to \infty$, creates an instability that drives the theory toward $\rho = \infty$. If instead $V(\rho)$ is negative for large ρ, then the leading correction to the free theory creates an instability that drives the theory toward smaller ρ, and hence toward stronger coupling. Either way, the leading correction term creates an instability, and a (meta)stable vacuum can arise only if higher-order corrections make comparably important contributions that counterbalance the instability. But once two[23] consecutive terms in the weak coupling expansion are comparable, one expects that the entire series must be included. While it could happen that the first and second non-vanishing terms are competitive because the second is accidentally large, verifying that this leads to a consistent solution requires examinination of higher terms in the series to rule out unanticipated accidental enhancements at higher orders. Thus, metastable vacua are quite generally found at points in moduli space where the weak coupling expansions break down. This fact presents a major obstacle to the search for metastable string vacua, because in nearly all cases, at most the *first* non-vanishing correction in each expansion (α' or g_s) is known explicitly.

In the case of Kähler moduli stabilization in type IIB flux compactifications, no-scale structure ensures that the classical potential for the Kähler moduli vanishes, so the leading correction to the potential is in fact the dominant potential energy term overall. At *generic* points in the parameter space, one expects that the correction of leading importance will come from the first non-vanishing perturbative correction, which is necessarily the first correction[24] to K, because the superpotential is not renormalized in perturbation theory. Following [365], we

---

[23] When $V > 0$ for large ρ, three separate terms are required – see [364, 366].

[24] Whether the leading perturbative correction to the potential comes from the first α' correction to K, or instead from the first g_s correction to K, is not obvious a priori, and can depend on parameter values – see Section 5.5 for a detailed discussion.

conclude that vacua at generic points in the parameter space are the result of competition among terms at different perturbative orders. Because of the absence of perturbative computations beyond leading order, it has proved very difficult to find controllable vacua in this regime (however, see e.g. [334, 367, 368]).

The two leading ideas for Kähler moduli stabilization, the KKLT scenario [369] and the Large Volume Scenario (LVS) [370], succeed by targeting regions of parameter space where vacua result from competition among *known* correction terms. To anticipate slightly, the KKLT mechanism involves competition between a classical flux superpotential (3.95), made small by fine-tuning fluxes, and the nonperturbative superpotential (3.110). The LVS construction works in a region of Kähler moduli space where some cycles are exponentially larger than others, so that the leading α' correction (3.101) involving the large overall volume $\mathcal{V}$ competes with nonperturbative superpotential terms (3.110) involving the small cycles. In both cases, one can argue that the unknown higher corrections do not spoil the vacuum structure. We now turn to explaining these mechanisms in more detail.

KKLT scenario

The seminal KKLT proposal [369] for constructing stabilized vacua bypasses all perturbative corrections and instead makes use of nonperturbative contributions to the superpotential.

In the presence of three-form flux the complex structure moduli and dilaton acquire supersymmetric masses via the classical superpotential (3.95), cf. Section 3.3.1. If we denote the typical mass scale by m_{flux}, then at energies $E \ll m_{\text{flux}}$ the complex structure moduli and dilaton can be integrated out (see discussion below), and the classical superpotential W_0 becomes a constant. The fields remaining in the low-energy effective theory are the Kähler moduli,[25] which do not appear in the classical superpotential.

As shown in Section 3.3.2, nonperturbative effects can generate superpotential interactions for the Kähler moduli, either through strong gauge dynamics (such as gaugino condensation) on D7-branes, or through instanton contributions from Euclidean D3-branes. The combination of the constant flux superpotential (3.95) with the nonperturbative terms (3.110) or (3.111) leads to

$$W = W_0 + \sum_{i=1}^{h_+^{1,1}} \mathcal{A}_i \, e^{-a_i T_i} + \cdots , \qquad (3.114)$$

where the ellipsis denotes higher-order nonperturbative effects. In writing (3.114), we have assumed that there is a nonperturbative term for each of the Kähler moduli T_i. The status of this important assumption is not completely understood;

---

[25] If spacetime-filling D3-branes are present, their positions are also light fields in the effective theory, as explained in detail in Section 5.1.

while examples do exist in which there is a nonperturbative term for each Kähler modulus [371, 372], it has not been shown that this situation is generic.[26]

For an arbitrary Kähler potential K, the superpotential (3.114) leads to the scalar potential

$$V_{(np)} = e^K K^{j\bar{\imath}} \left[a_j \mathcal{A}_j a_{\bar{\imath}} \bar{\mathcal{A}}_{\bar{\imath}} e^{-(a_j T_j + a_{\bar{\imath}} \bar{T}_{\bar{\imath}})} \right.$$
$$\left. - \left(a_j \mathcal{A}_j e^{-a_j T_j} \bar{W} \partial_{\bar{\imath}} K + a_{\bar{\imath}} \bar{\mathcal{A}}_{\bar{\imath}} e^{-a_{\bar{\imath}} \bar{T}_{\bar{\imath}}} W \partial_j K \right) \right] . \qquad (3.115)$$

Taking $K = K_0 = -2\ln(\mathcal{V})$, cf. Eq. (3.94), and considering the single-modulus case ($h_+^{1,1} = 1$), $\mathcal{V} = (T + \bar{T})^{3/2}$, one finds, in terms of $\sigma \equiv \frac{1}{2}(T + \bar{T})$,

$$V_{(np)} = \frac{a\mathcal{A} e^{-a\sigma}}{2\sigma^2} \left[\left(1 + \frac{a}{3}\sigma \right) \mathcal{A} e^{-a\sigma} + W_0 \right] . \qquad (3.116)$$

This potential is plotted in Fig. 3.3 (dashed line). It is easy to see that the vacuum solution is supersymmetric anti-de Sitter space. Letting $\sigma_\star$ be the value of the Kähler modulus at the minimum, we find $(\partial_T V_{(np)})_\star = (D_T W)_\star = 0$ and

$$W_0 = -\mathcal{A} e^{-a\sigma_\star} \left(1 + \frac{2}{3}a\sigma_\star \right) . \qquad (3.117)$$

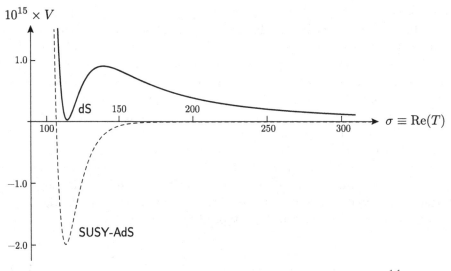

Fig. 3.3 Potential for the Kähler modulus T in a KKLT scenario with $h_+^{1,1} = 1$. The dashed line shows the potential in the absence of a supersymmetry-breaking anti-D3-brane. The figure was generated for $\mathcal{A} = 1$, $a = 0.1$, and $W_0 = -10^{-4}$.

---

[26] The stabilization scenario of [373] is very similar to the KKLT construction, but requires only one nonperturbative term, arising on a four-cycle Σ_4 that is *ample*. Roughly speaking, Σ_4 is ample if it is a positive linear combination of a basis of four-cycles of positive volume – see [373] for further background and a precise definition.

Control over the instanton expansion of the superpotential, corresponding to neglecting the ellipsis in (3.114), requires that $a\sigma_\star \gg 1$. Moreover, perturbative (α' and g_s) corrections to the Kähler potential (3.94) may be neglected if $\sigma_\star \gg 1$.[27] We see from (3.117) that the volume is stabilized in a controlled limit only for an exponentially small value of the flux superpotential, $W_0 \ll \mathcal{A}$. This can be achieved through a fine-tuned choice of quantized flux, following [197].

A number of authors have critically examined the two-step procedure of integrating out the complex structure moduli and dilaton, and then studying the effective theory for the Kähler moduli, instead of analyzing all moduli on the same footing [374–378]. The underlying justification for a two-step procedure is that the mass scale m_{flux} is set by the flux quantization condition, and does not diminish as W_0 is fine-tuned to be small, whereas the mass of the Kähler modulus T at the minimum is proportional to W_0. To understand this, we expand the flux superpotential around the supersymmetric minimum,[28]

$$W_0 = W_0|_{Z=0} + \ell_A Z^A + m_{AB} Z^A Z^B + \cdots , \tag{3.118}$$

where $Z^A \equiv \{\tau, \zeta^a\}$, and $W_0|_{Z=0}, \ell_A, m_{AB}$ are constants dictated by the quantized three-form fluxes. Via a fine-tuned choice of fluxes, one can arrange for $W_0|_{Z=0}$ to be small, and this contrivance does *not* render m_{AB} atypically small at the same time – in fact, a further fine-tuning would be needed to reduce m_{AB}. This fact is true in generic configurations, but can fail in simple examples with special structures. For example, because W_0 involves the dilaton only through the definition (3.20) of G_3, which is linear in τ, we see that $m_{\tau\tau} = 0$. As a result, the dilaton acquires a mass from W_0 only by mixing with the complex structure moduli ζ^a, through couplings $m_{\tau\zeta^a}$. In a compactification with $h_-^{1,2} = 0$, where there are no complex structure moduli, no such coupling is possible, the dilaton does not acquire a mass of order m_{flux}, and it is not consistent to integrate out τ before studying the Kähler moduli [374, 375]. However, this example is of limited interest because the mechanism of [197] is inoperative there. In summary (see [378]), integrating out the complex structure moduli and dilaton is consistent when these fields have large supersymmetric masses, which is the generic situation.[29] One can therefore treat W_0 as a constant, taking $W_0 = W_0|_{Z=0}$, as we have done in the remainder of this section.

---

[27] String loop corrections to K are suppressed at large volume, and not only by factors of g_s, because K_0 involves $\mathcal{V}$, so that any additive correction to K is subleading in volume; see Section 5.5.

[28] To be precise, we mean the minimum determined by $D_{\zeta_a} W_0 = D_\tau W_0 = 0$, where we stress that W_0 is the flux superpotential (3.117), not the full superpotential (3.114).

[29] Some care is needed to ensure that holomorphy is maintained in this process, as explained in [375] and described in an explicit example in [377].

Large Volume Scenario

The Large Volume Scenario [370] achieves stabilization of the Kähler moduli by balancing the leading α' correction (3.101) to K against the nonperturbative superpotential (3.110). The success of this approach rests on stabilizing the overall volume $\mathcal{V}$ at such large values that one can consistently neglect the (unknown) α' and g_{s} corrections that are formally subleading in $\mathcal{V}$ compared to (3.101).

Combining the constant flux superpotential (3.95) with the α'-corrected Kähler potential (3.101) leads to the following contribution to the scalar potential:

$$\delta V_{(\alpha')} = 3\hat{\xi}e^K \frac{(\hat{\xi}^2 + 7\hat{\xi}\mathcal{V} + \mathcal{V}^2)}{(\mathcal{V} - \hat{\xi})(2\mathcal{V} + \hat{\xi})^2} W_0^2 \approx \frac{3}{4}\hat{\xi}W_0^2 \frac{1}{\mathcal{V}^3}, \qquad (3.119)$$

where we have defined $\hat{\xi} \equiv \xi/g_{\mathrm{s}}^{3/2}$, cf. (3.101). Adding (3.119) to (3.115), one finds[30]

$$
\begin{aligned}
V_{(np)} + \delta V_{(\alpha')} = e^K \Bigg\{ & K^{j\bar{\imath}} \Big[a_j \mathcal{A}_j \, a_{\bar{\imath}} \bar{\mathcal{A}}_{\bar{\imath}} \, e^{-(a_j T_j + a_i \bar{T}_i)} \\
& - \Big(a_j \mathcal{A}_j \, e^{-a_j T_j} \bar{W} \partial_{\bar{\imath}} K + a_{\bar{\imath}} \bar{\mathcal{A}}_{\bar{\imath}} \, e^{-a_i \bar{T}_i} W \partial_j K \Big) \Big] \\
& + \frac{3}{4}\hat{\xi}W_0^2 \frac{1}{\mathcal{V}} \Bigg\}.
\end{aligned}
\qquad (3.120)
$$

At very large volume, the perturbative term (3.119) dominates over the nonperturbative terms (3.115). Competition between (3.119) and (3.115) can occur if one or more cycles are exponentially smaller than the largest cycles. Denoting the small cycle volumes by $\tau_s \equiv \frac{1}{2}(T_s + \bar{T}_s)$, the idea is to take the limit

$$\mathcal{V} \to \infty, \quad \text{with } a_s \tau_s = \ln \mathcal{V}. \qquad (3.121)$$

Along the ray in the Kähler moduli space defined by (3.121), the exponentials $e^{-a_s T_s}$ in (3.115) are proportional to $1/\mathcal{V}$, and all terms in (3.120) are of the same order in $1/\mathcal{V}$. Notice that the hierarchy (3.121) is only possible for $h_+^{1,1} > 1$ — we will therefore take $h_+^{1,1} > 1$ for the remainder of this discussion.

The sign of $\hat{\xi}$ is determined by the topology of the compactification, with $\hat{\xi} > 0$ corresponding to $\chi(X_6) < 0$. In this section we will assume that $\hat{\xi} > 0$, which implies that the contribution (3.119) is positive, so that the potential (3.120) approaches zero from below at large $\mathcal{V}$ along the ray (3.121). To establish the existence of a minimum, one then needs to argue, first, that the potential along (3.121) becomes positive at sufficiently small $\mathcal{V}$, so that by continuity the potential restricted to (3.121) is minimized at an intermediate point $\mathcal{V}_*$. Second, one must show that at $\mathcal{V}_*$, (3.120) is non-decreasing in the $h_+^{1,1} - 1$ directions in the Kähler moduli space that are perpendicular to the ray (3.121).

---

[30] For the α' and string loop expansions to be valid, we require $\mathcal{V} \gg \hat{\xi} \gg 1$, as discussed further below; see [379] for a systematic exposition of the α' expansion in this setting.

A useful heuristic argument that is valid in certain simple cases (with provisos enumerated below) goes as follows. If the term (3.119) is dominant over the exponential terms at small volume, this establishes that (3.120) likewise becomes positive at small volume. Next, if the leading exponential terms in (3.120) are positive, and all $h_+^{1,1} - 1$ Kähler moduli appear in the nonperturbative superpotential, the potential increases in the directions transverse to the ray (3.121). In combination, these assumptions imply the existence of a minimum at exponentially large volume. This minimum has negative vacuum energy, so the spacetime solution is AdS_4. Because the F-terms are non-vanishing in the minimum, supersymmetry is spontaneously broken [370].

Let us now discuss the conditions for a minimum in more detail, following [380]. We divide the Kähler moduli into two classes: those corresponding to big and small cycles,

$$\{T_i\} = \{T_b^\rho\} \cup \{T_s^r\} \,, \tag{3.122}$$

where $r = 1, \dots, N_s$ and $\rho = 1, \dots, N_b = h_+^{1,1} - N_s$. We consider the large volume limit

$$\mathcal{V} \to \infty \,, \quad \text{with} \quad a_s^r \tau_s^r = \ln \mathcal{V} \quad \text{for all} \quad r = 1, \dots, N_s \,. \tag{3.123}$$

To check for the existence of a minimum in the limit (3.123), one needs to examine in detail the inverse Kähler metric $K^{i\bar{j}}$, and in particular the block corresponding to the small cycle moduli T_s^r. A systematic treatment for $N_s = 1$ and $N_s = 2$ appears in [380].

To understand the results of [380], one piece of geometrical background is necessary. Suppose that $\mathcal{M}$ is a complex manifold (potentially containing singularities) of complex dimension n, and let p be a point in $\mathcal{M}$. The *blow-up* of $\mathcal{M}$ at a non-singular point p replaces p with a copy of $\mathbb{P}^{n-1}$, known as the *exceptional divisor*. The blow-up of a singular point can result in more general exceptional divisors. When the blow-up of $\mathcal{M}$ is a Calabi–Yau threefold, the exceptional divisor is a four-cycle, with size parameterized by one of the Kähler moduli. When the exceptional divisor satisfies the rigidity condition (3.113), the corresponding Euclidean D3-brane superpotential term is non-vanishing [348].

A necessary condition for an LVS minimum is that at least one of the $N_s \geq 1$ small cycles is a rigid exceptional divisor arising from blowing up a singular point [380]. When $N_s = 1$, this condition guarantees that (3.120), *restricted to the ray* (3.123), has a minimum at exponentially large volume. Whether this is a minimum of the full potential depends on the curvature in the $N_b - 1$ directions perpendicular to (3.123), as we discuss further below. For the case $N_s = 2$, if the two small cycles correspond to blow-ups of distinct points, then (3.120) restricted to (3.123) again has a minimum at exponentially large volume, with the same caveat about transverse directions. If instead the two small cycles are two independent resolutions of the same singular point, then an LVS minimum along (3.123) exists only if there is a basis in which the volume $\mathcal{V}$ is symmetric in the two Kähler moduli T_s^1 and T_s^2. For a discussion of the necessary conditions on $K^{i\bar{j}}$ in the context of a survey of a class of Calabi–Yau manifolds, see [381].

A canonical class of examples of LVS vacua arise in what are known as "Swiss-cheese" Calabi–Yau manifolds, whose volumes can be written as[31]

$$\mathcal{V} = \alpha \tau_b^{3/2} - p^{(3/2)}(\tau_s^r) \, , \tag{3.124}$$

where $\alpha > 0$, and $p^{(3/2)}$ is a homogeneous polynomial of degree $3/2$ in the small cycle moduli τ_s^r, $r = 1, \ldots, N_s$. A proper subset of Swiss-cheese Calabi–Yau manifolds take the "strong" form

$$\mathcal{V} = \alpha \left(\tau_b^{3/2} - \sum_{r=1}^{N_s} \lambda_r (\tau_s^r)^{3/2} \right), \tag{3.125}$$

with $\lambda_r > 0$. This compactification has a single large four-cycle, with volume τ_b, and $N_s = h_+^{1,1} - 1$ small four-cycles, with volumes τ_s^r. Increasing one of the τ_s^r with all else fixed *decreases* $\mathcal{V}$, so the small cycles act like holes in a large cheese. The structure (3.125) can arise if the N_s small cycles correspond to the blow-ups of N_s distinct singular points. In the case of a compactification of strong Swiss-cheese form (3.125), the necessary conditions described in [380] are readily met, for any $N_s > 0$.

The final, critical question is whether the potential is stable or unstable in the $N_b - 1$ directions perpendicular to (3.123). In fact, (3.120) per se, which includes only the leading α' correction to K, namely (3.101), has $N_b - 1$ flat directions. The exact moduli potential, incorporating all perturbative and non-perturbative effects in g_s and α', very plausibly depends on the $N_b - 1$ fields that are unlifted by (3.120).[32] However, appealing to an unknown potential to lift these remaining moduli is problematic, not least because there is no evidence that the resulting masses-squared will all be positive. That is, further perturbative corrections beyond (3.101) could well introduce instabilities along one or more of the $N_b - 1$ flat directions of (3.120), leading to an LVS saddle point rather than a minimum.[33] Indeed, as we argue in Section 3.5.3, in certain ensembles of supergravity theories it is overwhelmingly improbable that all $N_b - 1$ flat directions are stabilized rather than destabilized; the probability of stability is

---

[31] See [381] for a study of the incidence of the form (3.124) in a class of Calabi–Yau manifolds.

[32] As we will explain in Section 5.5.2, it has been suggested [44, 380] that the leading g_s correction to K, with form conjectured in [335] following computations in [333, 334], can stabilize the $N_b - 1$ flat directions. However, a more detailed demonstration of stability would be valuable.

[33] However, if it can be established that the potential increases as one moves toward each of the boundaries of the moduli space, then one can again make a continuity argument for the existence of a minimum.

exponentially small in N_b. Whether the assumptions of Section 3.5.3 are applicable to the moduli potential in LVS is an important open question (see [382] for recent work).

In summary, the necessary conditions for an LVS minimum are the following: $\hat{\xi} > 0$; $h_+^{1,1} \equiv N_s + N_b > 1$; $N_s \geq 1$ Kähler moduli corresponding to the blow-ups of points. For $N_s > 1$, further conditions on the blow-ups are necessary [380], while for $N_b > 1$, it is necessary that further corrections, beyond (3.120), render stable the $N_b - 1$ flat directions of (3.120). Explicit examples with $N_b = 1$ that meet all other necessary criteria are now well known [370, 379, 380].

Several differences between LVS and the KKLT scenario should be emphasized. In LVS, some cycles are exponentially larger than others, while in KKLT the cycles are not hierarchically different in size. In KKLT, the classical flux superpotential W_0 is fine-tuned to be exponentially small, while in LVS W_0 is of order unity. In KKLT, the AdS_4 vacuum is supersymmetric, whereas in LVS the AdS_4 vacuum is non-supersymmetric. However, in both scenarios some form of "uplifting" effect is required to achieve a de Sitter vacuum, as we now explain.

3.4 De Sitter vacua

The KKLT and LVS vacua just described are fully stabilized, in the sense that there are no remaining instabilities and no flat directions of the potential. Even so, these vacua have negative energy and are therefore unsuitable for a realistic cosmology. To describe the early universe (inflation) and the late universe (dark energy) requires vacua with positive energy, i.e. de Sitter solutions. Constructing metastable de Sitter vacua in string theory turns out to be far more difficult than constructing stable anti-de Sitter vacua. As a first step toward appreciating the problem, one can ask what it is about AdS vacua that makes them a natural endpoint of the moduli stabilization procedure. In the KKLT scenario, supersymmetry guarantees the stability of the AdS solution. In LVS, the AdS vacua are not supersymmetric, but their stability can be established by asymptotic arguments, in particular by the fact that $V \to 0$ from below for $\mathcal{V} \to \infty$. In contrast, dS vacua are much more susceptible to instabilities. This becomes apparent when one tries to construct explicit de Sitter solutions in string compactifications.

3.4.1 Uplifting to de Sitter

The leading paradigm for constructing metastable de Sitter solutions from stabilized AdS solutions is known as *uplifting*. The stable vacuum is interpreted as a background solution to which the effects of supersymmetry breaking in some new sector, not considered in the original stabilization, may be added. Although the steps of stabilization and uplifting are conveniently described as sequential, in reality the full set of equations of motion, for all fields, must of course be solved simultaneously. This presents a difficulty, because the vacuum energy contribution from the uplifting sector cannot be a perturbatively small correction to the original vacuum energy. In most approaches the stabilization in AdS is analyzed in a supersymmetric effective action, and one must take care that the large supersymmetry breaking from the uplifting sector does not invalidate this treatment. In summary, the task in uplifting is to identify a sector that breaks supersymmetry dynamically, in a parametrically controlled manner, and makes a positive contribution to the vacuum energy without disrupting the physics that led to a stabilized AdS vacuum. As we explain in the following, these requirements are very challenging, even taken in isolation.

First of all, one must engineer a sector of fields that breaks supersymmetry. As a concrete example, consider placing multiple D-branes at the singular apex of a Calabi–Yau cone, leading to a supersymmetric gauge theory in four dimensions. Some of the resulting gauge theories have metastable vacua in which supersymmetry is dynamically broken [383–389], while in other cases, such as [390–392], there are runaway instabilities in directions parameterized by Kähler moduli [391, 393, 394]. But even if one finds a configuration of D-branes on a noncompact Calabi–Yau cone leading to a flat space gauge theory that dynamically breaks supersymmetry, establishing that metastability survives compactification is highly nontrivial (but see [387]). The essential issue is that in the low-energy Lagrangian of a compactification, all parameters are determined by the vevs of fields, and are therefore dynamical at sufficiently high energies. Any gauge theory construction relying on a non-dynamical parameter – for example, the mass of a quark flavor, as in [395] – is potentially vulnerable, upon compactification, to an instability along which this parameter evolves. Often a second stage of model-building is required in which one generates the desired vev dynamically and establishes the absence of instabilities – see e.g. [386].

After identifying a supersymmetry-breaking sector, one must compute the effects of supersymmetry breaking on the remaining fields. A pervasive but potentially deceptive picture for uplifting is that the uplifting sector exists "somewhere else" in the compactification; stabilization in AdS is imagined to result from sources and fields in one region, while supersymmetry breaking arises in another region, and the vacuum energy contributions are therefore approximately additive, by locality in the extra dimensions. One problem with this modular picture, as we explain in detail in Section 4.2, is that geometric separation does not imply complete decoupling of two sectors. At the very least, the supersymmetry-breaking sector interacts with the remaining fields by its coupling to the overall compactification volume $\mathcal{V}$: any source $\mathcal{S}$ of positive energy[34] in the four-dimensional theory must be negligible in the limit $\mathcal{V} \to \infty$, and so must enter the Lagrangian as

---

[34] A ten-dimensional cosmological constant would be an exception, but this is excluded by ten-dimensional supersymmetry.

$$\rho_S = \frac{D}{\mathcal{V}^\alpha} , \tag{3.126}$$

asymptotically at large $\mathcal{V}$, with D and α being positive constants. The potential (3.126) contributes to the equation of motion for the Kähler modulus parameterizing the volume $\mathcal{V}$; because $D > 0$, there is a force toward larger volume, cf. Section 3.2.1. This force can substantially change the vev of $\mathcal{V}$, or even drive runaway decompactification. The net result is that a computation of physical parameters in the original AdS vacuum will not necessarily give an accurate prediction for these quantities in the dS solution. Accurate determination of the effective action in metastable de Sitter solutions remains a core challenge for inflationary model-building in string theory, as we discuss further in Chapter 4.

Many constructions of uplifting to de Sitter vacua along the lines of [369], as well as alternatives to uplifting, have been proposed: see e.g. [368, 396–413]. Analyses in type IIA string theory include [302, 303, 414–425], while for proposals in the heterotic string, see [426–430]. See [431] for an early construction of de Sitter vacua in supercritical string theory, i.e. for total spacetime dimension $D > 10$. Discussions of de Sitter vacua of M-theory, and of supergravity theories with $N > 1$ supersymmetry in four dimensions, can be found in e.g. [432–439]. Cosmological solutions in constructions modeled on [369] can be found in [440, 441]; see [442] for related earlier work.

3.4.2 SUSY breaking from antibranes

The archetypal configuration [443] for uplifting to de Sitter space consists of p anti-D3-branes placed at the tip of a *Klebanov–Strassler (KS) throat* [444], which is a smooth, asymptotically conical supergravity solution described in detail in Section 5.1.1. The tip of the KS throat is a three-sphere threaded by three-form flux:

$$\frac{1}{(2\pi)^2\alpha'} \int_{S^3} F_3 \equiv M , \tag{3.127}$$

with M an integer. The KS solution preserves $\mathcal{N} = 1$ supersymmetry in four dimensions, but the anti-D3-branes are incompatible with these supersymmetries, so the total configuration is non-supersymmetric.

Because the anti-D3-branes carry negative D3-brane charge, cf. Eq. (3.23), while the fluxes in the KS solution carry positive D3-brane charge, annihilation of anti-D3-branes and flux is possible in some circumstances. For a given background there is a critical value $p_\star \approx 0.08\,M$ such that for $p > p_\star$, rapid classical annihilation can occur, while for $p < p_\star$ the leading annihilation instability involves quantum tunneling [443], and is nonperturbatively slow. It was therefore argued in [443] that a collection of $p < p_\star$ anti-D3-branes in a KS throat is a metastable, supersymmetry-breaking configuration. In [369], and in many subsequent works, this configuration was used as a module for effecting uplifting: see Fig. 3.3.

The idea of antibrane uplifting has been challenged [445]. In particular, it was observed that the known, approximate solutions for p anti-D3-branes in a KS background are singular. If one could establish that the corresponding full, exact solution manifests *unphysical* singularities, this would imply that anti-D3-branes in a KS throat do not provide a consistent metastable super-symmetry-breaking configuration. To discuss this important point [445–460], we first have to explain the sense in which the known solutions are approximations (see also Section 4.1.4).

The meaning of exact and approximate

By an exact solution of string theory, we mean a configuration of the massless fields that solves the exact equations of motion, i.e. the equations of motion that incorporate all perturbative and nonperturbative corrections in the α' and g_s expansions. In contrast, an exact solution of classical, two-derivative[35] super-gravity – generally abbreviated as an "exact supergravity solution" – solves the equations of motion expressed to leading order in α' and g_s. These are the equations of motion determined by the two-derivative, ten-dimensional actions (3.12), (3.15) for type IIA and type IIB string theory, respectively. Next, we recall exact and approximate solutions involving D-brane sources (see Section 3.1.2 for more details). Consider, for example, a stack of N coincident D3-branes placed in ten-dimensional Minkowski space. The D3-branes warp the space; comparing with (3.32), the characteristic radius of curvature R is

$$R^4 = 4\pi g_s N (\alpha')^2 \, , \qquad (3.128)$$

so that corrections in the α' expansion can be ignored for $g_s N \gg 1$, while as usual, corrections in the string loop expansion can be ignored for $g_s \ll 1$. Thus, the exact supergravity solution determined by the D3-brane sources is an approx-imation to an underlying exact string theory solution, and the small expansion parameters governing the approximation are $g_s \ll 1$ and $(g_s N)^{-1} \ll 1$. Notice that for any fixed N, the string loop and α' expansions cannot both be arbitrarily accurate. In particular, if one imagines sending $g_s \to 0$ for N fixed, a curvature singularity develops, and the α' expansion becomes invalid near the source. For one or more D3-branes in flat space, this singularity is not surprising, and is not indicative of any sickness; at weak string coupling a D3-brane is a heavy source whose transverse thickness is of order $\sqrt{\alpha'}$. This system is well behaved and can be defined by referring to the conformal field theory describing open strings ending on the D-branes. Less practically, one could imagine incorporat-ing all α' corrections in order to obtain a solution that does not break down near the source. Summarizing, a single D3-brane in flat space is a singular source in supergravity; this is the expected and allowable singularity that arises from

---

[35] The "two-derivative" qualifier refers to omission of higher-curvature contributions, and is usually assumed implicitly.

a localized source, just as for a point charge in classical electromagnetism. For N coincident D3-branes, the curvature of the supergravity solution is small at large $g_s N$.

Singular antibranes

In view of the above remarks, it should come as no surprise that a single anti-D3-brane placed in a KS throat is a singular source in supergravity. More generally, for p anti-D3-branes, there is no reason to expect a smooth supergravity solution if $g_s p \ll 1$; this would amount to better behavior than that of D3-branes, which are supersymmetric in the KS background and hence are "maximally innocuous." On the other hand, for $g_s p \gg 1$ it is reasonable to expect that a smooth, exact supergravity solution exists, but none has been constructed to date; only singular approximate solutions have been obtained. The important question is whether the singularities are a signal of unphysical behavior, or instead merely reflect our technical limitations.

In the following, we will discuss two aspects of the singularity problem; first, we will ask whether the singularities could be artifacts of the approximations involved in the analysis. Even if one can argue that approximations are not the cause of the singularity, one still has to ask whether the singularities are unexpected and signal an inconsistency for antibranes in KS throats.

Approximate treatments

Determining the supergravity solution for antibranes in KS is extremely complicated, and two further approximations, beyond the fundamental expansions described above, have been employed to simplify the task: these are *linearization* and *smearing*. Linearization refers to an expansion of the supergravity equations of motion to first order in the strength of the source. The smearing approximation replaces anti-D3-branes at a specific location on the S^3 with an equivalent charge and tension uniformly distributed over the S^3. This reduces the equations of motion from PDEs to ODEs. One may wonder whether either of these approximations could be the source of the apparent singularity.

Linearization

The linearized supergravity solution for p anti-D3-branes smeared around the S^3 has been obtained in [447, 448] (for related earlier work see [461, 462]), and passes nontrivial consistency checks [446, 457]. The characteristic radius of curvature near the source is $R_p = (4\pi g_s p)^{1/4} \alpha'^{1/2}$, so the linearized solution can be trusted at radial distances $r \gg R_p$ away from the tip; nearer to the tip it is inconsistent to neglect α' corrections, and some of the background fields become singular. In particular, the three-form fluxes are singular near the source. It has been argued in [453] that the singularity in the flux is not a consequence of linearization: the nonlinearly backreacted, but still smeared, solution displays singularities.

(See e.g. [451, 456, 463] for related work on the problem of singularities from localized sources.)

Smearing and brane polarization

What sort of smooth supergravity solution might one expect for $p \gg g_s^{-1}$ non-smeared anti-D3-branes? As noted in [443], anti-D3-branes that are initially coincident are driven to redistribute themselves along an S^2 in the S^3, manifestly breaking some of the symmetries preserved by a configuration smeared on the S^3. This process can be viewed as *polarization* of the branes [298] by the flux background, as in the related solution found by Polchinski and Strassler [464], where brane polarization resolves the singularity present in the unpolarized configuration. In [447], it was conjectured that a smooth anti-D3-brane solution can be modeled on the system in [464], with polarization of the anti-D3-branes along an $S^2 \subset S^3$ being responsible for removing the singularities.[36] Such a solution is clearly incompatible with a smearing approximation, but solving the equations of motion in this setting is a formidable technical challenge, and at present it is not known whether brane polarization will resolve the singularities.

Expected and unexpected singularities

One further issue in the study of singularities from anti-D3-branes concerns the nature of the singular behavior. No one should be surprised by the fact that the electric field sourced by a pointlike electron in classical Maxwell theory is singular near the electron. The corresponding potential Φ obeys

$$\nabla^2 \Phi = 4\pi e\, \delta(\boldsymbol{x}) \,, \tag{3.129}$$

for an electron at position $\boldsymbol{x}$, which is solved by

$$\Phi(\boldsymbol{x}') = -\frac{e}{|\boldsymbol{x} - \boldsymbol{x}'|} \,. \tag{3.130}$$

The singularity of (3.130), and of the corresponding electric field, is expected, because the electron is a singular, perfectly localized source for the electric field. Of course, the divergence in the energy of the electric field is removed in the quantum theory.

The question, then, is whether the singularities seen in [445, 447, 448] are expected, and therefore plausibly resolved in the exact solution. A central concern raised by [445] is that the singularities in three-form flux "do not appear to have a distinct physical origin" [445]. That is, according to [445] it is not obvious

---

[36] D-branes can generally polarize in multiple ways, and an alternative to the polarization identified in [443], where the anti-D3-branes spread along an $S^2 \subset S^3$, is for the anti-D3-branes to spread along a different S^2, namely the S^2 that shrinks toward the tip of the throat (see Section 5.1.1). This process moves the anti-D3-branes radially outward, away from the tip, a direction of motion that is opposed by the classical potential from fluxes. It was shown in [454] that this alternative radial polarization is not possible, but this does not exclude the expected polarization of [443].

how the anti-D3-brane can serve as a source for singular three-form flux, and correspondingly these singularities are unexpected.

It is certainly true that the only flux sourced by an anti-D3-brane in empty flat space is five-form flux F_5, just as for a D3-brane: see the Chern–Simons coupling Eq. (3.23). For an anti-D3-brane in a classical flux background, the problem is more subtle: the supergravity equations of motion are nonlinear, and the various fluxes are coupled to each other, as we will explain below. To understand this case, we begin by developing intuition in a simpler example.

Let us see how a point source of one field $\mathcal{A}$, in a classical background of a second field $\mathcal{B}$, can source a singular profile of a third field $\mathcal{C}$, even if the source does not have a direct coupling to $\mathcal{C}$ in the Lagrangian. Consider classical four-dimensional electromagnetism coupled to an axion ϕ, with Lagrange density

$$\mathcal{L} = -\frac{1}{2}(\partial\phi)^2 - \frac{1}{4}F_{\mu\nu}F^{\mu\nu} - \frac{\phi}{f}F_{\mu\nu}F_{\rho\sigma}\epsilon^{\mu\nu\rho\sigma} , \qquad (3.131)$$

where f is the axion decay constant. Suppose that there is a constant classical background magnetic field $\boldsymbol{B} = B\,\hat{\boldsymbol{z}}$, and place an electron in this field, at rest at the origin. The electron sources an electric field $\boldsymbol{E} = -e\hat{\boldsymbol{r}}/r^2$, so that in spherical polar coordinates (r, θ, φ) one has

$$\boldsymbol{E} \cdot \boldsymbol{B} = -\frac{eB}{r^2}\cos\theta . \qquad (3.132)$$

The equation of motion for ϕ is therefore

$$\nabla^2\phi = \frac{eB}{fr^2}\cos\theta . \qquad (3.133)$$

Thus, the axion ϕ effectively has a local source, even though the electron alone does not couple to ϕ. This example illustrates that in theories with local sources and multiple coupled fields, not every singular field profile arises from a "one field, one source" coupling in the Lagrangian; classical background fields can also play a role.

In the case of an anti-D3-brane in a KS throat, the classical background field analogous to $\boldsymbol{B}$ is the three-form flux G_3 of the KS solution. Schematically, the anti-D3-brane sources singular five-form flux, which couples to the non-singular background three-form flux, and thereby sources a singular three-form flux. To see this more explicitly, we consider the equation of motion of the three-form flux. In the KS solution the dilaton is constant and the imaginary anti-self-dual component of the flux, $G_- \equiv (\star_6 - i)G_3$, satisfies

$$\mathrm{d}G_- = -\mathrm{d}\left(\frac{\Phi_- G_+}{\Phi_+}\right) , \qquad (3.134)$$

where $G_+ \equiv (\star_6 + i)G_3$ and $\Phi_\pm \equiv e^{4A} \pm \alpha$. The anti-D3-brane is a singular source for Φ_-, while $G_+ \neq 0$ and $\Phi_+ \neq 0$ in the KS background. Thus, solving (3.134) requires that G_- be singular. As a result, divergences in three-form flux are to be expected when anti-D3-branes are placed in a KS throat.

Summary

Let us summarize the key facts and questions about singularities from anti-D3-branes. The linearized solution describing p anti-D3-branes smeared around the tip of a Klebanov–Strassler throat has been obtained in [445, 447, 448], and passes multiple consistency checks [446, 457]. The three-form flux in this solution is singular near the source [445], and the singularity is not an artifact of linearization [453]. We have argued that singularities in flux should in fact be expected in this setting, but to show definitively that the singularities found in the solutions of [445, 447, 448] are (or are not) physical, the most compelling course is to exhibit the corresponding non-singular solution (or show that none exists). A leading proposal for a non-singular resolution, for $g_s p \gg 1$, involves the anti-D3-branes polarizing [298], as in the Polchinski–Strassler solution [464], but this necessarily breaks the symmetries used to smear the anti-D3-branes, and obtaining the corresponding solution is a difficult open problem.

3.5 Statistics of string vacua

I would be happy personally if the multiverse interpretation is not correct, in part because it potentially limits our ability to understand the laws of physics. But none of us were consulted when the universe was created.

Edward Witten [465].

At the fundamental level, string theory contains no continuously adjustable dimensionless parameters, but the theory has an astronomical number of solutions, or vacua. These solutions are distinguished from each other by the vevs of continuously adjustable moduli fields, and also by discrete data, consisting of topological invariants of the compactification itself, such as Hodge numbers; topological properties of any branes wrapping internal cycles, and of gauge bundles on these branes; and the number of units of quantized flux threading each cycle. The number of distinct choices of integer data is extremely large, because many compactifications have hundreds of independent cycles on which flux can be placed.

3.5.1 Landscape of stabilized vacua

For the purposes of cosmology, it is important to understand solutions whose effective theories contain no massless scalar fields, i.e. solutions without moduli. Such vacua are necessarily isolated: classical transitions from one to another require energy input. The *string landscape* is the collection of all consistent solutions of string theory that have four large spacetime dimensions[37] and do not have moduli. The expectation that string theory has a vast array of isolated solutions dates back to the early days of the theory [466, 467], but detailed understanding of flux compactifications in recent years has brought the landscape into focus and has made explicit investigation possible. In the same period, the discovery of dark energy [1, 2] has made the understanding of de Sitter solutions of string theory an urgent question.

What are the prospects for understanding the structure of the string landscape? There are two overarching challenges: accurately characterizing the effective theories whose isolated solutions comprise the landscape, and then exploring their innumerable vacua. At present, there is some degree of understanding of the effective theories resulting from Calabi–Yau compactifications of type II, type I, and heterotic strings, in the regime of weak coupling and large volume. Certain compactifications of M-theory and F-theory are likewise understood. However, despite prolonged study, non-Calabi–Yau compactifications (even if supersymmetric) are less well understood, in part because fewer geometric and topological tools are applicable. It would be premature to declare that the properties of the effective theories of presently understood compactifications are in fact general characteristics of string theory. Indeed, we find it plausible that most of the landscape remains to be discovered. Even so, in the absence of an alternative, one can begin by surveying the part of the landscape that rests on known compactifications.

This brings us to the second difficulty, working out the characteristics of the set of vacua of a fully specified ensemble of effective theories. Understanding through enumeration is inconceivable for systems with 10^{500} vacua, which strongly motivates a *statistical* approach, initiated by Douglas in [468]. Instead of computing all physically relevant quantities (also known as "observables") – such as gauge groups, coupling constants, and mass spectra – in a small number of actual vacua, one can instead determine the statistical distribution of a given observable, or the correlations among observables, in a broad class of vacua. We stress that the motivation for a statistical treatment of observables in the landscape goes beyond the practical difficulty of computing observables in explicit examples. Few now believe that string theory has a unique vacuum consistent with all observations, and the pressing task is not so much to find "the" vacuum describing our universe, but rather to understand the characteristic properties of realistic vacua. Solving the Schrödinger equation for one single microstate of the ocean is of much less practical use than understanding thermodynamic and hydrodynamic quantities; the statistical description is simpler, but also more important as a description of the phenomena of the system. Equally, in the landscape, the distributions of

---

[37] The four-dimensional spacetime is often assumed to be maximally symmetric, i.e. de Sitter space, Minkowski space, or anti-de Sitter space. Isolated solutions without four large spacetime dimensions could also be considered to be part of the landscape, but we will focus on the class of vacua that are directly relevant for cosmology.

observables can display emergent simplicity. Examples of simple patterns seen in the distributions of observables can be found in [45, 325, 326, 469]. For a comprehensive account of the statistics of flux vacua, we refer the reader to the excellent review [45].

3.5.2 Counting vacua

There is a general consensus[38] that the number of vacua in the landscape is immense, but it will be worthwhile to review key aspects of the argument. For concreteness, we will consider type IIB flux compactifications on Calabi–Yau orientifolds (or more generally, compactifications of F-theory).

Consider an orientifold of a Calabi–Yau threefold, with a specified choice $\mathfrak{F}$ of quantized three-form fluxes; that is, for each independent three-cycle Σ_3, one chooses $\int_{\Sigma_3} F_3 \in (2\pi)^2 \alpha' \mathbb{Z}$ and $\int_{\Sigma_3} H_3 \in (2\pi)^2 \alpha' \mathbb{Z}$. The result is a potential on the complex structure moduli space $\mathcal{M}_C$,

$$V = V_{\mathfrak{F}}(\zeta_1, \ldots, \zeta_{h^{2,1}}) . \tag{3.135}$$

As explained in Section 3.3, this flux-induced potential is responsible for the stabilization of the complex structure moduli; the local minima of $V_{\mathfrak{F}}$ are generally isolated points $\{p_1, \ldots, p_K\}$ in $\mathcal{M}_C$, and the complex structure moduli masses are generically nonvanishing at such minima. However, the number K of local minima of $V_{\mathfrak{F}}$ is *not* the primary large number responsible for the scope of the landscape; instead, the large number of choices $\mathcal{N}_{\mathfrak{F}}$ of quantized flux $\mathfrak{F}$, corresponding to distinct possibilities for the elementary topological data of the compactification, is the origin of the diversity of vacua. As explained in [45], $\mathcal{N}_{\mathfrak{F}}$ is large in Calabi–Yau compactifications because there are many – typically, hundreds – of independent three-cycles that the two fluxes can thread. Each choice $\mathfrak{F}$ creates a distinct potential $V_{\mathfrak{F}}$ on $\mathcal{M}_C$, and the number $\mathcal{N}_{\mathfrak{F}}$ of such choices is inarguably stupendous, at least of order 10^{500}.

Let us now describe more carefully how the number of choices of flux $\mathcal{N}_{\mathfrak{F}}$ is related to the number of vacuum solutions. As a first step toward understanding the statistics of string vacua, one can count *supersymmetric* vacua in type IIB flux compactifications. More precisely, following [473], we will discuss configurations in which the F-terms[39] $D_{\zeta_i} W_0$ of the complex structure moduli ζ_i vanish. At this stage the Kähler moduli sector is ignored completely, so one must bear in mind that what we term "vacua" here are merely solutions to the equations of motion in one sector, not full-fledged solutions of the total theory.

---

[38] Limitations and weaknesses of the current evidence have been described in e.g. [445, 454, 455, 470–472].

[39] The F-terms described here are those due to the classical flux superpotential W_0, but nonperturbative contributions to the superpotential – for example, from Euclidean D3-branes – introduce further dependence on the complex structure moduli.

To count vacua, a natural object to consider is the density of local minima as a function of the location ζ in moduli space:

$$d\mathcal{N}_{\min}(\zeta) \equiv \sum_i \delta(\zeta - \zeta_i) \,. \tag{3.136}$$

In practice, $d\mathcal{N}_{\min}$ is far more challenging to study than the related *index density* $d\mathcal{I}_{\min}$, defined by

$$d\mathcal{I}_{\min}(\zeta) \equiv \sum_i \delta(\zeta - \zeta_i)\,(-1)^{F_i} \,, \tag{3.137}$$

where $(-1)^{F_i}$ is the sign of the determinant of the fermion mass matrix (see [45]). The integral of $d\mathcal{I}_{\min}$ over the moduli space is manifestly not the total number of vacua; it is instead a sum weighted by signs. The advantage of considering $d\mathcal{I}_{\min}$ is that it is computable: one can obtain the elegant Ashok–Douglas formula [473],[40]

$$\sum_{L \leq L_{\max}} d\mathcal{I}_{\min} = \frac{(2\pi L_{\max})^{b_3}}{\pi^{b_3/2} b_3!} \det(-\mathcal{R} - \omega) \,, \tag{3.138}$$

where ω is the Kähler form on the moduli space, $\mathcal{R}$ is the curvature two-form, b_3 is the third Betti number of the compactification, and the number $L_{\max}$ represents a tadpole constraint on the flux. Equipped with (3.138), one can estimate the actual number of vacua by attempting to place bounds on the degree of difference between $d\mathcal{I}_{\min}$ and $d\mathcal{N}_{\min}$. One pivotal observation is that the number of vacua is *exponential* in b_3.

There are two critical caveats that prevent one from concluding at this stage that type IIB string theory compactified on a Calabi–Yau manifold with large b_3 has an exponentially large number of metastable de Sitter vacua. First, we have thus far described only the complex structure moduli, and a local minimum of the potential on $\mathcal{M}_C$ may or may not correspond to a local minimum of the exact potential on the full moduli space $\mathcal{M}_{\text{total}}$, which also includes the Kähler moduli and the positions of D-branes. Second, $\mathcal{M}_C$ is *noncompact*, as is $\mathcal{M}_{\text{total}}$; in particular, the Kähler moduli space $\mathcal{M}_K$ can be continued toward infinite volume, where one recovers ten-dimensional flat space. Noncompactness of $\mathcal{M}_C$ implies that $V_{\mathfrak{F}}$ may not have a minimum in $\mathcal{M}_C$.[41] Thus, one is not strictly guaranteed *any* vacua for a given choice of flux. Equation counting does certainly suggest that $V_{\mathfrak{F}}$ will generically have one or more minima inside $\mathcal{M}_C$, but topology does not necessitate this.

With this background, we emphasize that the celebrated counting of 10^{500} vacua in the landscape (cf. [45]) does not refer to a counting of metastable vacua

---

[40] Evidence supporting the result (3.138) in explicit flux compactifications on Calabi–Yau threefolds was obtained in [474, 475], building on [476].

[41] For $\mathcal{M}_K$, one manifestation of the corresponding fact is that the potential can have its minimum at infinite compactification volume.

of the full potential for all moduli (at any level of approximation); it is a counting of *supersymmetric vacua of the complex structure moduli sector*, neglecting the Kähler moduli and postponing the question of metastable supersymmetry breaking.

Let us therefore ask whether one can extrapolate from this result to estimate the number of de Sitter vacua in type IIB flux compactifications. One might be tempted to argue as follows: suppose that one single metastable de Sitter vacuum is found, e.g. a KKLT solution on a particular Calabi–Yau with a particular choice $\mathfrak{F}_\star$ of quantized flux. As famously explained by Bousso and Polchinski [197], the many possible choices of quantized p-form flux in compactifications with many p-cycles lead to a "discretuum" of closely spaced vacuum energy densities. Can one then apply this logic and appeal to the existence of many fluxes $\mathfrak{F}'_\star, \mathfrak{F}''_\star, \ldots$ that differ (by discrete quanta) from $\mathfrak{F}_\star$, but lead to a very similar cosmological constant, in order to replicate the single de Sitter vacuum into $\mathcal{O}(\mathcal{N}_{\mathfrak{F}})$ de Sitter vacua? No: the fact that $V_{\mathfrak{F}_\star}$ has a metastable local minimum in no way implies that $V_{\mathfrak{F}'_\star}$ has a local minimum. This fact can also be understood in concrete examples: a change of quantized fluxes that leads to a small change in the cosmological constant generally involves large changes in the individual flux quanta, and correspondingly makes an order-unity change to the effective action, entirely changing the distribution of extrema (if any exist).

One must therefore be cautious when using the vast number of supersymmetric (or no-scale supersymmetry-breaking) vacua in the complex structure moduli sector, cf. (3.138), to argue for the existence of a comparable number of metastable de Sitter vacua of the full potential on the total moduli space: $\mathcal{N}_{dS} \neq \mathcal{N}_{\mathfrak{F}}$ in general. We will now discuss this issue in detail.

3.5.3 Random supergravity

As a practical matter, it is far easier to find *critical points* of $V_{\mathfrak{F}}$, i.e. points where $\partial_a V_{\mathfrak{F}} = 0$ for all fields ϕ_a, than it is to find minima of $V_{\mathfrak{F}}$. For the problem of counting metastable vacua, one can therefore employ a strategy of counting the number $\mathcal{N}_{\text{c.p.}}$ of critical points and estimating $\mathcal{N}_{dS}$ via

$$\mathcal{N}_{dS} = \mathcal{N}_{\text{c.p.}} \times f_{dS} , \qquad (3.139)$$

where f_{dS}, defined by (3.139), is the fraction of all critical points that are in fact metastable de Sitter vacua. (Precisely analogous logic applies for vacua with any other property – for example, one could estimate the number of vacua with Standard Model gauge group by computing $\mathcal{N}_{\text{c.p.}}$ and the associated fraction f_{SM}.) To further simplify the analysis, one can first ask what fraction f_{min} of all critical points are local minima, without demanding that the cosmological constant at the minimum be positive:

$$\mathcal{N}_{\text{min}} = \mathcal{N}_{\text{c.p.}} \times f_{\text{min}} . \qquad (3.140)$$

The number of local minima, $\mathcal{N}_{\min}$, obviously provides an upper bound on $\mathcal{N}_{dS}$.

The problem of counting de Sitter vacua therefore hinges on determining the probability that a randomly chosen critical point is in fact a metastable minimum. Let us be very precise about the notion of probability that is relevant here. The intent is to begin with a compactification of fixed topology – for example, a Calabi–Yau with specified Hodge numbers – and consider all consistent choices of quantized flux $\mathfrak{F}$. For each choice $\mathfrak{F}_\star$, one imagines finding all the critical points $\{p_i^{(\mathfrak{F}_\star)}\}$ of $V_{\mathfrak{F}_\star}$ in the moduli space $\mathcal{M}$ (rather than in its compactification $\overline{\mathcal{M}}$), and assembling the ensemble $\mathfrak{C}$ of all critical points,

$$\mathfrak{C} \equiv \bigcup_{\mathfrak{F}_\star} \left\{ p_i^{(\mathfrak{F}_\star)} \right\} , \tag{3.141}$$

for any choice of flux. Equation counting suggests that for a generic choice of flux, there will be at least one critical point, so we expect[42]

$$\mathcal{N}_{\text{c.p.}} \gtrsim \mathcal{N}_{\mathfrak{F}} . \tag{3.142}$$

However, it still remains to estimate f_{dS}. In [49] it was shown that for broad classes of supergravity theories with $N \gg 1$ scalar fields, f_{dS} is spectacularly small, and can even be as small as $\mathcal{O}(\mathcal{N}_{\mathfrak{F}}^{-1})$. We will now summarize the argument of [49].

Consider an $\mathcal{N} = 1$ supergravity theory with N chiral superfields. The F-term potential (3.39), in units with $M_{\text{pl}} = 1$, is

$$V = e^K \left(K^{a\bar{b}} D_a W \overline{D_b W} - 3|W|^2 \right) . \tag{3.143}$$

The object of primary interest is the *Hessian matrix* $\mathcal{H}$ at a critical point p of the potential,

$$\mathcal{H} = \begin{pmatrix} \partial^2_{a\bar{b}} V & \partial^2_{ab} V \\ \partial^2_{\bar{a}\bar{b}} V & \partial^2_{\bar{a}b} V \end{pmatrix} . \tag{3.144}$$

At a local minimum of the potential, the eigenvalues $\lambda_1 \leq \lambda_2 \cdots \leq \lambda_N$ of $\mathcal{H}$ are non-negative, so

$$f_{\min} = P(\lambda_1 > 0) , \tag{3.145}$$

where as explained above, the probability P is computed in the ensemble consisting of the Hessian matrices at each of the critical points in $\mathfrak{C}$.

To express $\mathcal{H}$ in a convenient form [49, 477], we perform a coordinate transformation to set $K_{a\bar{b}} = \delta_{a\bar{b}}$ at p, and a Kähler transformation to set $K = 0$ at p. We denote the geometrically-covariant and Kähler-covariant derivative by $\mathcal{D}_a$, and define the first three covariant derivatives of the superpotential as

$$F_a \equiv \mathcal{D}_a W , \quad Z_{ab} \equiv \mathcal{D}_a \mathcal{D}_b W , \quad U_{abc} \equiv \mathcal{D}_a \mathcal{D}_b \mathcal{D}_c W . \tag{3.146}$$

---

[42] In some circumstances one can show that the number of critical points per choice of flux is exponentially large. We thank Edward Witten for this observation.

The Hessian then takes the form [49, 477]

$$
\mathcal{H} = \begin{pmatrix} Z_a{}^{\bar{c}}\,\bar{Z}_{\bar{b}\bar{c}} - F_a\bar{F}_{\bar{b}} - R_{a\bar{b}c\bar{d}}\bar{F}^c F^{\bar{d}} & U_{abc}\bar{F}^c - Z_{ab}\overline{W} \\ \overline{U}_{\bar{a}\bar{b}\bar{c}}F^{\bar{c}} - \bar{Z}_{\bar{a}\bar{b}}W & Z_{\bar{a}}{}^c\,Z_{bc} - F_b\bar{F}_{\bar{a}} - R_{b\bar{a}c\bar{d}}\bar{F}^c F^{\bar{d}} \end{pmatrix}
$$

$$
+\, \mathbb{1}\left(F^2 - 2|W|^2 \right), \tag{3.147}
$$

where indices are raised with $\delta^{a\bar{b}}$, $\mathbb{1}$ is the $2N \times 2N$ identity matrix, and $R_{a\bar{b}c\bar{d}}$ is the Riemann tensor of the metric on field space.

The idea at this stage is to recognize that the large dimension N of the field space need not remain an obstacle, but can instead be an expansion parameter. The Hessian is a large matrix, and *random matrix theory* [478] provides a powerful tool for determining its eigenvalue spectrum. The foundational insight in random matrix theory [479] is that one can make sharp predictions about the statistical properties of the eigenvalues of a large ($N \times N$) diagonalizable matrix given very limited information about the actual entries of the matrix. The guiding principle here is *universality*, which states that for $N \gg 1$, the statistics of the eigenvalues have little dependence on the statistics of the matrix entries. This may be thought of as central limit behavior for matrices.

Universality will be essential to the argument, so we pause for a brief illustration; see [480–482] for in-depth discussions. Consider a real, symmetric $N \times N$ matrix M, whose independent entries M_{ij} ($i \geq j$) are independent stochastic variables drawn from a normal distribution $\mathcal{N}(0,\sigma)$ with mean zero and standard deviation σ. Compare with this a real, symmetric $N \times N$ matrix $\tilde{M}$ that has the same symmetries as M, but whose independent entries $\tilde{M}_{ij}$ ($i \geq j$) are stochastic variables that are not necessarily independent – i.e. the entries may have some correlations – and are drawn from diverse non-Gaussian distributions. The magic of universality is that for $N \gg 1$, M and $\tilde{M}$ have the same eigenvalue spectrum; the correlations and non-Gaussianities disappear[43] at large N.

To apply random matrix theory to vacuum statistics, following [49, 468, 473, 477], we first define a *random supergravity* as a four-dimensional $\mathcal{N} = 1$ supergravity theory whose superpotential W and Kähler potential K are random functions, in the sense that the components of their covariant derivatives, such as F_a, Z_{ab}, and U_{abc}, are stochastic variables drawn from one or more statistical distributions. (See [485] for related work.) In view of universality, it will suffice to take all the independent tensor components to be drawn from normal distributions (though this choice is not central to the analysis), but with the possibility of distinct *scales* $F_{\mathrm{rms}}, Z_{\mathrm{rms}}$, and U_{rms} for the components of F_a, Z_{ab}, and U_{abc},

---

[43] The fine print is that the correlations cannot be too numerous [483], and the statistical distributions must have appropriately bounded moments. Universality has been formulated and established rigorously in many settings – cf. [480, 484] – but to simplify the discussion we will continue to omit the associated technicalities. More details can be found in [49].

respectively. These relative scales control the degree of supersymmetry break-ing: the soft supersymmetry-breaking masses are of order $F/M_{\rm pl}$, while the size $m_{\rm susy}$ of the supersymmetric mass terms is determined by the eigenvalues of Z_{ab}. Strictly unbroken supersymmetry would imply vacuum stability, while for

$$F \ll m_{\rm susy} M_{\rm pl} \tag{3.148}$$

supersymmetry breaking is a small effect, and supersymmetry may be expected to increase the likelihood of stability. If instead one has

$$F \gtrsim m_{\rm susy} M_{\rm pl} \, , \tag{3.149}$$

the supersymmetry-breaking masses are at least as large as the supersymmet-ric masses, and supersymmetry has little protective effect. When the scales appearing in the input statistical distributions are taken to be comparable, i.e. when

$$F_{\rm rms} \sim Z_{\rm rms} \sim U_{\rm rms} \, , \tag{3.150}$$

then (3.149) holds for a typical member of the ensemble, while approximate supersymmetry as in (3.148) can occur via a rare fluctuation. We will refer to the ensemble of critical points generated via (3.150) as *generic* critical points; it was argued in [49] that the overwhelming majority of critical points are in fact of this form.

In a random supergravity theory with N chiral superfields, the Hessian (3.147) is a $2N \times 2N$ matrix whose entries are stochastic variables. Considerable struc-ture is evident in (3.147), and the next step, following [49], is to decompose (3.147) into a sum of constituent random matrices with simple properties. The eigenvalue spectrum of $\mathcal{H}$ – which is the quantity controlling the probability of metastability – can then be obtained by appropriately convolving the spectra of the constituents.

To this end, we briefly outline the properties of two classic ensembles of random matrices. The (complex) *Wigner ensemble*, also known as the *Gaussian unitary ensemble*, consists of $N \times N$ Hermitian matrices M of the form

$$M = A + A^\dagger \, , \tag{3.151}$$

where the entries A_{ij} are stochastic complex variables with uniformly distri-buted phase and normally distributed magnitude, $|A_{ij}| \in \mathcal{N}(0, \sigma)$. The eigenvalue density $\rho(\lambda)$ of a typical member of the Wigner ensemble is given by the Wigner semicircle law,

$$\rho(\lambda) = \frac{1}{2\pi N \sigma^2} \sqrt{4N\sigma^2 - \lambda^2} \, . \tag{3.152}$$

Next, the *complex Wishart ensemble* consists of matrices of the form

$$M = AA^\dagger \, , \tag{3.153}$$

where A is a complex $N \times P$ matrix ($P \geq N$), and again A_{ij} are stochastic vari-ables with magnitude drawn from $\mathcal{N}(0, \sigma)$. From the form (3.153) it is clear that the eigenvalues of a Wishart matrix are necessarily non-negative. The eigenvalue spectrum of a typical member of the Wishart ensemble is given by the Marčenko–Pastur law,

$$\rho(\lambda) = \frac{1}{2\pi N \sigma^2 \lambda} \sqrt{(\eta_+ - \lambda)(\lambda - \eta_-)} \,, \qquad (3.154)$$

with $\eta_\pm \equiv N\sigma^2(1 \pm \sqrt{P/N})^2$.

We are now prepared to use random matrix theory to analyze the eigenvalue spectrum of the supergravity Hessian matrix (3.147), in a random supergravity theory. We will begin by studying generic critical points, as defined by (3.150). One recognizes (3.147) as the sum of constituent matrices with simple structures; for example,

$$\mathcal{H}_Z \equiv \begin{pmatrix} Z_a{}^{\bar{c}}\, \bar{Z}_{\bar{b}\bar{c}} & 0 \\ 0 & \bar{Z}_{\bar{a}}{}^{c}\, Z_{bc} \end{pmatrix} \qquad (3.155)$$

is manifestly positive-definite, and is well approximated by a Wishart matrix. By continuing along these lines, one finds [49] that the eigenvalues of the Hessian (3.147) are well approximated by those of

$$\mathcal{H}_{WWW} \equiv \mathcal{H}_{\text{Wigner}} + \mathcal{H}_{\text{Wishart}}^{(I)} + \mathcal{H}_{\text{Wishart}}^{(II)} \,, \qquad (3.156)$$

where $\mathcal{H}_{\text{Wigner}}$ is a Wigner matrix, and $\mathcal{H}_{\text{Wishart}}^{(I),(II)}$ are Wishart matrices. To obtain the spectrum of $\mathcal{H}_{WWW}$, one convolves the spectra of the constituents, which are given in (3.152) and (3.154). Because the matrices in question do not commute with each other, this must be what is known as a *free convolution* [486], denoted by $\boxplus$:

$$\rho(\mathcal{H}_{WWW}) = \rho(\mathcal{H}_{\text{Wigner}}) \boxplus \rho(\mathcal{H}_{\text{Wishart}}^{(I)}) \boxplus \rho(\mathcal{H}_{\text{Wishart}}^{(II)}) \,. \qquad (3.157)$$

An analytic expression for $\rho(\mathcal{H}_{WWW})$ was obtained in [49] (we omit it here for brevity). In Fig. 3.4 we compare a histogram of the eigenvalues of the full Hessian matrix (3.147) in random supergravity, making no approximation, to the analytic result of the Wigner $\boxplus$ Wishart $\boxplus$ Wishart (WWW) model (3.157). The model has no freely adjustable parameters: we take $N = 200$ fields in both the simulations and the analytic model. The agreement is excellent; the slight tail at the right edge is a consequence of finite N. Although formal results in the subject often require the limit $N \to \infty$, Fig. 3.4 makes it clear that $N = 200$, which is an entirely reasonable number of fields in Calabi–Yau compactifications, is a sufficiently large value of N. We conclude that the Hessian matrix (3.147) at a generic critical point of a random supergravity theory is very well approximated by the analytic model (3.157).

At first glance, the eigenvalue spectrum (3.157) depicted in Fig. 3.4 may appear to determine f_{min} as follows: according to the probability density $\rho(\lambda)$, a single eigenvalue is positive with probability

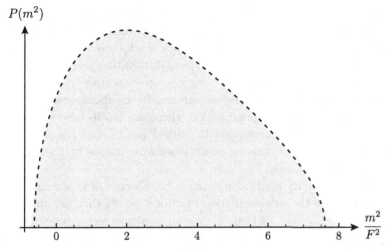

Fig. 3.4 The histogram shows the spectrum of eigenvalues of the full Hessian matrix in random supergravity (for $N = 200$ fields), while the curve gives the analytic result from the WWW model (3.157) [49]. The curve is a parameter-free prediction of the model, not a fit. (Figure adapted from [49].)

$$f_> \equiv \frac{\int_0^\infty \rho(\lambda)}{\int_{-\infty}^\infty \rho(\lambda)} \,, \tag{3.158}$$

suggesting that $f_{\min} = (f_>)^N$. This is *not* correct: $\rho(\lambda)$ describes the eigenvalue density for a typical Hessian matrix in the ensemble, and can be used to compute the probability that a single eigenvalue λ_i falls in some given interval, provided that the remaining $N-1$ eigenvalues are unconstrained. However, the eigenvalues of a random matrix are *strongly correlated*, and manifest eigenvalue repulsion (see [478]); if $N-1$ eigenvalues happen to be positive, the probability density for the final eigenvalue λ_i is very different from (3.157).

What is needed, beyond knowledge of the typical eigenvalue spectrum, is a means of computing the probability of finding a very atypical matrix $\mathcal{H}_{WWW}$ in the WWW ensemble (3.156), one that has only positive eigenvalues. Fortunately, there is a well-developed theory, pioneered by Tracy and Widom [487], describing fluctuations of the extreme (i.e. largest and smallest) eigenvalues of a random matrix. For a large class of random matrix ensembles, including the WWW ensemble [488], one finds [489–491]

$$P(\lambda_1 > \zeta) = \exp\left(-N^2\, \Psi(\zeta)\right) \,, \tag{3.159}$$

where the function $\Psi(\zeta)$, which depends on the particular ensemble and is computable in simple cases, is N-independent at leading order in large N.

In summary, eigenvalue repulsion dictates that the probability that a Hessian matrix in the ensemble defined by (3.150) has only positive eigenvalues is given by

$$f_{\min} \equiv P(\lambda_1 > 0) = \exp\left(-c\, N^2\right) \,, \tag{3.160}$$

for a constant c. Thus, at large N, an overwhelming fraction of generic critical points are unstable saddle points, not metastable minima.[44] This is to be contrasted with the naive estimate $f_{min} = (f_>)^N$, which is "only" exponentially small at large N. Furthermore, we recall from (3.138) that the number of choices of flux is exponential in N, $\mathcal{N}_{\mathfrak{F}} \propto e^{dN}$, for a positive constant d. Comparing with (3.160), we conclude that when the assumptions of the above random supergravity analysis hold to good approximation, there are hardly any vacua at large N – at least, none that correspond to critical points that are generic in the sense of (3.150), with soft supersymmetry-breaking masses at least as large as the supersymmetric masses.

Because metastability is so improbable in the absence of supersymmetry, it is natural to examine the sub-population of critical points that are *approximately supersymmetric*, obeying (3.148). Unbroken supersymmetry would guarantee stability,[45] and one expects greatly increased likelihood of stability in the approximately supersymmetric regime [477].[46] Detailed investigation [500] shows that at an approximately supersymmetric critical point,

$$f_{min} \equiv P(\lambda_1 > 0) = \exp\left(-dN\right) , \tag{3.161}$$

where $d \approx 0.35$ is a constant. We conclude that most metastable flux vacua arise in regions of approximate supersymmetry.

An important proviso is that the WWW model (3.157) describes the spectrum of eigenvalues of $\mathcal{H}$ in a random supergravity theory, in which K and W are random functions (as defined above). Although universality blunts the effect of non-random correlations on the eigenvalue spectrum, there are well-motivated supergravity theories in which K and W have so much structure that (3.160) must be modified. The simplest example[47] consists of two sectors, heavy and light, with additively separable K and W:

$$K = K_l(\phi, \bar{\phi}) + K_h(\Sigma, \bar{\Sigma}) , \tag{3.162}$$

$$W = W_l(\phi) + W_h(\Sigma) . \tag{3.163}$$

If the N_h heavy fields Σ receive large supersymmetric masses at a high scale Λ_h, and supersymmetry is spontaneously broken in the sector of N_l light fields ϕ at

---

[44] The extreme scarcity of minima in a landscape whose Hessian matrices are governed by Wigner's Gaussian orthogonal ensemble was first discussed in the context of cosmology in [492]. An analysis of uplifting of supersymmetric AdS vacua of type IIA string theory, leading to the same conclusion, appears in [493].

[45] Unbroken supersymmetry in AdS does not guarantee the absence of tachyons allowed by the Breitenlohner–Freedman bound [494], but the techniques described above can be used to compute the probability that there are no tachyons [495]. See also [496].

[46] The principal instability corresponds to the scalar partner of the Goldstino – see [497] for an analysis of geometric conditions that ensure stability in this direction, and [498, 499] for discussions of inflation along this direction.

[47] A second important example is the Large Volume Scenario, cf. Section 3.3.3, which leads to non-supersymmetric AdS_4 vacua that have been argued to be automatically tachyon-free in certain cases [370, 380]. The incidence of instabilities in LVS has been analyzed in [382].

a much lower scale Λ_l, then only the light fields will be vulnerable to instabilities caused by supersymmetry breaking. One therefore finds

$$f_{\min} \equiv P(\lambda_1 > 0) = \exp\left(-c\,N_l^2\right) , \qquad (3.164)$$

which for $N_h \gg N_l$ is a vastly increased probability of stability, compared with the estimate $P(\lambda_1 > 0) = \exp(-c\,(N_l + N_h)^2)$ which overlooks the fact that the heavy fields are robustly stabilized. The lesson is that the number N appearing in (3.160) is the number of fields that are dynamically accessible at the energy scale of the critical point in question.[48]

The principal reason for using caution in applying the results (3.160) and (3.161) has already been noted above: we have only a rudimentary understanding of the array of effective theories that emerge from string theory, so it is too early to give a complete account of the vacua of string theory, by any means. Nevertheless, we would like to stress that the assumption of a random superpotential and Kähler potential that underpinned the discussion of random supergravity is *not* tantamount to assuming that the supergravities arising in string theory have "no structure." Instead, universality ensures that in the large N limit, the eigenvalue spectrum of $\mathcal{H}$ takes the universal form, unless the correlations in K and W are extremely strong. In other words, many sorts of underlying patterns in the $\mathcal{N} = 1$ data are compatible with (3.160) and (3.161); such patterns are obscured in the eigenvalue spectrum, and the only patterns that do survive in the spectrum are those determined by the macroscopic structure of (3.147), not by the statistical properties of K and W themselves. For this reason, random matrix theory is actually a conservative approach to the problem of counting vacua; it serves to expose structure that is inherent in supergravity, through the form of (3.147), while blurring out detailed – and presently unknown – microphysics.

The techniques described above have a wealth of applications, most notably to the problem of characterizing inflation in a potential with many fields [492, 502–507], which we will briefly discuss in Section 4.4, Section 5.1.2, Section 5.1.6, and Section 5.4.1.

---

[48] See also [501], where the probability of metastability in a Gaussian landscape was shown to be anticorrelated with the magnitude of the vacuum energy.

4

What is string inflation?

Inflationary scenarios constructed in effective field theory have limitations stemming from incomplete knowledge of the ultraviolet completion. Because the inflationary dynamics is extraordinarily sensitive to Planck-suppressed operators in the effective theory, merely parameterizing our ignorance of quantum gravity is untenable; predictions obtained in this approach amount to reflections of implicit or explicit assumptions about the characteristics of quantum gravity. This fundamental problem motivates pursuing a more complete understanding of inflation in the context of string theory.

In Chapter 5, we will discuss an array of attempts to derive inflation in string theory. Before grappling with model-dependent details, however, it is worthwhile to take a broad overview of the subject. Many of the technical challenges that arise in string inflation are cognate across a range of models, and the phenomenological characteristics are likewise parallel. In this chapter, we provide a schematic account of the essential aspects of inflation in string theory. We will sharpen these considerations with detailed case studies in Chapter 5.

4.1 From strings to an inflaton

The aim of most work on the subject can be summarized by the simple expression

$$S_{10}[\mathcal{C}] \ \mapsto \ S_4[\Phi(t)] \ , \tag{4.1}$$

where the configuration $\mathcal{C}$ refers to the ten-dimensional data of geometry, fluxes, localized sources, and quantum effects, while $\Phi(t)$ represents a time-dependent configuration of scalar fields in the four-dimensional effective theory. The task is to specify compactification data $\mathcal{C}$ that lead, upon dimensional reduction, to an effective theory S_4 with interesting cosmology. To describe inflationary solutions, we require that S_4 has a positive vacuum energy contribution and one or more light moduli Φ whose time-dependent vevs describe a controlled instability of the vacuum.

4.1.1 Energy scales

Understanding the primary energy scales that are involved provides a useful perspective on the problem. Observations of the CMB directly probe energies of order the inflationary expansion rate H when modes cross the horizon and freeze (see Chapter 1). However, as shown in Chapter 2, inflation is sensitive to physics at higher energy scales. When inflation is formulated in effective field theory, these scales parameterize unknown ultraviolet physics, but in string theory they are computable and have specific meanings, as we now explain.

The fundamental scale of string theory is the string scale $M_{\rm s} = (\alpha')^{-1/2}$. At energies below $M_{\rm s}$, only the massless states of the string are excited, and the theory reduces to an effective supergravity in ten dimensions. Most models of inflation in string theory are formulated in this island of theoretical control. The drawback of being in the regime $H \ll M_{\rm s}$ is that truly stringy effects are highly suppressed as far as CMB observables are concerned. To describe situations with $H > M_{\rm s}$, one would need to use the full string theory; the time dependence of the background would create excited string states. Quantitative analysis of such a regime is out of reach at present.

Compactification on an internal space of volume $\mathcal{V}M_{\rm s}^{-6}$ introduces one or more additional scales, the Kaluza–Klein scales $M_{\rm KK} \sim M_{\rm s}\mathcal{V}^{-1/6}$. We will usually work in the regime where $M_{\rm KK} \ll M_{\rm s}$, so that the theory is a ten-dimensional supergravity for intermediate energies, $M_{\rm KK} < E < M_{\rm s}$, while it reduces to a four-dimensional effective theory at low energies, $E < M_{\rm KK}$. The four-dimensional theory will itself be supersymmetric if the compactification preserves some of the ten-dimensional supersymmetries, for example by having suitably reduced holonomy. Most models of string inflation satisfy $H < M_{\rm KK}$, and as a result it is hardly surprising that many such models reduce to well-known EFT models. Formulating inflation (or alternatives to inflation) as a truly higher-dimensional phenomenon would be interesting, but requires rethinking many of the fundamental aspects of the problem, such as the horizon problem and the generation of primordial perturbations.

The four-dimensional Planck scale becomes a derived scale in string theory. It is related to the string scale, the Kaluza–Klein scale, and the string coupling via (3.48), which has the schematic form

$$M_{\rm pl} \sim g_s^{-1} \left(M_{\rm s}/M_{\rm KK}\right)^3 M_{\rm s} \gg M_{\rm s} \,. \tag{4.2}$$

We note that applying the standard inflationary slow-roll analysis requires that one works in the four-dimensional Einstein frame and normalizes all fields with respect to the fixed Planck scale.

Finally, we have the scale of supersymmetry breaking in the early universe, $M_{\rm SUSY}$, by which we mean the highest scale of supersymmetry breaking that is unrelated to inflation. The fact that no superpartners have been observed to date plausibly puts the scale of supersymmetry breaking in the present vacuum at or above the TeV scale, but the breaking of supersymmetry may well have been different at the time of inflation. For $M_{\mathrm{SUSY}} < H$, supersymmetry is only spontaneously broken during inflation, and can partially protect against radiative corrections (see [249] for a recent discussion). The associated theoretical control provides crucial underpinning for most models of inflation in string theory. However, supersymmetry could be much more badly broken; indeed, in non-supersymmetric compactifications $M_{\mathrm{SUSY}} \gtrsim M_{\mathrm{KK}}$.

To sum up, most controlled treatments of string compactifications, and of inflation within it, rely on the hierarchy of scales

$$M_{\mathrm{SUSY}} < H < M_{\mathrm{KK}} < M_{\mathrm{s}} < M_{\mathrm{pl}} \,. \tag{4.3}$$

As our understanding of string theory improves, it may be possible to move away from the comfort of this particular hierarchy of scales and explore a wider parameter space of string cosmologies.

4.1.2 Spectrum of states

Fields with masses that are smaller than the Hubble scale, $m < \frac{3}{2}H$, are both classically and quantum-mechanically active during inflation. To characterize effective theories of inflation, we need to determine the spectrum and the interactions of these light fields. The simplest toy models of inflation assume that only one field is light and the rest have masses far above the Hubble scale (see Fig. 4.1); the heavy fields can then be integrated out, and one is left with a model of *single-field inflation*. If several fields are light one speaks of *multi-field inflation*. In both scenarios, the heavy fields only affect the couplings of the low-energy theory, but do not participate actively in the generation of the primordial perturbations.

However, this hierarchy of mass scales is rarely the situation one encounters in actual constructions of inflation in string theory. Even if one manages to arrange for one or more very light fields, it is hard to avoid having extra fields with intermediate masses. In particular, most models of inflation in string theory constructed to date involve spontaneously broken supersymmetry ($M_{\mathrm{SUSY}} < H$), which generally leads to moduli fields with masses of order H [249].[1] These fields fluctuate quantum-mechanically during inflation and therefore have to be included in the computation of the primordial perturbations. The phenomenology of these models of *quasi-single-field inflation* [509] has been explored in [121, 211, 212, 249, 265, 275, 277, 510–516].

4.1.3 Inflaton candidates

Models of string inflation can be classified by the nature of the field that serves as the inflaton. A few of the leading candidates are as follows.

---

[1] Lighter moduli, with $m \ll H$, may be natural in certain circumstances: see e.g. [508].

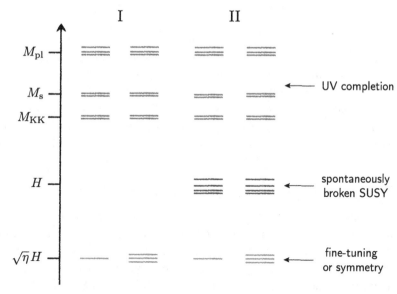

Fig. 4.1 Mass spectra of inflationary models. Phenomenological models of inflation frequently assume a large hierarchy between one or more light inflaton fields and the extra states of the UV completion (I). On the other hand, concrete examples of inflation in string theory often contain fields with masses of order the Hubble scale (II) arising from the spontaneous breaking of supersymmetry. Robust symmetries, or fine-tuning, are required to explain the presence of scalars with masses $m \sim \sqrt{\eta}H$.

Brane moduli

The positions of mobile, spacetime-filling branes[2] in the internal space can be moduli in the four-dimensional effective theory. Many leading models of string inflation are built on the time dependence of these brane position moduli. The complex interactions of a brane with other sources in the compactification create the inflaton potential. If the forces on the brane are weak enough, it moves non-relativistically and may source slow-roll evolution in the four-dimensional spacetime. In Sections 5.1 and 5.2, we will analyze several examples of this sort. If instead the brane moves relativistically, kinetic effects dominate the dynamics. This leads to the interesting possibility of inflation being driven not by a flat slow-roll potential, but by the nonlinear interactions in the kinetic part of the DBI action (3.27) for the brane. Models of DBI inflation are explored in Section 5.3; we will see that string theory plays an important role in explaining the radiative stability of these theories.

---

[2] The primary examples are Dp-branes with $p \geq 3$, NS5-branes, or M5-branes, wrapping suitable cycles. Orientifold planes, in contrast, are non-dynamical: their positions are not parameterized by light scalars.

Kähler moduli

Models of Kähler moduli inflation identify the inflaton with time-dependent deformations of the volumes of even-dimensional cycles. Some of the most promising configurations involve changes in the volume of one or more four-cycles, keeping the overall volume fixed. In constructions in type IIB string theory, the inflaton potential arises from the leading effects that violate no-scale structure – typically a combination of α' corrections, string loop corrections, and nonperturbative effects. We describe various realizations of Kähler moduli inflation in Section 5.5.

Complex structure moduli

In the best-understood moduli stabilization scenarios in type IIB string theory, the complex structure moduli are stabilized at a high scale by flux and can be integrated out at the time of inflation. Unsurprisingly, time dependence of complex structure moduli has played a limited role in models of inflation in type IIB flux compactifications. In stabilized flux compactifications of type IIA string theory [517–519], in contrast, complex structure moduli could a priori be natural inflaton candidates. However, in the best-understood classes of type IIA compactifications, there are no-go theorems for inflation [417, 418, 520] (see also [521]). As a result, it appears difficult to construct explicit scenarios for inflation driven by complex structure moduli.

Axions

String compactifications typically contain a plenitude of axion fields. These axions are particularly attractive inflaton candidates because they enjoy shift symmetries to all orders in perturbation theory. Such symmetries are a key ingredient in technically natural models of inflation [36], including the large-field models required for significant primordial gravitational waves. String theory offers the opportunity to determine which low-energy shift symmetries are compatible with quantum gravity. Inflation driven by a single axion field requires a super-Planckian decay constant, a feature that is difficult to realize in controlled string compactifications [47]. In Section 5.4, we show how models of axion inflation in string theory can overcome this obstacle, and we discuss the rich phenomenology of axion inflation.

4.1.4 Approximations

In an ideal world, one would derive the inflaton action from first principles, beginning with fundamental integer data $\mathcal{C}$ for a compactification, solving the equations of motion of the ten-dimensional effective supergravity theory S_{10}, order by order in α' and in g_s, and integrating out massive degrees of freedom (including the Kaluza–Klein modes of the compactification) to arrive at a four-dimensional effective theory S_4.[3] Unfortunately, computing the effective action in a metastable non-supersymmetric compactification is a formidable technical challenge; a direct approach, making no approximations, would require unforeseen advances in our understanding of string theory. Indeed, even in compactifications that preserve $\mathcal{N} = 2$ supersymmetry in four dimensions, such as compactifications of type II string theory on Calabi–Yau threefolds, the metric on the internal space cannot be computed analytically; in non-supersymmetric solutions, the difficulties are far greater. In practice, a four-dimensional effective theory is deduced based on a partial specification of the compactification data $\mathcal{C}$, and an arsenal of approximation schemes is used in place of a complete calculation.

We will outline some of the most important expansion parameters, systematic (and non-systematic!) approximation schemes, and simplifying assumptions that are used in determining four-dimensional effective theories and extracting their dynamics. It will be important to remember these limitations when we present case studies in Chapter 5.

α' expansion

The α' expansion is reliable when the gradients of the background fields are small in units of α'. However, the compactification volume is finite, and is typically restricted by the desire to achieve the hierarchy $H < M_{\mathrm{KK}}$, so that inflation is inherently four-dimensional. As a result, the α' expansion is often a barely controlled approximation scheme, rather than a convergent parametric expansion, in the regions of interest. This issue is particularly severe in models of high-scale (equivalently, large-field) inflation, because the lower limits on the Kaluza–Klein mass become more stringent.

String loop expansion

The weak coupling approximation retains only the leading terms in $g_s \ll 1$. One might hope to make $g_s = e^{\Phi}$ small in a stabilized vacuum through appropriate choices of flux, but because the dilaton Φ couples to most fields and localized sources, taking $g_s \ll 1$ very often disrupts the delicate balance of energies responsible for moduli stabilization. As a result, arbitrarily weak coupling is rarely achievable in practice.

Probe approximation

Localized objects, such as D-branes or orientifold planes, are often treated as probes, meaning that they are not included as sources in the ten-dimensional equations of motion. For sources that respect some of the supersymmetries

---

[3] More ambitiously, one might pursue solutions that are exact in α' and g_s, but there has been very little progress in this direction. For attempts to drive inflation through α' corrections, see e.g. [522].

preserved by the background, the backreaction can be restricted to a limited set of fields, and sometimes obeys a superposition principle, e.g. in the case of D3-branes in a background of ISD flux. However, for non-supersymmetric sources this approximation is tenable only at considerable distances, which are not always available in a compact space. For lack of an alternative, an unjustified probe approximation is occasionally made for low-codimension objects such as D7-branes and D8-branes, whose effects can be felt at arbitrarily large distances. Many instances have been found in which the probe approximation misses crucial aspects of the inflationary dynamics.

Large charge approximation
The polar opposite of the probe approximation takes the number of backreacting localized sources, or the total corresponding charge, to be so large that the radii of curvature of the resulting geometry are large in units of α'. While this limit has proved to be very fruitful in noncompact geometries, e.g. taking a large D3-brane charge leads to the large N limit in the AdS/CFT correspondence [523], Gauss's law presents difficulties in compact models. In many compact examples the large charge approximation is used despite being marginally valid at best.

Smeared approximation
When computing the backreaction of localized sources is unmanageably complex, considerable simplifications can be achieved by imagining that the sources are distributed in a highly symmetric manner, or are distributed throughout the entire space. For example, D-brane sources in Calabi–Yau cones are often treated as smeared over one or more angular directions of the cone, so that a problem that is properly posed as a system of PDEs is approximated by a system of ODEs. Similarly, negative tension contributions from orientifolds are sometimes taken to be uniformly distributed in the compactification, postponing the question of possible singularities near an actual localized orientifold. This approximation is very effective at reducing the difficulty of a problem, but its accuracy is poorly characterized.

Linear approximation
Linearization in the strength of a source for some of the supergravity fields is very common, but not always self-consistent. For example, in [43] it was shown that the leading contribution to the action for a D3-brane in a warped region arises at quadratic order in an expansion in the strength of sources in the bulk of the compactification.

Noncompact approximation
Although analytic expressions for metrics on compact Calabi–Yau threefolds remain unavailable, metrics are known for many noncompact Calabi–Yau manifolds, i.e. Calabi–Yau cones. The noncompact approximation attempts to

represent a region in a Calabi–Yau compactification as a finite portion of a non-compact Calabi–Yau cone, subject to boundary conditions in the ultraviolet that represent the effects of compactification.

Large volume expansion

Although the α' expansion corresponds to an expansion in inverse volumes, there is a special sort of large volume expansion that deserves separate mention. When the compactification volume is *exponentially* large, as in the Large Volume Scenario [330, 370], then a considerable number of corrections to the effective action in the α' and string coupling expansions can be ignored. In particular, it has been argued that only a subset of the terms arising at order $(\alpha')^3$ make leading contributions.

Adiabatic approximation

It is often assumed that heavy fields adiabatically follow their instantaneous minima as a light field evolves. The detailed form of the inflaton potential can depend on the precision with which heavy fields are integrated out (see Section 5.1 for an example). As discussed in Section 2.1.5, a heavy mode of frequency ω can be integrated out in this way only when the adiabatic condition $\dot{\omega}/\omega^2 < 1$ holds; more rapid evolution leads to excitation of the heavy modes. (Analyses of the effects of heavy fields include [512, 515, 524, 525]. See also the discussion of resonance in Section 5.4.3.)

Truncation

Often the low-energy effective theory is truncated by omitting one or more fields that would be integrated out in a more sophisticated analysis. Omitting fields with $m \ll H$, which evolve and fluctuate during inflation, is widely understood to be inconsistent. A more reasonable – but not always justified – approximation omits fields with $m \gtrsim H$.

Moduli space approximation

An unjustified and misleading oversimplification asserts that the existence of a moduli space for some field ϕ in a supersymmetric compactification "suggests" that ϕ will have a relatively flat potential, even after supersymmetry breaking. This serves only to mask the actual problem, which is understanding the effective action of the non-supersymmetric theory.

Zealous application of the approximation schemes described above can eventually lead to a well-characterized four-dimensional effective action, but inflation is not an automatic consequence. In the following sections, we will describe the most common obstacles that arise after the effective theory has been determined.

4.2 The eta problem

The eta problem is omnipresent in realizations of inflation in string theory, but
it takes various guises in different models. In this section, we will summarize the
causes of the eta problem in string inflation at a conceptual level, to provide
a framework for understanding the detailed incarnations of the problem in the
examples of Chapter 5.

4.2.1 Compactness and non-decoupling

A pivotal insight about inflation in string theory is that the effects of compactification and moduli stabilization do *not* decouple from the inflationary dynamics.
The problem of stabilizing the moduli and the problem of computing the inflaton potential cannot be treated independently, and the inflaton sector cannot be
understood in isolation from the other sectors of the theory [42]. The importance
of moduli stabilization in string inflation is widely appreciated in the recent literature, but achieving control of the moduli potential remains one of the main
technical challenges of the subject. Moreover, the failure of decoupling of different
sectors in string compactifications has important consequences for the dynamics
of inflation in the presence of moduli-stabilizing ingredients.

An instructive picture of decoupling is possible in compactifications with
D-branes, where one often constructs distinct sectors of the theory on collections
of D-branes located in different parts of the compactification. These sectors are
said to *decouple* if the details of one sector are irrelevant for the dynamics in
another, i.e. if the sectors serve as non-interacting modules for the purpose of
computing some four-dimensional observable. Complete decoupling is not always
desirable: the interaction between two sectors could be responsible for inflationary evolution, as in the example of a well-separated brane–antibrane pair – see
Sections 5.1 and 5.2. However, the problem is that hardly any sector decouples
from inflation, so detailed understanding of all hidden sectors is necessary [526].

A common but dangerous assumption is that sufficient geometric separation
of two sectors, A and B, makes the couplings between the sectors negligible.
(More refined criteria involve separation along a warped direction, or separation
without any branes stretched between the sectors, but the principle is the same.)
To check this assumption, one has to compute the couplings between the sectors
by integrating out massive fields that couple to both A and B. In particular, open
strings with one end on A and another end on B lead to massive fields in the four-
dimensional theory. Integrating out these strings leads to operators of the form[4]

---

[4] In many settings it is more efficient to compute the couplings between two separated
sectors by working in supergravity, rather than by integrating out stretched open strings.
In this closed string approach, one finds a supergravity solution that incorporates the
backreaction of sector A, and then evaluates the probe action for sector B at the
appropriate location in this solution, in order to determine the effect of sector A on

$$\Delta \mathcal{L} \supset \frac{1}{M_{AB}^{\delta_A + \delta_B - 4}} \mathcal{O}_A^{(\delta_A)} \mathcal{O}_B^{(\delta_B)} \,, \tag{4.4}$$

where $\mathcal{O}_A^{(\delta_A)}$ is an operator of dimension δ_A consisting of the fields of sector A, and similarly for $\mathcal{O}_B^{(\delta_B)}$, while M_{AB} is the mass of the strings stretched between the sectors. For example, if $\mathcal{O}_B^{(4)} \equiv V_0$ is a constant contribution to the vacuum energy originating in sector B, and taking ϕ to be a scalar field in sector A, then with $\mathcal{O}_A^{(2)} \equiv \phi^2$ we find the coupling

$$\Delta \mathcal{L} \supset \frac{V_0}{M_{AB}^2} \phi^2 \,. \tag{4.5}$$

This is precisely the dimension-six ultraviolet-sensitive inflaton mass term discussed in Section 2.3. The problematic interaction (4.5), and kindred couplings, will be negligible if $M_{AB} \gg M_{\mathrm{pl}}$, but will otherwise alter the inflationary dynamics. Notice that two sectors decouple, for the purposes of inflation, if the interactions between the sectors are *more than Planck-suppressed*. The general expectation in effective field theory is that Planck-mass degrees of freedom that participate in the ultraviolet completion of gravity will induce Planck-suppressed interactions; the absence of such couplings requires a special structure or symmetry in the quantum gravity theory. We will see that this expectation is borne out in string theory.[5]

The erroneous intuition that supports decoupling is that M_{AB} is dictated by the distance d between the sectors, via $M_{AB} \sim d/\alpha'$, so that the effects of A on B can be made negligible by taking d to be large. Of course, in a compactification, the distance d is bounded by the diameter L of the compact space. Moreover, in a roughly isotropic compactification, the total volume scales as $\mathcal{V} \propto L^6$, so that $M_{\mathrm{pl}} \propto L^3/(g_{\mathrm{s}}(\alpha')^2)$ and hence[6]

$$\frac{M_{AB}}{M_{\mathrm{pl}}} \lesssim g_{\mathrm{s}} \left(\frac{\ell_{\mathrm{s}}}{L}\right)^2 \,. \tag{4.6}$$

Thus, $M_{AB}/M_{\mathrm{pl}} < 1$ when the volume is controllably large: the stretched string mass cannot parametrically exceed the Planck mass in an isotropic compactification. Consequently, the couplings between spatially separated D-brane sectors will generically be at least gravitational in strength; the corresponding operators will be suppressed by no more than the Planck mass.

---

sector B. This method has been used, for example, to determine the coupling between D3-branes and quantum effects stabilizing the Kähler moduli [346].

[5] If B is taken to be a supersymmetry-breaking sector, and A is the visible sector, then the notion of decoupling described here corresponds to what is called *sequestering* [527] in the literature on supersymmetry breaking. Investigations of sequestering in string theory [528–533] have confirmed that complete decoupling is extremely rare, but partial suppression of some couplings can occur in certain cases [531].

[6] We display only the parametric scaling; factors of 2π can be important for this relation, but depend on the precise geometry and must be analyzed on a case-by-case basis.

Isotropy is a strong assumption, and it is important to check whether decoupling arises automatically in suitably anisotropic compactifications. If the compactification has p large directions of size L and $6 - p$ small directions of size S, then

$$\frac{M_{AB}}{M_{\mathrm{pl}}} \lesssim g_{\mathrm{s}} \left(\frac{\ell_{\mathrm{s}}}{L}\right)^{\frac{1}{2}p-1} \left(\frac{\ell_{\mathrm{s}}}{S}\right)^{\frac{1}{2}(6-p)} , \tag{4.7}$$

so that for $p > 1$ the coupling is again at least gravitational in strength at large volume. (For the case $p = 1$, see Section 4.3.) A significant example consists of a warped throat geometry: a warped cone over an angular manifold X_5 is an example of a highly anisotropic space, if X_5 is chosen appropriately – for example, one might consider $X_5 = S^5/\mathbb{Z}_k$ for $k \gg 1$. Rather surprisingly, it was shown in [254] that for *any* X_5, the stretched string mass is less than the Planck mass (see Section 5.1.1). Thus, "slender" warped throats do not evade the general argument that gravitational-strength couplings are unavoidable.

The fact that compactness prevents decoupling leads to important constraints on the interactions between localized sources. We will illustrate the issues in the example of a D3-brane/anti-D3-brane pair in a general unwarped six-manifold X_6, though the problem is more general (see Section 5.1).

The Coulomb potential of a D3-brane/anti-D3-brane pair separated by a distance r is

$$V(r) = 2T_3 \left(1 - \frac{1}{2\pi^3} \frac{T_3 g_{\mathrm{s}}^2 \kappa^2}{r^4}\right) , \tag{4.8}$$

where T_3 is the D3-brane tension (3.29), and κ is the gravitational coupling defined in (3.11). The canonically normalized field ϕ is related to r by $\phi = \sqrt{T_3}\, r$. Computing the slow-roll parameter η, we find

$$\eta \approx -\frac{10}{\pi^3} \frac{V}{r^6} , \tag{4.9}$$

where we have used (3.48) and $T_3^2 g_{\mathrm{s}}^2 \kappa^2 = \pi$. The Coulomb potential (4.8) is evidently steep at small separations and grows flatter at large separations. However, the brane–antibrane pair cannot be separated by a distance greater than the diameter of the compactification, so unless X_6 is highly anisotropic, the potential (4.8) is too steep to support inflation [534].[7] This is one of the simplest examples of the phenomenon of non-decoupling described above.

---

[7] When the background is warped, the Coulomb potential (4.8) takes the modified form (5.30) [42], and is extremely flat even at modest separations; see Section 5.1.

4.2.2 Compactness and backreaction

In a warped background, the version of the eta problem that we have just discussed seems to disappear [42]. However, a more subtle issue arises: the backreaction of the D3-branes on the compact geometry leads to instabilities and to a recurrence of the eta problem. We will briefly sketch the argument.

Going beyond the probe approximation, a D3-brane located at the position y_b in a six-dimensional space with coordinates y acts as a point source for a perturbation δe^{-4A} of the geometry (3.37):

$$\nabla_y^2 \left(\delta e^{-4A(y_b;y)} \right) = -\mathcal{C} \left(\frac{\delta(y_b - y)}{\sqrt{g(y)}} - \bar{\rho}(y) \right) , \tag{4.10}$$

with $\mathcal{C} \equiv 2g_s^2 \kappa^2 T_3 = (2\pi)^4 g_s(\alpha')^2$. In order to satisfy Gauss's law on the compact space [305], we have included a background charge density $\bar{\rho}(y)$, with $\int d^6 y \sqrt{g}\, \bar{\rho}(y) = 1$. To be precise, the tadpole in question is gravitational, so that $\bar{\rho}(y)$ corresponds to a negative tension source, as in Section 3.3.1. The solution to (4.10) can be written as [346]

$$\delta e^{-4A(y_b;y)} = \mathcal{C} \left(\mathcal{G}(y_b; y) - \int d^6 y' \sqrt{g}\, \mathcal{G}(y; y')\, \bar{\rho}(y') \right) , \tag{4.11}$$

where the function $\mathcal{G}(y; y')$ satisfies

$$\nabla_{y'}^2 \mathcal{G}(y; y') = \nabla_y^2 \mathcal{G}(y; y') = -\frac{\delta(y - y')}{\sqrt{g}} + \frac{1}{V} . \tag{4.12}$$

Acting with $\nabla_{y_b}^2$ on (4.11), we find

$$\nabla_{y_b}^2 \left(\delta e^{-4A(y_b;y)} \right) = -\mathcal{C} \left(\frac{\delta(y_b - y)}{\sqrt{g(y_b)}} - \frac{1}{V} \right) , \tag{4.13}$$

which does not depend on the background charge distribution $\bar{\rho}(y)$.[8] The leading term in the scalar potential for a D3-brane is therefore (see Section 5.1 for more details)

$$V(y_b) = 2T_3 e^{4A(y_b)} \approx 2T_3 \left(1 - \delta e^{-4A(y_b)} \right) . \tag{4.14}$$

Computing the trace of the Hessian, we find

$$\mathrm{Tr}(\eta) \approx -\frac{M_{\mathrm{pl}}^2}{T_3} \nabla_{y_b}^2 \left(\delta e^{-4A(y_b;y)} \right) = -2 , \tag{4.15}$$

where we have used (4.13) and (3.48). Thus, the potential for a D3-brane in the presence of an anti-D3-brane, with no other sources beyond those required by tadpole cancellation, necessarily has a steep unstable direction, preventing sustained inflation [42].

---

[8] For discussions of the effects of the background charge, see [535].

Although we have presented the problem in the example of D3-branes, parallel considerations apply to any scenario in which the backreaction of a source creates a potential for the motion of some object within the compactification; the instabilities that arise will quickly end inflation. On the other hand, all realistic models involve additional sources of stress energy – at the very least, to stabilize the moduli – and the moduli-stabilizing contributions can in principle lead to a potential suitable for inflation. This almost always requires some degree of fine-tuning. To make this fine-tuning *explicit*, and thus to obtain a complete inflationary scenario in string theory, rather than a plausibility argument for inflation, requires computing the moduli potential in extraordinary detail.

4.2.3 Eta problem in supergravity

Most contemporary scenarios for string inflation preserve supersymmetry down to the scale $H < M_{\mathrm{KK}}$, and hence can be described in four-dimensional $\mathcal{N} = 1$ supergravity. The positive vacuum energy during inflation spontaneously breaks supersymmetry. Inflation then often suffers from a particular form of the eta problem that arises from couplings in supergravity [248].

We sketched the basics of $\mathcal{N} = 1$ supergravity in four dimensions in Section 3.2. Let us take the inflaton φ to be a complex[9] scalar in a chiral multiplet. Assuming that φ is a gauge singlet, its interactions are determined by the Kähler potential $K(\varphi, \bar{\varphi})$ and the superpotential $W(\varphi)$. The Lagrangian for the inflaton is

$$\mathcal{L} = -K_{\varphi\bar{\varphi}} \partial_\mu \varphi \partial^\mu \bar{\varphi} - e^{K/M_{\mathrm{pl}}^2} \left[K^{\varphi\bar{\varphi}} D_\varphi W \overline{D_\varphi W} - \frac{3}{M_{\mathrm{pl}}^2} |W|^2 \right] . \qquad (4.16)$$

In (4.16) we have omitted the F-terms $D_\chi W$ of additional moduli χ; including these terms is straightforward and does not change our conclusions. We have also omitted a possible D-term contribution, which we will comment on below.

Expanding the Kähler potential around a reference location $\varphi \equiv 0$,

$$K = K(0) + K_{\varphi\bar{\varphi}}(0)\varphi\bar{\varphi} + \cdots , \qquad (4.17)$$

the Lagrangian (4.16) becomes

$$\mathcal{L} \approx -\partial_\mu \phi \partial^\mu \bar{\phi} - V(0) \left(1 + \frac{\phi\bar{\phi}}{M_{\mathrm{pl}}^2} + \cdots \right) , \qquad (4.18)$$

where we have defined the canonically normalized field $\phi\bar{\phi} \equiv K_{\varphi\bar{\varphi}}(0)\varphi\bar{\varphi}$. The ellipsis in (4.18) corresponds to terms arising from the expansion of K and W inside the square brackets in (4.16). These terms are model-dependent and can be of the same order as the model-independent term that we have shown explicitly. However, without fine-tuning the model-dependent terms against the universal

---

[9] The actual inflationary instability will generally involve one real component of φ, e.g. the real or imaginary part, phase, or magnitude of φ.

term, we get a large contribution to the inflaton mass and hence to the eta parameter:

$$m_\phi^2 = \frac{V(0)}{M_{\rm pl}^2} + \cdots = 3H^2 + \cdots \qquad \Rightarrow \qquad \eta = 1 + \cdots . \qquad (4.19)$$

Thus, a generic inflationary model in $\mathcal{N} = 1$ supergravity suffers from the eta problem [248].

An instructive special case is the theory of a spacetime-filling D3-brane in a compactification with a single Kähler modulus T. Parameterizing the D3-brane position in the compact space with three complex scalars z_α, $\alpha = 1, 2, 3$, the Kähler potential takes the DeWolfe–Giddings [331] form (3.100):

$$K = -3 \ln \left[T + \bar{T} - \gamma k(z_\alpha, \bar{z}_\alpha) \right] \equiv -2 \ln \mathcal{V} , \qquad (4.20)$$

in units where $M_{\rm pl} \equiv 1$. In the second equality we have indicated the dependence on the *physical volume* $\mathcal{V}$, as contrasted with the *holomorphic volume* T. The latter is the proper Kähler coordinate on the moduli space, and can appear in the superpotential. On the other hand, the rescaling to four-dimensional Einstein frame entering (3.40) involves a power of $\mathcal{V}$, and so all sources of positive energy in four dimensions contribute to a runaway potential for $\mathcal{V}$.

In the absence of a superpotential for T and z_α, all four fields have vanishing potential. One might hope that T could be stabilized by superpotential interactions, leaving z_α as flat directions. However, in the presence of a superpotential for T, the F-term potential (4.16) depends on both T, through the superpotential, and $\mathcal{V}$, through the prefactor e^K. Displacement of the D3-brane changes $k(z_\alpha, \bar{z}_\alpha)$, and hence alters either T or $\mathcal{V}$. As a result, superpotential stabilization of T leads to a mass for z_α, through the mixing in (4.20). This is another manifestation of the eta problem [42].

It has been suggested that the eta problem in supergravity may be evaded if inflation is driven by a D-term potential [536]; the argument given above is then inapplicable. Moreover, the D-term potential has been argued to be less sensitive than the F-term potential to inflaton-dependent corrections to the Kähler potential. A significant difficulty[10] is that all known scenarios for complete moduli stabilization involve some F-term potential for the moduli, and in general $V_F \gtrsim V_D$. Expanding V_F as in (4.18), the eta problem reappears, because of the inflaton dependence of the F-term contribution to moduli stabilization. See [538] and Section 5.2.1 for discussions of this effect in an explicit string inflation scenario.

By particle physics standards, the fine-tuning required to go from $\eta \sim \mathcal{O}(1)$ to $\eta \sim \mathcal{O}(0.01)$ is not extreme. Nevertheless, it would certainly be preferable if a symmetry principle made inflation technically natural. A simple way to achieve

---

[10] A criticism of D-term inflation based on consistency conditions in supergravity can be found in [537].

this in the present context [539][11] is to impose a shift symmetry on one of the real components of the complex scalar ϕ, e.g. $(\phi + \bar{\phi}) \mapsto (\phi + \bar{\phi}) + const$. If this symmetry is exact, then the superpotential is independent of ϕ and the Kähler potential can only be a function of the imaginary part $\phi - \bar{\phi}$, i.e. at lowest order we have

$$K = (\phi - \bar{\phi})^2 \,. \tag{4.21}$$

Now the e^K factor in (4.16) is independent of $\phi + \bar{\phi}$, and the real part of ϕ is protected from a dangerous mass term. This time only the unprotected field $\phi - \bar{\phi}$ receives a mass of order H. Examples of supersymmetric inflation models with these structures can be found in [229, 230, 249, 251, 541, 542]. Further work on inflation in supergravity includes [499, 543–548].

A fundamental limitation of simply assuming a shift symmetry in the low-energy supergravity is that couplings to Planck-mass degrees of freedom can readily spoil the symmetry (see Section 2.1.4). Thus, asserting an exact shift symmetry in supergravity is untenable,[12] and the question is how badly the symmetry is lifted in string theory. In Sections 5.2.1 and 5.4, we will encounter examples of inflation in string theory that try to exploit shift symmetries to construct natural models of slow-roll inflation. This is a prime example of the utility of string theory in assessing ultraviolet-sensitive questions: the nature of the remnant symmetry can be determined by direct calculation within string theory. A fair summary is that approximate symmetries are ubiquitous in string theory, but symmetries that are powerful enough to resolve the eta problem and make inflation natural are considerably less common.

4.3 Super-Planckian fields

The prospect of detecting primordial gravitational waves makes it essential to understand inflationary scenarios involving super-Planckian inflaton displacements, $\Delta\phi \gtrsim M_{\rm pl}$. As we explained in Section 2.3, such large-field models are exquisitely sensitive to Planck-scale physics: at least naively, an infinite series of non-renormalizable terms should be incorporated in the inflaton action. Examining large-field inflation in string theory sharpens and refines the problem; the task of understanding and controlling the effective inflaton action becomes a matter of explicit computation.

---

[11] See also [540], in which an assumed Heisenberg symmetry protects the flatness of the potential.

[12] In certain field theories with special structures, it is possible to suppress all dangerous symmetry breaking terms to the necessary level. For example, in [251], it was shown that if the inflaton is the phase of a baryonic operator in SUSY QCD theories with gauge groups $SU(N \geq 5)$, symmetry breaking operators only arise at dimension seven or larger. In this case, the inflaton shift symmetry is an accidental symmetry and symmetry breaking effects are controlled by gauge symmetry. (The same mechanism controls proton decay in the Standard Model.) Similarly, coupling the inflaton to a conformal field theory can suppress the Wilson coefficients of the dangerous operators by RG flow [542].

It is useful to divide constraints on super-Planckian displacements into two classes, kinematic and dynamic. Kinematic constraints on the field range are purely geometric: if the field space has a finite diameter, then by definition there is a maximum possible geodesic distance between two points, although the path length between an initial and final configuration can still be arbitrarily large. For fields in string theory that have restricted ranges for purely geometrical reasons, one can make very strong statements about the impossibility of using those fields to construct inflationary models with observable tensors. Even for the fields that kinematically allow super-Planckian vevs, one must consider the dynamical question of whether inflation can persist over such a displacement, i.e. whether controllably flat potentials can extend over such large distances in field space. This requires careful study of corrections to the inflaton potential.

In this section we will describe some of the general aspects of the kinematic and dynamic problems. A definitive treatment of dynamics requires detailed information about the geometry and potential energy in a metastable compactification, and is therefore deferred to the examples of Chapter 5.

4.3.1 Geometric constraints

First, we will examine the size of the moduli space for a Dp-brane in a simple toroidal compactification. Consider a Dp-brane that fills the four-dimensional spacetime and wraps a $(p-3)$-cycle of volume $\mathcal{V}_{p-3} = (2\pi L)^{p-3}$ on an isotropic six-torus of volume $\mathcal{V} = (2\pi L)^6$. The dynamics of the brane is then that of a point particle in $9-p$ compact dimensions. We will derive a kinematic constraint on the canonical range of this particle. Suppose that the Dp-brane moves along one of the circles in the T^6, with coordinate y; the maximum possible distance from its starting point is then $\Delta y = \pi L$. Dimensional reduction of the DBI action defines the canonically-normalized field as $\phi^2 = T_p \mathcal{V}_{p-3} y^2$, so that the maximal displacement is

$$\Delta\phi^2 < \frac{1}{8\pi} \frac{M_s^2}{g_s} \left(\frac{L}{\ell_s}\right)^{p-1} . \qquad (4.22)$$

It may appear that we can make this field range arbitrarily large by choosing $L \gg \ell_s$ and/or $g_s \ll 1$. However, what is relevant for the Lyth bound is the canonical field range in units of the four-dimensional Planck mass (3.48),

$$M_{\text{pl}}^2 = \frac{1}{\pi} \frac{M_s^2}{g_s^2} \left(\frac{L}{\ell_s}\right)^6 . \qquad (4.23)$$

We find

$$\frac{\Delta\phi^2}{M_{\text{pl}}^2} < \frac{g_s}{8} \left(\frac{\ell_s}{L}\right)^{7-p} . \qquad (4.24)$$

For $p < 8$, the Planck mass grows faster with L than $\Delta\phi$ does, so that in the limit of theoretical control ($L > \ell_s$ and $g_s < 1$), the field excursion is sub-Planckian.

The constraint (4.24) is clearly weakest for a high-dimensional brane on an anisotropic compactification with one large dimension. Consider the spacetime $\mathbb{R}^{1,3} \times S^1/\mathbb{Z}_2 \times X_5$, where X_5 is a compact manifold of volume $\mathcal{V}_5$ and the interval $S^1/\mathbb{Z}_2$ has length πL. A D8-brane that fills $\mathbb{R}^{1,3}$ and wraps X_5 is then a point particle on $S^1/\mathbb{Z}_2$. Going through the same logic as above, one finds

$$\frac{\Delta\phi^2}{M_{\mathrm{pl}}^2} < \frac{g_{\mathrm{s}}}{4\pi}\frac{L}{\ell_{\mathrm{s}}} , \tag{4.25}$$

with no dependence on $\mathcal{V}_5$. The field range now becomes parametrically large for $L \gg \ell_{\mathrm{s}}$. This result closely parallels the finding in Section 4.2.1 that stretched string masses can become super-Planckian in compactifications with one large dimension and five small dimensions.

4.3.2 Backreaction constraints

Although the kinematic range (4.25) accessible to a *probe* D8-brane can be very large, the low codimension of the D8-brane makes backreaction a serious problem. In fact, backreaction by the D8-brane restricts the range to be sub-Planckian.[13]

First, we note that the D8-brane charge and tension lead to tadpoles that must be canceled. For a consistent compactification on $\mathbb{R}^{1,3} \times S^1/\mathbb{Z}_2 \times X_5$, we introduce a pair of O8-planes that sit at each end of the interval and wrap X_5, and take the total number of D8-branes to be 16, initially situated in two groups of eight on top of the orientifold planes.[14] Now the inflaton candidate is the position y of a single D8-brane, leaving the remaining D8-branes at the endpoints of the interval. The backreaction problem is that the moving D8-brane has charge and tension, and sources corrections to the metric and dilaton once it is removed from the O8-plane: cf. (3.33). Because of the low codimension of the source, it turns out that the dilaton *diverges* before the D8-brane can be displaced by $\Delta\phi = M_{\mathrm{pl}}$. Thus, consistently incorporating backreaction prevents super-Planckian displacements.

A rather different example where the would-be inflaton induces corrections that limit its own field range arises in N-flation [549], as detailed in Section 5.4.1. The essential idea is that, as in assisted inflation [550], the inflaton Φ is a collective excitation of $N \gg 1$ elementary fields ϕ_i. If the ϕ_i each have kinematic range $\Delta\phi$, the total range is

$$\Delta\Phi = \sqrt{N}\Delta\phi . \tag{4.26}$$

The backreaction problem in this scenario is that the Planck mass is renormalized by loops of the N light fields. Without detailed knowledge of the ultraviolet completion, one can estimate this correction as

---

[13] We thank Juan Maldacena for discussions on this point.

[14] This is known as a compactification of the type I$'$ theory – see for example the discussion in [292].

$$\delta M_{\rm pl}^2 \sim \frac{N}{16\pi^2} \Lambda_{\rm UV}^2 \,, \tag{4.27}$$

in terms of an ultraviolet cutoff $\Lambda_{\rm UV}$. Because the correction (4.27) has the same scaling with N as the displacement (4.26), taking N large does not parametrically increase the field range in a theory where the quantum corrections take the form (4.27). Overcoming this problem requires replacing the estimate (4.27) with a precise computation in an ultraviolet completion, and then identifying circumstances in which the scaling differs from (4.27): see Section 5.4.1.

4.3.3 Stability constraints

Given a field space in which super-Planckian displacements are possible, sustained large-field inflation requires a potential energy source that varies slowly over this distance. A significant obstacle to constructing a gently sloped potential in a string compactification is that the inflationary energy itself backreacts on the geometry, and can disrupt the stabilization of the moduli, as we now explain.

Many sources contribute to the moduli potential in a general string compactification: p-form fluxes, localized D-branes and orientifold planes, and perturbative and nonperturbative quantum effects are among the best-studied examples. A single source generally induces an instability, as explained in Section 3.3.3, and the characteristic of solutions with stabilized moduli is a delicate – and often precarious – balance among multiple contributions to the potential energy, leading to a moduli potential $U_{\rm mod}$ with a local minimum. The inflationary potential energy itself is one such contribution, but, crucially, this energy V necessarily diminishes as inflation proceeds, with initial and final energies differing by $V_i - V_f \equiv \Delta V$. In scenarios where $V \ll U_{\rm mod}$ and $\Delta V \ll U_{\rm mod}$, the inflationary energy poses a limited risk to stability. When instead $V \gtrsim U_{\rm mod}$, the initial inflationary energy may overcome the barriers in the moduli potential, driving runaway evolution. Even worse, when $\Delta V \gtrsim U_{\rm mod}$ the inflationary contribution changes so dramatically during the course of inflation that instabilities are unavoidable unless the remaining sources for the moduli potential provide precisely compensating energies with just the right time dependence.

Destabilization is a particular difficulty for large-field inflation in string theory, because the inflationary energy density V is necessarily large, of order $M_{\rm GUT}^4$, and changes significantly during inflation.[15] With only two decades of energy between the inflationary energy and the Planck scale, there is little room for a hierarchy of the form

$$V^{1/4} \ll M_{\rm KK} \ll M_{\rm s} \ll M_{\rm pl} \tag{4.28}$$

---

[15] While the precise change is model-dependent, the ratio of initial to final energies is generally sizable: for example, $V_i/V_f \sim 10^2$ in $m^2\phi^2$ chaotic inflation.

that would underpin theoretical control, as in (4.3). The Kaluza–Klein scale M_{KK} sets the maximal scale of the moduli potential, $U_{\mathrm{mod}} \lesssim M_{\mathrm{KK}}^4$, so the first relation in (4.28), $V \ll M_{\mathrm{KK}}^4$, indicates the separation of scales that could be compatible with $V \ll U_{\mathrm{mod}}$.

While destabilization that leads to runaway decompactification is ruinous, more controllable backreaction of the inflationary energy on the moduli potential can alter the character of an inflationary model without preventing sustained inflation. In particular, given sufficiently high barriers around a local minimum of the moduli potential, a time-dependent inflationary energy can induce evolution of the moduli within the basin of attraction of the minimum. Incorporating the motion of the moduli can then change the form of the inflaton potential, as in the rather general *flattening* mechanism of [551].[16] Thus, although shifts of the moduli do not necessarily end inflation, their effects must be taken into account.

The twin issues of limited parametric separation and of backreaction by the inflationary energy are common to all scenarios for large-field inflation in compactifications of string theory. The problem is simply an outcome of the high energy scale (1.38), combined with the existence of extra dimensions with radii greater than the Planck length. Even so, these fundamental problems take many different guises in explicit constructions, and can be subtle to identify and extirpate. In Chapter 5, we will encounter these challenges explicitly: e.g. backreaction by relativistic D-branes in the DBI model (Section 5.3), and by induced charge on NS5-branes in axion monodromy models (Section 5.4.2).

4.4 Multi-field dynamics

Moduli fields are ubiquitous in string compactifications, as we explained in Section 3.2. After integrating out ultraviolet degrees of freedom, incorporating the effects of fluxes, localized sources, and quantum corrections to the action, one generally finds a complicated potential for the moduli. Although a subset of the moduli may acquire large supersymmetric masses, $m \gg H$ – e.g. complex structure moduli in type IIB flux compactifications, cf. Section 3.3 – the generic outcome is that a significant number of moduli have masses $m \lesssim H$, and are therefore dynamically active during inflation. The resulting inflationary models are quite complex, and are just beginning to be explored in detail.

The challenge of analyzing a model with multiple light moduli can be divided into two principal tasks: (i) determining the effective Lagrangian, and (ii) computing the observational signatures. We will address these issues in turn.

---

[16] See also [225, 552], where the slight shift of the overall volume induced by motion of a D3-brane leads to important corrections to the D3-brane potential.

4.4.1 Ensembles of effective theories

As explained in Section 2.1, the effective Lagrangian for N scalar fields $\Phi \equiv \{\phi_1, \ldots, \phi_N\}$ can be written in the form (2.14),

$$\mathcal{L}_{\text{eff}}[\Phi] = \mathcal{L}_l[\Phi] + \sum_i c_i \frac{\mathcal{O}_i[\Phi]}{\Lambda^{\delta_i - 4}}, \tag{4.29}$$

where $\mathcal{O}_i[\Phi]$ stands for operators of dimension δ_i constructed from $\phi_1, \ldots, \phi_N$ and their derivatives,[17] and c_i are the associated Wilson coefficients.[18] The Wilson coefficients depend on unknown details of the compactification, and computing them is impractical. Moreover, we lack any principle that could select a single compactification, and are therefore obliged to marginalize over the unconstrained details of the bulk. Stated differently, scenarios for inflation in flux compactifications of string theory lead not to one fully specified Lagrangian, but to an *ensemble* of possible inflationary Lagrangians, each with the same operator content $\{\mathcal{O}_i\}$ but with different sets of associated Wilson coefficients $\{c_i\}$. Fine-tuning the parameters of a model – implicitly, by adjusting quantized fluxes and other integer data – ultimately involves selecting an appropriate Lagrangian from the ensemble.

How can anything be learned if the Wilson coefficients are unknown? One strategy is to take the c_i to be elements of some statistical distribution Ω, and then determine only the statistical properties of the ensemble of effective Lagrangians. A natural concern is that the conclusions might depend on Ω, which, just like the values of the individual c_i, is usually not computable. Fortunately, in effective theories with *many fields* – and therefore a large number of operators with[19] $\delta \lesssim 6$ – the potential is a sum of many terms, with the consequence that central limit behavior can wash out most of the dependence on the shape of Ω. Universality therefore restores some degree of predictivity. Concretely, one can approximate Ω by a Gaussian distribution with zero mean and standard deviation σ,[20] even if the true distribution of the individual Wilson coefficients c_i is highly non-Gaussian.

In this approach, inflation can arise from accidental cancelations among two or more terms in the potential. A primary goal for a statistical analysis is then to determine how frequently inflation occurs, and when it does, what the characteristic properties of the evolution are.

---

[17] Curvature invariants are also allowed in principle, but can usually be neglected during an inflationary phase with $H \ll M_{\text{pl}}$.

[18] Symmetries of the high-scale theory may forbid certain operators, or suppress different Wilson coefficients to varying degrees, as detailed in Section 2.1. Incorporating these effects is straightforward, cf. e.g. [43, 250, 553].

[19] See Section 2.3.2 for an explanation of the cutoff value $\delta \sim 6$ in small-field inflation.

[20] The standard deviation σ controls the rms size of non-renormalizable contributions to the potential, and is therefore physical; one can estimate σ by the general logic of Section 2.1.

Although numerical experiments in the particular example of warped D-brane inflation (see Section 5.1) give strong evidence that *six* fields can be large enough for universality to take hold [250, 554], much remains to be learned about the statistics of general multi-field models – see [265, 507, 555–561] for related work.

4.4.2 Multi-field perturbations

Extracting the cosmological signatures of an effective theory with multiple light fields is challenging. We will briefly describe the qualitative problems (and opportunities), deferring details to Appendix C.

Super-horizon evolution

The essential difference between a model with one light field and a model with two or more light fields is that in the former case there is only one clock, so that the evolution of the perturbations is captured by the Goldstone action (1.16) for π. The resulting curvature perturbations, $\mathcal{R} = -H\pi$, are purely adiabatic, and are conserved outside the horizon. In multi-field models, the vevs of additional fields ψ provide additional clocks, whose fluctuations correspond to *entropy* perturbations.

Entropy fluctuations can evolve outside the horizon, and also couple to the curvature perturbations in such a way as to permit the latter to evolve outside the horizon, so that the late-time curvature perturbation can be a complicated function of all the fluctuations at horizon crossing,

$$\mathcal{R} = f(\pi_\star, \psi_\star) \, . \tag{4.30}$$

In some cases, the entropy perturbations eventually decay and the evolution reaches an *adiabatic limit*, where the curvature perturbation can again be expressed as $\mathcal{R} = -H\pi$. After that time, the superhorizon curvature perturbations are conserved. If instead reheating occurs before an adiabatic limit is reached, the late-time curvature perturbations are extremely sensitive to the details of reheating, leading to a loss of predictivity.

Single-field slow-roll models automatically predict curvature perturbations that are adiabatic, approximately scale-invariant, and approximately Gaussian, in excellent agreement with observations. None of these properties is automatic in a general multi-field model. For $m_\psi \sim H$, the entropy fluctuations have a strongly scale-dependent spectrum. If these fluctuations give the dominant contribution in (4.30), this can destroy the scale-invariance of the spectrum of curvature perturbations. Scale-invariance can be preserved, however, if the couplings of the inflaton to the additional fields preserve the approximate shift symmetry of the inflaton [249, 509].

Alternative sources for curvature perturbations

Models with multiple light fields offer alternative mechanisms for generating the observed density perturbations, including *modulated reheating* [174–176, 266] and the *curvaton scenario* [134, 135, 177, 562].

In modulated reheating, superhorizon fluctuations in one or more light spectator fields ψ modulate the end of inflation, or the decay rate of the inflationary energy density. In other words, if the decay rate Γ is a function of the fields, $\Gamma = \Gamma(\psi)$, then the decay rate inherits the spatial variations of the ψ fields. This converts fluctuations of ψ to density fluctuations in the post-inflationary universe. The observed curvature perturbations can be non-Gaussian if the function $\Gamma(\psi)$ is nonlinear.

In the curvaton scenario, a light spectator field ψ, the "curvaton," survives until after reheating. Once the Hubble rate drops below the mass m_ψ, the curvaton begins to oscillate, evolving as non-relativistic matter. The energy density associated with the curvaton therefore redshifts more slowly than the post-inflationary radiation background, and eventually the curvaton makes a significant contribution to the total energy density of the universe. When ψ ultimately decays, its superhorizon fluctuations are imprinted into density fluctuations in the visible sector. The fluctuations may be non-Gaussian if the potential $V(\psi)$ is anharmonic and/or if the decay rate $\Gamma(\psi)$ is nonlinear.

4.5 Reheating

Any complete model of inflation must explain how the energy stored in the inflaton eventually reaches the visible sector and initiates the hot Big Bang. There are two basic requirements for the process of *reheating*: Standard Model degrees of freedom must be heated to a temperature sufficient for baryogenesis, and the cosmic history must not be spoiled by overproduction of relic particles in other sectors. The rich structure of inflationary models in string theory leads to significant challenges for successful reheating, as well as a range of novel phenomena, as we now review.[21]

4.5.1 Heating the visible sector

The universal feature of string constructions that complicates reheating is the existence of fields beyond the inflaton, the Standard Model fields, and the four-dimensional graviton. Light, long-lived hidden-sector fields, such as moduli, have long been known to threaten the successes of the standard thermal history. Moduli decays occurring after baryogenesis can dilute the baryon asymmetry, while decays occurring during or after Big Bang nucleosynthesis can photodissociate the light elements, ruining the prediction of their abundances.

---

[21] See [563] for a review of reheating in field theory.

On the other hand, cosmologically long-lived relic particles can yield too much dark matter or even overclose the universe. String theory provides a plethora of candidates for dangerous relics, including compactification moduli, Kaluza–Klein modes, excited strings, and axions, as well as hidden sector matter and radiation.[22]

In conventional field-theoretic studies of reheating, as well as of the related nonperturbative process known as *preheating*, a primary question concerns the efficiency with which the inflaton transfers its energy into other degrees of freedom. The difficulty in many string-theoretic constructions is rather different [564]; the inflaton readily liberates its energy into hidden sector fields, and the question is whether a sufficiently large fraction ends up in the Standard Model rather than in harmful relics. The challenge of reheating after inflation in string theory can be compared with that of keeping a house warm in a cold winter; a furnace alone is insufficient, and one must also have insulation to direct a large fraction of the energy output to the desired region.

Reheating crucially involves the Standard Model, so to discuss reheating in a string construction one cannot remain agnostic about how the visible sector is realized. In D-brane models in type II and type I string theory, as well as in the strongly coupled heterotic string, the visible sector is generally localized on one or more branes (or at the intersections of branes). When the inflationary energy is also localized on a brane, one can take a modular approach, in which the inflationary sector and the visible sector are constructed separately, in local geometries approximating regions of some unspecified compactification, and their interactions are then computed or parameterized. This strategy has been fruitful in extensive explorations [564–575] in the context of warped D-brane inflation [42], as we review in Section 5.1. Reheating in other models involving D-branes has been studied in e.g. [576–578].

In models where the inflaton is a closed string modulus (see Section 5.5), new challenges arise, as described in [579, 580]. Investigations in models where the inflaton is a closed string modulus include [581–585].

The phenomenology of reheating in string theory is quite rich. A violent end to inflation, e.g. through brane–antibrane annihilation, provides a setting in which fields that can otherwise be omitted from the effective theory play a role: strong violations of the adiabatic approximation allow very massive fields to contribute to the dynamics, as further discussed in Section 5.6. Furthermore, the existence of multiple light fields can lead to modulated reheating, in which the dominant contribution to the temperature anisotropies arises from spatial variations in the couplings between the inflaton and the visible sector, or to the conversion of entropic perturbations to curvature perturbations; investigations of these effects in string theory include [582, 585] and [265, 554, 571], respectively. Finally,

----

[22] Dark radiation, corresponding to relativistic species in a hidden sector, can have distinctive signatures; see e.g. [314–316, 318].

condensation of a complex tachyon produces a network of topological defects, *cosmic strings*, which have striking signatures, as we now explain.

4.5.2 Cosmic strings

Symmetry-breaking phase transitions can lead to the formation of topological defects classified by the topology of the vacuum manifold. Cosmologically important examples include zero-dimensional defects, such as magnetic monopoles; one-dimensional defects, known as *cosmic strings*; and two-dimensional defects, i.e. domain walls. Magnetic monopoles from a GUT phase transition could overclose the universe [586], and one of the early successes of inflation was explaining how monopoles could be diluted [18]. Domain walls likewise come to dominate the energy density of the universe, and are ruled out. Cosmic string networks, on the other hand, evolve so that their density tracks the density of the dominant component (radiation or matter); this is called *scaling*. As a result, cosmic strings are constrained, but not excluded, and they produce spectacular, unmistakable signatures that could be detected in coming experiments. Moreover, cosmic strings arise very naturally in constructions of inflation in string theory. Here, we will review key facts about cosmic strings, referring the reader to the textbook [587] and the reviews [588–591] for many more details. We will begin with generalities that apply to all cosmic strings, and then describe the special aspects of the cosmic superstrings that arise in string theory, following [589].

Cosmic strings arise whenever a $U(1)$ symmetry is broken: the winding number of the $U(1)$ around the core of the string is the topological conserved quantity responsible for stability. The minimum cosmic string density produced in a cosmological phase transition in which a $U(1)$ symmetry is broken is set by the *Kibble mechanism* [592]. Causality prevents the phase of the complex scalar order parameter from being correlated on super-horizon distances, so that at least one horizon-spanning string defect is produced per horizon volume.

The evolution of cosmic strings involves a few elementary processes: stretching along with the expansion of the universe; intersection and reconnection, including loop formation; and energy loss through emission of gravitational radiation. Reconnection of a string after intersection is known as *intercommutation*, and the probability P of intercommutation is a key phenomenological parameter. The self-intersection and intercommutation of a long string leads to the formation of a loop, which breaks off from the long string and gradually decays by emitting gravitational waves. Thus, the network of strings involves a number of long, horizon-crossing strings, as well as populations of loops in different stages of decay.

The signatures of cosmic strings are distinctive. A string produces a conical defect geometry; denoting the string tension by μ, the deficit angle is $8\pi G\mu$, where G is Newton's constant. The associated gravitational lensing of background objects can lead to double images.[23] Moreover, a moving cosmic string generates a temperature contrast in the CMB [597] – this is known as the Kaiser–Stebbins effect (for related signatures in 21cm radiation, see [598]). Stochastic contributions to the CMB anisotropy are also important; cosmic strings with high tension could produce density perturbations sufficient to seed large-scale structure. However, the corresponding anisotropies lack phase coherence, and so do not manifest acoustic peaks. Thus, cosmic strings can at most contribute a subdominant component [599–603] of the primordial perturbations. The continual emission of gravitational radiation produces a stochastic background of gravitational waves, which could be detected directly by LIGO or Virgo [604], or indirectly by inducing stochastic fluctuations in the arrival of pulsar signals (see e.g. [605]). Finally, smooth loops of string develop one or more sharp *cusps* in each period of oscillation. Near the cusp, the string is extremely relativistic, and emits an intense burst of gravitational waves in a cone pattern [606–608]. A cusp event directed toward a gravitational wave detector such as LIGO could allow detection of strings with comparatively low tension.[24] Bursts can also occur if strings break following the formation of monopole–antimonopole pairs [610, 611].

For many years, the study of cosmic strings focused exclusively on strings arising in quantum field theory – Nielsen–Olesen strings [612], also called vortex lines – rather than on the fundamental strings of superstring theory. Witten had observed in [613] that in perturbative constructions, the tension of fundamental strings was large enough so that cosmic F-strings were excluded by the isotropy of the CMB.[25] Furthermore, he showed that heterotic cosmic strings form boundaries for axion domain walls, whose tension causes the strings to contract rapidly and disappear.

A renewed study of cosmic superstrings was initiated by Tye and collaborators in [614–616]. The essential new insight was that if the Standard Model arises on D-branes, the visible sector couplings and the string tension in Planck units can be adjusted independently, by changing the string coupling g_s and the compactification volume. Thus, the string tension can be low enough to satisfy observational constraints. (A similar argument applies in the strongly coupled heterotic string [617].) Moreover, in inflationary scenarios involving moving D-branes, reheating typically proceeds by the condensation of a complex tachyon [618], leading to cosmic string defects via the Kibble mechanism.

---

[23] Most searches for cosmic string lensing involve extragalactic objects (cf. e.g. [593]), but microlensing of stars within the galaxy [594] (rather than of distant quasars [595]) could probe very low tensions, particularly if the string loops cluster substantially [596].

[24] Cosmic strings of even lower tension might be detectable if they passed through the Earth [609], causing devastating earthquakes while simultaneously providing a window on Planck-scale physics.

[25] The isotropy of the CMB gives an upper limit on the inflationary scale, and hence on the tension of cosmic strings that could be produced in a phase transition after inflation. Moreover, high-tension strings can be excluded by searches for lensing and for the Kaiser–Stebbins effect.

Cosmic superstrings have several important characteristics that distinguish them from strings arising as topological defects in perturbative quantum field theories [619–621]. In type IIB string theory, there are two elementary one-dimensional objects: the fundamental string, or "F-string," and the D1-brane, or "D-string." These strings can form bound states involving p F-strings and q D-strings, if p and q are relatively prime [622, 623]. The resulting (p, q) string has tension [623]

$$\mu_{p,q} = \frac{1}{2\pi\alpha'}\sqrt{(p - C_0 q)^2 + e^{-2\Phi}q^2} \ . \tag{4.31}$$

Networks of (p, q) strings yield scaling solutions [624], just like simpler cosmic strings. The intercommutation probabilities of cosmic F-strings and D-strings can be much smaller than those for field theory cosmic strings, as carefully examined in [621]. In particular, a colliding pair of strings can miss each other in the compact dimensions [616, 619, 621], and the string coupling g_s also suppresses the intercommutation probability.

Perhaps the most compelling setting for cosmic superstring production is warped D-brane inflation [42], in which annihilation of a D3-brane/anti-D3-brane pair via condensation of a complex tachyon automatically produces a collection of cosmic strings, and warping provides a natural parametric mechanism through which the tension can be small enough to obey observational bounds. The stability and tension of these strings depend on the details of the model [620, 625, 626], and we defer further discussion to Section 5.1.6.

One might hope that cosmic superstrings can be distinguished from strings arising as topological defects in field theory – see [589] for a thorough discussion of this point. This hope is not entirely unjustified; cosmic superstrings with $P < 1$ can be told apart from strings in a *perturbative* field theory, which have $P \approx 1$. Furthermore, the spectrum of tensions (4.31) appears distinctive. On the other hand, a field theory with $SL(2, \mathbb{Z})$ invariance would reproduce (4.31). More generally, the duality between string theory and field theory makes it difficult, even in principle, to distinguish F-strings, D-strings, or (p, q) strings of string theory from corresponding defects in strongly coupled field theories [589]; for example, the (p, q) strings produced in warped D-brane inflation can also be viewed as strings of the dual gauge theory. Even so, the detection of a network of cosmic (p, q) strings would be an unsurpassed opportunity to probe high-scale physics!

4.6 Inflation in string theory: a checklist

An ideal model of the early universe in quantum gravity would begin from fundamental topological data, arrive at an effective theory via an explicit and well-controlled computation, and make definitive, distinctive predictions that are consistent with current data but could be falsified or verified with future experiments. There is little prospect of deriving such a model in the near future. A more realistic hope is to specify some integer data (for example, the topology of a Calabi–Yau orientifold) and explicitly solve some equations of motion (e.g. those of the Kähler moduli) while appealing to the existence of generic solutions for the remaining equations (e.g. the complex structure moduli and dilaton equations of motion given a choice of quantized three-form flux).

Let us summarize the essential requirements for a successful model of inflation derived in string theory.

- The inflaton action should be computed in an expansion around a metastable de Sitter vacuum, with all approximations under good control.
- For every physical effect contributing to the moduli potential, one must know the corresponding correction to the inflaton potential.
- All assertions about ultraviolet-sensitive quantities must be justified through controlled calculations.
- If a dimensionless parameter needs to be large or small in order for inflation to succeed, one should know whether the required value can be achieved in a consistent compactification.
- For each field with a mass $m \ll H$, the small mass should be explained either by fine-tuning of explicitly known, fully specified operators in the effective theory, or by a symmetry that can be shown to survive in string theory.
- All quantum-mechanically active fields, i.e. fields with $m < \frac{3}{2}H$, must be included in the phenomenology.
- The model should contain a mechanism to produce density fluctuations that are nearly scale-invariant, Gaussian, and adiabatic.
- The inflationary phase must end, and then transition to successful reheating of the Standard Model, without overproduction of relics.

Distinctive observational signatures, while obviously desirable, are ultimately optional.

In the next chapter, we will review some of the leading examples of string inflation. We will see that no model is completely successful on all points of the above checklist.

5

Examples of string inflation

In this chapter, we will survey a number of representative examples of inflation in string theory. We will try to be reasonably complete in our discussion of inflationary mechanisms, within the limitations of space and expertise, but we will not be able to present all of the results in the subject. Our focus will be on extracting a few important lessons from the collective works of many researchers.

The individual sections are largely self-contained and can be read in any order. In Section 5.1, we consider the motion of a D3-brane [627] in a warped throat region [42] as a source for inflation. We present a number of interrelated perspectives on the potential energy of a D3-brane in a warped flux compactification, and then discuss the challenge of achieving slow-roll behavior in this setup. In Section 5.2, we study a few examples of brane inflation in unwarped compactifications, including D3/D7 inflation [628–632], fluxbrane inflation [633, 634], and M5-brane inflation [635, 636]. In Section 5.3, we discuss relativistic brane motion as a source of non-slow-roll inflation. We describe DBI inflation [39] as an effective field theory and highlight microphysical constraints imposed by compactification. In Section 5.4, we argue that string axions are promising inflaton candidates. We give detailed analyses of N-flation [549] and axion monodromy inflation [35]. In Section 5.5, we describe models in which the inflaton is a Kähler modulus (or the associated axion), including racetrack inflation [637, 638] and inflationary scenarios in large volume compactifications [44, 639, 640]. Finally, in Section 5.6, we look at dissipative effects as a source of inflation and critically assess the prospects for dissipative inflation in string compactifications [641–644].

5.1 Inflating with warped branes

The positions of localized sources in a string compactification correspond to scalar fields in the four-dimensional effective theory. In [627], Dvali and Tye proposed that the separation between two branes could serve as an inflaton candidate. This idea was made more precise in [534, 645], where the two branes were taken to be a D3-brane and an anti-D3-brane, respectively. These objects attract each other gravitationally, and also through the R-R four-form potential C_4, under which they carry opposite charges; moreover, at small separations a tachyon appears in the spectrum, and the brane and antibrane annihilate, providing a natural end to inflation. (See [646] for a proposal in which the annihilation itself drives inflation.)

In [534, 645], the Coulomb interaction (4.8) of the brane–antibrane pair was computed and identified with the inflaton potential. The Coulomb force diminishes with increasing distance, suggesting that for sufficiently large separations, the Coulomb interaction could drive slow-roll inflation. However, Burgess et al. [534] demonstrated that the branes would have to be separated by a distance that is larger than the size of the compact space to give a potential that can source slow-roll inflation (see Section 4.2).

The character of the problem changed when Kachru et al. (KKLMMT) [42] made two pivotal observations about D-brane inflation. First, they established that *warping* of the extra dimensions suppresses the Coulomb force between the brane–antibrane pair, flattening the potential even for modest brane separations. However, building on advances in moduli stabilization (cf. Section 3.3.3), they also showed that the inflaton potential for a D-brane system is *not* given by the Coulomb potential alone; the leading contributions to the curvature of the inflaton potential come from the physical effects that stabilize the moduli. This was the first of many manifestations of the eta problem in the context of stabilized string compactifications. The task is therefore to specify the moduli-stabilizing effects and derive the complete inflaton potential. We pick up the story at this stage.[1]

5.1.1 D3-branes and warped geometries

The scenario of [42] operates in the context of flux compactifications of type IIB string theory (see Section 3.3.3), which can naturally contain warped throat regions (see e.g. [647]). In this section, we will introduce some geometrical facts about these spacetimes. We will first approximate the warped region by five-dimensional anti-de Sitter space, and then upgrade to the warped deformed conifold geometry [444, 648]. In Section 5.1.2, we will derive the D3-brane potential in these warped backgrounds.

D3-branes in anti-de Sitter space

Consider a stack of N D3-branes in ten-dimensional Minkowski space. The D3-branes source a nontrivial background for the massless fields of type IIB supergravity. In string frame, the solution for the metric is

---

[1] This section is based mostly on refs. [42, 43, 224].

$$ds^2 = e^{2A(r)}\eta_{\mu\nu}dx^\mu dx^\nu + e^{-2A(r)}\left(dr^2 + r^2 d\Omega_{S_5}^2\right) , \tag{5.1}$$

where $d\Omega_{S_5}^2$ is the metric on a five-sphere and $e^{4A(r)}$ is a harmonic function of the transverse coordinates,

$$e^{-4A(r)} = 1 + \frac{L^4}{r^4} , \qquad \text{with} \qquad \frac{L^4}{(\alpha')^2} = 4\pi g_s N . \tag{5.2}$$

This is a simple example of a warped solution, as in (3.37). The solution has constant dilaton[2] and a nontrivial four-form potential

$$\alpha(r) \equiv (C_4)_{tx^i} = e^{4A(r)} . \tag{5.3}$$

Equation (5.3) leads to the self-dual five-form flux $\tilde{F}_5 = (1 + \star_{10})dC_4$. Recalling the line element of five-dimensional anti-de Sitter space, AdS_5, in Poincaré coordinates,

$$ds^2_{AdS_5} = \frac{L^2}{r^2}dr^2 + \frac{r^2}{L^2}\eta_{\mu\nu}dx^\mu dx^\nu , \tag{5.4}$$

we see that (5.1) reduces to $AdS_5 \times S^5$ for $r \ll L$.

We now consider the dynamics of a mobile D3-brane in the $AdS_5 \times S^5$ background (see Fig. 5.1). The action for a D3-brane in Einstein frame is[3]

$$S_{D3} = -T_3 \int d^4\sigma \sqrt{-\det(G^E_{ab})} + \mu_3 \int C_4 . \tag{5.5}$$

To preserve four-dimensional Poincaré symmetry, the D3-brane is spacetime-filling, i.e. its worldvolume coordinates σ^a coincide with the spacetime coordinates x^μ. The brane is pointlike in the extra dimensions. We denote its radial location in anti-de Sitter space by r. Since the angular isometries of S^5 are unbroken, we can (for now) assume that the D3-brane has a fixed location along the angular coordinates. Evaluating the action (5.5) in the background (5.1) gives the following Lagrangian for the brane position:

$$\mathcal{L} = -T_3 e^{4A(r)}\sqrt{1 + e^{-4A(r)}g^{\mu\nu}\partial_\mu r\partial_\nu r} + T_3\alpha(r) . \tag{5.6}$$

For small velocities, $\dot{r}^2 \ll e^{4A(r)}$, we can expand the square root to get

$$\mathcal{L} \approx -\frac{1}{2}(\partial\phi)^2 - T_3\left(e^{4A(\phi)} - \alpha(\phi)\right) , \tag{5.7}$$

---

[2] Recall from Section 3.1.2 that D3-branes decouple from fluctuations of the dilaton. Moreover, their backreaction on the metric of an ISD compactification (cf. Section 3.3.1) is completely captured by an overall warp factor, as in (5.1). D3-branes are therefore considerably simpler to treat than branes of other dimensionality.

[3] We have taken the gauge field strength $\mathcal{F}_2$ on the D3-brane worldvolume to vanish, which corresponds to considering a D3-brane without dissolved D1-brane charge.

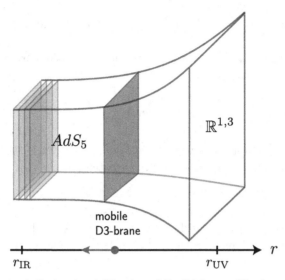

Fig. 5.1 Brane inflation in AdS_5. A mobile D3-brane fills four-dimensional spacetime and is pointlike in the extra dimension.

where we have defined the canonically normalized field $\phi^2 \equiv T_3 r^2$. From (5.3), we see that a single D3-brane experiences no force in the anti-de Sitter background: electrostatic repulsion from the four-form background exactly cancels the gravitational attraction.

D3-branes on the conifold

An anti-de Sitter background is not a realistic setting for D-brane inflation. First of all, the spacetime is not compact, but ranges from $r = 0$ to $r = \infty$. Furthermore, the metric becomes singular, with infinite redshift, at $r = 0$. A more promising scenario for D-brane inflation[4] involves a D3-brane in a finite warped throat region of a flux compactification [42]. We will now review a few geometric prerequisites for a discussion of this model.

Singular conifold

The singular conifold is a six-dimensional Calabi–Yau cone X_6 that can be presented as the locus in $\mathbb{C}^4$ defined by

$$\sum_{A=1}^{4} z_A^2 = 0 \, , \tag{5.8}$$

---

[4] *Mirage cosmology* [649] is an alternative to inflation in which the spacetime metric is the induced metric on a D-brane moving through a background supergravity solution. Discussions of mirage cosmologies involving D3-branes in warped throat regions include [650–652].

where $A \in \{1, 2, 3, 4\}$. This describes a cone over a base Y_5, which is topologically – but not metrically – equivalent to $S^2 \times S^3$. To see this, note that if z^A is a solution to (5.8) then so is λz^A, with $\lambda \in \mathbb{C}$. Writing $z^A = x^A + iy^A$, the complex equation (5.8) may be recast as three real equations,

$$x \cdot x = \frac{1}{2}\rho^2 , \qquad y \cdot y = \frac{1}{2}\rho^2 , \qquad x \cdot y = 0 . \tag{5.9}$$

The first equation defines a three-sphere S^3 with radius $\rho/\sqrt{2}$, while the last two equations describe a two-sphere S^2 fibered over the S^3. More precisely, the base Y_5 of the cone is the Einstein manifold[5] $T^{1,1}$, which is the coset space

$$T^{1,1} = [SU(2) \times SU(2)]/U(1) , \tag{5.10}$$

with isometry group $SU(2) \times SU(2) \times U(1)$. The metric on $T^{1,1}$ is

$$d\Omega^2_{T^{1,1}} \equiv \frac{1}{9}\left(d\psi + \sum_{i=1}^{2} \cos\theta_i d\phi_i\right)^2 + \frac{1}{6}\sum_{i=1}^{2}\left(d\theta_i^2 + \sin^2\theta_i d\phi_i^2\right) , \tag{5.11}$$

where $\theta_i \in [0, \pi]$, $\phi_i \in [0, 2\pi]$ and $\psi \in [0, 4\pi]$. The metric on the conifold can then be written as

$$ds^2 = dr^2 + r^2 d\Omega^2_{T^{1,1}} , \tag{5.12}$$

where $r \equiv \sqrt{3/2}\,\rho^{2/3}$. To express (5.12) as a manifestly Kähler metric, we introduce three complex coordinates z^α, $\alpha \in \{1, 2, 3\}$. The Ricci-flat Kähler metric on the singular conifold,

$$ds^2 = k_{\alpha\bar{\beta}}\, dz^\alpha d\bar{z}^{\bar{\beta}} , \tag{5.13}$$

then follows from the Kähler potential [653]

$$k(z_\alpha, \bar{z}_\alpha) = \frac{3}{2}\left(\sum_{A=1}^{4}|z^A|^2\right)^{2/3} , \tag{5.14}$$

via $k_{\alpha\bar{\beta}} = \partial_\alpha \partial_{\bar{\beta}} k$.

Deformed conifold

In the singular conifold, the base manifold $T^{1,1}$ shrinks to zero size at $z_A = 0$, and the metric on the cone has a curvature singularity. To remove the singularity, we consider a small modification of the embedding condition (5.8),

$$\sum_{A=1}^{4} z_A^2 = \epsilon^2 . \tag{5.15}$$

---

[5] An Einstein manifold satisfies $R_{ab} \propto g_{ab}$.

This defines the deformed conifold. The deformation parameter ε can be made real by an appropriate phase rotation. Equation (5.15) can then be written as

$$x \cdot x - y \cdot y = \varepsilon^2 \, , \tag{5.16}$$

$$x \cdot x + y \cdot y = \rho^2 \, . \tag{5.17}$$

At the tip of the cone, $\rho^2 = \varepsilon^2$, the S^3 remains finite ($x \cdot x = \varepsilon^2$), while the S^2 shrinks to zero size ($y \cdot y = 0$). Sufficiently far from the tip, the right-hand side of (5.15) can be ignored and the metric of the deformed conifold is well-approximated by that of the singular conifold. Most models of D-brane inflation operate in this regime.

D3-branes on the conifold

Now consider placing a stack of N D3-branes at the singular tip, $z_A = 0$, of the singular conifold. As before, the branes backreact on the geometry, producing the warped ten-dimensional line element [648]

$$ds^2 = e^{2A(r)}\eta_{\mu\nu}dx^\mu dx^\nu + e^{-2A(r)}\left(dr^2 + r^2 d\Omega_{T^{1,1}}^2\right) \, , \tag{5.18}$$

where

$$e^{-4A(r)} = 1 + \frac{L^4}{r^4} \, , \quad \text{with} \quad L^4 \equiv \frac{27\pi}{4}g_{\rm s}N(\alpha')^2 \, . \tag{5.19}$$

For $r \ll L$, the solution is $AdS_5 \times T^{1,1}$ [654].

Warped deformed conifold

Finally, we describe the warped deformed conifold, or *Klebanov–Strassler* (KS) geometry [444]. This is a noncompact, smooth solution of type IIB supergravity in which warping is supported by background fluxes. The KS solution can be obtained by considering the backreaction of N D3-branes at the tip of the singular conifold, together with the backreaction of M D5-branes wrapping the collapsed S^2 at the tip, but we will find it useful to give an alternative presentation in which all D-branes are replaced by fluxes carrying the associated charges (cf. [655]).

The geometric substrate for the solution is the deformed conifold (5.15), which contains two independent three-cycles: the S^3 at the tip, known as the *A-cycle*, and the Poincaré dual three-cycle, known as the *B-cycle*. The background three-form fluxes of the KS solution are quantized

$$\frac{1}{(2\pi)^2\alpha'}\int_A F_3 = M \quad \text{and} \quad \frac{1}{(2\pi)^2\alpha'}\int_B H_3 = K \, , \tag{5.20}$$

where $M \gg 1$ and $K \gg 1$ are integers. These fluxes give rise to nontrivial warping. The line element for the KS solution takes the form

$$ds^2 = e^{2A(r)}\eta_{\mu\nu}dx^\mu dx^\nu + e^{-2A(r)}d\tilde{s}^2 \, , \tag{5.21}$$

where $\mathrm{d}\tilde{s}^2$ is the metric of the deformed conifold defined by (5.15). As in the deformed conifold, the infrared geometry is smooth; the A-cycle is finite in size, with radius $r_A = \sqrt{g_s M \alpha'}$, so the supergravity approximation remains valid near the tip provided that $g_s M \gg 1$. For our purposes, it will suffice to cut off the radial coordinate at a minimum value r_{IR}, and work at $r \gg r_{\mathrm{IR}}$ (but see [444] for a precise description of the tip geometry). Far from the tip, the line element is well-approximated by (5.8), with

$$
e^{-4A(r)} = \frac{L^4}{r^4}\left(1 + \frac{3g_s M}{8\pi K} + \frac{3g_s M}{2\pi K}\ln\frac{r}{r_{\mathrm{UV}}}\right), \qquad (5.22)
$$

where

$$
L^4 \equiv \frac{27\pi}{4}g_s N(\alpha')^2\,, \qquad N \equiv MK\,. \qquad (5.23)
$$

Here, r_{UV} is an ultraviolet cutoff, discussed further below. The logarithmic running of the warp factor corresponds to that seen in the singular warped conifold solution of [648]. The warp factor $e^{A(r)}$ in (5.21) reaches a minimal value $e^{A(r_{\mathrm{IR}})} \equiv e^{A_{\mathrm{IR}}}$ at the tip, and is given in terms of the flux quanta by [305]

$$
e^{A_{\mathrm{IR}}} = \exp\left(-\frac{2\pi K}{3g_s M}\right)\,. \qquad (5.24)
$$

The exponential hierarchy is a consequence of the logarithmic running in (5.22). The KS solution given in (5.21) and (5.22) is the canonical example of a *warped throat* geometry, and provides the basis for the most explicit studies of warped D-brane inflation.

Before proceeding, we should emphasize that the ten-dimensional KS solution, involving a noncompact warped deformed conifold, does not give rise to dynamical gravity upon dimensional reduction to four dimensions; the compactification volume, and hence the four-dimensional Planck mass, are infinite. For model-building purposes, one considers instead a flux compactification containing a finite warped throat region that is well-approximated by a finite portion of the KS solution, from the tip $r = r_{\mathrm{IR}}$ to some ultraviolet cutoff $r = r_{\mathrm{UV}}$. Beyond this, the throat attaches to a *bulk* space, corresponding to the remainder of the compactification (see Fig. 5.2). The metric of the bulk is poorly characterized in general, but the influence of the bulk supergravity solution on dynamics in the throat region can be parameterized very effectively. The validity of the finite throat approximation was systematically investigated in [43] – see Section 5.1.2.

A field range bound

The total compactification volume is the sum of the throat volume,

$$
\mathcal{V}_T \equiv \int \mathrm{d}\Omega^2_{T^{1,1}} \int_{r_{\mathrm{IR}}}^{r_{\mathrm{UV}}} r^5 \mathrm{d}r\, e^{-4A(r)} = 2\pi^4 g_s N(\alpha')^2 r_{\mathrm{UV}}^2\,, \qquad (5.25)
$$

and the volume $\mathcal{V}_{\mathcal{B}}$ of the bulk space. The Planck mass (3.48), $M_{\mathrm{pl}}^2 = \mathcal{V}/g_s^2 \kappa^2$, is finite, with $\mathcal{V} \equiv \mathcal{V}_{\mathcal{T}} + \mathcal{V}_{\mathcal{B}}$. Ignoring the bulk volume gives a lower bound on the Planck mass,

$$M_{\mathrm{pl}}^2 > \frac{N}{4} \frac{r_{\mathrm{UV}}^2}{(2\pi)^3 g_s (\alpha')^2} \ . \tag{5.26}$$

The amount of canonical field range available to a D3-brane in the throat region (the region of controlled evolution) is bounded from above by

$$\Delta\phi^2 < T_3 r_{\mathrm{UV}}^2 = \frac{r_{\mathrm{UV}}^2}{(2\pi)^3 g_s (\alpha')^2} \ . \tag{5.27}$$

Combining (5.26) and (5.27), we arrive at the remarkably simple formula [254]

$$\frac{\Delta\phi}{M_{\mathrm{pl}}} \leq \frac{2}{\sqrt{N}} \ . \tag{5.28}$$

Since the validity of the supergravity approximation requires $N \gg 1$, this result precludes super-Planckian field ranges in models of inflation based on D3-branes in warped throats. The geometric bound (5.28) implies that warped D3-brane inflation does not allow for observable gravitational waves. Note that this argument is purely kinematic, and does not involve the D3-brane potential.

5.1.2 D3-brane potential

Equation (5.7) gives the potential for a D3-brane in the warped backgrounds (5.1), (5.8), and (5.21) as

$$V(\phi) = T_3 \left(e^{4A(\phi)} - \alpha(\phi) \right) \ . \tag{5.29}$$

This vanishes for compactifications with imaginary self-dual (ISD) fluxes [305]. However, generic string compactifications contain various sources that break the ISD condition and generate a nontrivial potential for the D3-brane.

Coulomb potential

In [42], an anti-D3-brane was added to the compactification, following [369, 443]. The antibrane minimizes its energy in regions of maximal warping, and is therefore stabilized at the tip of the conifold, $r = r_{\mathrm{IR}}$. The anti-D3-brane perturbs the background supergravity solution, and the D3-brane experiences a corresponding force. This is described by the Coulomb potential [42, 446, 461]

$$V_{\mathcal{C}}(\phi) = D_0 \left(1 - \frac{27}{64\pi^2} \frac{D_0}{\phi^4} \right) \ , \tag{5.30}$$

where the scale of the potential, $D_0 \ll 2T_3$, is set by the warped tension of the antibrane

$$D_0 \equiv 2T_3 e^{4A(r_{\mathrm{IR}})} \ . \tag{5.31}$$

The potential (5.30) is extremely flat, even for small values of the field ϕ. If this were the end of the story, warped D-brane inflation would be a strikingly natural scenario, but life is not so simple.

Curvature coupling

To source inflation, the system has to be coupled to dynamical gravity. Besides the Einstein–Hilbert term, the four-dimensional effective action contains a curvature coupling [42]

$$V_{\mathcal{R}}(\phi) = \frac{1}{12} R \phi^2 . \tag{5.32}$$

In de Sitter space, the four-dimensional spacetime curvature R equals $12H^2$. During inflation, the coupling in (5.32) therefore induces a dangerous mass term for the inflaton

$$
\begin{aligned}
V(\phi) &= V_{\mathcal{C}}(\phi) + V_{\mathcal{R}}(\phi) + \cdots \\
&\approx V_0 + H^2 \phi^2 + \cdots \qquad \Rightarrow \qquad \eta \approx \frac{2}{3} + \cdots .
\end{aligned} \tag{5.33}
$$

This is an incarnation of the eta problem. The flatness of the Coulomb potential has been completely destroyed by the curvature coupling. However, this is still not the final answer [42]. In all stabilized string compactifications there are additional contributions to the D3-brane action, and these must be included in order to determine whether inflation can occur.

Beyond the probe approximation

To compute these corrections we have to go beyond the probe approximation and allow the D3-brane to backreact on the geometry. In fact, the curvature coupling (5.32) can be interpreted as such a backreaction effect [43]. The presence of the D3-brane perturbs the overall volume of the compactification, $\mathcal{V}$. Moreover, this perturbation will depend on the position of the brane. As the brane moves through the warped region, its effect on the volume varies. The compactification volume therefore develops a dependence on the brane position, $\mathcal{V} = \mathcal{V}(\phi)$. As a result, a potential that is flat in string frame need not stay flat in Einstein frame, since the transformation between the frames involves a factor of the volume. The eta problem in (5.33) arises from precisely this effect; see Section 4.2.3.

However, it is easy to see that there will be further corrections. In Section 3.3.3, we explained that Kähler moduli stabilization in the KKLT scenario involves nonperturbative effects on D7-branes (or from Euclidean D3-branes) wrapping certain four-cycles. The volumes $\mathcal{V}_4$ of these four-cycles will also depend on the D3-brane position, $\mathcal{V}_4(\phi)$. As the D3-brane moves, the four-cycle volume adjusts. This changes the gauge coupling on the wrapped D7-branes (or the Euclidean D3-brane action) and hence the strength of the nonperturbative effects. This leads to important corrections to the D3-brane potential.

In the following, we will describe the complete D3-brane potential from two different perspectives: first we will derive the potential in four-dimensional supergravity, and then we will provide an equivalent treatment in ten-dimensional supergravity.

4D perspective

The four-dimensional effective theory can be described by the F-term potential of $\mathcal{N} = 1$ supergravity,

$$V_F = e^K \left[K^{I\bar{J}} D_I W \overline{D_J W} - 3|W|^2 \right] , \qquad (5.34)$$

where I, J runs over all moduli. We make the standard KKLT assumption that the complex structure moduli and the dilaton are stabilized at sufficiently high energies. The remaining moduli (assuming $h^{1,1}_- = 0$) are then the Kähler moduli T_i and the brane position moduli z^α ($\alpha = 1, 2, 3$). For simplicity of presentation, we restrict to compactifications with only a single Kähler modulus T, but all our considerations generalize to $h^{1,1}_+ > 1$. We define $Z^I \equiv \{T, z^\alpha\}$. The tree-level Kähler potential is the logarithm of the compactification volume

$$K = -2 \ln(\mathcal{V}) , \qquad (5.35)$$

where $\mathcal{V}$ is an implicit function of the Z^I. Corrections to (5.35) are important in many other contexts, cf. Section 5.5, but can be neglected in D3-brane inflation.

Backreaction on the volume

As mentioned above, a D3-brane with finite energy density backreacts on the overall compactification volume, which therefore depends on the brane position z^α [225, 331]:

$$\mathcal{V} = \left(T + \bar{T} - \gamma k(z_\alpha, \bar{z}_\alpha) \right)^{3/2} , \qquad (5.36)$$

where $k(z_\alpha, \bar{z}_\alpha)$ is the Kähler potential (5.14) and γ is a constant. In Appendix B of [225], the parameter γ was related to the stabilized value of the Kähler modulus,

$$\gamma \equiv \frac{T_3}{6} \left(T + \bar{T} \right)_{\text{IR}} . \qquad (5.37)$$

Here, $T_{\text{IR}} \equiv T(r_{\text{IR}})$ stands for the value of the Kähler modulus when the D3-brane is near the tip of the throat. In [225], it was shown that the minimum of the potential for the Kähler modulus T shifts slightly as the D3-brane moves, and the effect of this shift was further examined in [552].

F-term potential

Combining (5.35) and (5.36), we find that the Kähler potential is of the form postulated by DeWolfe and Giddings [331], cf. (3.100):

$$K(Z^I, \bar{Z}^I) = -3 \ln \left[T + \bar{T} - \gamma k(z_\alpha, \bar{z}_\alpha) \right] \equiv -3 \ln \left[U(Z^I, \bar{Z}^I) \right] . \qquad (5.38)$$

The F-term potential for (5.205) combined with a general superpotential $W(Z^I)$ was determined in [225, 656, 657],

$$
V_F(T, z_\alpha) = \frac{1}{3U^2} \left[\left(T + \bar{T} + \gamma \left(k_\gamma k^{\gamma\bar{\delta}} k_{\bar{\delta}} - k \right) \right) |W_{,T}|^2 - 3 \left(\overline{W} W_{,T} + c.c. \right) \right.
$$

$$
\left. + \underbrace{\left(k^{\alpha\bar{\delta}} k_{\bar{\delta}} \overline{W_{,T}} W_{,\alpha} + c.c. \right) + \frac{k^{\alpha\bar{\beta}}}{\gamma} W_{,\alpha} \overline{W}_{,\beta}}_{\Delta V_F} \right] , \tag{5.39}
$$

where $k_\alpha \equiv \partial_\alpha k$ and $k_{\alpha\bar{\beta}} \equiv \partial_\alpha \partial_{\bar{\beta}} k$. The label ΔV_F has isolated terms that arise exclusively from the dependence of the superpotential on the brane position z^α. The remainder is the standard KKLT F-term potential [369].

First consider the situation in which the superpotential does not depend on the brane coordinate, $W = W(T)$. In this case, $\Delta V_F = 0$ and the remaining terms in the square brackets in (5.39) depend only weakly on the inflaton. The potential can therefore be written as

$$
V_F(r) \approx \frac{V_0}{(1 - \frac{1}{6}\phi^2)^2} \approx V_0 + \frac{1}{3} \frac{V_0}{M_{\text{pl}}^2} \phi^2 , \tag{5.40}
$$

where in the second equality we have made the dependence on the Planck mass explicit. We see that the inflaton has a mass of order the Hubble scale, $H^2 \approx V_0/(3M_{\text{pl}}^2)$. This is how the curvature coupling (5.32) arises in the effective supergravity description.

Backreaction on D7-branes

Gaugino condensation on a stack of N_c D7-branes leads to

$$
|\Delta W| \propto \exp\left(-\frac{2\pi}{N_c} \mathcal{V}_4 \right) , \tag{5.41}
$$

where $\mathcal{V}_4$ is the "warped volume" (3.108) of the D7-branes (see Fig. 5.2). Changing the position ϕ of a spacetime-filling D3-brane alters the warp factor $A(\phi)$, and hence $\mathcal{V}_4(\phi)$, so that $\Delta W = \Delta W(\phi)$. To quantify this effect, one computes the backreaction of the D3-brane on the four-cycle wrapped by the D7-branes [346]. For a four-cycle defined by a holomorphic embedding

$$
f(z_\alpha) = 0 , \tag{5.42}
$$

the result can be written as

$$
W(T, z_\alpha) = W_0 + \mathcal{A}(z_\alpha) e^{-aT} , \qquad a \equiv \frac{2\pi}{N_c} , \tag{5.43}
$$

where the function $\mathcal{A}(z_\alpha)$ is defined in terms of the embedding (5.42),

$$
\mathcal{A}(z_\alpha) = \mathcal{A}_0 \left(\frac{f(z_\alpha)}{f(0)} \right)^{1/N_c} . \tag{5.44}
$$

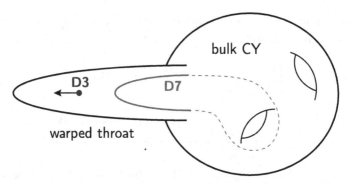

Fig. 5.2 Schematic of a finite warped throat containing D7-branes wrapping a compact four-cycle. A portion of the four-cycle extends into the throat region. Gaugino condensation on the D7-branes leads to a D3-brane potential.

Fine-tuning to produce a flat potential

Which embedding functions $f(z_\alpha)$ lead to forces that can balance the curvature coupling? This question has been addressed in a number of papers [224, 225, 656, 657]. An important no-go result was proved in [225, 656]. The infinite class of embeddings studied in [658] does not allow any inflationary solutions. In fact, to date only a single explicit embedding is known in which inflation can occur [225, 657]. This is the so-called Kuperstein embedding [659]

$$f(z_1) = \mu - z_1 . \tag{5.45}$$

In this example, the scalar potential (5.40) receives a correction scaling as $\phi^{3/2}$ ($\propto z_1$),

$$V_F(\phi) \approx V_0 + \cdots + \lambda\phi^{3/2} + \frac{V_0}{M_{\rm pl}^2}\phi^2 + \cdots . \tag{5.46}$$

The last two terms shown in (5.46) contribute to η with opposite signs. Let ϕ_0 be the point in field space where the second slow-roll parameter vanishes, $\eta(\phi_0) = 0$. Near this point we have $|\eta| \ll 1$. This is not even a fine-tuning, but arises dynamically. What does involve fine-tuning are the requirements that ϕ_0 be in the region of control (i.e. inside the warped throat) and that the potential be monotonic, with a small first derivative (small ϵ) at the same point. If this can be arranged, the result is inflation near an approximate *inflection point*. Figure 5.3 shows an example of a successful scan in the parameter space of warped D3-brane inflation [225].

10D perspective

The above example provides an existence proof for inflation in warped throat geometries, but the setup is too special to provide a good sense of the range of possibilities. Moreover, this analysis implicitly assumes that the physics inside the

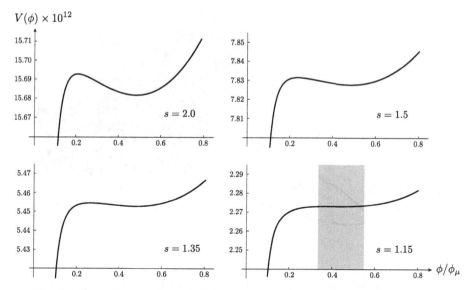

Fig. 5.3 Example scan through the parameter space of warped D3-brane inflation (figure adapted from [225]). The scan parameter s is the ratio of the antibrane energy to the F-term energy before uplifting. Successful inflation occurs in the gray shaded region.

throat decouples completely from the physics of the bulk, which as we stressed in Section 4.2 is rarely the case. Finally, we have modeled the warped throat region by a finite portion of a noncompact warped Calabi–Yau cone. This approximation fails where the finite throat is attached to the remainder of the compactification. In this section, we describe a more general analysis that addresses these deficiencies.

The essential idea is that all "compactification effects" – i.e. all information about moduli stabilization and supersymmetry breaking in the remainder of the compactification – can be expressed as non-normalizable perturbations of the noncompact solution [43, 660] (see Fig. 5.4),

$$\delta\Phi(r) = \delta\Phi(r_{\rm UV}) \left(\frac{r}{r_{\rm UV}}\right)^{\Delta-4} . \tag{5.47}$$

Here, $\delta\Phi$ is the deviation of some supergravity field Φ from its profile in the noncompact solution, $r_{\rm UV}$ is the radial location of the ultraviolet end of the throat, and the constant Δ is related to the scaling dimension $\Delta_{\mathcal{O}}$ of the operator that is dual, by the AdS/CFT correspondence, to the perturbation $\delta\Phi$.[6] By determining the spectrum of perturbations of the warped conifold, we will be able to identify the leading corrections to the D3-brane potential.

---

[6] When Φ is canonically normalized, we have $\Delta = \Delta_{\mathcal{O}}$, while for the field Φ_- given in (5.48) below, $\Delta = \Delta_{\mathcal{O}} + 4$.

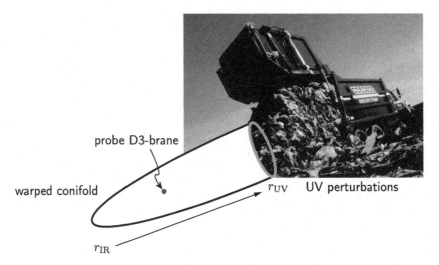

probe D3-brane

warped conifold

r_{UV} UV perturbations

r_{IR}

Fig. 5.4 Compactification induces complicated UV perturbations to the warped conifold solution. In the IR the lowest-dimension perturbations dominate in the D3-brane potential.

Locality in the internal space dictates that the effective action for a D-brane probe at some point is specified by the supergravity fields at that point. This suggests the following strategy: find the most general supergravity solution for a finite warped throat that asymptotes in the infrared to the Klebanov–Strassler solution, by classifying all possible perturbations $\delta\Phi$. Far from the ultraviolet region, the solution is given to good approximation by retaining the subset of modes with the lowest values of Δ, i.e. the modes dual to the most relevant perturbations of the dual field theory Lagrangian.

In a general six-dimensional cone, it would be challenging to determine the spectrum of dimensions Δ. However, the conifold is a cone over the coset space $T^{1,1}$, which is amenable to harmonic analysis via group theory techniques. Thus, by approximating a finite warped region as a portion of the warped conifold and using the spectroscopy of $T^{1,1}$, one can determine the leading non-normalizable modes. Correspondingly, one obtains the form of the leading contributions to the potential of a D3-brane in a KS throat. We now give a few details of this analysis.

10D Supergravity

To determine the D3-brane potential (5.29), we need solutions for the warp factor $e^{4A(r)}$ and the four-form potential $\alpha(r)$. In particular, we will be interested in the solution for the field

$$\Phi_- \equiv e^{4A} - \alpha \ . \tag{5.48}$$

Taking the metric ansatz

$$\mathrm{d}s^2 = e^{2A(y)} g_{\mu\nu}\mathrm{d}x^\mu\mathrm{d}x^\nu + e^{-2A(y)} g_{mn}\mathrm{d}y^m\mathrm{d}y^n \ , \tag{5.49}$$

where $g_{\mu\nu}$ is the metric of a maximally symmetric four-dimensional spacetime, the field equations of ten-dimensional type IIB supergravity imply the master equation[7]

$$\nabla^2 \Phi_- = R_4 + \frac{g_s}{96}|\Lambda|^2 + e^{-4A}|\nabla\Phi_-|^2 + \mathcal{S}_{\text{loc}} , \qquad (5.50)$$

where ∇^2 is the Laplacian constructed using the conifold metric (5.12), $\mathcal{S}_{\text{loc}}$ is a localized source due to anti-D3-branes, and

$$\Lambda \equiv \Phi_+ G_- + \Phi_- G_+ , \qquad (5.51)$$

with

$$G_\pm \equiv (\star_6 \pm i)G_3 \qquad \text{and} \qquad \Phi_\pm \equiv e^{4A} \pm \alpha . \qquad (5.52)$$

At the same time, the three-form flux must satisfy the equation of motion,

$$d\Lambda + \frac{i}{2}\frac{d\tau}{\text{Im}\tau} \wedge (\Lambda + \bar{\Lambda}) = 0 . \qquad (5.53)$$

The solutions to (5.50) can be organized as follows:

$$V(x, \Psi) = V_0 + V_C(x) + V_R(x) + V_B(x, \Psi) , \qquad (5.54)$$

where $x \equiv r/r_{\text{UV}}$ and Ψ stands for all five angular coordinates. We will describe each of the terms in (5.54) in turn.

Constant contributions

The constant V_0 represents possible contributions from distant sources of supersymmetry breaking – in the bulk of the compactification, or in other throats – that exert negligible forces on the D3-brane, and only contribute to the inflationary vacuum energy. This situation corresponds to maximal decoupling of the source of supersymmetry breaking from the D3-brane action: the two sectors communicate only through four-dimensional curvature. As explained in Section 4.2, complete decoupling of this sort is very rare. We have in fact made an artificial but convenient division, using V_0 to represent the sum of all[8] constant contributions to the potential, from diverse sources, each of which will in general also contribute non-constant terms in other categories described as follows.

Local sources

As before, $V_C(x)$ is the Coulomb potential sourced by $\mathcal{S}_{\text{local}}$,

$$V_C(x) = D_0 \left(1 - \frac{27}{64\pi^2} \frac{D_0}{T_3^2 r_{\text{UV}}^4} \frac{1}{x^4} \right) . \qquad (5.55)$$

---

[7] In comparison to (3.89), we have now allowed the four-dimensional curvature R_4 to be nonvanishing; compare (3.83) and (5.49).

[8] In fact, one constant contribution is grouped in V_C rather than in V_0; this is the vacuum energy contributed by the brane–antibrane pair, denoted D_0 in (5.55).

In the inflationary regime (far from the tip), the dependence on the D3-brane position x is a subdominant effect. This is a restatement of the fact that warping – captured by the smallness of D_0 in (5.55) – makes the Coulomb potential extremely flat.

The eta problem revisited

The Friedmann equation relates the Ricci curvature in four dimensions, $R_4 = 12H^2$, to the inflationary energy density, $V \approx V_0 + D_0$. Integrating (5.50), we find a curvature-induced mass term,

$$V_R(x) = \frac{1}{3}\mu^4 x^2 + \cdots , \qquad \text{where} \quad \mu^4 \equiv (V_0 + D_0)\frac{T_3 r_{\text{UV}}^2}{M_{\text{pl}}^2} . \tag{5.56}$$

This is how the curvature-coupling aspect of the eta problem arises in ten-dimensional supergravity.

Bulk contributions

Finally, we have a term that characterizes all possible contributions from stress–energy in the bulk of the compactification,

$$V_B(x, \Psi) = \mu^4 \sum_{LM} c_{LM}\, x^{\Delta(L)}\, f_{LM}(\Psi) , \tag{5.57}$$

where c_{LM} are constant coefficients, $L \equiv (j_1, j_2, R)$ and $M \equiv (m_1, m_2)$ label the $SU(2) \times SU(2) \times U(1)$ quantum numbers under the isometries of $T^{1,1}$, and the functions $f_{LM}(\Psi)$ are angular harmonics on $T^{1,1}$ (whose explicit forms can be found in [43]). The exponents $\Delta(L)$ have been computed in detail in [43], building on a spectroscopic analysis of perturbations on $AdS_5 \times T^{1,1}$ [661]. We briefly summarize the results. We split the bulk contributions into homogeneous solutions of the six-dimensional Laplace equation [328]

$$\nabla^2 \Phi_h = 0 , \tag{5.58}$$

and inhomogeneous contributions sourced by flux [660],

$$\nabla^2 \Phi_f = \frac{g_s}{96}|\Lambda|^2 . \tag{5.59}$$

The solutions are characterized by their scaling dimensions $\Delta(L)$. Solutions to (5.58) satisfy

$$\Delta_h(L) \equiv -2\sqrt{H(j_1, j_2, R) + 4} , \tag{5.60}$$

where

$$H(j_1, j_2, R) \equiv 6\left[j_1(j_1 + 1) + j_2(j_2 + 1) - \frac{1}{8}R^2\right] . \tag{5.61}$$

Taking into account selection rules [660, 661] for the angular quantum numbers, the first few scaling dimensions are

$$\Delta_h = \frac{3}{2}, \, 2, \, 3, \, \sqrt{28} - 2, \, \cdots . \tag{5.62}$$

The flux contributions in (5.59) lead to the following solutions:

$$\Delta_f(L) = \delta_i(L) + \delta_j(L) - 4, \tag{5.63}$$

where

$$\delta_1(L) \equiv -1 + \sqrt{H(j_1, j_2, R+2) + 4}, \tag{5.64}$$

$$\delta_2(L) \equiv \sqrt{H(j_1, j_2, R) + 4}, \tag{5.65}$$

$$\delta_3(L) \equiv 1 + \sqrt{H(j_1, j_2, R-2) + 4}. \tag{5.66}$$

Incorporating the selection rules, we find [553, 660]

$$\Delta_f = 1, \, 2, \, \sqrt{28} - 3, \, \frac{5}{2}, \, \sqrt{28} - \frac{5}{2}, \, \cdots . \tag{5.67}$$

The total bulk potential (5.57) therefore contains terms with the scaling dimensions

$$\begin{aligned}
\Delta &= \{\Delta_h, \Delta_f\} \\
&= 1, \, \frac{3}{2}, \, 2, \, \sqrt{28} - 3, \, \frac{5}{2}, \, \sqrt{28} - \frac{5}{2}, \, 3, \, \sqrt{28} - 2, \, \frac{7}{2}, \, \cdots .
\end{aligned} \tag{5.68}$$

A few remarks about the analysis leading to (5.68) are necessary. One should recognize that (5.59) is nonlinear in perturbations of the background; a linear treatment would capture only the homogeneous solutions solving (5.58), with dimensions given in (5.62), while the *leading* term at small r, corresponding to $\Delta_f = 1$ in (5.67), actually arises at quadratic order in perturbations of three-form flux. This is possible because the perturbations corresponding to various supergravity fields do not enter on an equal footing: some perturbations are allowed by the ISD background, and hence have order-unity perturbations $\delta\Phi$ at $r = r_{\mathrm{UV}}$, while other perturbations are forbidden[9] in the ISD solution, and have perturbations $\delta\Phi \sim e^{-aT}$ at $r = r_{\mathrm{UV}}$. These hierarchies can be captured by a careful spurion analysis [43, 553].

Notice that we again have a contribution scaling as $\phi^{3/2}$, just as in the four-dimensional analysis. This suggests that the basic phenomenology is again that

---

[9] The classically forbidden perturbations of three-form flux are sourced by the quantum effects that stabilize the Kähler moduli, i.e. by nonperturbative effects on D7-branes or by Euclidean D3-branes [43, 662].

of inflection point inflation, and a number of numerical investigations [250, 265, 554, 663, 664] have confirmed this expectation.

5.1.3 Multi-field dynamics

The effective theory describing an inflating D3-brane in a conifold region attached to a stabilized compactification has a natural mass scale: the inflationary Hubble parameter, H. Moreover, all continuous global symmetries are broken by the compactification. The general arguments reviewed in Section 2.1 then suggest that the six real scalar fields parameterizing the D3-brane position should have masses $m \sim \mathcal{O}(H)$. Inflation will not occur naturally, and some accidental cancelations among terms in the potential are required in order for one of the scalars to have a mass $m \ll H$. Once such a cancelation has occurred, it is quite unlikely that all five of the other fields will have masses $m \gg H$; a more probable outcome is that one or more of these fields will be light enough to evolve and fluctuate during inflation. Thus, the warped D3-brane inflation scenario generically gives rise to models of multi-field inflation, or more precisely of *quasi-single-field* inflation [509].

To understand the phenomenology of these models, neither a slow-roll approximation nor a single-field truncation is appropriate, and one must solve the equations of motion for the perturbations numerically, without making any approximations. The exact power spectra for more than 10^4 realizations from the ensemble of [250] were obtained in [265], with key results summarized in Section 5.1.6.

One intriguing finding of [265] is that the spectrum of scalar masses is predicted to good accuracy by a very simple matrix model inspired by [49], cf. Section 3.5. The model for the 6×6 mass matrix $\mathcal{M}$ takes the form[10]

$$\mathcal{M} = \begin{pmatrix} A\bar{A} + B\bar{B} & C \\ \bar{C} & \bar{A}A + \bar{B}B \end{pmatrix}, \qquad (5.69)$$

in terms of 3×3 complex symmetric matrices A, B, and C whose entries are assumed to be random complex numbers drawn from a Gaussian distribution. The eigenvalue spectrum of $\mathcal{M}$ agrees surprisingly well with the empirical mass spectrum found in [265], even though the methods of random matrix theory are formally applicable only to large matrices; evidently three is a sufficiently large number in the present context.

---

[10] The physical relevance of the matrix model (5.69) can be understood by comparing it with the Wigner+Wishart+Wishart model (3.157) of [49]. The positive-definite blocks $A\bar{A}$ and $B\bar{B}$ are consequences of spontaneously broken four-dimensional supersymmetry; in the limit of unbroken supersymmetry the mass matrix must be positive definite. The methods used in [43] to construct the ensemble of effective Lagrangians were inherently ten-dimensional, and made no direct connection with the structure of four-dimensional $\mathcal{N} = 1$ supersymmetry. The fact that the stability properties enjoined by four-dimensional supersymmetry nevertheless emerge after the intricate analysis described above is encouraging evidence that the entire construction is self-consistent.

The procedure described above led to an EFT for six real fields, the coordinates of the D3-brane. This captures completely general contributions to the action for these fields that stem from *heavy* degrees of freedom in the remainder of the compactification. However, the open string EFT constructed in this way can differ from the complete EFT that arises from dimensional reduction of all open and closed string fields. We have implicitly truncated the spectrum (see Section 4.1.4), assuming that the closed string moduli have masses $m \gg H$. For complex structure moduli and the axiodilaton, which acquire large supersymmetric masses from three-form flux, truncation is generally justified; but without special model building (cf. [665]) the typical mass scale of the Kähler moduli is $m \sim H$. As a result, the EFT may include a number of relatively light Kähler moduli, in addition to the six open string fields studied above, and there is comparatively little hope of determining the precise form of the potential for these closed string moduli.[11] However, in view of the successes of universality and random matrix theory in characterizing the six-field effective theory [250, 265], and bearing in mind that having more fields makes these methods more robust, we find it plausible that the statistical signatures of scenarios with dynamical Kähler moduli can be obtained in like manner.

5.1.4 Reheating

The reheating stage of warped D-brane inflation has been carefully examined in [564, 565, 567, 570, 572], revealing a complex cascade of energy from the inflaton to the visible sector and to invisible relics – see Fig. 5.5. To set the stage, we remark that a modular approach to reheating is very natural in this context; because the inflaton sector involves a D3-brane in a local geometry, it is reasonable to identify the warped throat where inflation occurs as one module, and to situate the Standard Model on D-branes in a different region of the geometry, either in another warped throat or in the unwarped bulk region. These model-building choices critically affect the success of reheating. We will not review all possibilities here, and will emphasize the interesting "two-throat" scenario in which the visible sector resides in a warped throat distinct from that in which inflation occurs. This choice affords much latitude in model-building, as well as leading to novel phenomenology for reheating. Moreover, if the Standard Model D-branes were inside the inflationary throat, any relic cosmic strings would quickly disintegrate through contact with these D-branes; in the bulk or in another throat, the D-branes are at a safe distance, and long-lived cosmic strings, with the associated interesting signatures, are possible.

The outline of the end of inflation, and of reheating, is as follows. Inflation occurs while the D3-brane passes through the vicinity of an inflection point

---

[11] As noted above, some effects of a single light Kähler modulus were considered in [552], and the response of the overall volume to the displacement of the D3-brane played a key role in the stability analysis of [225].

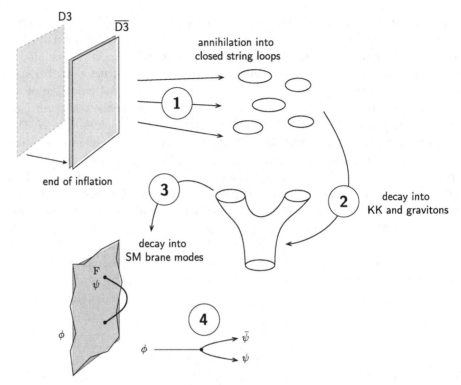

Fig. 5.5 The stages of reheating after warped D-brane inflation (figure adapted from [564]).

in its potential, and accelerated expansion ends once the D3-brane reaches a steeper portion of the potential. The D3-brane then falls rapidly toward the anti-D3-brane at the tip of the throat. Eventually, the separation of the brane–antibrane pair becomes small enough that a tachyon develops. The tachyonic instability causes the D3-brane pair to fragment, and to decay into highly excited, non-relativistic closed string modes [666–668], which quickly decay into massive Kaluza–Klein excitations of the supergravity fields (i.e. massless string modes) in the inflationary throat.

A few words about interactions in warped throats are necessary. The Kaluza–Klein modes of a warped throat have wavefunctions that peak exponentially in the infrared, and their mutual interactions are suppressed by the infrared scale $m_{\mathrm{IR}} \sim e^{A_{\mathrm{IR}}} M_{\mathrm{pl}} \ll M_{\mathrm{pl}}$, where $e^{A_{\mathrm{IR}}}$ is the warp factor at the tip of the throat. On the other hand, their couplings to Kaluza–Klein *zero modes*, including the graviton, are suppressed by M_{pl}. The warping creates a gravitational potential barrier that confines massive particles to the infrared region; access to other throats is via tunnelling[12] through the bulk of the compactification, which is

---

[12] See [669] for an analysis of energy transfer in warped reheating via induced motion of D-branes.

very slow compared to perturbative decays [564, 565, 567, 569, 670]. As a result, the characteristic timescales typically obey

$$\tau_{\text{therm}} \ll \tau_{\text{graviton}} \ll \tau_{\text{tunnel}} , \qquad (5.70)$$

where τ_{therm} denotes the thermalization time for Kaluza–Klein modes of the inflationary throat, τ_{graviton} is the timescale for decay to gravitons, and τ_{tunnel} is the tunnelling timescale.

Shortly after the decay of excited strings to excited Kaluza–Klein modes, the energy previously stored in the inflaton condensate is still largely confined to the inflationary throat. The success of reheating depends on channeling a sufficiently large fraction of this energy into Standard Model degrees of freedom, rather than into four-dimensional gravitons; long-lived relic particles protected by approximate isometries; or matter or radiation in other sectors. We now discuss these challenges in turn.

Overproduction of gravitons

KK modes decay to four-dimensional gravitons with a rate set by M_{pl}. If no other channels extract energy more quickly from the inflationary throat, the universe will be dominated by gravitational radiation, ruining Big Bang nucleosynthesis. Tunneling can transfer energy to other throats, but because generically $\tau_{\text{graviton}} \ll \tau_{\text{tunnel}}$ (cf. [565]), additional mechanisms may be needed to dilute the graviton abundance.

The heaviest Kaluza–Klein modes have the wavefunctions that reach farthest into the ultraviolet, and so have the largest tunneling probability. Efficient tunneling therefore requires that the lifetime τ_{KK} of the heaviest Kaluza–Klein modes obeys $\tau_{\text{KK}} \gtrsim \tau_{\text{tunnel}}$. This presents a further constraint on the parameters [565].

Kaluza–Klein relics from angular isometries

Suppose that one of the throats in the compactification enjoys approximate angular isometries, such as the $SU(2) \times SU(2)$ isometry of the Klebanov–Strassler solution. The associated angular momentum is approximately conserved, and Kaluza–Klein modes carrying this charge can only decay[13] through symmetry-violating interactions. Charged Kaluza–Klein modes produced during reheating will be long-lived Kaluza–Klein relics [564], and can readily overclose the universe.

To determine whether Kaluza–Klein relics decay sufficiently quickly for successful cosmology, one can examine the isometry-breaking perturbations sourced by the compactification [564, 572, 671], as explained in Section 5.1.2. A detailed analysis of this issue appeared in [572], where it was concluded that for

---

[13] Annihilation can in principle reduce the relic density [570], but only for problematically small values of the warp factor [572].

the relics to decay before nucleosynthesis, irrelevant[14] perturbations that break supersymmetry at a sufficiently high scale must be introduced. The conclusion obtained in [570] is more positive; the effects of warping and of the compact bulk were argued to lead to a much smaller relic abundance than that found in [564, 572].

Excitation of other sectors

Some scenarios consider an additional "intermediate" throat (for example, where supersymmetry is broken) whose warp factor falls between those of the inflationary and visible-sector throats. In that case, tunneling leads to Kaluza–Klein excitations of this throat. These are only very slowly depopulated by transfer to the visible-sector throat, presenting a serious problem [564]. More generally, if light moduli associated with other sectors become populated, these can come to dominate the energy density of the universe, with consequences discussed in e.g. [314–319].

Reheating above the local string scale

The reheating temperature can exceed the warped string scale $e^{A_{IR}}/\sqrt{\alpha'}$ in a throat that is much more strongly warped than the inflationary throat – for example, if the electroweak hierarchy is addressed by warping of the visible-sector throat. Reheating can then induce copious production of excited strings in the strongly warped throat [566]. Analyzing this process in detail remains challenging.

5.1.5 Fine-tuning

Considerable effort has been directed at finding mechanisms that can alleviate the fine-tuning of the potential in warped D-brane inflation – see [39, 672–678]. Here, we will outline a few of the leading approaches. The *DBI mechanism*, which turns a steep potential from a liability into an asset, will be discussed in Section 5.3. Discrete symmetries can be used to forbid problematic mass terms [673], and in some cases have been shown to be compatible with moduli stabilization [679]. Dynamical mechanisms have also been found: it was shown in [674, 676] that if N D3-branes become trapped in a metastable minimum of the potential in the throat, and sequentially tunnel out, the barrier diminishes with each tunneling event. For favorable parameter values, the potential for the final D3-brane is an inflationary inflection point.

The second fine-tuning problem of warped D-brane inflation – indeed, of most scenarios for inflation in string theory – is that rather special initial conditions are

---

[14] The restriction to irrelevant perturbations in [572] rests on the requirement that the background throat solution is a good approximation in the infrared. However, [43, 553] showed that certain *relevant* perturbations are necessarily present in Klebanov–Strassler regions of KKLT compactifications. Approximate no-scale symmetry ensures that these perturbations have exponentially small coefficients and do not destroy the throat. The effects of such perturbations on Kaluza–Klein relic decays have not been assessed.

required for successful inflation to occur. When the potential is approximately flat in a small fraction of the field space, and is steep elsewhere, then generic trajectories passing through the would-be inflationary region will *overshoot* the flat portion without initiating an inflationary phase, as emphasized long ago in [680]. The DBI kinetic term (see Section 5.3) has been argued to ameliorate the overshoot problem [681], though this conclusion was challenged by [682]. Negative spatial curvature resulting from tunneling entirely removes overshooting in certain classes of potentials, and reduces its severity in general [683, 684]. Finally, it was argued in [685] that the overshooting of an inflection point is ameliorated by particle production near points in field space where new species become light [686–689].

A different perspective on overshooting was given in [250], in which inflationary solutions were found by Monte Carlo sampling of the ensemble of potentials obtained in [43], followed by numerical solution of the six-field equations of motion. In this setting, the potential was fine-tuned by chance, rather than by hand. Surprisingly, the overshoot problem was absent: for each inflationary trajectory that was found for a given potential V and for some fixed initial conditions, an $\mathcal{O}(1)$ fraction of the space of possible initial positions likewise led to prolonged inflation. Thus, while inflation was not a generic outcome in the joint space of Lagrangians and initial conditions, for each successful Lagrangian that was found, inflation occurred for generic initial positions[15] of the D3-brane.

5.1.6 Phenomenology

The phenomenology of warped D-brane inflation has been the subject of intense investigation (e.g. [225, 250, 265, 552, 554, 657, 690–692]). In this section, we will summarize some of the main conclusions.[16] We will start with the simplified single-field treatment [225, 657] in which the angular degrees of freedom are integrated out using an adiabatic approximation. This is not always consistent, as the angular fields can have masses smaller than the inflationary Hubble scale, but it serves to develop intuition for the more complex multi-field dynamics studied in [250, 265, 552, 554, 695, 696]. We will then present the multi-field results of [250, 265, 554], which incorporate the complete potential derived in [43], and follow the full six-field[17] dynamics numerically, making no approximation.

Single-field expectations

In [225, 657], the six-dimensional field space was analyzed analytically. The potential was minimized in the angular directions and an effective potential for

---

[15] The initial kinetic energy of the D-brane was required to be somewhat smaller than the initial potential energy.

[16] We will emphasize the original scenario [42] in which a D3-brane falls toward the tip of a Klebanov–Strassler throat. Scenarios involving D-branes moving on the S^3 at the tip of the throat include [693, 694].

[17] As explained above, light Kähler moduli may also evolve during inflation.

the radial direction was determined. As expected, the potential for the effective radial coordinate has an inflection point. Near the inflection point, we can write the potential as

$$V(\phi) \approx V_0 \left[1 + \lambda_0 \frac{\phi}{M_{\text{pl}}} + \frac{1}{3!} \mu_0 \frac{\phi^3}{M_{\text{pl}}^3} + \cdots \right] , \qquad (5.71)$$

where the constants V_0, λ_0, and μ_0 can be related to microscopic parameters of the model [225]. A slow-roll analysis of this potential leads to the following predictions [225].

Power spectrum

The spectral index derived from (5.71) has the analytic solution [225, 227]

$$n_s - 1 \approx -\frac{4\pi}{N_{\text{tot}}} \cot\left(\pi \frac{N_\star}{N_{\text{tot}}}\right) \approx -\frac{4}{N_\star}\left(1 + \mathcal{O}\left(\frac{N_\star^2}{N_{\text{tot}}^2}\right)\right) , \qquad (5.72)$$

where $N_\star$ corresponds to the number of e-folds, between the horizon exit of the pivot scale and the end of inflation, and N_{tot} denotes the total number of e-folds, defined as

$$N_{\text{tot}} = \int_{-\infty}^{\infty} \frac{1}{\sqrt{2\epsilon}} \frac{\mathrm{d}\phi}{M_{\text{pl}}} = \pi \sqrt{\frac{2}{\lambda_0 \lambda_1}} . \qquad (5.73)$$

The number of e-folds from some initial vev ϕ until the end of inflation at ϕ_{end} is

$$N_e(\phi) = \int_{\phi_{\text{end}}}^{\phi} \frac{1}{\sqrt{2\epsilon}} \frac{\mathrm{d}\phi}{M_{\text{pl}}} = \frac{N_{\text{tot}}}{\pi} \arctan\left(\frac{\eta(\phi) N_{\text{tot}}}{2\pi}\right)\bigg|_{\phi_{\text{end}}}^{\phi} . \qquad (5.74)$$

For N_{tot} not much greater than $N_\star \approx 60$ the spectrum is strongly blue and the model is hence ruled out by observations (see Fig. 5.6). For $N_{\text{tot}} \approx 2N_\star$, the spectrum on CMB scales is exactly scale-invariant, while for $N_{\text{tot}} > 2N_\star$, the spectrum is red and asymptotes to the lower limit $n_s \to 1 - 4/N_\star \approx 0.93$ for $N_{\text{tot}} \gg 2N_\star$.

The running of the spectral index follows from (5.72),

$$\alpha_s = -\frac{4\pi^2}{N_{\text{tot}}^2} \sin^{-2}\left(\pi \frac{N_\star}{N_{\text{tot}}}\right) \approx -\frac{4}{N_\star^2}\left(1 + \mathcal{O}\left(\frac{N_\star^2}{N_{\text{tot}}^2}\right)\right) . \qquad (5.75)$$

Notice that both the tilt n_s and the running α_s are determined by N_{tot} alone (for fixed $N_\star$).

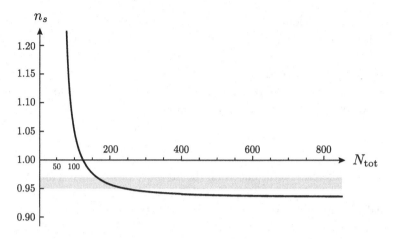

Fig. 5.6 Prediction for n_s as a function of the total number of e-folds. The gray band shows the range of n_s allowed by Planck.

Absence of tensors

Combining the geometric bound (5.28) on the inflaton field range[18] with the Lyth bound (2.105), we find [254]

$$r < \frac{4}{N} \times 0.01 \ll 0.01 \,. \tag{5.76}$$

This is a conservative bound that assumes that inflation occurs over the entire length of the throat, and that the bulk makes a negligible contribution to the total compactification volume. In all known examples, inflation is confined to a small part of the throat (the part where the potential is tuned to be flat) and the tensor amplitude is much smaller than the maximum allowed by the geometric bound. This implies that gravitational waves are unobservable in warped brane inflation.[19]

Multi-field effects

As explained in Section 5.1.3, a proper description of warped D-brane inflation involves all six D3-brane coordinates (and ultimately any light Kähler moduli). Here, we summarize a few key phenomenological results that emerge from an intensive Monte Carlo investigation of the dynamics and signatures of the six-field effective theory [250, 265, 554]. A few words about the methodology are necessary: in [250, 265, 554], scalar potentials were drawn at random from the ensemble described in Section 5.1.3, and the equations of motion were solved numerically

[18] See [697] for another discussion of geometric constraints in warped D-brane inflation, with implications for eternal inflation.

[19] If inflation is driven by the motion of a Dp-brane wrapping a $(p-3)$-cycle, the field range can be larger than that for a D3-brane [698, 699]. However, arranging for a nearly-flat potential is challenging, and backreaction of the moving brane can be important.

beginning from a random initial condition. The cosmological signatures were then evaluated in the subset of trials that led to $N_e \geq 60$ e-folds of inflation.

Inflationary probabilities

First, one can compute the relative probability $P(N_e)$ of N_e e-folds of inflation in the ensemble. In [250], it was shown that

$$P(N_e) = P(N_\star) \left(\frac{N_\star}{N_e} \right)^3 , \qquad (5.77)$$

where $N_\star$ is a reference value encoding the absolute probability. Thus, the probability of N_e e-folds of inflation is proportional to $1/N_e^3$. This result can be derived analytically in an inflection point model [250] (see [683] for earlier work in a slightly different model), and is consistent with the simpler analytic arguments of Section 5.1.2, which focused on the appearance of the term $\phi^{3/2}$ in an effective single-field description. Of course, the total number of e-folds is not in itself an observable, but whether or not $N_e \gg 60$ strongly influences the likelihood of observing relics of a pre-inflationary stage, such as traces of bubble collisions [700–704].

Violations of slow roll

A useful measure of violations of the slow-roll approximation is the ratio m_σ^2/H^2, where m_σ is the mass of fluctuations in the adiabatic direction (see Appendix C for a precise definition and further discussion). Slow-roll violations are strongly correlated with the total number of e-folds of inflation; realizations with $N_e \gg 100$ have $m_\sigma^2 \approx -0.1 H^2$ at the moment when the CMB exits the horizon, agreeing with the analytic result for single-field inflection point inflation. However, realizations with $N_e \approx 60$ have $m_\sigma^2 \approx H^2$, so that the slow-roll approximation is marginally valid at best. The effect on the spectrum is a slight increase in n_s compared with the slow-roll result [265], see Fig. 5.7.

Bending of the trajectory

Characteristic trajectories leading to prolonged inflation begin by spiraling in the angular directions, and then settle down to an inflection point that is approximately parallel to the radial direction. As a result, multi-field effects are generically significant during the first 5–10 e-folds of inflation, but are subsequently exponentially suppressed. See Appendix C for background on multi-field effects from bending trajectories.

Decay of entropic perturbations

Although all six open string scalars have masses that are very roughly $\mathcal{O}(H)$, the precise distribution of masses is important. By directly evaluating the Hessian matrix, or using the matrix model given in Section 5.1.2, one can show that in nearly all realizations the lightest field is tachyonic, the second-lightest field has

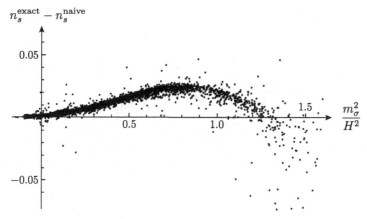

Fig. 5.7 Multi-field effects on the spectral index in warped D-brane inflation, versus the mass m_σ of the adiabatic fluctuation (figure adapted from [265]). The exact tilt, n_s^{exact}, is the result of a six-field numerical calculation making no slow-roll approximation, while the naive tilt, n_s^{naive}, follows from simply evaluating (2.44) at horizon exit.

$m^2 \sim H^2$, and the four remaining fields have $m^2 > \frac{9}{4}H^2$. Thus, there is at most one instability, and only two fields fluctuate. Moreover, the five entropic perturbations decay exponentially after exiting the horizon [265], i.e. an "adiabatic limit" [705] is reached. This is important, for if one or more entropic perturbations were to persist until the time of reheating, predicting the scalar power spectrum would become extremely difficult [705]. A detailed treatment of related multi-field effects at the end of D-brane inflation appears in [695].

Scalar power spectrum
In models producing $N_e \lesssim 60$–70 e-folds in total, multi-field effects dictate the observable anisotropies, while in models yielding $N_e \gg 70$ e-folds, a single-field approximation is valid and the analytic treatment given above applies without modification. In light of (5.77), models with multi-field effects are much more common than approximately single-field models, within the class of all realizations yielding $N_e \geq 60$ e-folds. However – see Fig. 5.6 and the related discussion – the scalar power spectrum computed *in the single-field approximation* is unacceptably blue in models producing $N_e \lesssim 120$ e-folds. Multi-field effects quite generally shift the spectrum toward the red, i.e. $n_s^{\text{exact}} - n_s^{\text{naive}} < 0$, but the magnitude of the effect is only occasionally large enough to produce models consistent with observations, which fall in the gray band in Fig. 5.8.

Tensor amplitude
The inflationary inflection points arising in the ensemble are extremely small in Planck units; for the parameters explored, $r \lesssim 10^{-12}$, which is far below the

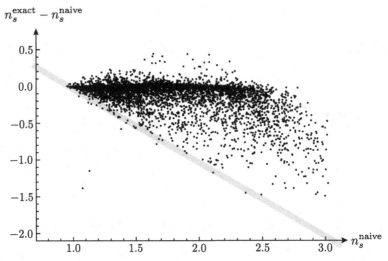

Fig. 5.8 The spectral index in realizations of warped D-brane inflation with significant multi-field effects (figure adapted from [265]). The gray band shows the region allowed at 2σ by WMAP7; the Planck constraints are slightly more stringent.

upper limit allowed by the Lyth bound (2.105) combined with the geometric bound (5.28).

Non-Gaussianity

Although conversion of entropic perturbations to curvature perturbations is commonplace in models yielding $N_e \lesssim 60\text{--}70$ e-folds, this does not automatically lead to large non-Gaussianity, because the cubic couplings in the D3-brane potential can be quite small [265]. More importantly, in the subset of models allowed by constraints on the tilt, multi-field effects, including non-Gaussianity, are extremely rare.[20]

Cosmic strings

Cosmic superstrings are one of the most striking signatures of D-brane inflation. Following [589], we recall the conditions for cosmic strings to be cosmologically relevant: the strings must be produced after inflation, remain stable over cosmological times, and be observable without already being excluded. Finally, one may also hope that the strings have distinctive signatures revealing their origin in string theory. All four conditions can be met in warped D-brane inflation, as we now explain.

Condensation of the D3-brane/anti-D3-brane tachyon at the end of inflation automatically produces a population of cosmic F-strings and D-strings, as well as

---

[20] Sharp features in the radial profile of the warp factor were argued in [706] to produce observable signatures in the power spectrum and bispectrum.

the more general (p, q) string bound states. Whether these strings are stable depends on whether there are D-branes in the warped throat where inflation occurs – see [620] for a detailed treatment. First of all, (p, q) strings (including the $(1, 0)$ F-string and $(0, 1)$ D-string) are not BPS in this setting; the two-forms $B_{\mu\nu}$ and $C_{\mu\nu}$ whose charges the strings carry are projected out by the orientifold action [620]. Correspondingly, a string can break apart by coming into contact with its orientifold image. However, in the generic situation in which there are no orientifold fixed planes within the throat itself, a string has to fluctuate out of the throat to meet its image in the image throat. This is an exponentially slow process, as the potential due to the warp factor confines the strings to the bottom of their respective throats, and for practical purposes breakage via the orientifold image can be ignored [620]. A more significant risk comes from D3-branes or anti-D3-branes in the inflationary throat, which could serve as the substrate for the Standard Model [707] or as a source of supersymmetry-breaking energy [443]. If any D3-branes or anti-D3-branes are present, cosmic strings fragment immediately and are cosmologically irrelevant. If D7-branes are present but D3-branes and anti-D3-branes are not, the D-string remains stable [620].

The spectrum of tensions of (p, q) strings in a warped throat was obtained in [708]:

$$T_{(p,q)} \approx \frac{e^{2A_{\mathrm{IR}}}}{2\pi\alpha'} \sqrt{\frac{q^2}{g_s^2} + \left(\frac{bM}{\pi}\right)^2 \sin^2\left(\frac{\pi(p - qC_0)}{M}\right)}, \qquad (5.78)$$

where $e^{A_{\mathrm{IR}}}$ is the warp factor at the tip of the throat, M is the flux on the A-cycle, and $b \approx 0.93$ is a constant arising in the Klebanov–Strassler solution. This result is primarily governed by the warp factor, which can be exponentially small. Hence, if the warp factor were a free parameter, it would be easy to ensure that the cosmic string tension was low enough to satisfy any conceivable observational bound. However, the warp factor in the inflationary throat determines the scale of the inflaton potential, and is therefore constrained by the normalization of the scalar fluctuations. Recalling from (2.42) that the amplitude of the scalar power spectrum involves both V and ϵ, we conclude that once ϵ is known, the warp factor and hence the cosmic string tension are predicted. The distribution of values of ϵ in a simple model for the potential was studied in [626]. Tensions that satisfy present constraints but can be detected in the coming generation of observations are achievable, but the associated fine-tuning has not yet been quantified completely.

5.2 Inflating with unwarped branes

We have just seen, in Section 5.1, that D3-branes in warped throat regions of type IIB flux compactifications lead to a class of highly computable inflationary models with rich phenomenology. At the same time, D-branes in more general geometries – in which warping may be present but is not a dominant effect –

provide an array of interesting models with some theoretical advantages. In this section, we will discuss a few examples of inflation driven by branes in unwarped regions.

5.2.1 D3/D7 inflation

An interesting and uniquely explicit scenario for D-brane inflation in an unwarped compactification is the D3/D7 model [628–632, 709]. This model has close parallels with the warped brane inflation scenario detailed in the previous section, so we will be brief, emphasizing the distinctive features of the D3/D7 construction.

The background geometry is a compactification of type IIB string theory on the orientifold $K3 \times T^2/\mathbb{Z}_2$. At each of the four fixed points there are four D7-branes atop an O7-plane, all of which wrap the $K3$ manifold. This configuration corresponds to M-theory on $K3 \times T^4/\mathbb{Z}_2$, to the $T^4/\mathbb{Z}_2$ being the orbifold limit of $K3$. Displacing the D7-branes from the orientifold planes leads to a geometry that lifts to M-theory on $K3 \times K3$ (see [710] for an analysis of moduli stabilization in this regime).

Now we add a spacetime-filling D3-brane, which sits at a point in the internal space, and in particular on $T^2/\mathbb{Z}_2$ (see Fig. 5.9). The position of the D3-brane on $T^2/\mathbb{Z}_2$, relative to the stack of D7-branes, was proposed to be the inflaton [628]. The inflaton sector therefore consists of two real fields describing the D3-brane location on the torus.

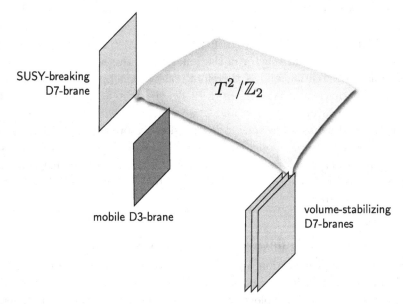

Fig. 5.9 Schematic of D3/D7 inflation (figure adapted from [632]).

A spacetime-filling D3-brane in the $K3 \times T^2/\mathbb{Z}_2$ orientifold actually preserves $\mathcal{N} = 2$ supersymmetry in four dimensions, in the absence of flux, so there is no potential for D3-brane motion, and the would-be inflaton is perfectly massless. However, introducing two-form flux $\mathcal{F}_2$ in the D7-brane worldvolume modifies the situation; when the flux is not self-dual in the worldvolume (i.e. $\star_4 \mathcal{F}_2 \neq \mathcal{F}_2$, with $\star_4$ the Hodge star in the four compact directions of the D7-brane worldvolume), then supersymmetry is broken, and the D3-brane feels a force. As explained in [628], the worldvolume flux corresponds to a field-dependent Fayet–Iliopoulos term[21] ξ, so the D3/D7 model described so far is a model of D-term inflation.[22] Specifically, D-term supersymmetry breaking by D7-brane fluxes introduces mass splittings in the supermultiplets of strings stretched between the D3-brane and the D7-brane, and integrating out these fields leads to the Coleman–Weinberg potential (cf. [632] for an updated discussion),

$$V_D(\phi) = \frac{g^2 \xi^2}{2} \left(1 + \frac{g^2}{16\pi^2} U(x) \right) , \qquad (5.79)$$

where we have defined $x \equiv \phi/\sqrt{\xi}$, and

$$U(x) \equiv (x^2 + 1)^2 \ln(x^2 + 1) + (x^2 - 1)^2 \ln(x^2 - 1) - 4x^4 \ln(x) - 4\ln(2) , \quad (5.80)$$

with g the coupling of the $U(1)$ gauge field.

At this stage we have specified the geometric data of a non-supersymmetric compactification, but without further ingredients this configuration will be unstable to decompactification: the $\mathcal{N} = 2$ supersymmetric compactification has unfixed moduli, and supersymmetry-breaking positive energy from the D3-brane potential creates an instability. Fixing all the Kähler moduli of the supersymmetric compactification may be achievable [710], but we will first focus on a scenario in which a single overall volume modulus is stabilized by gaugino condensation in the gauge theory living on the D7-branes.

As explained in detail in Section 5.1.2, mixing of D3-brane position moduli with Kähler moduli, via the DeWolfe–Giddings Kähler potential (3.100), implies that stabilization of the volume generically leads to stabilization of the D3-brane position.[23] We will now examine this crucial point more closely and determine whether there are non-generic exceptions.

The Kähler potential takes the form

$$K(Z^I, \bar{Z}^I) = -3\ln \left[T + \bar{T} - \gamma k(z_\alpha, \bar{z}_\alpha) \right] , \qquad (5.81)$$

---

[21] See [711–713] for analyses of consistency conditions for Fayet–Iliopoulos terms in supergravity, and [714] for the implications for cosmic strings.

[22] An important alternative means of breaking supersymmetry is the addition of an anti-D3-brane in a warped region – see [709] for a comprehensive discussion of the D3/D7 model with antibrane supersymmetry breaking.

[23] A very different perspective on this fact was recently given in [715].

with $k(z_\alpha, \bar{z}_\alpha)$ the Kähler potential for the metric on the internal space. Let us restrict attention to a single complex field $z_1 \equiv x + iy$, corresponding to the D3-brane position on $T^2/\mathbb{Z}_2$. Suppose for the moment that the superpotential W is independent of z_1, so that the only dependence of the potential energy on x and y comes through the appearance of these fields in K. If $k(x, y)$ depends nontrivially on both x and y, then both real fields will have nontrivial masses in the stabilized vacuum, rather than corresponding to flat directions.

An influential proposal is to invoke a shift symmetry in the Kähler potential [539], so that k – and hence K – is independent of one of the fields. For example, if

$$k = \frac{1}{2}(z_1 + \bar{z}_1)^2 = x^2 \,, \tag{5.82}$$

then y corresponds to a flat direction of the F-term potential for the moduli. (The dependence of the D-term potential (5.79) on z_1 is mild enough to be suitable for inflation.) Although this approach appears reasonable in supergravity, in a string construction one is not free to write down an effective action with a desired form: the action follows from dimensional reduction of a specified configuration. Moreover, some shift symmetries do not admit ultraviolet completions. It is therefore essential to determine whether the D3/D7 model actually enjoys a shift symmetry that allows Kähler modulus stabilization to coexist with a flat direction for D3-brane motion.

It was shown in [629, 631] that the tree-level Kähler potential is indeed shift-symmetric, so that before accounting for additional terms in the effective action, the D3-brane potential takes the form of a nearly flat trough oriented along the symmetry direction. However, the nonperturbative superpotential, which is critical in the stabilization of the volume, necessarily depends on the D3-brane position, contrary to our assumption above. The one-loop correction to the gauge kinetic function for the D7-brane gauge theory was computed explicitly in [347], and was found to depend on the D3-brane position, so that the gaugino condensate superpotential likewise depends on the D3-brane location. For D7-branes with gauge group $SU(N_c)$ at position $z_{D7} = \mu$ in $T^2/\mathbb{Z}_2$, one finds [347]

$$W = W_0 + \left[\vartheta_1\big(\sqrt{2\pi}(z_1 + \mu), \zeta\big)\, \vartheta_1\big(\sqrt{2\pi}(z_1 - \mu), \zeta\big) \right]^{-1/N_c} e^{-2\pi T/N_c} \,, \tag{5.83}$$

where ϑ_1 is a Jacobi theta function, and ζ is the complex structure of the T^2 [347]. It was then shown in [538, 716] that the appearance of the D3-brane position in the nonperturbative superpotential (5.83) spoils the shift symmetry and prevents inflation from occurring naturally.

The gauge theory description of this effect is simple and instructive: strings stretching from the D3-brane to the D7-branes ("3-7 strings") correspond to flavors in the condensing theory, and their masses depend on the D3-brane's separation from the D7-branes. The dependence of the low-energy condensate on the mass of the flavors implies that the superpotential depends on the D3-brane position. As a simple example, consider $\mathcal{N} = 1$ supersymmetric Yang–Mills theory with gauge group $SU(N_c)$ (for $N_c > 2$) and a single flavor Q with mass parameter m. The gaugino condensate superpotential below the scale m, which results from integrating out Q, takes the form

$$W = \Lambda^{3-1/N_c} m^{1/N_c} , \qquad (5.84)$$

where Λ is the dynamical scale of the high-energy theory. In a string theory realization of this gauge theory, $m = m_{37}$ is the mass of the stretched strings, which depends on the D3-brane position ϕ; for sufficiently small separations, $m \propto \phi$. Thus, $W \propto \phi^{1/N_c}$, and the gaugino condensate superpotential depends on the D3-brane position.

One of the virtues attributed to models of D-term inflation is the absence of inflaton mass terms from Kähler potential couplings. The D3/D7 model is arguably the best-studied model of D-term inflation in string theory, and an important lesson from this model is that moduli stabilization by superpotential terms introduces F-term energy, which itself may depend on the inflaton, even if the "intended" inflaton potential comes from a D-term. In other words, a model of D-term inflation in a compactification stabilized by superpotential terms for the moduli is not purely a D-term scenario, and the moduli sector introduces masses in the inflaton sector.

Although the global form of the moduli potential is readily computed from (5.83), in practice one can expand in ϕ; the leading contribution to the inflaton potential from moduli stabilization is an inflaton mass term. The total potential $V = V_F + V_D$ then takes the form

$$V(\phi) = V_D(\phi) - \frac{m^2}{2}\phi^2 + \frac{\lambda}{4}\phi^4 , \qquad (5.85)$$

where $V_D(\phi)$ is given in (5.79). The resulting phenomenology is discussed in Section 5.2.4.

5.2.2 Fluxbrane inflation

An influential idea for achieving inflation with D-branes is to consider a pair of branes that are separated in the compact space, and are almost parallel, but misaligned by a small relative angle θ [717–719].[24] The small angle leads to controllably small breaking of supersymmetry, resulting in a force that draws the branes together, at which point they merge and reheat the universe. Brane–antibrane inflation [534, 645] can be viewed as a special case in which the branes are precisely antiparallel.

---

[24] A T-dual configuration, in which the inflationary coordinate is a Wilson line, has been investigated in [720] (see also [721]).

Just as in the cases of brane–antibrane inflation and the D3/D7 model, the approach taken in the literature was to begin by analyzing the interaction potential V_{int} of the misaligned D-brane pair, assuming that the closed string moduli were stabilized by some mechanism, and then later to attempt to incorporate (or minimize) the effects of the moduli potential V_F, which we may take to be an F-term potential. This approach was a pragmatic one, because methods for computing the interaction potential were developed long before techniques for computing the moduli potential. However, from the present perspective we must emphasize that the division into interaction potential and moduli potential is somewhat arbitrary, and is often very misleading; the essence of the eta problem described in Section 4.2 is that the moduli potential *is not subleading* as a contribution to the inflationary dynamics. Bearing this in mind, we will nevertheless briefly describe the properties of the interaction potential V_{int}.

The interaction potential for a brane–antibrane pair in an unwarped compact space is generally too steep for successful inflation [534], except possibly for certain antipodal configurations (see e.g. [614]). The proposal of [717] was that weak supersymmetry breaking by a small angle $\theta \ll 1$ would diminish the Coulomb force to the extent that V_{int} could drive slow-roll inflation. Compactness of the internal space introduces a crucial difficulty: the potential between branes with $\theta \ll 1$ is indeed small (compared to the vacuum energy) if the computation is performed with the internal directions taken to be noncompact, but the result is quite different for compact internal dimensions [42]. As explained in Section 4.2, the effect of compactification is to make the interaction potential for the branes be of the same order as the vacuum energy, ruining the favorable hierarchy obtained by taking the branes to be noncompact.

More recently, the relative position of two D7-branes has been proposed as an inflationary direction [633, 634]. Consider type IIB string theory compactified on an O3/O7 orientifold of a Calabi–Yau threefold X_6. Suppose that there is a continuous family Σ_4 of four-cycles in X_6, on any representative of which a D7-brane can be wrapped. Wrap two D7-branes a, b on distinct representatives in Σ_4; the D7-branes can then be separated to some extent, although they generally intersect along a two-cycle (see Fig. 5.10). If gauge flux $\mathcal{F}$ is introduced on the shared two-cycle, the D7-branes feel a force that tends to make them coincide.[25] Because a key part of the inflaton potential arises from worldvolume flux, this scenario is called *fluxbrane inflation*.

The potential for the canonically normalized D7-brane coordinate takes the form

$$V(\phi) = V_D(\phi) + V_F(\phi) , \tag{5.86}$$

---

[25] This setup is T-dual to a configuration of branes at angles; to see this, take the compactification to be a torus and T-dualize along a circle in the two-cycle threaded by the flux $\mathcal{F}$.

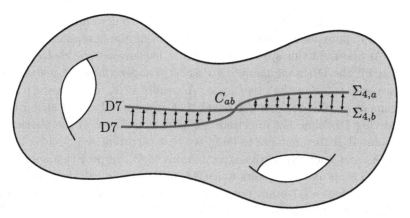

Fig. 5.10 In fluxbrane inflation [633, 634], the inflaton coordinate is the effective separation of a pair of intersecting D7-branes (figure adapted from [633]).

where[26]

$$V_D(\phi) = V_0\left(1 + \alpha \ln(\phi/\phi_0)\right) , \qquad (5.87)$$

and V_0, α, ϕ_0 are constants. The D-term potential $V_D(\phi)$ is a consequence of the supersymmetry-breaking flux, and can be thought of as a Coulomb potential, while the F-term potential $V_F(\phi)$ results from moduli stabilization and has not been computed in detail.

Although the similarities to D3/D7 inflation are apparent, several distinctive features of fluxbrane inflation were identified in [633]. First of all, the D7/D7 interaction potential due to flux can be made flat enough for inflation, evading the well-known difficulty [534] of achieving a sufficiently flat Coulomb interaction within an unwarped compact space. Of course, as in warped D-brane inflation [42], the Coulomb potential is only a small part of the final story; one must compute the moduli-stabilizing potential $V_F(\phi)$ – which generically contributes $\Delta\eta \gtrsim 1$ – and determine whether it spoils inflation. Moduli stabilization in the type IIB orientifold (and F-theory) context is comparatively well understood, so detailed study of the moduli potential is possible. Careful investigations of this question appear in [633, 634], but the issue is not yet settled.

The second notable feature of fluxbrane inflation is that the range of the canonically normalized inflaton corresponding to a wrapped D7-brane can be much larger than in D3-brane inflation, as first recognized in [698, 699]. Moreover, the D-term potential is readily adjusted to avoid cosmic string constraints, by arranging for a hierarchy in the stabilized values of the Kähler moduli [633].

---

[26] In an alternative parameter regime for fluxbrane inflation, the dominant nonconstant term in the potential is sinusoidal [261], instead of logarithmic as in (5.87).

A fundamental challenge in fluxbrane inflation is to determine whether there might be an approximate shift symmetry protecting the potential for D7-brane motion. At first sight this appears implausible, because *generic* choices of three-form flux lift the D7-brane moduli, giving large supersymmetric masses to the D7-brane scalars.[27] On the other hand, Appendix E of [633] gives a plausibility argument for the existence of fluxes that stabilize all closed string moduli while leaving D7-brane flat directions, at least at the level of the classical flux superpotential. It then remains to be shown that perturbative (g_s and α') corrections to K, and nonperturbative contributions to W, respect this approximate symmetry. This is an open question; Euclidean D3-branes carrying worldvolume flux may introduce a D7-brane potential [634] (cf. [722]), and there are intricate interactions with the stabilization of other moduli, including the dilaton and the Kähler moduli. Determining the moduli potential in detail will be a crucial step toward obtaining the phenomenological signatures of the model.

5.2.3 M5-brane inflation

Although until now we have discussed D3-branes in type IIB string theory, closely related proposals exist in the context of M-theory compactifications on $S^1/\mathbb{Z}_2 \times X_6$, with X_6 a Calabi–Yau threefold. In this case, it was proposed that the inflaton could correspond to the position of one or more M5-branes along the interval, with inflation ending as the M5-branes collide with and dissolve into the "end-of-the-world" brane. The single-M5-brane scenario was proposed in [635], while a multiple-brane model appeared in [636] (see also [723]).[28]

A fundamental difficulty in realizing inflation along these lines is the absence of parametrically controlled constructions of de Sitter vacua in heterotic string theory, at weak or at strong coupling. Extensive efforts building on sophisticated studies of heterotic compactifications have led to scenarios for stabilization of the geometric[29] moduli in anti-de Sitter vacua – see for example [725]. Even so, de Sitter constructions remain challenging (but see the recent work [726]). A general obstacle to parametric control, in both supersymmetric and non-supersymmetric vacua, is that the quantized three-form flux H_3 in heterotic string theory is *real*, and hence cannot be adjusted in the same manner as the complex flux G_3 of type IIB string theory.[30]

---

[27] D3-branes, in contrast, do enjoy a moduli space in the leading order no-scale compactifications of [305], but at this same order the Kähler moduli are unstabilized. The challenge described in Section 5.1 is that the nonperturbative effects that lift the Kähler moduli inevitably spoil the flatness of the D3-brane potential.

[28] It was argued in [724] that the tensor-to-scalar ratio r can be large in multi-M5-brane inflation.

[29] Vector bundle moduli are not necessarily stabilized, but are sometimes assumed to be absent.

[30] The difficulties inherent in constructing parametrically controlled heterotic vacua with H_3 were appreciated many years ago [342, 343], and have been only partially overcome: see [726, 727].

Furthermore, the eta problem appears in a predictable manner in models with moving M5-branes. The Kähler potential for the volume modulus S of the Calabi–Yau, the length modulus T of the interval, and the position Y of a single M5-brane includes the term [636]

$$K \supset -\ln\left(S + \bar{S} - \frac{(Y + \bar{Y})^2}{T + \bar{T}}\right) . \tag{5.88}$$

This is evidently analogous to the DeWolfe–Giddings Kähler potential for a D3-brane, and leads to a mass term for the M5-brane position in the presence of effects stabilizing S. More generally, it is difficult to arrange for an M5-brane to have a potential suitable for inflation while simultaneously stabilizing the geometric moduli. It was suggested in [636] that the M5-brane potential would be satisfactory if the effects of gaugino condensation and H_3 flux could be neglected during inflation, becoming relevant only later in cosmic history, but it is not clear that such a scenario, if realizable, can avoid destabilization of the geometric moduli.

5.2.4 Phenomenology

Once moduli stabilization is properly incorporated, inflation in the D3/D7 model remains possible, but necessarily involves fine-tuning. Equipped with the global form of the nonperturbative superpotential thanks to the worldsheet calculation of [347], the authors of [632, 709] systematically analyzed the potential in search of inflationary regions. Two qualitatively different scenarios were found.

Saddle-point inflation

If the gaugino condensate responsible for the stabilization of the Kähler moduli is assumed to arise exclusively on a stack of D7-branes near a single fixed point of $T^2/\mathbb{Z}_2$, then after fine-tuning of the parameters, the potential for a D3-brane at an approximately antipodal location in the torus can develop an unstable saddle point. (For this scenario, it is essential that the primary source of supersymmetry breaking is an anti-D3-brane in a warped region.) The resulting model has $r \ll 1$ and $n_s \lesssim 0.95$. The characteristic redness of the spectrum in saddle-point models of this form is discussed in [728].

Inflection point inflation

If the dominant force on the D3-brane comes from interactions with supersymmetry-breaking fluxes on a D7-brane, as in [628], then the potential can be fine-tuned to have an inflationary inflection point, with phenomenology broadly similar to that described in Section 5.1.6. The potential takes the form (5.85), incorporating a Coleman–Weinberg term, as well as quadratic and quartic terms from moduli stabilization. When the quartic terms are significant, the fine-tuning for inflation is extreme, and was argued in [709] to be at the level of one part in 10^6. On the other hand, [632] exhibit parameter ranges in which the moduli contribution is approximately quadratic and the fine-tuning is milder.

It is worth remarking that the kinematical field range $\Delta\phi$ of the canonically normalized inflaton in D3/D7 inflation can be super-Planckian, $\Delta\phi > M_{\rm pl}$, if the $T^2/\mathbb{Z}_2$ is highly anisotropic [632]. For a rectangular torus with side lengths L_1 and L_2, we have

$$M_{\rm pl} \propto \sqrt{{\rm Vol}(K3)L_1L_2} \, , \tag{5.89}$$

while the field range along the side of length L_1 has the parametric dependence

$$\Delta\phi_1 \propto \sqrt{L_1/L_2} \, . \tag{5.90}$$

In [632], it was argued that one can take L_1/L_2 to be large enough so that $\Delta\phi_1 > M_{\rm pl}$, without compromising computability. This fact is consistent with the general arguments made in Section 4.3, where anisotropy of the compactification was the only plausible route to a parametrically controlled super-Planckian field range for a D-brane. Even so, this observation has not led to a full-fledged inflationary scenario with large tensor-to-scalar ratio r, because of the difficulty of arranging that the potential remains flat over a distance $\Delta\phi > M_{\rm pl}$.

Cosmic strings in models of D-term inflation have been the subject of much discussion – see [630, 632] for summaries with original references. In early versions of the D3/D7 model, long-lived cosmic strings were thought to be present in problematic numbers, in conflict with upper limits from measurements of the CMB temperature power spectrum. This problem has been avoided in two ways. First, in extensions of the scenario with additional charged fields [630, 729], the vacuum manifold is simply connected and the resulting cosmic strings are "semilocal," i.e. non-topological, leading to weaker constraints. Second, it was recognized in [632] that contributions to the inflaton potential from moduli stabilization introduce an additional parameter, allowing separation of the amplitude of inflationary density perturbations from the density perturbations due to cosmic strings.

The signatures of fluxbrane inflation are broadly similar to those of D3/D7 inflation, although as noted above, the observational constraints from cosmic strings are readily satisfied through hierarchical stabilization of the Kähler moduli. Detailed statements, for example about the spectral index, will have to await a complete computation of the moduli potential.

The phenomenology of M5-brane inflation is more difficult to characterize, because it depends on presently unknown properties of the moduli potential, as well as on the intricacies of the M5-brane collision with the end-of-the-world brane. We will leave the signatures of this scenario as a question for the future.

5.3 Inflating with relativistic branes

As we have learned in the previous two sections, inflation in systems driven by slowly moving D-branes suffers from the eta problem. In [39], Silverstein and Tong observed that this problem may be alleviated if the D-branes move relativistically. The model relies on the nontrivial structure of the DBI action for the D-branes and is called *DBI inflation*. In this section, we will explore this idea.[31] In Section 5.3.1, we present the DBI mechanism, followed, in Sections 5.3.2 and 5.3.3, by explanations of the symmetries that control quantum corrections. Microphysical constraints arising in the embedding in string theory are presented in Section 5.3.4, and the observational signatures of DBI inflation are summarized in Section 5.3.5.

5.3.1 Dirac–Born–Infeld inflation

We begin with the same system as in Section 5.1: a spacetime-filling D3-brane probing $AdS_5 \times S^5$,

$$ds^2 = \left(\frac{r}{L}\right)^2 \eta_{\mu\nu} dx^\mu dx^\nu + \left(\frac{L}{r}\right)^2 \left(dr^2 + r^2 d\Omega_{S^5}^2\right) , \tag{5.91}$$

where $L^4 = 4\pi g_s N(\alpha')^2$, with N the total D3-brane charge of the background. To arrive at a simple model with four-dimensional gravity, we cut off the AdS space in the infrared and ultraviolet, taking $r_{\rm IR} < r < r_{\rm UV}$ (as in Randall–Sundrum scenarios [730]). The D3-brane Lagrangian takes the form

$$\mathcal{L} = -\frac{\phi^4}{\lambda} \left(\sqrt{1 + \frac{\lambda}{\phi^4}(\partial\phi)^2} - 1\right) - V(\phi) , \tag{5.92}$$

where $\phi \equiv \sqrt{T_3}\, r$ and[32]

$$\lambda \equiv T_3 L^4 = \frac{1}{2\pi^2} N . \tag{5.93}$$

The potential $V(\phi)$ in (5.92) requires some explanation. In $AdS_5 \times S^5$ a probe D3-brane feels no force, but physical effects associated with the infrared and ultraviolet deformations of the spacetime typically generate a potential for D3-brane motion. Indeed, as we have seen in Section 5.1, the cutoff AdS geometry is most naturally viewed as an approximation to a region of a finite Klebanov–Strassler throat[33] that is attached to a flux compactification – the IR cutoff at $r_{\rm IR}$ corresponds to the tip of the throat, while the UV cutoff at $r_{\rm UV}$ corresponds to the remainder of the compactification. Supersymmetry breaking in the infrared

---

[31] This section is based mostly on [39, 41].

[32] The constant λ is proportional, but not equal, to the standard 't Hooft coupling in $AdS_5 \times S^5$, $\lambda_t \equiv g_{\rm YM}^2 N = 4\pi g_s N$.

[33] Such a region actually corresponds to a section of $AdS_5 \times T^{1,1}$, up to logarithmic corrections, but the angular manifold is immaterial at present.

(e.g. by an anti-D3-brane), as well as supersymmetry breaking and moduli stabilization in the ultraviolet then lead to a potential for the D3-brane position, as explained in detail in Section 5.1.2. Any such effect can be captured by a suitable change to the supergravity background, cf. Eq. (5.29).

The action (5.92) is adapted straightforwardly to more general warped backgrounds,

$$\mathcal{L} = -T(\phi) \left(\sqrt{1 + \frac{(\partial \phi)^2}{T(\phi)}} - 1 \right) - V(\phi) \, , \qquad (5.94)$$

where we have introduced the warped tension of the brane, $T(\phi) \equiv T_3 e^{4A(\phi)}$. The precise functional forms of the warp factor $e^{4A(\phi)}$ and the potential $V(\phi)$ determine the details of the phenomenology of DBI inflation (see [690, 731]). As a generalization of the AdS background (5.91), we will use Calabi–Yau cones that approximate finite warped throat regions attached to type IIB flux compactifications. Although the KS throat provides a rare example of a warped throat that is smooth in the infrared, here we will also consider more general warped Calabi–Yau cones, with ten-dimensional line element

$$ds^2 = e^{2A(r)} \eta_{\mu\nu} dx^\mu dx^\nu + e^{-2A(r)} \left(dr^2 + r^2 d\Omega_{X_5}^2 \right) \, , \qquad (5.95)$$

where X_5 is an arbitrary Einstein manifold. Just as the KS solution can be approximated by $AdS_5 \times T^{1,1}$ for $r_{\mathrm{IR}} \ll r \ll r_{\mathrm{UV}}$, up to logarithmic corrections, many solutions of the form (5.95) can be approximated by $AdS_5 \times X_5$: the warp factor is then[34]

$$e^{-4A(r)} \approx \frac{L^4}{r^4} \, , \qquad \text{with} \quad L^4 \equiv \frac{4\pi^4 g_{\mathrm{s}}}{\mathrm{Vol}(X_5)} N(\alpha')^2 \, , \qquad (5.96)$$

where $\mathrm{Vol}(X_5)$ denotes the volume of X_5 (in string units). For a throat of the form (5.95), the range of the canonically normalized D3-brane position is given, as in (5.28), by [254]

$$\frac{\Delta\phi}{M_{\mathrm{pl}}} \leq \frac{2}{\sqrt{N}} \, , \qquad (5.97)$$

and in particular is independent of X_5. We will see in Section 5.3.5 that if one manages to achieve a DBI phase, (5.97) provides a stringent upper bound on the tensor-to-scalar ratio.

UV model

We will refer to the UV model as the situation in which a D3-brane moves *into* the warped region, i.e. toward small ϕ, from the ultraviolet [39, 41].

---

[34] For a generalization of DBI inflation to arbitrary warp factor, assuming an appropriate potential, see [732, 733].

IR model

An interesting alternative is the IR model, in which inflation occurs as the D3-brane leaves the tip region and moves toward the ultraviolet end of the throat [734–736]. The initial conditions for the IR model are very appealing [734]. Suppose that p anti-D3-branes are introduced into a KS throat region. If $p \lesssim 0.08\,M$, with M the flux quantum number defined in (5.20), the anti-D3-branes form a metastable configuration at the tip [443]. Over an exponentially long timescale, this state can decay; the anti-D3-branes annihilate against flux, liberating $(M-p)$ D3-branes. The observation of [734] is that the D3-brane potential arising from moduli stabilization may drive some or all of these D3-branes to move out of the throat region, and during this process a phase of DBI inflation can occur. A simple model of the radial potential is

$$V(\phi) = V_0 - \frac{1}{2}\beta H^2 \phi^2 \,, \tag{5.98}$$

where V_0 is a constant, and in generic configurations,[35] $-1 \lesssim \beta \lesssim 1$. The IR model corresponds to $\beta > 0$. For a discussion of obstacles to a computable realization of the IR model in string theory, see Section 5.3.4.

Relativistic dynamics

For a spatially homogeneous D3-brane, i.e. for $\phi = \phi(t)$, it is natural to define a "Lorentz factor," by analogy to relativistic particle dynamics:

$$\gamma \equiv \left(1 - \frac{\dot{\phi}^2}{T(\phi)}\right)^{-1/2} \,. \tag{5.99}$$

The requirement that γ be real enforces a speed limit on the motion of the probe D3-brane,

$$\dot{\phi}^2 < T(\phi) \,. \tag{5.100}$$

Notice that the bound is independent of the properties of the potential and becomes stronger in regions of strong warping, where $e^{4A(\phi)} \ll 1$ and hence $\dot{\phi}^2 \ll T_3$.

In Section 5.1, we studied non-relativistic D3-brane motion, corresponding to $\gamma \approx 1$; expanding the square root in (5.6) then led to the two-derivative action (5.7). DBI inflation operates in the regime of relativistic brane dynamics, with $\gamma \gg 1$, where higher-derivative terms in (5.92) cannot be neglected. Varying (5.92) with respect to the four-dimensional metric gives the stress–energy tensor sourced by the D3-brane. This corresponds to the stress–energy of a perfect fluid, with energy density and pressure given by

---

[35] Potentials that tend to expel a D3-brane from the infrared are quite common in the ensemble obtained in [43], but are far more complicated than (5.98), involving all five angular directions and an array of competing terms (see [250]).

$$\rho = (\gamma - 1)T + V \,, \tag{5.101}$$

$$P = (\gamma - 1)\frac{T}{\gamma} - V \,. \tag{5.102}$$

Coupling to gravity gives the Friedmann equation

$$3M_{\mathrm{pl}}^2 H^2 = (\gamma - 1)T + V(\phi) \,, \tag{5.103}$$

and the continuity equation

$$\dot{\phi} = -\frac{2M_{\mathrm{pl}}^2 H'}{\gamma} \,, \tag{5.104}$$

where $H' \equiv dH/d\phi$. Using (5.104) in (5.99), we find

$$\gamma = \sqrt{1 + \frac{(2M_{\mathrm{pl}}^2 H')^2}{T}} \,. \tag{5.105}$$

The Hubble slow-roll parameters are [39]

$$\varepsilon = -\frac{\dot{H}}{H^2} = \frac{2M_{\mathrm{pl}}^2}{\gamma}\left(\frac{H'}{H}\right)^2 \,, \tag{5.106}$$

$$\tilde{\eta} = \frac{\dot{\varepsilon}}{H\varepsilon} = \frac{2M_{\mathrm{pl}}^2}{\gamma}\left[2\left(\frac{H'}{H}\right)^2 - 2\frac{H''}{H} + \frac{H'}{H}\frac{\gamma'}{\gamma}\right] \,. \tag{5.107}$$

Notice the factors of γ^{-1} in both ε and $\tilde{\eta}$. For large γ, the slow-roll parameters are therefore *suppressed* relative to the expectation derived from the non-relativistic limit. This leads to the intriguing possibility of achieving inflation even for potentials that naively seem to be too steep to drive prolonged inflation.

Accelerated expansion occurs if the potential energy dominates over the kinetic energy. Demanding that $V(\phi)$ is the leading term on the right-hand side of (5.101) gives the condition

$$\frac{V}{\gamma T} \gg 1 \,. \tag{5.108}$$

Thus, DBI inflation can occur near the location ϕ only if the potential is *large in local string units*. Next, a defining requirement for a DBI phase is that the D3-brane be relativistic.[36] Using (5.108) in (5.103) and (5.105), we find

$$\gamma^2 = \frac{2}{3}\epsilon\frac{V}{T} \gg 1 \,, \tag{5.109}$$

where $\epsilon \equiv \frac{1}{2}M_{\mathrm{pl}}^2(V'/V)^2$. We observe that although (5.109) involves ϵ and can plausibly be satisfied by making the potential very steep, the condition (5.108) is independent of the functional form of the potential. We will see below that (5.108) presents a serious obstacle in the search for a consistent embedding in string theory.

---

[36] In the remainder of this section, we will work to leading order in $\gamma^{-1} \ll 1$.

Analytical [39, 41] and numerical [690, 731] studies have shown that for suitable potentials $V(\phi)$, the DBI Lagrangian (5.92) can indeed support an inflationary phase in which the nontrivial kinetic term plays a crucial role. However, before describing the phenomenology of DBI inflation, one should first ask whether the DBI Lagrangian (5.92) gives an accurate and consistent representation of the physics of a relativistic D3-brane. There are several important questions. First, do quantum corrections, either in EFT or from Planck-scale physics, lead to significant changes to the very special kinetic terms in (5.92)? Second, do potentials $V(\phi)$ satisfying the necessary conditions (5.109) and (5.108) arise naturally, in the same setting where the kinetic terms take the necessary form? Finally, does backreaction of the D3-brane energy, which is large when (5.108) holds, modify the background or the dynamics? We will discuss these issues in turn.

5.3.2 DBI as an EFT

The action in (5.92) may be viewed as an effective theory with cutoff $\Lambda^4(\phi) \equiv \phi^4/\lambda$. The relativistic limit corresponds to the limit $(\partial\phi)^2 \to \Lambda^4$. Naively, this suggests a breakdown of the effective theory and a loss of predictivity. In particular, one might worry that the form of the DBI action in the ultra-relativistic limit is unstable to quantum corrections. Moreover, one may ask whether it is consistent to work to arbitrary order in single derivatives $\partial_\mu\phi$ and yet neglect all terms involving higher derivatives, such as $\Box\phi$. In this section, we address both of these concerns.

Quantum corrections

We will first explain why the action in (5.92) does not receive large quantum corrections in the ultra-relativistic limit. In particular, we will show that the DBI action in anti-de Sitter space is uniquely fixed by the nonlinearly realized conformal symmetries inherited from the isometries of the background spacetime.[37]

The isometry group of five-dimensional anti-de Sitter space, $SO(2,4)$, contains Poincaré transformations, spacetime dilatations and special conformal transformations (SCTs). The D3-brane action is invariant under the four-dimensional Poincaré subgroup $ISO(1,3) \subset SO(2,4)$, but some of the five-dimensional symmetries are only nonlinearly realized. In particular, the D3-brane position modulus ϕ transforms nonlinearly under the conformal symmetries of the AdS spacetime [523]:

$$\text{dilatations}: \qquad x^\mu \mapsto \tilde{x}^\mu \equiv (1+c)x^\mu \,,$$

$$\phi(x) \mapsto \phi(\tilde{x}) + c \,, \tag{5.110}$$

---

[37] This argument was first made in [523] and further elaborated in [737, 738].

$$\text{SCTs}: \qquad x^\mu \mapsto \tilde{x}^\mu \equiv x^\mu + (b \cdot x)x^\mu - \frac{1}{2}\left(x^2 + \frac{\lambda}{\phi^2}\right)b^\mu ,$$

$$\phi(x) \mapsto \phi(\tilde{x})\big(1 - (b \cdot x)\big) , \qquad\qquad (5.111)$$

where c and b^μ are infinitesimal transformation parameters. We see that dilatations and SCTs shift the field value and its gradient, respectively. After gauge fixing, these transformations of ϕ become global symmetries [738], which constrain the form of the action. First, we note that the unbroken four-dimensional Lorentz symmetry and the nonlinearly realized dilatation symmetry (5.110) imply

$$S = \int \mathrm{d}^4x\; \phi^4 f\big((\partial\phi)^2/\phi^4\big) + \cdots , \qquad\qquad (5.112)$$

where f is an arbitrary function and the ellipsis denotes corrections involving at least two derivatives acting on ϕ. In order for this action to be invariant under the SCTs (5.111), the function f must take the form [523]

$$f(z) = \alpha\left[\sqrt{1 + \lambda z} - \beta\right] , \qquad\qquad (5.113)$$

where z is shorthand for $(\partial\phi)^2/\phi^4$. The coefficients α and β can be fixed using supersymmetry;[38] first, recall that in the absence of a supersymmetry-breaking potential, a D3-brane feels no force in an AdS background (it is BPS). This implies that $\beta = 1$. Second, the kinetic term $(\partial\phi)^2$ is not renormalized in the supersymmetric limit. This fixes the normalization, $\alpha = -\lambda^{-1}$. Purely on the basis of symmetries, we have therefore arrived at the DBI action in the form (5.92), implying that the action is protected by symmetry. Moreover, the argument is nonperturbative and so applies to all orders in the quantum theory. This is the famous non-renormalization theorem of the DBI action: quantum corrections can only arise at higher order in derivatives. Similar non-renormalization arguments apply to the generalized backgrounds (5.94).

Higher-derivative corrections

But what about the higher-derivative terms? In the limit $(\partial\phi)^2 \to \Lambda^4$, the full square-root structure of the kinetic term is important and the dynamics cannot be described by the first few orders in an expansion in $(\partial\phi)^2/\Lambda^4$. This is somewhat unconventional from a low-energy effective field theory point of view, so it deserves a bit more discussion. For example, is it really consistent to go to all orders in $(\partial\phi)^2$, but ignore all operators with higher derivatives such as those involving $\Box\phi$? In fact, we do the same when we study a point particle in the ultra-relativistic limit, $\dot{x}^2 \to 1$. In that case, we also trust the full square-root action, $L = -m^2\sqrt{1 - \dot{x}^2}$, but neglect higher time derivatives such as terms involving the acceleration $\ddot{x}$. The justification is that the equations of motion enforce all

---

[38] Here, we assume that supersymmetry is only broken spontaneously during inflation, so it still constrains the form of the action.

higher derivatives to vanish in the ultra-relativistic limit, i.e. as $\dot{x}^2 \to 1$. In the absence of warping, an identical argument holds for the DBI action, i.e. the DBI action is valid for arbitrarily high velocities $\dot{\phi}$, as long as the proper acceleration is smaller than the string scale. One expects the same conclusion to hold in a warped background, provided that the warp factor changes sufficiently slowly. As shown in [39], this is indeed the case in the AdS background as long as $\lambda \gg 1$, which is precisely the limit in which the supergravity description is a good approximation.

5.3.3 DBI as a CFT

The AdS/CFT correspondence [523, 739, 740] (for a review, see [300]) provides an alternative viewpoint on the inflaton action. To describe this, we will briefly sketch the essential elements of the correspondence in its simplest incarnation, which is the duality between $\mathcal{N} = 4$ super-Yang–Mills (SYM) theory with gauge group $U(N)$, in flat four-dimensional spacetime, and type IIB string theory in $AdS_5 \times S^5$ [523].

The essential idea of the correspondence is that there are two equivalent descriptions of the region near to a stack of N D3-branes in flat ten-dimensional spacetime: the gauge theory description involving open strings on the D-branes, and the gravitational description involving the curved spacetime sourced by the branes. Near a stack of N D3-branes, the background takes the $AdS_5 \times S^5$ form (5.91). The asymptotic symmetry group of this spacetime (at the boundary, $r \to \infty$) is $SO(4,2) \times SO(6)$. We recognize $SO(4,2)$ as the *conformal group* in four spacetime dimensions. Recalling that $\mathcal{N} = 4$ super-Yang–Mills theory is a conformal theory and has an R-symmetry group $SO(6) \simeq SU(4)$, one finds a perfect match between the global symmetries of the gauge theory and the asymptotic symmetries of the gravitational solution.

For the present purpose, the relevant application of the duality is to the *Coulomb branch*[39] of $\mathcal{N} = 4$ SYM. This is the moduli space corresponding to the positions of D3-branes along the radial coordinate r of AdS_5, and in the five angles on S^5. The D3-brane positions are parameterized by scalars that transform in the adjoint of $SU(N)$. Let us give a vev to one eigenvalue ϕ of the adjoint scalar. This induces the symmetry breaking $U(N) \to U(N-1) \times U(1)$. (In the bulk this corresponds to separating one of the branes from the stack.) Modes Ψ that are charged under the $U(1)$ symmetry obtain masses proportional to ϕ. (In the bulk picture these correspond with the masses of strings

---

[39] The terminology is arguably more appropriate to $\mathcal{N} = 2$ theories, where the Coulomb branch is distinguished from the Higgs branch because the former is parameterized by scalars in vector multiplets and the latter is parameterized by scalars in hypermultiplets. The important point is just that motion on the Coulomb branch does not change the rank of the gauge group, but can change the rank of the non-Abelian part of the gauge group, i.e. $U(N) \to U(N-1) \times U(1)$, while motion on the Higgs branch can change the total rank.

stretching from the mobile brane to the stack.) Integrating out the fields Ψ generates higher-dimension operators suppressed by powers of ϕ itself. The vev of ϕ can be extremely small compared to the string scale and, thus, higher-dimension contributions are more important in the DBI model than one might naively expect. The first correction is protected by supersymmetry and takes the form $\lambda(\partial\phi)^4/\phi^4$. The CFT is strongly coupled, so all higher-order terms are important and need to be resummed. This is difficult to do in the field theory, but the AdS/CFT correspondence tells us that the answer will be the DBI action (5.92).

5.3.4 Microphysical constraints

In Section 5.3.2, we have seen that DBI inflation is natural in the bottom-up sense, in that loop corrections in the EFT are under control. We now turn to a discussion of top-down naturalness, including the question of whether DBI inflation arises in a consistent string compactification. For concreteness, we will focus our discussion on relativistic branes in warped Calabi–Yau cones, cf. Eq. (5.95).[40]

Achieving a steep potential

Slow-roll inflation requires a potential that is flat in Planck units, with $\eta \ll 1$, whereas DBI inflation requires a potential that is steep enough to drive the moving D-brane to have large kinetic energy, i.e. to obey (5.109). These two options are *not exhaustive*: a potential can easily be too steep for slow-roll inflation and yet too gentle for DBI inflation.

To understand whether (5.109) is readily satisfied for D3-branes in Calabi–Yau cones, we can make use of the potential derived in [43], which describes the forces on a D3-brane in a KS throat attached to a KKLT compactification (see Section 5.1). This potential incorporates the Coulomb interaction with an anti-D3-brane, as well as the full spectrum of contributions from moduli stabilization. In a Monte Carlo study [250] based on [43], DBI inflation did not occur by chance: in the full set of more than 10^7 trials, $\gamma - 1$ never exceeded 10^{-8}. The reason for this finding is that the D3-brane potential from moduli stabilization is small in local string units, $V(\phi) \ll T(\phi)$, and is not parametrically steep, so that $\epsilon \lesssim 1$. It follows from (5.109) that $\gamma - 1 \ll 1$. While it is plausible that somewhat larger Lorentz factors could arise very near the tip of the throat, which was not directly studied in [250], the problem of backreaction becomes severe in this regime, as discussed below. The earlier analysis [731] worked with a simpler model of the potential, but arrived at compatible conclusions; in realizations consistent with

---

[40] Identifying an alternative setting for DBI inflation in a string compactification would be very interesting, and would undoubtedly lead to modified microphysical constraints, but we are not aware of any complete example.

the field range bound (5.97) of [254], and with observational constraints on n_s, the Lorentz factor was bounded by $\gamma - 1 < 10^{-7}$.

A remark about the Coulomb potential is relevant here. The potential $V_C(\phi)$ given in (5.30) applies to a KS throat (in the $AdS_5 \times T^{1,1}$ approximation). For a cone over X_5, one finds instead

$$V_C(\phi) = D_0 \left(1 - \frac{\pi}{4 \, \text{Vol}(X_5)} \frac{D_0}{\phi^4} \right) . \tag{5.114}$$

For small $\text{Vol}(X_5)$ – for example, if $X_5 = S^5/Z_p$ with $p \gg 1$ – the Coulomb force is increased in strength; heuristically, the field lines are collimated along a narrow throat. In extreme cases, this increased force can compel the D3-brane to be relativistic; [731] found that a DBI phase arose for $\text{Vol}(X_5) \lesssim 10^{-17}$. However, it is highly implausible that such a throat can be embedded in a consistent compactification, and moreover, even if this issue were overlooked, the examples of [731] are incompatible either with observations or with the microphysical bound of [254].

Backreaction

The requirement (5.108) for accelerated expansion, $V(\phi) \gg \gamma T(\phi)$, implies that the D3-brane potential energy must substantially exceed the local string scale. This carries the risk that whatever physics generates the potential will simultaneously distort the background supergravity solution (i.e. the AdS geometry (5.91) or a related warped throat geometry). We will offer two related perspectives on this problem.

First, we present an observation due to Maldacena (also described in [23]). Suppose that the D3-brane potential arises from coupling the warped throat sector to a hidden sector that breaks supersymmetry, and take the potential to be quadratic, $V(\phi) = \frac{1}{2} m^2 \phi^2$, so that

$$\frac{V(\phi)}{\gamma T(\phi)} = \frac{\lambda}{2\gamma} \frac{m^2}{\phi^2} . \tag{5.115}$$

Consider the effect of hidden sector supersymmetry breaking on the Kaluza–Klein spectrum of the throat. Barring an efficient sequestering mechanism, the Kaluza–Klein modes will acquire masses $M_{KK} \sim m$. Since the lightest Kaluza–Klein modes in an undistorted throat have

$$M_{KK} \sim \frac{1}{L} e^{A_{IR}} \sim \frac{r_{IR}}{L^2} , \tag{5.116}$$

supersymmetry breaking will typically cut off the throat at

$$r_{IR} \sim mL^2 . \tag{5.117}$$

Thus, the canonical field ϕ parameterizing D3-brane motion obeys

$$\phi^2 \gtrsim \phi_{IR}^2 \sim T_3 (mL^2)^2 = \lambda m^2 . \tag{5.118}$$

Combining (5.115) and (5.118), we find that

$$\frac{V(\phi)}{\gamma T(\phi)} \lesssim \frac{1}{2\gamma} \ll 1 . \tag{5.119}$$

We conclude that unless the source of supersymmetry breaking couples much more strongly to the D3-brane than to the Kaluza–Klein modes of the background, the throat is truncated in the infrared, excluding the region where DBI inflation could occur.

One can argue for the cutoff (5.117) in a slightly different way, beginning with the potential for a D3-brane probe of a general supergravity background. From (5.29), we have $V(\phi) = T_3\Phi_-$, where the scalar Φ_-, defined in (5.52), involves the warp factor and the four-form potential. Thus,

$$\frac{V(\phi)}{\gamma T(\phi)} = \frac{1}{\gamma}\frac{e^{4A} - \alpha}{e^{4A}} . \tag{5.120}$$

DBI inflation therefore requires

$$|\alpha| \gg \gamma e^{4A} . \tag{5.121}$$

In the noncompact, supersymmetric Klebanov–Strassler background, $e^{4A} = \alpha$, i.e. $\Phi_- = 0$, but in a finite warped throat, supersymmetry breaking and moduli stabilization in the bulk source perturbations of Φ_- and of G_- [43, 328] – see (5.52). In turn, these perturbations source corrections to the metric. The condition (5.121) requires strong deviations from the ISD background, which lead to correspondingly large corrections to the metric that eventually terminate the throat in the infrared. One can then show [553, 741] that $V(\phi) \lesssim T(\phi)$ in the accessible region, in agreement with (5.119).

Super-Planckian fields

Another important microphysical difficulty is that in many simple examples (e.g. for a quadratic potential) a field range $\Delta\phi$ of order $M_{\rm pl}$ is necessary [41]. In view of (5.97), this is at best marginally possible, but only if $N \lesssim 4$: at large N, $\Delta\phi \ll M_{\rm pl}$. On the other hand, the requirement of a nearly-Planckian range in DBI inflation is a common finding, not a theorem, and it may be possible to find a compactification in which the potential satisfies (5.109) and (5.108), without violating the immutable bound (5.97).

Bremsstrahlung

A further consistency requirement is that the speed limit felt by the D3-brane be dictated by the nontrivial kinetic term, not by other modes of dissipation.[41] Because the D3-brane is necessarily relativistic, and is accelerated by

---

[41] While alternative means of shedding energy could give rise to interesting cosmological scenarios, the dynamics would not be governed by (5.92) alone, and much further analysis would be required. See Section 5.6 for several examples of dissipative models.

the potential, it will emit gravitational and scalar synchrotron radiation into the compact dimensions. This process was analyzed in [742], where it was shown that losses due to bremsstrahlung dominate the dynamics in a significant fraction of parameter space – including the regime of weak string coupling and large volume – but can be neglected in the remainder.

Excitation of massive strings

A very significant obstacle to predictivity in the IR model is that in the initial stage of expansion, the Hubble scale exceeds the local string scale, and massive open strings can be excited [734–736] (see the summary in [743]). One can try to estimate the corresponding perturbations [735, 736], but a reliable computation is inaccessible with current tools. This problem cannot be relegated to unobservably large angular scales; the maximum number of e-folds N_e^{EFT} that can be produced after the EFT computation of the perturbations becomes valid is [735]

$$N_e^{\text{EFT}} \approx \frac{N^{1/8}}{\sqrt{\beta}} \,, \tag{5.122}$$

where as usual N is the D3-brane charge of the throat and β was defined in (5.98). Unless $N_e^{\text{EFT}} \gg 60$, the observed CMB perturbations will be dictated by uncomputable fluctuations of massive strings. While this is a fascinating possibility that escapes the confines of the effective field theory of a finite number of fields, no meaningful predictions are possible at present. For generic potentials with $\beta \sim 1$, solving the horizon problem without encountering uncomputable perturbations requires $N \gtrsim 10^{14}$; moreover, because $\beta N_e \gg 1$ (with N_e the number of e-folds) is required for a consistent DBI phase [743], we must have $N \gtrsim 10^7$ regardless of the value of β. D3-brane charges of this magnitude are difficult to realize in compact spaces.

5.3.5 Phenomenology

In spite of the many apparent obstacles to realizing DBI inflation in a consistent string compactification, the DBI scenario is a leading example of a field-theoretic inflationary mechanism that is underpinned by the symmetries of an ultraviolet theory. Many authors have deferred the question of explicit ultraviolet completion, and directly investigated the rich phenomenology that follows from (5.94) – or a generalization, e.g. with additional fields – in the regime where $\gamma \gg 1$. We will briefly sketch the signatures of DBI inflation that emerge from this approach.

DBI inflation is a special case of so-called $P(X)$ theories (see Section 2.2.3) whose action is given by

$$S = \int d^4x \sqrt{-g} \left[\frac{M_{\text{pl}}^2}{2} R + P(X, \phi) \right] \,, \tag{5.123}$$

where $P(X, \phi)$ is an arbitrary function of $X \equiv -\frac{1}{2}(\partial \phi)^2$ and ϕ. DBI inflation is recovered for

$$P(X, \phi) = -T(\phi) \left(\sqrt{1 - \frac{2X}{T(\phi)}} - 1 \right) - V(\phi) . \tag{5.124}$$

The phenomenology of $P(X)$ theories has been explored in [124].

Scalar modes

The theory for the fluctuations in $P(X)$ theories maps to the effective Goldstone action of Section 1.2.1 with sound speed given by

$$c_s^2 = \frac{P_{,X}}{P_{,X} + 2X P_{,XX}} . \tag{5.125}$$

The scalar power spectrum therefore follows from (1.28). Substituting (5.124) into (5.125), we get

$$c_s^2 = 1 - \frac{2X}{T(\phi)} = \frac{1}{\gamma^2(\phi)} . \tag{5.126}$$

We see that a nontrivial warp factor can lead to a field dependence of the sound speed, $c_s(\phi)$. On CMB scales, the scale-invariance of the spectrum constrains the variation of the sound speed. On smaller scales, a significant evolution of the spectrum and hence of the sound speed is still allowed. Large-scale structure constraints on the running of the spectrum were analyzed in [744].

Tensor modes

The tensor-to-scalar ratio in $P(X)$ theories is given by

$$r = 16 c_s \varepsilon , \tag{5.127}$$

and a generalized Lyth bound can be derived [254]:

$$\frac{\Delta \phi}{M_{\rm pl}} = \int_0^{N_\star} \sqrt{\frac{r(N)}{8} \frac{1}{c_s P_{,X}}} \, \mathrm{d}N . \tag{5.128}$$

We notice a nontrivial generalization of the slow-roll result (2.101) through the factor $c_s P_{,X}$. However, the Lagrangian of DBI inflation (5.124) is algebraically special, satisfying

$$c_s P_{,X} = 1 . \tag{5.129}$$

The correspondence between $\Delta \phi$ and r in DBI inflation is thus the same as in slow-roll inflation. The geometric field range bound (5.28) therefore also forbids large tensors in DBI inflation [254].[42]

---

[42] The bound $r < 10^{-7}/\mathrm{Vol}(X_5)$ can be derived by combining (5.28) with the *assumption* that $f_{\rm NL}^{\rm equil} \gtrsim 1$. In other words, DBI inflation in a Calabi–Yau cone cannot have both detectable equilateral non-Gaussianity and detectable tensors [745].

Equilateral non-Gaussianity

The cubic Goldstone action takes the form of (1.71), with

$$A = c_s^2 \left(-1 - \frac{2}{3} \frac{X P_{,XXX}}{P_{,XX}} \right) . \tag{5.130}$$

Substituting (5.124) into (5.130), we get[43]

$$A = -1 . \tag{5.131}$$

Mapped onto the equilateral and orthogonal templates (see Section 1.4.2), the size of the bispectrum for $P(X)$ theories is

$$f_{\mathrm{NL}}^{\mathrm{equil}} = (- 0.27 + 0.08 A) \frac{1 - c_s^2}{c_s^2} , \tag{5.132}$$

$$f_{\mathrm{NL}}^{\mathrm{ortho}} = (+ 0.02 + 0.02 A) \frac{1 - c_s^2}{c_s^2} . \tag{5.133}$$

For the DBI Lagrangian, the orthogonal component is small and the amplitude of equilateral non-Gaussianity becomes

$$f_{\mathrm{NL}}^{\mathrm{equil}} = -\frac{35}{108} \left(\frac{1}{c_s^2} - 1 \right) \approx -\frac{35}{108} \gamma^2 , \tag{5.134}$$

where in the second equality we have assumed the relativistic limit. The Planck constraint, $-117 < f_{\mathrm{NL}}^{\mathrm{equil}} < 33$ (68% C.L.) [10], implies

$$\gamma \lesssim 24 \quad (95\% \text{ C.L.}) . \tag{5.135}$$

Joint constraints for a quadratic potential

In the important special case where $V(\phi) = \frac{1}{2} m^2 \phi^2$, the above results can be combined to place strong limits on the model parameters [254] (see also [745]). For a quadratic potential, one has [39]

$$2 \left(\frac{M_{\mathrm{pl}}}{\phi} \right)^2 = \varepsilon \gamma = \frac{1}{16} r \gamma^2 , \tag{5.136}$$

where in the final equality we used (5.125) and (5.127). From (5.134), one then finds [254]

$$2 \left(\frac{M_{\mathrm{pl}}}{\phi} \right)^2 = \frac{27}{140} r \left| f_{\mathrm{NL}}^{\mathrm{equil}} \right| . \tag{5.137}$$

Incorporating the geometric bound (5.97), we find the upper limit,

$$N < \frac{27}{70} r \left| f_{\mathrm{NL}}^{\mathrm{equil}} \right| \lesssim 9 \quad (95\% \text{ C.L.}) . \tag{5.138}$$

---

[43] This result is independent of the potential and the warp factor, unlike other observables such as n_s.

This small value of N is in tension with the required amplitude of the scalar perturbations; using (5.137) in (1.28) leads to [254]

$$\Delta_{\mathcal{R}}^2(k_\star) = \left(\frac{32}{3\pi}\right)^2 \frac{3}{r^4 (f_{\rm NL}^{\rm equil})^2} \frac{{\rm Vol}(X_5)}{N} \gtrsim 15 \frac{{\rm Vol}(X_5)}{N} , \qquad (5.139)$$

where the inequality uses the Planck upper limits on r and $f_{\rm NL}^{\rm equil}$. Because $\Delta_{\mathcal{R}}^2(k_\star) = 2.2 \times 10^{-9}$, we conclude that

$$N \gtrsim 7 \times 10^9 \; {\rm Vol}(X_5) . \qquad (5.140)$$

Quadratic DBI inflation in a Calabi–Yau cone is therefore virtually excluded. One would need an extremely small angular manifold X_5, e.g. by orbifolding by a large discrete group, while also keeping $N \lesssim 9$, which almost certainly renders the supergravity approximation invalid.

Multi-field effects

So far, our discussion of DBI inflation has been restricted to purely radial evolution of the D3-brane, but in general the potential will depend on the angular coordinates. Including these effects leads to multi-field models of DBI inflation,[44] whose phenomenology has been studied comprehensively in [747–755]. When one or more of the angular fields are light during inflation, their quantum fluctuations lead to entropy perturbations, which propagate with the same speed of sound c_s as the adiabatic mode. In the limit $c_s \ll 1$, the amplitude of the entropy perturbations is boosted relative to the adiabatic fluctuations. The amplitude of the bispectrum can be suppressed,

$$f_{\rm NL}^{\rm equil} \approx -\frac{35}{108} \gamma^2 \cos^2 \Theta , \qquad (5.141)$$

where the angle Θ parameterizes how much of the final curvature perturbation arises from entropy perturbations,[45] with $\Theta = 0$ if there is no transfer of entropy modes and $\Theta = \pi/2$ if the final curvature perturbation is mostly of entropic origin. This can relax the constraint (5.135) on γ.

5.4 Inflating with axions

String axions are promising inflaton candidates. Equipped with a continuous shift symmetry to all orders in perturbation theory, the axion potential is stable against radiative corrections. Weakly breaking the symmetry – either spontaneously by nonperturbative effects, or explicitly through the presence of branes – can lead to inflationary theories that are natural in the bottom-up sense

---

[44] For an analysis of DBI inflation with N D3-branes, see [746].

[45] The transfer from entropy perturbations to curvature perturbations can depend on the physics of reheating. Investigations of the distinctive features of reheating in DBI inflation include [578, 756, 757].

of Section 2.1.2. In *natural inflation* [36, 758], the role of the inflaton is played by a single axion ϕ with the Lagrangian (3.78),

$$\mathcal{L}(\phi) = -\frac{1}{2}(\partial\phi)^2 - \Lambda^4 \left[1 - \cos\left(\frac{\phi}{f}\right)\right] + \cdots , \tag{5.142}$$

where f is the axion decay constant, and Λ is a dynamically generated scale. To facilitate comparison with the literature we have redefined the decay constant by a factor of 2π compared with Section 3.2.3, i.e. $f_{\text{here}} = 2\pi f_{\text{there}}$.

For sufficiently large values of f/M_{pl}, the model (5.142) gives rise to prolonged inflation, and for $f \gtrsim 10 M_{\text{pl}}$, the spectral index n_s is compatible with the constraints from the Planck mission.[46] However, as we have reviewed in Section 3.2.3, super-Planckian decay constants have not been obtained to date in a controlled string compactification. Natural inflation is therefore an interesting example of the importance of explicit ultraviolet completion. Although inflation with a single axion and super-Planckian decay constant is natural from the bottom-up perspective, it does not seem to be natural from the top down, unless additional structures are present. In the rest of this section, we will discuss a few of the leading proposals for what these extra structures might be.

5.4.1 Inflation with multiple axions

One strategy is to extend (5.142) by introducing one or more additional axions [549, 759], each with a sub-Planckian decay constant, and arrange that a combination of these fields effectively enjoys a super-Planckian decay constant. We will describe two mechanisms: "alignment" of two axions [759], and assisted inflation [550] with $N \gg 1$ axions [549].

Two axions

Consider two axion fields ϕ_1 and ϕ_2 that couple to linear combinations of two confining non-Abelian gauge groups a and b, with the Lagrangian density [759]

$$\mathcal{L} \supset \sum_{i=1}^{2} \frac{\phi_i}{f_i} \left(\frac{c_{ia}}{32\pi^2} \text{Tr}\left[F^{(a)} \wedge F^{(a)}\right] + \frac{c_{ib}}{32\pi^2} \text{Tr}\left[F^{(b)} \wedge F^{(b)}\right] \right) , \tag{5.143}$$

where the coefficients $c_{ia} = \{c_{1a}, c_{2a}\}$ and $c_{ib} = \{c_{1b}, c_{2b}\}$ are dimensionless, and the decay constants f_1, f_2 have the dimensions of mass.

In terms of the dynamical scales Λ_a and Λ_b of the two gauge groups, the potential for the axions is

$$V = \Lambda_a^4 \left[1 - \cos\left(c_{1a}\frac{\phi_1}{f_1} + c_{2a}\frac{\phi_2}{f_2}\right)\right] + \Lambda_b^4 \left[1 - \cos\left(c_{1b}\frac{\phi_1}{f_1} + c_{2b}\frac{\phi_2}{f_2}\right)\right] . \tag{5.144}$$

---

[46] The constraint on the decay constant depends on the choice of prior. The result we have cited here is for a uniform prior on $\log(f)$ [9].

The central observation of [759] is that if

$$\frac{c_{1a}}{c_{2a}} = \frac{c_{1b}}{c_{2b}} , \qquad (5.145)$$

then one linear combination ξ of the axions is unlifted, and effectively has infinite range.[47] When (5.145) is approximately satisfied, ξ has a decay constant f_ξ that can be much larger than f_1 and f_2. In particular, for sufficiently precise alignment, one can have $f_\xi > M_{pl}$ with f_1, $f_2 \ll M_{pl}$. For the simple case where $c_{1a} = c_{1b} = 1$ and $\Lambda_a^4 \gg \Lambda_b^4$, one finds

$$f_\xi = \frac{\left(c_{2a}^2 f_1^2 + f_2^2\right)^{1/2}}{|c_{2b} - c_{2a}|} , \qquad \text{with} \quad \xi = \frac{\phi_2 f_2 - c_{2a}\phi_1 f_1}{c_{2a}^2 f_1^2 + f_2^2} . \qquad (5.146)$$

The relation (5.145) is plausibly radiatively stable. However, for inflation to proceed along the ξ direction, driven by the potential (5.144), other nonperturbative contributions to the potential for ξ must be negligible. It would be valuable to construct an explicit example in a stabilized string compactification and ascertain whether nonperturbative effects associated with moduli stabilization, which could be unrelated to the confining gauge groups in (5.143), might spoil the flatness of the inflaton direction. (For a closely related idea, see [264], as discussed in Section 5.4.2 below.)

Many axions: N-flation

A different idea for axion inflation takes advantage of the fact that string compactifications often come with large numbers of axion fields; perhaps the collective excitations of hundreds of axions, each with a sub-Planckian decay constant, can yield an effective decay constant of super-Planckian size.[48] This idea is called *N-flation* [549] and is based on the earlier proposal of *assisted inflation* [550] (see also [761, 762]).[49]

Consider N axions whose Lagrangian is simply N copies of (5.142),

$$\mathcal{L} = \sum_{i=1}^{N} \left[-\frac{1}{2}(\partial\phi_i)^2 - V_i(\phi_i) \right] , \qquad (5.147)$$

where $V_i(\phi_i) \equiv \Lambda_i^4 \left[1 - \cos(\phi_i/f_i)\right]$. We have assumed that cross-couplings in the axion potential are negligible, as discussed below. Each individual axion therefore experiences a force only from its own potential V_i, but Hubble friction from the sum of all potentials $\sum_i V_i$,

$$\ddot{\phi}_i + 3H\dot{\phi}_i = -\partial_i V_i , \qquad \text{where} \quad 3M_{pl}^2 H^2 \approx \sum_{i=1}^{N} V_i . \qquad (5.148)$$

---

[47] A related mechanism is used in racetrack constructions – see Section 5.5.1.

[48] This section is based on [502, 549, 760].

[49] A different approach to assisted inflation in string theory is M-flation [763–765].

Compared with the single axion case, each individual axion feels enhanced Hubble friction, suggesting that one might be able to achieve a friction-dominated situation even without any axion being at a super-Planckian distance from the minimum.

Let us begin by considering the simple case in which all the axions have equal masses $m_i = \Lambda_i^2/f_i \equiv m$. Moreover, let us consider small displacements from the minimum, $\phi_i \ll f_i$. The collective excitation $\Phi^2 \equiv \sum_i \phi_i^2$ then has the potential

$$V(\Phi) = \frac{1}{2}m^2\Phi^2 . \tag{5.149}$$

Successful inflation requires $\Phi > M_{\mathrm{pl}}$, but this does not mean that the individual axion vevs ϕ_i have to be large; for sufficiently large N, the individual displacements can be sub-Planckian, $\phi_i \approx \Phi/\sqrt{N} < M_{\mathrm{pl}}$. In typical examples the required number of axions is $N \gtrsim \mathcal{O}(10^3)$ [549].

One might object that quantum gravity constraints should restrict the range of Φ to be sub-Planckian, just like the ranges of the ϕ_i, because in moving from individual displacements ϕ_i to the collective field Φ, one has merely changed from Cartesian to spherical polar coordinates, which should be immaterial unless some physical effect is sensitive to the change of coordinates. This concern is unfounded; the N discrete axionic shift symmetries $\phi_i \mapsto \phi_i + 2\pi f_i$ that persist at the nonperturbative level do in fact define a preferred Cartesian coordinate system. (After the periodic identifications, the axion field space is an N-torus, with individual radii $f_i < M_{\mathrm{pl}}$.) In string theory constructions (see below), only the f_i are directly constrained.

A much graver concern is that loops of the N light axion fields renormalize the Planck mass; on general grounds one expects [549][50]

$$\delta M_{\mathrm{pl}}^2 \sim \frac{N}{16\pi^2} \Lambda_{\mathrm{UV}}^2 , \tag{5.150}$$

where Λ_{UV} is the ultraviolet cutoff. Because the collective field displacement scales as $\Phi \propto \sqrt{N}\phi$, with ϕ the mean of the individual displacements, while the correction to the Planck mass in (5.150) also scales as $\sqrt{N}$, we conclude that one cannot obtain a parametrically super-Planckian displacement purely by working at large N. Instead, one must grapple with the ultraviolet-sensitive details, e.g. by refining the field-theoretic estimate (5.150) through a computation in string theory. To learn more, we now turn to a string theory realization of N-flation [502, 549, 760].

N-flation in type IIB string theory

As an explicit realization of N-flation in string theory, we consider a KKLT compactification in which $h_+^{1,1} \equiv N \gg 1$ complexified Kähler moduli T_i are

---

[50] A counterpoint to this finding appears in [766], where it is argued that the portion of the eta problem arising from loops of light fields is actually suppressed at large N.

stabilized by nonperturbative effects. As explained in Section 3.2, the associated axions ϑ_i, $i = 1, \ldots, N$, correspond to the integrals of C_4 over orientifold-even four-cycles, cf. Eq. (3.70). For simplicity of presentation, we take $h_-^{1,1} = 0$, so that by (3.68), $T_i = \tau_i + i\vartheta_i$, with τ_i a real four-cycle volume.[51]

The superpotential takes the form (3.114),

$$W = W_0 + \sum_{i=1}^{N} \mathcal{A}_i\, e^{-a_i T_i} \,, \tag{5.151}$$

where $a_i = 2\pi$ (for Euclidean D3-branes) or $a_i = 2\pi/N_i$ (for gaugino condensation in a gauge group with dual Coxeter number N_i). Note that the axions ϑ_i appear in the phase of each nonperturbative term. In view of (3.94), the Kähler potential has a complicated dependence on the Kähler moduli T_i:

$$K = -2\ln\left[\mathcal{V}(T_i, \bar{T}_i)\right] \,, \tag{5.152}$$

where $\mathcal{V}$ is the volume of the compact Calabi–Yau manifold. The $\mathcal{N} = 1$ supergravity theory defined by (5.151) and (5.152) has N light chiral superfields,[52] and generically admits supersymmetric AdS_4 solutions. Introducing an anti-D3-brane following [369], one can find a solution with a small positive cosmological constant. The anti-D3-brane potential energy has negligible dependence on the axions ϑ_i, and will be treated as a constant for the purpose of axion inflation.

We now expand around the minimum, in small fluctuations of the (dimensionless) axions ϑ_i:

$$\mathcal{L} = -\frac{1}{2}M_{\mathrm{pl}}^2 K_{ij}\partial_\mu\vartheta^i\partial^\mu\vartheta^j - \frac{1}{2}M_{ij}\vartheta^i\vartheta^j + \cdots \,. \tag{5.153}$$

The mass matrix M_{ij} is determined by the superpotential (5.151), the Kähler potential (5.152), and their derivatives – see [502] for the explicit expression – and depends on the vevs of the (real) Kähler moduli τ_i. In a generic basis, the axions will be cross-coupled both in their kinetic terms and in the potential; neither K_{ij} nor M_{ij} will be diagonal. Because of the axion shift symmetry, the Kähler metric K_{ij} is independent of the ϑ^i, up to nonperturbatively small corrections. One can always perform a change of coordinates to set $K_{ij} \mapsto \delta_{ij}$, which rotates and rescales the axion fields. Denoting the new, canonically normalized axion fields by ϕ_i and performing a further orthogonal rotation to diagonalize the mass matrix, we arrive at

---

[51] Another interesting and rather explicit construction with similar qualitative features works with $h_-^{1,1} \equiv N \gg 1$, taking the inflationary axions to arise from the dimensional reduction of the R-R two-form potential, cf. Eq. (3.63) [760]. One advantage of the model of [760] is that it is comparatively straightforward to arrange for the Kähler moduli masses to be larger than the inflaton mass.

[52] At the energy scales of interest, the dilaton and the complex structure moduli are already stabilized, and W_0 and $\mathcal{A}_i$ are constants.

$$\mathcal{L} = \sum_{i=1}^{N} \left[-\frac{1}{2}(\partial\phi_i)^2 - \frac{1}{2}m_i^2\,\phi_i^2 \right] . \qquad (5.154)$$

The masses m_i of the decoupled fields ϕ_i have a rather complicated dependence on $W_0, \mathcal{A}_i, \tau_i$. Fortunately, at large N, random matrix theory yields a simple approximate expression for the axion mass spectrum [502]. The mass matrix belongs to the Wishart ensemble described in Section 3.5.3, and the mass spectrum is given by the Marčenko–Pastur law (3.154),

$$\rho(m^2) = \frac{1}{2\pi N\sigma^2 m^2}\sqrt{(\eta_+ - m^2)(m^2 - \eta_-)} . \qquad (5.155)$$

Here, $\eta_\pm \equiv N\sigma^2(1 \pm \sqrt{P/N}\,)^2$, where, as above, $N = h_+^{1,1}$, while $P = h_-^{2,1}$ is the number of complex structure moduli, and σ controls the typical scale of terms in the moduli potential – see [502] for a detailed explanation.[53] The eigenvalue spectrum (5.155) is shown in Fig. 5.11.

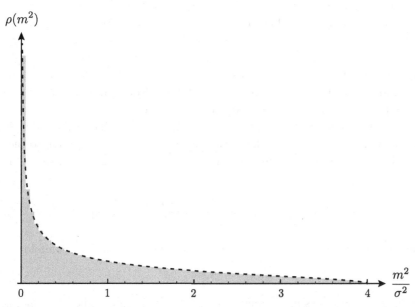

Fig. 5.11 The eigenvalue spectrum of a Wishart matrix, given by the Marčenko–Pastur law (5.155) (figure adapted from [49]). The spectrum of axion masses around a KKLT minimum is well described by this law when $N \equiv h_+^{1,1} \gg 1$. The curve is the analytic result, while the histogram is the result of simulations, both for $N = P = 200$.

[53] One further assumption implicit in obtaining (5.155) in a KKLT compactification is that the gravitino mass $m_{3/2}$ is small compared to the scale of supersymmetric masses; see [49, 495] for discussions of how this could be achieved.

Because the axion decay constants f_i are proportional to the eigenvalues of the Kähler metric K_{ij}, the distribution of decay constants in a class of string compactifications could be determined in much the same way that the distribution of masses m_i was obtained above.

Microscopic challenges

Two significant issues make it difficult to realize N-flation in a stabilized string compactification.

Light Kähler moduli One obstacle faced by any realization of N-flation in a string compactification with spontaneously broken supersymmetry is that the axions are partnered with real scalar fields, by $\mathcal{N} = 1$ supersymmetry; in the KKLT example given above, these are the four-cycle volumes τ_i. Arranging that the real moduli have masses above the Hubble scale, $m \gtrsim H$, and hence are (barely) frozen during inflation, is challenging [767]. Highly contrived configurations, e.g. involving many racetrack terms in the superpotential, could in principle give large masses to these fields (see the discussion in [49]), but no examples have been constructed. If supersymmetry is not spontaneously broken during inflation, but is instead broken at a higher scale, then the real partners of the axions can be decoupled, but control of the potential becomes more difficult [768].

Renormalization of the Planck mass The problem of the renormalization of the Planck mass (5.150) is not automatically alleviated in string theory, but it takes a more precise form, as we now explain in the example of a KKLT realization of N-flation. The leading correction to the four-dimensional Planck mass that scales with the number of axions is (2.30), the four-loop σ-model correction to the ten-dimensional action. Upon dimensional reduction, one finds the Einstein–Hilbert term [549]

$$\mathcal{L} = \frac{M_{\text{pl}}^2}{2}\left(1 + \frac{\zeta(3)\chi(X_6)}{(2\pi)^3}\frac{(\alpha')^3}{\mathcal{V}} + \cdots\right)R_4\,, \tag{5.156}$$

where $\chi(X_6)$ is the Euler characteristic of the compactification manifold X_6,

$$\chi(X_6) = 2h^{1,1} - 2h^{2,1}\,, \tag{5.157}$$

and the omitted terms are higher order in α' and/or in g_s. The renormalization of the Planck mass encoded in (5.156) indeed has the same scaling with N as the field theory result[54] (5.150), *unless* $2h^{1,1} - 2h^{2,1} \ll h^{1,1}$. Thus, for suitable

---

[54] A potential confusion is that the leading correction in (5.156) arises as a loop effect on the worldsheet, not from loops of light moduli in spacetime, but nevertheless involves the number of moduli. As such, it corresponds to a non-renormalizable term in the effective theory, as in Section 2.3.2, rather than a radiative correction as in (5.150), cf. Section 2.3.1. However, corrections at higher order in g_s may be expected to involve actual loops of the light fields, in closer analogy with (5.150).

Hodge numbers, the correction to the Planck mass from (5.156) can be neglected. However, there are higher-order corrections in α', and in g_s, whose form is not known; if these are also proportional to $\chi(X_6)$, then renormalization of the Planck mass is harmless in compactifications with $2h^{1,1} - 2h^{2,1} \ll h^{1,1}$, but otherwise the problem plausibly reappears.[55] More detailed understanding of ultraviolet-sensitive quantum corrections would be needed to resolve this issue.

Phenomenology

A number of authors have studied the signatures of the phenomenological model (5.147), anticipating a microphysical realization; see e.g. [502, 769–778]. If a quadratic approximation is applicable, then the tensor-to-scalar ratio is given by $r \approx 0.13$ (for 60 e-folds of inflation).

5.4.2 Axion monodromy inflation

Another idea for extending the effective axion range uses the phenomenon of *monodromy*.[56] We speak of monodromy when a system reaches a new configuration after being transported around a closed loop in the (naive) configuration space. The classic example is a spiral staircase, where the naive configuration space is a circle, but the system changes upon transport by 2π: after each circuit we reach a higher level on the staircase. Something very similar occurs in the scalar potential for axions in certain string compactifications: the potential energy continues to increase as the axion traverses multiple circuits of its fundamental domain. The basic idea of monodromy inflation [779] is that inflation can persist through many cycles around the configuration space. The effective field range is then much larger than the fundamental period, but the axion shift symmetry protects the structure of the potential over each individual cycle [35, 779].

Monodromies are widespread in string compactifications, but constructing an explicit model of axion monodromy inflation is delicate, as we now explain.

Axion potentials

We will begin with a simple example: a D5-brane in type IIB string theory [35].[57] The brane fills the four-dimensional spacetime and wraps a two-cycle Σ_2 in the compact space. The axion $b \equiv \frac{1}{\alpha'} \int_{\Sigma_2} B_2$ will exhibit monodromy in the potential

---

[55] One cannot work at arbitrarily weak coupling and large volume, because in this limit the individual decay constants f_i are parametrically small compared to M_{pl}; see Section 3.2.3.

[56] This section is based on [35, 779].

[57] The first example of monodromy inflation constructed in string theory involves a D4-brane in a nilmanifold compactification of type IIA string theory [779], and relies on a scenario for moduli stabilization that is rather different from that presented in Section 3.3.3. We will discuss the scenario of [779] in Section 5.6.1.

energy, i.e. the potential energy of the wrapped brane is *not* a periodic function of the axion. To see this, consider the DBI action (3.27) for the D5-brane,

$$S_{\mathrm{D5}} = \frac{1}{(2\pi)^5 g_{\mathrm{s}}(\alpha')^3} \int_{\mathcal{M}_4 \times \Sigma_2} \mathrm{d}^6 \sigma \sqrt{-\det(G_{ab} + B_{ab})} \ . \tag{5.158}$$

Performing the integral over the two-cycle, we obtain the potential for the axion in the four-dimensional effective theory,

$$V(b) = \frac{\varrho}{(2\pi)^6 g_{\mathrm{s}}(\alpha')^2} \sqrt{(2\pi)^2 \ell^4 + b^2} \ , \tag{5.159}$$

where ℓ is the size of Σ_2 in string units, and the dimensionless number ϱ encodes a possible dependence on the warp factor. The presence of the brane has broken the axion shift symmetry, $b \mapsto b + (2\pi)^2$. We say that the brane has generated a monodromy for the axion. For large values of the axion vev, $b \gg \ell^2$, the potential is linear, $V(b) \propto b$.

A similar effect occurs if the D5-brane is replaced by an NS5-brane. The wrapped NS5-brane now produces a monodromy for the axion $c \equiv \frac{1}{\alpha'} \int_{\Sigma_2} C_2$. Dimensional reduction of the action for the NS5-brane introduces the following potential for the c axion:

$$V(c) = \frac{\varrho}{(2\pi)^6 g_{\mathrm{s}}^2(\alpha')^2} \sqrt{(2\pi)^2 \ell^4 + g_{\mathrm{s}}^2 c^2} \ . \tag{5.160}$$

Axion monodromy inflation

In both cases, inflation requires that the axion (b or c) has a large initial vev. The Lagrangian for the canonically normalized field is

$$\mathcal{L} = -\frac{1}{2}(\partial\phi)^2 - \mu^3 \phi \ , \qquad \text{with} \quad \mu^3 \equiv \frac{1}{f} \frac{\varrho}{(2\pi)^6 g_{\mathrm{s}}(\alpha')^2} \ , \tag{5.161}$$

where f is the decay constant of the corresponding axion, as defined in Section 3.2.3. During inflation the worldvolume flux on the wrapped fivebrane decreases, and the axion vev drops. For large initial vev, the axion moves a large effective distance in field space. Provided that the axion shift symmetry continues to protect the potential across super-Planckian displacements, the result is a natural model of chaotic inflation in string theory. (We critically discuss the stability of the potential in the remainder of this section.) Inflation ends at a small axion vev, at which point the inflaton starts to oscillate around the minimum of the potential. Couplings between the axion and other fields will drain energy from the inflaton sector. If this energy transfer happens predominantly to the visible-sector degrees of freedom, then the system successfully reheats and the hot Big Bang is initiated (see Section 5.4.2).

Although axion monodromy from an NS5-brane source yields the asymptotically linear potential (5.161), many other variants of large-field inflation can arise via monodromy. One can parameterize the resulting theories in terms of an exponent p as

$$\mathcal{L} = -\frac{1}{2}(\partial \phi)^2 - \mu^{4-p} \phi^p .$$ (5.162)

The nilmanifold monodromy scenario of [779] yields $p = 2/3$, while versions of axion monodromy inflation with $p = 2$ can arise from an appropriate coupling of an axion to a four-form field strength [780],[58] or on a pair of sevenbranes [783].

A general mechanism known as *flattening* [551] can affect the asymptotic form of the scalar potential for a light field ϕ in the presence of additional heavy fields Ψ. Given appropriate couplings of ϕ to Ψ, integrating out Ψ flattens $V(\phi)$ at large ϕ, in the sense of reducing the exponent p. It was argued in [551] that the linear potential of (5.161) is an example of flattening; the type IIB action includes terms proportional to

$$S \supset \int d^{10}X \, |C_2 \wedge H_3|^2 ,$$ (5.163)

which naively give rise to an energy that is quadratic in $\phi \propto \int C_2$, but the actual potential (5.161) is linear. The claim of [551] is that backreaction of localized D3-brane charge, which shifts the moduli vevs, is responsible for the flattening from $p = 2$ to $p = 1$. Further examples of flattening in axion monodromy models arising in type IIB compactifications on Riemann surfaces appear in [784].

Compactification and tadpole cancellation

The attentive reader will appreciate by now that it is essential to check that a proposed inflationary mechanism in string theory – and any symmetries that underlie it – survives compactification and moduli stabilization. We turn to a critical discussion of these issues.

A fundamental consistency requirement in a compact model is cancellation of all tadpoles. Changing the axion vev, b, in the presence of a D5-brane alters the D3-brane charge induced on the D5-brane, because of the Chern–Simons coupling (3.30):

$$S_{\text{CS}} \supset i\mu_5 \int_{\mathcal{M}_4 \times \Sigma_2} C_4 \wedge \mathcal{F}_2 .$$ (5.164)

That is, because a D3-brane has the Chern–Simons coupling

$$S_{\text{CS}}^{\text{(D3)}} = i\mu_3 \int_{\mathcal{M}_4} C_4 ,$$ (5.165)

a D5-brane wrapping Σ_2, with

$$\int_{\Sigma_2} \mathcal{F}_2 \neq 0 ,$$ (5.166)

---

[58] For related field-theoretic constructions, see [781, 782].

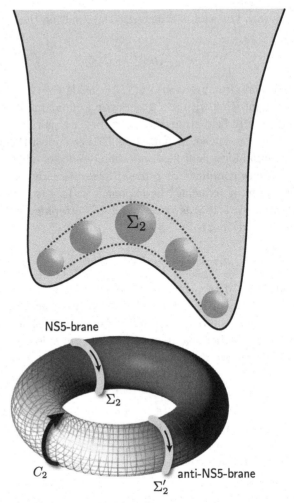

Fig. 5.12 The integral of the two-form C_2 over a two-cycle Σ_2 defines the c axion. In the presence of a wrapped NS5-brane this develops a monodromy. An anti-NS5-brane is required by Gauss's law on the compact space. The entire configuration should be situated in a warped region, and have a distant orientifold image (not shown). In the lower figure, the two-cycles are represented by circles.

carries non-vanishing D3-brane charge. Without a mechanism for absorbing or canceling this induced charge, Gauss's law would fix b to one definite value. However, there is a natural configuration in which the induced charge is canceled automatically: instead of a single D5-brane on a two-cycle Σ_2, consider a D5-brane and an anti-D5-brane, each wrapping Σ_2, but at different locations in the compact space,[59] as in the constructions of [785]. A similar construction applies for an NS5-brane pair (see Fig. 5.12), which, as we will see, yields a more promising inflationary model.

---

[59] That is, the D5-brane and anti-D5-brane wrap distinct, well-separated representatives of the same homology class.

Symmetry breaking from nonperturbative effects

To analyze the impact of moduli stabilization on axion monodromy inflation, it is necessary to specify a scenario for stabilization. Here, we will discuss the concrete example of axion monodromy inflation on fivebranes in type IIB compactifications with nonperturbatively stabilized Kähler moduli.

Eta problem for the b axion

To assess axion monodromy inflation with b as the inflaton, we consider a KKLT compactification with $h^{1,1}_+ = h^{1,1}_- = 1$, and correspondingly a single Kähler modulus T and a single complex "two-form scalar" $G = c - \tau b$. The $\mathcal{N} = 1$ supergravity data obtained from dimensional reduction takes the form

$$W = W_0 + \mathcal{A}e^{-2\pi T} , \tag{5.167}$$

$$K = -3\ln\left(T + \bar{T} + \gamma b^2\right) , \tag{5.168}$$

where $\gamma = e^{-\Phi}c_{+--}$, with c_{+--} the triple intersection number of the even and odd cycles.

Stabilization of T via the nonperturbative superpotential (5.167) breaks the shift symmetry for b, because of the way that b and T mix in the Kähler potential (5.168). One way to understand this effect is by considering the distinction between "physical volumes" and "holomorphic volumes." The proper Kähler coordinates on the moduli space are the complex scalar G and the holomorphic four-cycle volume T given in (3.65), but the physical volume of the compactification is $\mathcal{V} \propto (T + \bar{T} + \gamma b^2)^{3/2}$, which involves a nontrivial mixture of T with the inflaton b. If T is unstabilized, then the potential has a flat direction along which T and b shift, but $T + \bar{T} + \gamma b^2$ is invariant. Along this direction, the physical volume of the compactification is unchanged, but the holomorphic volume T is altered.

Kähler modulus stabilization through superpotential interactions involves the introduction of a potential for the holomorphic coordinate T. On the other hand, the presence of sources of positive or negative energy (such as fluxes and D-branes) leads to terms in the scalar potential that depend on physical volumes, and are proportional to powers of $e^K \propto \mathcal{V}^{-2}$. Because the potential energy therefore depends on both $T + \bar{T} + \gamma b^2$ and T, the flat direction along which b shifts is lifted. One finds that the canonically normalized field obtained from b has $\eta \sim 1$, so that inflation does not generically occur.

Notice that with the identification $b \to \phi$, $\gamma b^2 \to -\gamma k(\phi, \bar{\phi})$ the discussion above is exactly analogous to the eta problem in D3-brane inflation (see Section 5.1.2). The basic point is that a coordinate that appears in the Kähler potential almost certainly receives a potential on moduli stabilization.

Perturbative shift symmetry for the c axion

The NS-NS two-form B_2 is related to the R-R two-form C_2 by the S-duality subgroup of the full $SL(2,\mathbb{Z})$ duality, which also exchanges D5-branes and

NS5-branes. This immediately suggests an S-dual of the above scenario, in which an NS5-brane is wrapped on a two-cycle Σ_2, and the inflaton is proportional to $c \equiv \frac{1}{\alpha'} \int_{\Sigma_2} C_2$. The background flux compactification breaks $SL(2, \mathbb{Z})$, so the difficulties of the b axion model do not necessarily have to arise in the c axion model. The natural motivation for considering a c axion model is that c does not appear in the tree-level Kähler potential, so that the eta problem observed for b is absent. This is a direct consequence of the PQ symmetry for c, which is unbroken at the perturbative level, and constitutes one of the simplest realizations in string theory of the shift symmetry proposal of [539].

Nonperturbative symmetry breaking for the c axion
Let us now discuss the leading effects that break the shift symmetry of c. We begin at the level of the $\mathcal{N} = 1$ data, i.e. we first consider spontaneous breaking of the shift symmetry in the supersymmetric theory, before incorporating the explicit breaking due to the NS5-brane. In view of our general discussion in Section 3.2.3 of the breaking of axionic symmetries in string theory, the only possibilities involve nonperturbative effects.

Euclidean strings For a two-form symmetry, the first place to look is a two-dimensional worldvolume, corresponding to some form of Euclidean string. Although ordinary worldsheet instantons (Euclidean F-strings) do break the shift symmetry of b, fundamental strings do not carry charges under Ramond fields, so even worldsheet instantons will not break the shift symmetry of c. On the other hand, Euclidean D1-branes – or more generally, Euclidean (p, q) strings – do couple to C_2, and can break the symmetry. However, no such effect can be present in the superpotential, because of holomorphy; the real part of the action of any Euclidean string involves the volume of a two-cycle, which is not the real part of a holomorphic coordinate and therefore cannot appear in the superpotential [348]. (The corresponding superfields are linear superfields, not chiral superfields.) We conclude that symmetry-breaking terms from Euclidean D1-branes are confined to the Kähler potential. These contributions can easily be made negligible; they are nonperturbatively small, and – unlike Euclidean D3-brane terms in the super-potential – are unrelated to, and can be much smaller than, the terms appearing in the moduli potential. Indeed, recall that in a KKLT vacuum, perturbative corrections to K can be consistently neglected, and nonperturbative corrections to K are even smaller. See [37] for a more detailed discussion.

Euclidean D3-branes Because Euclidean D3-branes can make important contributions to the moduli potential, it is important to ask whether they also affect the potential for c. The action for a Euclidean D3-brane wrapping a four-cycle Σ_4 contains a term

$$S \supset i \int_{\Sigma_4} C_2 \wedge \mathcal{F}_2 , \qquad (5.169)$$

where $\mathcal{F}_2$ is the worldvolume flux. The path integral includes a sum over "magnetizations," i.e. over topologically distinct choices of $\mathcal{F}_2$, and if there exists a choice of magnetic flux such that (5.169) is nonvanishing, this will generally give rise to an eta problem for c. In short, unmagnetized Euclidean D3-branes do not affect c, but magnetized Euclidean D3-branes that intersect the NS5-brane can break the shift symmetry and lead to an eta problem.

Gaugino condensation on D7-branes Perhaps surprisingly, breaking of the shift symmetry of c by gaugino condensation on D7-branes is negligibly small. To see this, recall that the nonperturbative superpotential (3.110) from N_c D7-branes can be written in the form

$$W_{\lambda\lambda} = \mathcal{A} e^{-f/N_c} , \tag{5.170}$$

where the Wilsonian gauge kinetic function f – not to be confused with the axion decay constant – is a holomorphic function of the moduli, with $\mathrm{Re}(f) = 8\pi^2/g^2$. The gauge kinetic function is renormalized only at one loop and nonperturbatively:

$$f = f_0 + f_1 + f_{np} , \tag{5.171}$$

with $f_0 = 2\pi T$. The one-loop correction f_1 is independent of c, because perturbative strings do not carry R-R charge, so $W_{\lambda\lambda}$ can depend on c only through f_{np}. However, on general grounds $f_{np} \lesssim e^{-S}$, where S denotes the Euclidean action (for some appropriate extended object that couples to c, for example a magnetized Euclidean D3-brane) evaluated at a dominant saddle point. Thus,

$$W_{\lambda\lambda} = \mathcal{A} e^{-(f_0+f_1)/N_c} \left[1 + \mathcal{O}(e^{-S}) g(c) \right] , \tag{5.172}$$

where $g(c)$ is some function of c. We conclude that the dependence of the gaugino condensate superpotential on c is exponentially weaker than its dependence on T, so one can arrange for stabilization of T without inducing a large potential for c.

Let us summarize the nonperturbative effects that spontaneously break the shift symmetry of c. Euclidean (p, q) strings make innocuous periodic contributions to the Kähler potential, and do not introduce dangerously large masses for c. Gaugino condensation yields negligibly small dependence on c. Magnetized Euclidean D3-branes are problematic, and as a model-building criterion, one must ensure that any four-cycles stabilized by Euclidean D3-branes do not intersect the NS5-brane.[60]

---

[60] The precise condition on the triple intersection form is $c_{i\varphi\varphi} = 0 \ \forall i$, where φ denotes the orientifold-odd two-form corresponding to the inflaton, i.e. $c = G^\varphi \omega_\varphi$ (no sum), and i indexes all four-cycles stabilized by Euclidean D3-branes.

Symmetry breaking from backreaction

The arguments above establish that, given appropriate topology of the four-cycles involved in Kähler moduli stabilization, and taking the NS5-brane to be a *probe* of a fixed background compactification, the axion c has an all-orders shift symmetry for which the leading spontaneous breaking comes from Euclidean D1-brane contributions to the Kähler potential, and is easily made negligibly small. The leading explicit breaking originates in the NS5-brane DBI action (5.160); this is the candidate inflaton potential.

However, assuming that the NS5-brane is purely a probe of the compactification is not consistent, as originally noted in [35].

Backreaction from D3-brane charge

In the configuration with $c \gg 1$, a large D3-brane charge is induced on the NS5-brane. Upon solving the supergravity equations of motion with this charge as a source, one finds significant corrections to the warp factor[61] of the background that depend on c.[62] The critical question is whether a c-dependent warp factor leads to additional c-dependence in the potential energy, beyond that captured by the DBI action of the NS5-brane. We recall that the Kähler moduli T_i are assumed to be stabilized by a combination of Euclidean D3-branes and gaugino condensation on D7-branes, and the Euclidean D3-brane action (or D7-brane gauge coupling) depends on the warped volume of the corresponding four-cycle: cf. Eq. (3.108). Thus, a c-dependent modification of the warp factor entails a c-dependent correction to the exponentials appearing in the moduli potential. Schematically, one has

$$V(c) = \mu^3 f c + \widehat{V}[T_i(c)] , \qquad (5.173)$$

where the first term comes from the DBI action and $\widehat{V}$ stands for the moduli potential. The Kähler moduli T_i depend on c via the warp factor,

$$\mathrm{Re}(T_i) = \int_{\Sigma_4^i} \mathrm{d}^4 y \sqrt{g}\, e^{-4A(y;c)} . \qquad (5.174)$$

In generic configurations, the resulting c-dependence of the moduli potential is large enough to invalidate the derivation made in the probe approximation.

Fortunately, there is a mechanism that provides parametric suppression of this problematic backreaction effect. In order to satisfy Gauss's law, it was necessary to introduce an anti-NS5-brane, in addition to the NS5-brane; the induced charges on the brane and antibrane are equal and opposite. Thus, if the NS5-brane and anti-NS5-brane are relatively near to each other compared with their distance from the relevant four-cycles, the net flux of F_5 past the four-cycle will

---

[61] This is easily understood if one recalls that the backreaction of D3-branes in flat space leads to the $AdS_5 \times S^5$ geometry.

[62] The corresponding calculation for a D3-brane was performed in [346] – see also the discussion following Eq. (4.10).

be suppressed, and the correction to the warp factor will be correspondingly small. Instead of seeing a monopole D3-brane charge, the four-cycles see only a dipole. A concrete realization of this protective mechanism involves placing an NS5-brane and an anti-NS5-brane in a common warped throat, as discussed in detail in [37].[63]

To recap, backreaction of the induced D3-brane charge that is ultimately responsible for the inflationary energy leads to a correction to the warp factor that affects the scalar potential by modifying the Euclidean D3-brane action. This is consistent with the general arguments, in that the breaking of the axionic shift symmetry by backreaction effects is, strictly speaking, "nonperturbatively small," being proportional to $\exp(-T_p \mathrm{Vol}(\Sigma_p))$. However, it is essential to understand that the moduli potential, and the inflationary vacuum energy, are necessarily nonperturbatively small in the same sense. For the backreaction to be a small correction, the geometry must be arranged to respect an additional approximate symmetry, e.g. by situating the fivebrane pair at the bottom of a warped throat, as noted above. The original axion shift symmetry, on its own, does not suffice to guarantee a flat potential.

Backreaction from the NS5-branes

There is another backreaction effect that presents a possible concern [786]. The NS5-brane/anti-NS5-brane pair explicitly breaks supersymmetry, and moreover either member of the pair, in isolation, sources a fivebrane tadpole that is canceled by its counterpart. A toroidal orbifold computation presented in [786] shows that – in that unwarped setting – the backreaction of the fivebranes themselves, not including the induced D3-brane charge described above, grows logarithmically with their separation. The scale of the potential from backreaction is set by the mass of strings stretched between the branes. The conclusion of [786] is that for an NS5-brane/anti-NS5-brane pair that occupy well-separated throats, the potential from fivebrane backreaction is of order the unwarped bulk scale. It would be worthwhile to understand whether the findings of [786], especially the claim that the backreaction of a homologous fivebrane pair has real codimension two, are applicable in general warped backgrounds. One should bear in mind that the geometric configuration required to address the problem of D3-brane backreaction – namely, situating the fivebrane pair in a common warped region – also ameliorates the fivebrane backreaction described in [786].

The difficulties described above are substantially alleviated in the two-field generalization of axion monodromy known as *Dante's Inferno* [264], in which the inflaton trajectory is a gradual spiral in the two-dimensional axion field space. A hierarchy between the two axion decay constants – which is plausibly

---

[63] A precise computation of the axion decay constant in such a setting is an open problem. If the decay constant is significantly reduced by warping, the requisite induced charge increases, intensifying the problem of backreaction.

radiatively stable – provides the large number that serves to enlarge the effective axion field range, and correspondingly to suppress backreaction. One striking feature of this scenario is that the length of the inflationary trajectory can be parametrically larger than the diameter of the region in field space traversed during inflation. This makes it possible to produce a large primordial tensor signal without a super-Planckian *displacement*, although the arc length along the trajectory remains super-Planckian.

Reheating

Reheating in a string theory model with a shift-symmetric inflaton poses particular difficulties, as explored in [576, 579, 581, 583]. The general issue is that the shift symmetry that protects the inflaton simultaneously limits the couplings of the inflaton to the visible sector. In a naive model containing only the inflaton, the visible sector, and general relativity – with no additional sectors associated with the ultraviolet completion of gravity – the primary consequence would be slow reheating, which is not necessarily fatal. However, if the inflaton couples at least as strongly to hidden sector fields as it does to the visible sector, which is frequently the case in models of closed string inflation, then problematic reheating of the hidden sectors is difficult to avoid.

At present, no concrete results concerning reheating are available for the specific scenario of axion monodromy inflation with a c axion coupled to an NS5-brane pair, but we find it reasonable to expect that reheating of hidden sectors, including fields associated with the NS5-brane pair and the enveloping warped geometry, will present a challenge for the model.

5.4.3 Phenomenology

The phenomenology of axion inflation can be extremely rich.[64] Besides the model-independent gravitational wave signal, a host of additional model-dependent signatures have been explored, including oscillations in the power spectrum [37], deviations from scale-invariance [788], non-Gaussianity [38, 789], chiral gravitational waves [270, 788], and primordial black holes [790]. Since all of these effects are tied to the underlying axion shift symmetry, one has the hope of finding correlated signatures across different observational channels.

Signatures of nonperturbative effects

At leading order in the instanton expansion, the Lagrangian for axion monodromy inflation takes the form

$$\mathcal{L} = -\frac{1}{2}(\partial\phi)^2 - V_0(\phi) - \Lambda^4 \cos\left(\frac{\phi}{f}\right), \tag{5.175}$$

---

[64] See [787] for a recent review.

where $V_0(\phi) \equiv \mu^{4-p}\phi^p$. Since the scale Λ is generated nonperturbatively, the modulations can quite naturally be exponentially small, but it is also possible that the modulations could be large enough to spoil the monotonicity of the inflaton potential. Here, we assume that Λ happens to be large enough to be phenomenologically interesting, but not large enough to dominate the evolution. The monotonicity constraint is

$$b_\star \equiv \frac{\Lambda^4}{V_0'(\phi_\star)f} < 1 \, . \tag{5.176}$$

This parameter depends on the inflaton vev, unless the potential is linear. Here, we have evaluated it at $\phi = \phi_\star$, the value of the inflaton when the pivot scale $k = k_\star$ exits the horizon.

Before proceeding, we point out that $f/M_{\rm pl}$ is bounded from below, for two reasons. First, requiring the α' expansion to be under good control (in the NS5-brane construction [35] with $p = 1$) implies [37]

$$\frac{f^2}{M_{\rm pl}^2} > \frac{\sqrt{g_{\rm s}}}{(2\pi)^3 \mathcal{V}} \, . \tag{5.177}$$

Another bound on f arises from requiring the effective theory of the fluctuations to be weakly coupled for the parameter values of interest [791]. Control over the observational predictions requires the oscillation frequency $\omega = (-2M_{\rm pl}^2\dot{H})^{1/2}/f$ to be much smaller than the unitarity bound $4\pi f$. This implies

$$f \gg (2\Delta_\mathcal{R})^{-1/2} \frac{H}{2\pi} \, . \tag{5.178}$$

Using (1.37), we get

$$\frac{f}{M_{\rm pl}} \gg \sqrt{\frac{r}{16}} \Delta_\mathcal{R}^{1/2} \approx 4.5 \times 10^{-4} \left(\frac{r}{0.07}\right)^{1/2} \, . \tag{5.179}$$

The predictions for cosmological observables depend on the parameters $b_\star$ and f, as well as the parameters of the potential V_0 (e.g. $V_0(\phi_\star)$, $\epsilon_\star$ and $\eta_\star$).

Striking signatures of axion monodromy inflation arise from the periodic modulation in (5.175). In a general inflationary model, the solutions for the primordial perturbations inside the horizon are oscillatory. The periodic potential in (5.175) introduces a periodic driving force, which can resonate with the freely oscillating perturbations, modulating their amplitude as a function of wavenumber k [37, 792]. This resonance leads to modulations of the power spectrum, as well as to a specific type of non-Gaussianity.

Modulated power spectrum

For $b_\star \ll 1$, the oscillatory term in the potential can be treated as a perturbation. At first order in $b_\star$, the power spectrum is [37]

$$\Delta_{\mathcal{R}}^2(k) = \Delta_{\mathcal{R}}^2(k_\star) \left(\frac{k}{k_\star}\right)^{n_s-1} \left[1 + \mathcal{A}\cos\left(\frac{\phi_k}{f}\right)\right] , \tag{5.180}$$

where $\phi_k \simeq \phi_\star - \sqrt{2\epsilon_\star} \ln(k/k_\star)$ is the field value at horizon exit of the mode k. We have defined $n_s = 1 + 2\eta_\star - 6\epsilon_\star$ and

$$\mathcal{A} \equiv 3b_\star \left(\frac{2\pi f}{\sqrt{2\epsilon_\star}}\right)^{1/2} , \tag{5.181}$$

in units where $M_{\rm pl} \equiv 1$, and have worked to leading order in $f/\sqrt{2\epsilon_\star}$. For the special case $p = 1$, i.e. the linear potential that arises from NS5-branes, we have $1/\sqrt{2\epsilon_\star} = \phi_\star$, and $\Delta_{\mathcal{R}}^2(k)$ is known for arbitrary $f\phi_\star$:

$$\mathcal{A}_{p=1} = \frac{6b_\star}{\sqrt{1 + (3f\phi_\star)^2}} \sqrt{\frac{\pi}{2}} \coth\left(\frac{\pi}{2f\phi_\star}\right) f\phi_\star . \tag{5.182}$$

These oscillations in the power spectrum have recently been searched for in the WMAP data [793] and the Planck data [794]. So far, no signal has been detected.[65]

Resonant non-Gaussianity

At first order in $b_\star$, the bispectrum is [38]

$$B_{\mathcal{R}}(k_1, k_2, k_3) = f_{\rm NL}^{\rm res} \times \frac{(2\pi\Delta_{\mathcal{R}})^4}{k_1^2 k_2^2 k_3^2} \left[\sin\left(\frac{\sqrt{2\epsilon_\star}}{f} \ln\frac{K}{k_\star}\right)\right.$$
$$\left. + \frac{f}{\sqrt{2\epsilon_\star}} \sum_{i\neq j} \cos\left(\frac{\sqrt{2\epsilon_\star}}{f} \ln\frac{K}{k_\star}\right) + \cdots\right] , \tag{5.183}$$

where $K \equiv k_1 + k_2 + k_3$ and

$$f_{\rm NL}^{\rm res} \equiv \frac{3\sqrt{2\pi}}{8} b_\star \left(\frac{\sqrt{2\epsilon_\star}}{f}\right)^{3/2} . \tag{5.184}$$

The ellipsis in (5.183) stands for terms that are suppressed by higher powers of the slow-roll parameters or by positive powers of $f/\sqrt{2\epsilon_\star}$. Observable non-Gaussianity requires $f/\sqrt{2\epsilon_\star} \ll 1$. In this case, the second term in (5.183) is suppressed relative to the first term, except in the squeezed limit (where it ensures that Maldacena's consistency relation (1.70) holds). The unitarity bound (5.179) implies an upper bound on $f_{\rm NL}^{\rm res}$,

$$f_{\rm NL}^{\rm res} \ll \frac{3\sqrt{\pi}}{2}(2\Delta_{\mathcal{R}})^{-3/4} \approx 3\times 10^3 . \tag{5.185}$$

---

[65] Using WMAP9, it was found that a modulation with $\log_{10}(f/M_{\rm pl}) = -3.38$ improves the fit by $\Delta\chi^2 = 19$ [793], but the frequency of this signal coincides with the unitarity bound (5.179). Moreover, the signal does not seem to be present in the Planck data [794, 795].

Because the bispectrum is oscillating, it is nearly orthogonal to the standard bispectrum templates. It is therefore barely constrained by the bispectrum results (1.72)–(1.74), and a dedicated analysis is required to put meaningful constraints on the parameter $f_{\rm NL}^{\rm res}$.

Signatures of gauge field production

In order to reheat, the axion has to be coupled to extra fields, and one can ask whether these couplings affect the perturbations during inflation. Particularly interesting is the following dimension-five operator that couples the inflaton to a gauge field:[66]

$$\mathcal{L} \supset -\frac{\alpha}{4}\frac{\phi}{f}F\tilde{F} \,, \tag{5.186}$$

where F is the gauge field strength. This coupling respects the shift symmetry of the inflaton, as for constant ϕ the operator is a total derivative. When the inflaton has a time-dependent vev, $\phi(t)$, the conformal invariance of the gauge field is broken. This leads to production of gauge field quanta during inflation. To see this, consider the equation of motion for the two polarization modes of the gauge field (in Coulomb gauge):

$$\left(\frac{\partial^2}{\partial\tau^2} + k^2 \mp 2aHk\xi\right) A_\pm(\tau, k) = 0 \,, \qquad \text{where} \quad \xi \equiv \frac{\alpha\dot\phi}{2fH} \,. \tag{5.187}$$

We see that one of the helicities of the gauge field experiences tachyonic growth for $k/(aH) < 2\xi$. For $\xi > 0$, the unstable mode is A_+. Most of the power in the produced gauge field is in modes with $(8\xi)^{-1} < k/(aH) < 2\xi$. In this regime, the solution can be written as [796]

$$A_+(\tau, k) \simeq \frac{1}{\sqrt{2k}}\left(\frac{k}{2\xi aH}\right)^{1/4} e^{\pi\xi - 2\sqrt{2\xi k/(aH)}} \,. \tag{5.188}$$

The coupling of this solution to the inflaton, via (5.186), leads to a number of observational signatures.

Equilateral non-Gaussianity

The gauge field nonlinearities in the operator (5.186) source non-Gaussian inflaton fluctuations (this may be thought of as an inverse decay, $\delta A + \delta A \to \delta\phi$). The bispectrum is of equilateral type and has the amplitude [789]

$$f_{\rm NL}^{\rm equil} \simeq \frac{\Delta_{\mathcal{R},0}^6}{\Delta_{\mathcal{R}}^4} f_3(\xi) e^{6\pi\xi} \,, \tag{5.189}$$

---

[66] We will assume $\alpha \lesssim 1$. The case $\alpha \gg 1$ is discussed in [796], while estimates for α in type IIB string theory appear in [788]. Both bottom-up and top-down naturalness of the regime $\alpha \gg 1$ remain to be established.

where $\Delta_{\mathcal{R},0}$ stands for $\Delta_{\mathcal{R}}|_{\xi=0} = H^2/(2\pi\dot{\phi})$. The function $f_3(\xi)$ is determined numerically, but has the following limits:

$$f_3(\xi) = 2.8 \times 10^{-7}\,\xi^{-9} \quad \text{for} \quad \xi \gg 1 , \tag{5.190}$$

$$f_3(\xi) \approx 7.4 \times 10^{-8}\xi^{-8.1} \quad \text{for} \quad 2 < \xi < 3 . \tag{5.191}$$

We see that $f_{\rm NL}^{\rm equil}$ is exponentially sensitive to the model parameter ξ. Let us denote by $\xi_\star$ the value of ξ at the pivot scale $k_\star = 0.002\ \text{Mpc}^{-1}$. Using the seven-year WMAP data, [797] found $\xi_\star < 2.45$ (95% CL). In terms of the axion decay constant, this corresponds to

$$f > \frac{\alpha}{10\pi}\frac{H}{\Delta_{\mathcal{R}}} . \tag{5.192}$$

Using (1.37), this can be written as

$$\frac{f}{M_{\rm pl}} > \frac{\alpha}{10}\sqrt{\frac{r}{2}} \approx 2 \times 10^{-2}\left(\frac{\alpha}{1}\right)\left(\frac{r}{0.07}\right)^{1/2} . \tag{5.193}$$

For $\alpha \sim \mathcal{O}(1)$, this provides a strong constraint on the axion decay constant. This bound is similar in spirit to the bound in (2.117).

Non-scale-invariance

The power spectrum of curvature perturbations receives contributions both from the vacuum fluctuations of the inflaton and the fluctuations sourced by the gauge field,

$$\Delta_{\mathcal{R}}^2(k) = \Delta_{\mathcal{R},0}^2(k)\left[1 + \Delta_{\mathcal{R},0}^2(k)\,f_2(\xi)e^{4\pi\xi}\right] , \tag{5.194}$$

where

$$f_2(\xi) = 7.5 \times 10^{-5}\,\xi^{-6} \quad \text{for} \quad \xi \gg 1 , \tag{5.195}$$

$$f_2(\xi) \approx 3.0 \times 10^{-5}\xi^{-5.4} \quad \text{for} \quad 2 < \xi < 3 . \tag{5.196}$$

For large ξ, the sourced fluctuations can dominate over the vacuum fluctuations. Notice that $\xi \propto \sqrt{\epsilon}$ grows during inflation. Although ξ grows slowly, it appears in the exponent in (5.194), and can therefore lead to significant scale dependence of the power spectrum. Although this effect is constrained mainly by small-scale CMB and LSS data, it remains convenient to express the constraint as a bound on the parameter $\xi_\star$ (evaluated at $k_\star = 0.002\ \text{Mpc}^{-1}$). Assuming a quadratic inflaton potential and using WMAP and ACT data, [797] found $\xi_\star < 2.41$ (95% CL).

Primordial black holes

The growth of ξ can lead to the formation of primordial black holes. This can disturb the standard cosmology. Estimating the effects of strong backreaction at the end of inflation, [790] finds $\xi_\star < 1.5$. Although this is the strongest constraint on the parameter $\xi_\star$, it is plausibly subject to the largest theoretical uncertainties.

Chiral gravitational waves

The stress tensor associated with the produced gauge fields sources gravitational waves. An interesting property of the resulting tensor signal is that it is chiral [798], essentially because only one chirality of the gauge field is unstable. This parity violation could, in principle, be tested using the TB and EB correlators of the CMB [799]. However, to achieve a sufficiently large tensor amplitude requires values of ξ_* that are already ruled out by the bound on non-Gaussianity. On the other hand, the tensor modes become large on small scales, just like the scalars. It is therefore conceivable for the signal to be small on CMB scales, but detectable on scales accessible to terrestrial interferometers. In [800] it was estimated that $\xi_* > 2.2$ could be probed with Advanced LIGO.

5.5 Inflating with Kähler moduli

A very early idea for inflation in string theory was that a modulus could be the inflaton [801, 802]. In particular, the compactification volume is invariably present in the four-dimensional effective theories of string compactifications, and it is natural to ask if the volume modulus could be the field driving slow-roll inflation. In this section, we will describe scenarios in which the inflaton is a Kähler modulus, or an axion paired with a Kähler modulus.[67]

5.5.1 Racetrack inflation

Modular inflation in the context of flux compactifications was first realized in the *racetrack inflation* scenario [637]. As a concrete example, we consider a KKLT compactification with a single Kähler modulus T, and a superpotential of the "racetrack" form [637]

$$W = W_0 + \mathcal{A}\,e^{-aT} + \mathcal{B}\,e^{-bT} \,, \tag{5.197}$$

where W_0 is the constant flux superpotential, $\mathcal{A}$ and $\mathcal{B}$ are prefactors that depend on the vevs of the stabilized complex structure moduli, and a, b are constants. A superpotential of this form can be generated by gaugino condensation in a product gauge group; for gauge group $SU(N) \times SU(M)$ one has $a = 2\pi/N$ and $b = 2\pi/M$. The classical Kähler potential takes the form

$$K = -3\ln(T + \bar{T}) \,, \tag{5.198}$$

up to α' corrections that will be discussed momentarily. To complete the specification of the effective theory, we incorporate supersymmetry breaking by an anti-D3-brane in a warped throat region, which leads to a term in the potential of the form

$$\delta V = \frac{\varrho}{(T + \bar{T})^2} \,, \tag{5.199}$$

---

[67] This section is based mostly on [44, 370, 639]. For a recent review see [31].

with ϱ a constant that depends on the warp factor at the location of the anti-D3-brane.[68]

The authors of [637] showed that for suitable values of the parameters $W_0, \mathcal{A}, \mathcal{B}, a, b, \varrho$, the potential for T develops a saddle point[69] that is suitable for inflation. The evolution is primarily in the direction of the axion $\text{Im}(T)$, but $\text{Re}(T)$, corresponding to the compactification volume, does also evolve. In a related construction, [804] followed the Kallosh–Linde scenario [665] for Kähler moduli stabilization via a racetrack, and showed that for special values of the parameters, the axion $\text{Im}(T)$ is stabilized, but a single-field inflection point appears in the $\text{Re}(T)$ direction. The volume modulus then serves as the inflaton.

The parameter values that can lead to racetrack inflation in type IIB string theory are quite contrived.[70] This is not a fatal objection; because W_0, $\mathcal{A}$, $\mathcal{B}$ are determined by the vevs of complex structure moduli, which are in turn dictated by quantized fluxes, it is possible in principle to adjust their values rather precisely (as originally noted for the cosmological constant problem in [197]). On the other hand, large values of M, N require stacks of many D7-branes, which can be difficult to construct in explicit compactifications.

It is important to recognize that fine-tuning the leading-order classical effective action does not necessarily lead to a self-consistent model. Quantum effects, as well as curvature corrections from the α' expansion, inevitably contribute to the action, and it is not consistent to study solutions of the leading-order theory that require precision comparable to the size of these corrections. Indeed, it was shown in [805] that the higher-curvature term of (2.30) generically destabilizes $\text{Re}(T)$, and also renders the potential more steep in the $\text{Im}(T)$ direction, spoiling inflation. Although it is conceivable that new parameter values could be found for which the corrected action yields inflation, the complete set of contributing terms has not yet been determined, so that it is difficult to make the model more explicit and predictive.

A related idea is that the volume modulus T can serve as the inflaton if finely tuned combinations of corrections to the Kähler potential, cf. Section 3.3.2, lead to the appearance of an inflection point in the potential [806]. The scenario of [806] alleviates the tension between low-energy supersymmetry and high-scale inflation identified in [665] – see also [807–811].

---

[68] For an explanation of the exponent 2 in (5.199), which differs from the result given in [369], see [42].

[69] Racetrack-type superpotentials for Kähler moduli have also been argued to yield inflection point inflation [803].

[70] The example given in [637] has $N = 90$ and $M = 100$, as well as fine-tuned values for $W_0, \mathcal{A}, \mathcal{B}, \varrho$. In the more explicit construction of [638] (building on moduli stabilization results of [371]), successful inflation was found for $N = 40$ and $M = 258$, again with fine-tuning of the remaining parameters, while the scenario of [804] used $N = 58$ and $M = 60$, with B specified to 11 decimal places.

5.5.2 Large volume compactifications

The most concrete models of Kähler moduli inflation have been constructed in the Large Volume Scenario (LVS) [330]. We reviewed LVS compactifications in Section 3.3.3. For convenience, we will quickly summarize the aspects of that discussion that will be relevant for inflationary model-building.

The tree-level Kähler potential in the presence of the leading sigma model corrections, at order $(\alpha')^3$ [332] is

$$K = K_0 + \delta K_{(\alpha')} = -2\ln(\mathcal{V}) - \frac{\hat{\xi}}{\mathcal{V}} \,. \tag{5.200}$$

The contribution $\delta K_{(\alpha')}$ breaks the no-scale structure, but only lifts the volume modulus $\mathcal{V}$, leaving $h_+^{1,1} - 1$ complexified Kähler moduli massless (at this level of approximation). These are natural inflaton candidates. To assess whether such scenarios are viable, it is important to consider string loop corrections to the Kähler potential,

$$K = K_0 + \delta K_{(\alpha')} + \delta K_{(g_{\mathrm{s}})} \,, \tag{5.201}$$

but these corrections are difficult to compute. The only explicit results available are those obtained by Berg, Haack, and Körs for $\mathcal{N} = 1$ compactifications on the toroidal orientifold $T^6/(\mathbb{Z}_2 \times \mathbb{Z}_2)$ [333, 334], cf. Eqs. (3.103) and (3.104). Some progress has been made on extending these results to general Calabi–Yau manifolds [335, 812], and a specific functional form of the string loop corrections has been conjectured by Berg, Haack, and Pajer in [335] (see also [336] for a general discussion). Recall from Section 3.3.3 that these corrections can be separated into two types: those associated with the exchange of closed strings with Kaluza–Klein momentum, and those associated with the exchange of strings that wind non-contractible cycles.[71] The correction $\delta K_{(g_{\mathrm{s}})}^{\mathrm{KK}}$ is conjectured to be

$$\delta K_{(g_{\mathrm{s}})}^{\mathrm{KK}} \sim g_{\mathrm{s}} \sum_{i=1}^{h^{1,1}} \frac{\mathcal{C}_i^{\mathrm{KK}}(\zeta,\bar{\zeta}) M_{\mathrm{KK}}^{-2}}{\mathcal{V}} \sim g_{\mathrm{s}} \sum_{i=1}^{h^{1,1}} \frac{\mathcal{C}_i^{\mathrm{KK}}(\zeta,\bar{\zeta})(a_{ij}t^j)}{\mathcal{V}} \,, \tag{5.202}$$

where $a_{ij}t^j$ is some linear combination of the two-cycle size moduli t^j. The conjectured result for $\delta K_{(g_{\mathrm{s}})}^{\mathrm{W}}$ is

$$\delta K_{(g_{\mathrm{s}})}^{\mathrm{W}} \sim \sum_i \frac{\mathcal{C}_i^{\mathrm{W}}(\zeta,\bar{\zeta}) M_{\mathrm{W}}^{-2}}{\mathcal{V}} \sim \sum_i \frac{\mathcal{C}_i^{\mathrm{W}}(\zeta,\bar{\zeta})}{(b_{ij}t^j)\mathcal{V}} \,, \tag{5.203}$$

where the two-cycle $b_{ij}t^j$ corresponds to the curve of intersection of two D7-branes (see [335] for details). The unknown complex structure dependence has been absorbed into the functions $\mathcal{C}_i^{\mathrm{KK}}(\zeta,\bar{\zeta})$ and $\mathcal{C}_i^{\mathrm{W}}(\zeta,\bar{\zeta})$. Since we assume that

---

[71] The cycles in question are one-cycles within the curve of intersection of two D7-branes; the parent Calabi–Yau manifold does not have any non-contractible (non-torsion) one-cycles.

the complex structure moduli are stabilized at higher energies, we will treat them as constants, focusing on the dependence on the Kähler moduli.

The supergravity approximation holds when $t_i \gg 1$, in which case

$$\delta K_{(g_s)}^{\mathrm{KK}} \sim \sum_i \frac{t_i}{\mathcal{V}} > \delta K_{(\alpha')} \sim \frac{1}{\mathcal{V}} \,. \tag{5.204}$$

One might then worry that at large volume the g_s corrections (5.202) and (5.203) overwhelm the α' corrections $\delta K_{(\alpha')}$ and drastically change the vacuum structure, to say nothing of the inflationary phenomenology. However, what happens is a bit more subtle [335, 812]. Although the g_s corrections dominate over the α' corrections in the Kähler potential, they cancel to a certain degree in the scalar potential, so that the dominant contribution to the scalar potential actually comes from the α' corrections. This important phenomenon is called *extended no-scale structure* [812]. It arises because $\delta K_{(g_s)}^{\mathrm{KK}}$ is a homogeneous function, of degree -2, in the two-cycle volumes t_i [812].

To illustrate extended no-scale structure, we consider an example with a single modulus τ [31]. Schematically, we can write the Kähler potential as

$$K = -2\ln(\mathcal{V}) - \frac{\hat{\xi}}{\mathcal{V}} + \frac{\sqrt{\tau}}{\mathcal{V}} \,. \tag{5.205}$$

Taking the superpotential to be a constant, $W = W_0$, we find

$$V = \frac{W_0^2}{\mathcal{V}^3} \left[0 + \hat{\xi} + 0 \cdot \sqrt{\tau} + \frac{1}{\sqrt{\tau}} + \frac{1}{\tau^{3/2}} \right] \,. \tag{5.206}$$

The first zero in (5.206) corresponds to the famous no-scale structure, while the second vanishing contribution (namely, $0 \cdot \sqrt{\tau}$) is the consequence of what we have just referred to as extended no-scale structure. We see that the leading g_s contribution to the scalar potential scales as $1/\sqrt{\tau}$, and is *smaller* than the leading α' contribution proportional to $\hat{\xi}$. The g_s contribution to the scalar potential is therefore smaller than naively expected. Even so, we will find that g_s corrections can still make dangerously large contributions to the inflaton potential.

Extended no-scale structure can be understood from an alternative point of view [812]. In the low-energy effective field theory, we can interpret the g_s contribution as the one-loop Coleman–Weinberg potential,

$$\delta V_{\mathrm{CW}} \simeq 0 \cdot \Lambda^4 + \Lambda^2 \mathrm{STr}(M^2) + \mathrm{STr}\left[M^4 \ln\left(\frac{M^2}{\Lambda^2} \right) \right] \,, \tag{5.207}$$

where

$$\Lambda = M_{\mathrm{KK}} \simeq \frac{M_{\mathrm{pl}}}{\mathcal{V}^{2/3}} \,, \quad \mathrm{STr}(M^2) \simeq \frac{M_{\mathrm{pl}}^2}{\mathcal{V}^2} \,, \tag{5.208}$$

and STr denotes the supertrace.[72] The first term in (5.207) vanishes by supersymmetry. Substituting (5.208) into (5.207), we get

$$\delta V_{\rm CW} \simeq 0 \cdot \frac{1}{\mathcal{V}^{8/3}} + \frac{1}{\mathcal{V}^{10/3}} + \frac{1}{\mathcal{V}^4} , \qquad (5.209)$$

which, upon using $\tau \sim \mathcal{V}^{2/3}$, can be written as

$$\delta V_{\rm CW} \simeq \frac{1}{\mathcal{V}^3} \left[0 \cdot \sqrt{\tau} + \frac{1}{\sqrt{\tau}} + \frac{1}{\tau^{3/2}} \right] . \qquad (5.210)$$

This precisely matches the scaling in (5.206), implying that the extended no-scale feature of the potential is indeed related to supersymmetry [812].

In the rest of this section, we will use these results to construct inflationary solutions in LVS. We will describe three ways in which the inflaton potential is generated: (i) via nonperturbative effects [639] (Section 5.5.3), (ii) via string loops [44] (Section 5.5.4), and (iii) via poly-instanton effects [640, 813] (Section 5.5.5).

5.5.3 Blow-up inflation

The first models of Kähler moduli inflation [639] were constructed in Swiss-cheese compactifications of the Large Volume Scenario (see Section 3.3.3.) In order for one of the blow-up cycles to play the role of the inflaton, while keeping the overall volume fixed, at least three Kähler moduli are required (see Fig. 5.13). The compactification volume is then

$$\mathcal{V} = \alpha \left(\tau_b^{3/2} - \lambda_\phi \tau_\phi^{3/2} - \lambda_s \tau_s^{3/2} \right) . \qquad (5.211)$$

We will look at the part of the moduli space satisfying the hierarchies $\tau_b \gg \tau_\phi \gg \tau_s$. The field τ_b then determines the overall volume, while τ_ϕ and τ_s are blow-up cycles.

Inflaton potential

We first assume that string loop corrections can be ignored, so that the Kähler potential is given by (5.200). For the superpotential, we take

$$W = W_0 + \mathcal{A}_\phi e^{-a_\phi T_\phi} + \mathcal{A}_s e^{-a_s T_s} . \qquad (5.212)$$

---

[72] The supertrace is defined by $\mathrm{STr}(M^2) \equiv \sum_s (2s+1)(-1)^{2s} \mathrm{Tr}(M_s^2)$, where s is the spin and M_s^2 is the matrix of masses squared for particles of spin s. Note that bosonic and fermionic contributions enter with opposite sign.

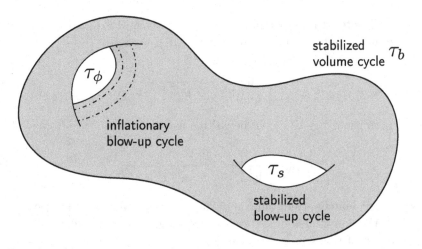

Fig. 5.13 Blow-up inflation in a three-modulus Swiss-cheese compactification. An evolving blow-up cycle $\tau_\phi(t)$ drives inflation, while a second blow-up cycle τ_s stabilizes the overall volume $\mathcal{V}(\tau_b)$.

This structure stabilizes the moduli[73] τ_b and τ_s, with $\mathcal{V} \sim \alpha \tau_b^{3/2} \sim e^{a_s \tau_s}$ and $\tau_s \gg 1$. Integrating out τ_b and τ_s leads to the potential [639][74]

$$V = W_0^2 \left(\mathfrak{a} \frac{\sqrt{\tau_\phi}\, e^{-2a_\phi \tau_\phi}}{\mathcal{V}} - \mathfrak{b} \frac{\tau_\phi\, e^{-a_\phi \tau_\phi}}{\mathcal{V}^2} + \mathfrak{c} \frac{\hat{\xi}}{\mathcal{V}^3} \right) , \qquad (5.213)$$

where $\mathfrak{a}$, $\mathfrak{b}$, and $\mathfrak{c}$ are order-one coefficients [639]. While $\mathcal{V}$ is fixed[75] during inflation, τ_ϕ will evolve, playing the role of the inflaton. For large τ_ϕ, the last two terms in (5.213) dominate and determine the inflaton potential

$$V(\phi) \simeq V_0 \left(1 - c_1 \mathcal{V}^{5/3} \phi^{4/3} \exp\left[-c_2 \mathcal{V}^{2/3} \phi^{4/3} \right] \right) , \qquad (5.214)$$

where we have defined the canonically normalized inflaton,

$$\phi \equiv \sqrt{4\lambda_\phi/(3\mathcal{V})}\, \tau_\phi^{3/4} , \qquad (5.215)$$

as well as the parameters $V_0 \equiv \mathcal{O}(1) \times W_0^2 \hat{\xi} \mathcal{V}^{-3}$, $c_1 \equiv \mathcal{O}(1) \times \hat{\xi}^{-1}$, and $c_2 \equiv \mathcal{O}(1) \times a_\phi$. It is instructive to expand (5.214) around the vev of ϕ in the minimum of (5.213), where

$$a_\phi \langle \tau_\phi \rangle \approx \mathcal{O}(1) \times \ln(\mathcal{V}) \qquad \Leftrightarrow \qquad \langle \phi \rangle \approx \mathcal{O}(1) \times \frac{\ln(\mathcal{V})^{3/4}}{\mathcal{V}^{1/2}} . \qquad (5.216)$$

---

[73] The axionic partner ϑ_ϕ of the inflaton τ_ϕ is not necessarily stabilized, and allowing ϑ_ϕ to have a nonvanishing initial velocity leads to the rich dynamics known as *roulette inflation* [814].

[74] This corrects a misprint in the corresponding formula (41) in [31].

[75] See [815] for numerical evidence supporting the validity of the single-field approximation.

Writing $\phi = \langle\phi\rangle + \hat\phi$ and using $\mathcal{V} \gg 1$, we find

$$V(\hat\phi) \simeq V_0 \left(1 - \kappa_1 e^{-\kappa_2\hat\phi}\right) , \tag{5.217}$$

where[76]

$$\kappa_1 \equiv c_1 \mathcal{V}^{5/3} \langle\phi\rangle^{4/3} \approx \mathcal{O}(\mathcal{V}\ln(\mathcal{V})) , \tag{5.218}$$

$$\kappa_2 \equiv \frac{4}{3} c_2 \mathcal{V}^{2/3} \langle\phi\rangle^{1/3} = \mathcal{O}(\mathcal{V}^{1/2}\ln(\mathcal{V})^{1/4}) . \tag{5.219}$$

Eta problem from string loops

The region of interest for inflation is given by

$$\mathcal{V}^{2/3}\phi^{4/3} \gg 1 \qquad \text{and} \qquad \phi \ll 1 , \tag{5.220}$$

where the first condition renders the potential (5.214) exponentially flat, while the second ensures that $\tau_\phi \ll \tau_b$, i.e. that the inflationary blow-up cycle makes a negligible contribution to the overall volume $\mathcal{V} \approx \alpha\tau_b^{3/2}$. However, at this point one should remember that we have not yet included string loop corrections. Using (5.205), we can estimate the string loop correction to the inflaton potential,

$$\delta V_{(g_s)} \sim \frac{1}{\sqrt{\tau_\phi}\,\mathcal{V}^3} \sim \frac{1}{\phi^{2/3}\mathcal{V}^{10/3}} . \tag{5.221}$$

The associated correction to the η parameter is

$$\delta\eta \sim \frac{\delta V''_{(g_s)}}{V_0} \sim \frac{1}{\phi^{8/3}\mathcal{V}^{1/3}} \sim \frac{\mathcal{V}}{\tau_\phi^2} , \tag{5.222}$$

where we are still using units with $M_{\rm pl} \equiv 1$. Using $\tau_\phi \approx \langle\tau_\phi\rangle$ and inserting (5.216) in (5.222), we find that

$$\delta\eta \approx a_\phi^2 \frac{\mathcal{V}}{\ln(\mathcal{V})^2} \gg 1 . \tag{5.223}$$

Thus, the leading string loop corrections to the Kähler potential – even after incorporating the cancelation from the extended no-scale structure – lead to *parametrically large* values of η.

  In conclusion, Kähler moduli inflation from a blow-up cycle suffers from a severe eta problem induced by string loop corrections associated with D7-branes wrapping the inflationary cycle. Note that nonperturbative effects on this cycle are required in order to generate the exponential in (5.214), which is central to the mechanism. One suggestion for evading the eta problem [44] is to arrange that only Euclidean D3-branes, not D7-branes, wrap the inflationary cycle; the desired superpotential is then generated, while the associated quantum corrections to the Kähler potential are not obviously determined by known and conjectured results

---

[76] Our volume scalings of κ_1 and κ_2 differ from [32].

[333–336, 812]. However, Euclidean D3-branes and gaugino condensation involve closely related physics. Indeed, there are examples where a quantum correction that was first computed as an open string loop effect in the D7-brane case, and appeared inaccessible in the corresponding Euclidean D3-brane case, was shown (by a closed string computation) to take precisely the same form for Euclidean D3-branes [346]. Whether the substitution of Euclidean D3-branes will address the eta problem of blow-up inflation remains an open question that could be resolved by direct computation.

5.5.4 Fiber inflation

A fundamental feature of the Large Volume Scenario is that the leading α' correction depends only on the overall volume, leaving the remaining Kähler moduli as flat directions. As we explained above, in blow-up inflation [639] nonperturbative effects generate an exponentially flat term in the potential; but it has proven difficult to prevent perturbative quantum corrections to the Kähler potential from introducing a parametrically larger – and unacceptably steep – contribution. Faced with this situation, it is natural to ask whether there exist compactifications in which the perturbative contributions, which are almost invariably significant, actually drive inflation.

The first proposal of this sort is *fiber inflation* [44]. The setting is a Calabi–Yau manifold that is a K3 fibration over a $\mathbb{P}^1$ base. In the simplest explicit example,[77] the volume can be written as

$$\mathcal{V} = \frac{1}{2}\sqrt{\tau_1}\tau_2 \,, \tag{5.224}$$

where we have chosen a convenient basis in which τ_1 is the volume of the K3 fiber [44]. As in blow-up inflation, a third blow-up cycle (whose volume we again denote by τ_s) turns out to be necessary. The volume is therefore assumed to take the form [44]

$$\mathcal{V} = \alpha\left(\sqrt{\tau_1}\tau_2 - \lambda_s\tau_s^{3/2}\right) \,, \tag{5.225}$$

where α and λ_s are model-dependent constants.

Inflaton potential

Before including string loop corrections, the Kähler potential is given by (5.200). If $\tau_1, \tau_2 \gg 1$, nonperturbative effects involving τ_1 and τ_2 can be neglected, and the superpotential takes the form

$$W = W_0 + \mathcal{A}_s e^{-a_s T_s} \,. \tag{5.226}$$

---

[77] The simplest example is a Calabi–Yau hypersurface in $\mathbb{P}_4^{(1,1,2,2,6)}$ – see [816] for details.

The scalar potential is then

$$V = a_s^2 \mathcal{A}_s^2 \frac{\sqrt{\tau_s}}{\mathcal{V}} e^{-2a_s \tau_s} - a_s \mathcal{A}_s W_0 \frac{\tau_s}{\mathcal{V}} e^{-a_s \tau_s} + \hat{\xi} W_0^2 \frac{1}{\mathcal{V}^3} . \tag{5.227}$$

The potential (5.227) depends only on τ_s and $\mathcal{V}$, which are stabilized at $\tau_s \sim g_s^{-1}$ and $\mathcal{V} \sim W_0 \sqrt{\tau_s} e^{a_s \tau_s}$. This leaves a flat direction in the (τ_1, τ_2) plane – namely, the direction along which $\mathcal{V}$ remains constant. This flat direction is plausibly lifted by string loop corrections to the Kähler potential. The main idea of fiber inflation is that these quantum corrections will provide the leading (non-constant) terms in the inflaton potential. Before proceeding, we must emphasize that the string loop corrections in question, Eqs. (5.202) and (5.203), are those *conjectured* in [335] (see also [336, 812]) as generalizations of the explicit computations of [333, 334] for the toroidal orientifold $T^6/(\mathbb{Z}_2 \times \mathbb{Z}_2)$. The viability of fiber inflation rests on the specific form assumed in [335], and it would be valuable to obtain more direct and detailed understanding of quantum corrections to the Kähler potential. Without further apologies, the potential from the conjectured string loop corrections is

$$\delta V_{(g_s)} = \frac{W_0^2}{\mathcal{V}^2} \left(\mathfrak{a} \frac{g_s^2}{\tau_1^2} - \mathfrak{b} \frac{1}{\sqrt{\tau_1} \mathcal{V}} + \mathfrak{c} \frac{g_s^2 \tau_1}{\mathcal{V}^2} \right) , \tag{5.228}$$

where $\mathfrak{a}, \mathfrak{b}, \mathfrak{c}$ are unknown order-one constants. This fixes the fiber modulus τ_1 at $\tau_1 \sim g_s^{4/3} \mathcal{V}^{2/3}$. An inflationary phase can arise if τ_1 is displaced far from this minimum, i.e. if the K3 fiber is initially large compared with the base, and then relaxes to smaller values.

As a simple first step, we suppose that τ_s and $\mathcal{V}$ remain fixed at their minima while τ_1 evolves, and can be integrated out. The resulting single-field potential takes the form

$$V(\phi) = V_0 \left(1 - \frac{4}{3} e^{-\phi/\sqrt{3}} + \frac{1}{3} e^{-4\phi/\sqrt{3}} + \frac{\mathfrak{C}}{3} e^{2\phi/\sqrt{3}} \right) , \tag{5.229}$$

where $V_0 \equiv \mathcal{O}(1) \times \mathcal{V}^{-10/3}$, $\mathfrak{C} \equiv 16 \mathfrak{a}\mathfrak{c}/\mathfrak{b}^2 \sim g_s^4 \ll 1$, and

$$\phi \equiv \frac{\sqrt{3}}{2} \ln \tau_1 . \tag{5.230}$$

The potential is plotted in Fig. 5.14. Successful inflation occurs in region II,[78] where the potential can be approximated as

$$V(\phi) \simeq V_0 \left(1 - \frac{4}{3} e^{-\phi/\sqrt{3}} \right) . \tag{5.231}$$

Interestingly, this form of the potential is similar to that obtained in the Starobinsky model and in Higgs inflation (see Section 2.2.2).

---

[78] The slow-roll conditions are also satisfied in region III, but constraints on the spectral index are violated there; see Section 5.5.6.

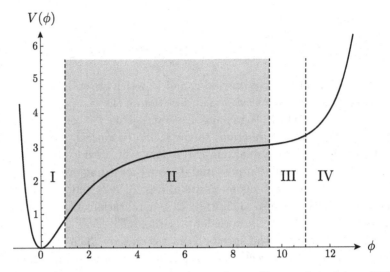

Fig. 5.14 Sketch of the potential for fiber inflation (figure adapted from [31]). The phenomenologically viable inflationary regime is the gray shaded region II. The slow-roll conditions are also satisfied in region III, but the spectrum of fluctuations is blue.

The single-field treatment presented above is not a priori justified, because the compactification volume $\mathcal{V}$ is light enough to evolve during inflation. Even so, [44] presents extensive numerical and analytical evidence showing that the single-field potential (5.229) gives an accurate picture of the two-field evolution. The motion of $\mathcal{V}$ is slow until the end of inflation, and moreover upon incorporating its evolution, i.e. setting $\mathcal{V} = \mathcal{V}(\phi)$, one finds negligible corrections to the slow-roll parameters of the effective single-field model. It would be interesting to know whether fluctuations of $\mathcal{V}$ contribute to the primordial perturbations in fiber inflation, along the lines of [582].

Naturalness and higher corrections

The structure of the potential (5.229) is ultimately dictated by the leading α' correction (5.200), which depends only on $\mathcal{V}$, and by the string loop corrections (5.202) and (5.203), which enter in (5.228) and lift all flat directions. Corrections from higher string loops, and at higher order in α', have not been computed, but are suppressed by additional (possibly fractional) powers of g_s and $\mathcal{V}^{-1}$.

To understand whether all such corrections can be neglected, we recall that the remarkable resilience of LVS rests in large part on the fact that at *exponentially* large $\mathcal{V}$, any unknown or unwanted corrections that are suppressed by any reasonable power (including a fractional power) of $\mathcal{V}$ are effectively negligible. As an example, a possible higher-derivative correction to the ten-dimensional action at order $(\alpha')^4$ would on dimensional grounds be suppressed compared to the leading term (3.101) by a factor $\mathcal{V}^{-1/3}$. In "traditional" LVS constructions,

$\mathcal{V}^{-1/3}$ is a very small number, justifying the omission of higher-derivative terms in ten dimensions.[79]

In fiber inflation, however, reproducing the normalization of the scalar power spectrum compels the volume to be modest in size. This follows because the potential (5.231) has only one free parameter; the scale is set by $V_0 \propto \mathcal{V}^{-10/3}$ and the slow-roll parameter ϵ is not parametrically adjustable. For the benchmark parameters given in [44], the scalar power spectrum has the right amplitude for $\mathcal{V} \approx 1700$. Neglecting terms suppressed by integer powers of $\mathcal{V}$ is clearly safe, but suppressions by $\mathcal{V}^{-1/3}$ are marginal, particularly in cases where the dimensionless prefactor is entirely unknown.[80]

Higher-loop corrections are a potentially important issue in a model driven by one-loop corrections. To understand higher-loop corrections in fiber inflation, we consider the limit $\tau_1 \to \infty$ at constant $\mathcal{V}$, corresponding to a K3 fiber that is large compared to the base $\mathbb{P}^1$, which becomes singular in the limit. The geometric singularity is reflected in a (power law) divergence in the one-loop corrections involving τ_2, which vanishes in the large fiber limit [44]. An obvious concern is that higher-loop corrections to the Kähler potential will become important in this regime. However, it was shown in [44] that slow-roll inflation also breaks down at small base volume, at a value of the base volume that is large enough so that higher-loop corrections are still small. As a result, higher-loop corrections are argued to be negligible during the inflationary phase.

Finally, the authors of [44] have argued that fibre inflation is robust because of a hidden symmetry that emerges in the limit of infinite volume. In four-dimensional terms, the problematic corrections to the inflaton potential are suppressed by powers of $\mathcal{V}$, and vanish in the decompactification limit $\mathcal{V} \to \infty$. This limit enjoys additional symmetries, most notably ten-dimensional general covariance, that could be related to cancelations in the four-dimensional action. Further analysis of this interesting possibility would be worthwhile.

In conclusion, fiber inflation is a promising inflationary scenario in LVS compactifications. If one grants the conjectured string loop corrections (5.202) and (5.203), which are the key to the inflaton potential, and also omits higher-order corrections, the dynamics is quite robust. We have argued above that at the comparatively small values of $\mathcal{V}$ relevant for fiber inflation, the validity of the approximations made to arrive at the inflationary potential merits further scrutiny. To be clear, the possible corrections are not parametrically large (as they are in blow-up inflation), but they could be important, and it would be interesting to have a sharper picture.

---

[79] See [379] for details of the α' expansion in LVS.

[80] However, because the leading α' correction (3.101) does not depend on the inflaton, one might conjecture that subleading α' corrections are likewise inflaton independent, and hence unimportant.

5.5.5 Poly-instanton inflation

In an effort to evade the eta problem of blow-up inflation, [640] (see also [813]) constructed a model in which *poly-instanton* terms in the superpotential make a critical contribution to the scalar potential. A poly-instanton[81] effect arises when the Euclidean action S_a of an instanton a receives corrections from a second instanton b [817] (see also the earlier work [818]), so that

$$W = \mathcal{A}_a \exp\left(-S_a + \mathcal{A}_b e^{-S_b}\right) , \tag{5.232}$$

where the moduli-dependent prefactors $\mathcal{A}_a$ and $\mathcal{A}_b$ are one-loop determinants.[82]

Setup

The compactification geometry assumed in [640] is the same as in Section 5.5.4; in particular, the compactification volume is given by (5.225),

$$\mathcal{V} = \alpha \left(\sqrt{\tau_1}\tau_2 - \lambda_s \tau_s^{3/2}\right) . \tag{5.233}$$

Building on explicit poly-instanton constructions in [819], [813] considered a slightly different model with

$$\mathcal{V} = \tau_b^{3/2} - \tau_s^{3/2} - (\tau_s + \tau_w)^{3/2} . \tag{5.234}$$

We will focus on (5.233), as realized in the explicit constructions of [820], but the issues and most of the phenomenology are very similar with the choice (5.234) [813].

We consider a stack of D7-branes wrapping the four-cycle associated with the modulus τ_s. We assume that the field theory on the D7-branes can be broken into two sectors that separately undergo gaugino condensation. The superpotential is then of the racetrack[83] form (5.197),

$$W = W_0 + \mathcal{A} \exp[-aT_s] - \mathcal{B} \exp[-bT_s] , \tag{5.235}$$

where the sign of the final term is a convenient phase choice. In addition, a Euclidean D3-brane is taken to wrap the fiber associated with τ_1. This leads to nonperturbative corrections to the gauge kinetic functions of the two condensing gauge groups. The poly-instanton corrected superpotential then takes the form

$$W = W_0 + \mathcal{A} \exp\left[-a\left(T_s + c_1 e^{-2\pi T_1}\right)\right] - \mathcal{B} \exp\left[-b\left(T_s + c_2 e^{-2\pi T_1}\right)\right] , \tag{5.236}$$

where c_1 and c_2 are constants.

---

[81] Poly-instantons should not be confused with *multi-instantons*, which are well known in field theory and correspond in string theory to multiple Euclidean branes wrapping the same cycle.

[82] Precisely this structure arises in axion monodromy inflation through instanton corrections to the holomorphic gauge coupling function f, cf. Eq. (5.171).

[83] The racetrack is unrelated to the existence of poly-instanton effects: it is a further model-building requirement. By introducing another adjustable parameter, the racetrack superpotential allows one to evade constraints that arise in a single-condensate model [640, 813].

Inflaton potential

In the absence of the poly-instanton corrections, i.e. for $c_1 = c_2 = 0$, the fields $\mathcal{V}$ and τ_s are stabilized as before. Again, one is left with a flat direction in the (τ_1, τ_2) plane. This time, the flat direction is lifted by the poly-instanton contributions in (5.236). As before, one can consistently integrate out $\mathcal{V}$ and τ_s, as well as the axion partners of all Kähler moduli (see [640] for details). The scalar potential for the distance from the minimum in the τ_1 direction, i.e. $\hat{\tau}_1 \equiv \tau_1 - \langle\tau_1\rangle$, is found to be

$$V = \frac{F_{\text{poly}}}{\mathcal{V}^{3+p}} \left(1 - (1 + 2\pi\hat{\tau}_1)e^{-2\pi\hat{\tau}_1} \right) , \tag{5.237}$$

where $F_{\text{poly}} \equiv \mathcal{O}(1) \times W_0$ and $p = \mathcal{O}(1)$. The order-one factors in the parameters F_{poly} and p depend in a complicated way on the microscopic parameters of the theory, and the precise expressions can be found in [640]. Using (5.230), we can write the potential in terms of the canonically normalized inflaton field,

$$V(\hat{\phi}) \simeq V_0 \left(1 - \kappa_2 \hat{\phi} e^{-\kappa_2 \hat{\phi}} \right) , \qquad \hat{\phi} \approx \frac{\sqrt{3}}{2} \frac{\hat{\tau}_1}{\langle\tau_1\rangle} , \tag{5.238}$$

where $V_0 \equiv F_{\text{poly}} \mathcal{V}^{-(3+p)}$ and $\kappa_2 \simeq \mathcal{O}(1) \times \ln(\mathcal{V})$.

Corrections

Because the setting (5.233) for poly-instanton inflation in [640] is precisely that of fiber inflation, while the geometry (5.234) in [813] is similar to that in blow-up inflation, one should ask about the string loop corrections to the Kähler potential that were crucial in Sections 5.5.3 and 5.5.4. In [640], it is argued that because D7-branes only wrap τ_s, not τ_1 or τ_2, with only Euclidean D3-branes wrapping τ_1, one does not expect open string loop corrections that depend on the inflaton τ_1. However, as we remarked in Section 5.5.3, it has not actually been shown that dangerous open string loop corrections are absent in this setting. Instead, a fair summary is that the calculation of [333, 334] that led to the conjecture [335] is not immediately applicable, and no first-principles computation of the quantum corrections has been presented. The absence of (a certain sort of) quantum corrections to an unprotected quantity such as the Kähler potential would be quite striking, and further investigation of this point is warranted.

In addition to corrections from loops of open strings ending on D7-branes, the Kähler potential can also be corrected by loops of closed strings. This quantum correction was estimated in [640], where it was found that closed string loops can significantly affect the shape of the inflaton potential. The size of the effect depends on an undetermined amplitude C_{loop} that depends on the complex structure moduli, and may be assumed to be of order unity in generic situations. In [640], it was assumed that for appropriate choices of flux one has $C_{\text{loop}} \lesssim 0.1$, in which case the loop corrections can be neglected.

5.5.6 Phenomenology

In the truncation to a single-field description, the models of inflation in LVS described in this section can all be written in terms of the approximate potential

$$V(\phi) \approx V_0 \left(1 - \kappa_1 e^{-\kappa_2 \phi}\right) . \tag{5.239}$$

In blow-up inflation, $\kappa_1 = \mathcal{O}(\mathcal{V} \ln \mathcal{V})$ and $\kappa_2 = \mathcal{O}(\mathcal{V}^{1/2} (\ln \mathcal{V})^{1/4})$, while in fiber inflation $\kappa_1 \sim \kappa_2 = \mathcal{O}(1)$, and in poly-instanton inflation $\kappa_1 \sim \kappa_2 = \mathcal{O}(\ln(\mathcal{V}))$. The slow-roll parameters derived from (5.239) are

$$\eta \simeq -\kappa_1 \kappa_2^2 e^{-\kappa_2 \phi} \qquad \text{and} \qquad \epsilon \simeq \frac{1}{2} \frac{\eta^2}{\kappa_2^2} . \tag{5.240}$$

This class of models therefore satisfies $\epsilon \ll \eta$ and hence (2.44) becomes

$$n_s \simeq 1 + 2\eta . \tag{5.241}$$

Given n_s, one predicts the tensor-to-scalar ratio:

$$r \simeq \frac{2}{\kappa_2^2}(n_s - 1)^2 \xrightarrow{n_s = 0.96} \frac{3 \times 10^{-3}}{\kappa_2^2} . \tag{5.242}$$

This prediction for r depends on the parameter κ_2, which differs for the different classes of Kähler moduli inflation scenarios.

Blow-up inflation

Because of the parametrically large string loop correction (5.222) to η in blow-up inflation, it is not necessarily well motivated to derive predictions from the uncorrected model of the form (5.239).[84] Here, we will only point out that in blow-up inflation without string loop corrections, $\kappa_2 = \mathcal{O}(\mathcal{V}(\ln \mathcal{V})^{1/4}) \gg 1$, cf. (5.217). Thus, by (5.242), the tensor-to scalar ratio r is extremely small. However, this feature relies on exponential flatness of the potential, which as explained above is very vulnerable to corrections.

Fiber inflation

Equation (5.231) is of the form (5.239), with $\kappa_2 = 1/\sqrt{3}$. This leads to a direct correlation between the scalar spectral index and the tensor-to-scalar ratio,

$$r \simeq 6(n_s - 1)^2 \xrightarrow{n_s = 0.96} 0.01 . \tag{5.243}$$

A word about predictions for n_s in fiber inflation is necessary. From (5.229) one readily sees that slow-roll inflation can occur in both regions II and III depicted in Fig. 5.14. In region III, $\eta > 0$, so that $n_s > 1$; the spectrum has a blue tilt, which

---

[84] For the same reason, in Section 5.1 we did not analyze the phenomenology that would arise in warped D3-brane inflation driven by a Coulomb potential with no corrections from moduli stabilization; although these predictions are widely quoted in the literature, they have little meaning.

is strongly disfavored by observations. The approach of [44] is to consider only inflationary dynamics in region II, but in fact the full model (5.229) can produce a blue *or* a red spectrum, depending on where on the potential the large-angle CMB fluctuations exit the horizon; see also [821, 822]. The situation is similar to that in inflection point inflation (cf. Section 5.1.6), which is unsurprising given the shape of the potential in Fig. 5.14.

Poly-instanton inflation

In (5.238), we found $\kappa_2 = \mathcal{O}(\ln(\mathcal{V}))$. A typical model [640] has $\kappa_2 = \ln(10^3) \sim 10$ and hence

$$r \sim 10^{-5} \ . \tag{5.244}$$

Such a low tensor amplitude is unobservable.

It seems quite generic that the inflaton field in Kähler moduli inflation couples to additional light degrees of freedom. This can modify the above results, which were based on a truncation to single-field inflation, and may lead to additional signatures. For example, variations of blow-up inflation have been proposed [582] that allow for large local non-Gaussianity via the curvaton mechanism [134, 135, 177, 562]: $f_{\rm NL}^{\rm loc} \sim \mathcal{O}({\rm few}) \times 10$. Similarly, extensions of the simplest fiber inflation models have been constructed [585] that produce relatively large local non-Gaussianity from modulated reheating [174–176, 266]: $f_{\rm NL}^{\rm loc} \sim \mathcal{O}({\rm few})$. Both possibilities are strongly constrained by the Planck bound (1.72).

5.6 Inflating with dissipation

In systems where the potential energy function is too steep to support slow-roll inflation, dissipation can provide an alternative source of accelerated expansion. Microscopically, dissipative effects arise if the inflaton is coupled to, and excites, additional degrees of freedom during inflation. To model this we add a direct coupling between the inflaton and the extra fields, collectively denoted ψ:

$$S = \int {\rm d}^4 x \sqrt{-g} \left[\frac{M_{\rm pl}^2}{2} R - \frac{1}{2}(\partial\phi)^2 - V(\phi) + \mathcal{O}(\phi, \psi) \right] \ . \tag{5.245}$$

Suitable couplings can lead to the production of ψ particles, which drains energy from the inflaton sector and leads to an enhanced effective friction that slows the evolution of the inflaton field. However, since the density of particles is diluted exponentially during inflation, it is difficult to maintain friction-dominated evolution. In this section, we present a few ideas on how this might nevertheless be achieved.[85]

---

[85] An effective field theory of dissipative inflation was constructed in [823]. Related work on *warm inflation* [824, 825] is reviewed in [826] (see also [796, 827]).

In Section 5.6.1, we describe *trapped inflation* [641], in which dissipative dynamics arises from repeated production of particles or strings. We explain how trapped inflation could plausibly arise in the class of string compactifications discussed in the context of axion monodromy inflation in Section 5.4.2, albeit in a slightly different parameter regime. Then, in Sections 5.6.2 and 5.6.3, we present two very recent ideas: inflation via flux cascades [642] and via magnetic drift [828]. Both are imaginative additions to the string inflation literature, so we include them here even though, at the time of writing, the models still lack explicit embeddings into fully specified string compactifications including moduli stabilization. We hope that our discussion will inspire the reader to determine whether these ideas can be realized in concrete compactifications. Finally, we close, in Section 5.6.4, with a summary of the phenomenology of trapped inflation.

5.6.1 Trapped inflation

The study of quantum-mechanical particle production during inflation was pioneered in the work of Kofman, Linde, and Starobinsky [829]. Here, we describe the basic elements of that analysis and then apply them to trapped inflation.[86]

Particle production

We start by computing the particle production for a simple field theory model [686, 829, 830]. The result of this computation will feed into the dynamics of the inflationary model. Consider a scalar field ψ coupled to the inflaton ϕ via the interaction

$$\mathcal{L}_{\text{int}} = -\frac{1}{2}g^2(\phi - \phi_0)^2\psi^2 \ . \tag{5.246}$$

Notice that the field ψ becomes massless at a specific point in field space, $\phi = \phi_0$. This is where the ψ particles are produced. Near this point, we can approximate the homogeneous inflaton evolution as

$$\phi(t) \approx \phi_0 + \dot{\phi}_0(t - t_0) \ , \tag{5.247}$$

which implies a time-dependent effective mass for the ψ particles,

$$m_\psi^2(t) \equiv g^2(\phi - \phi_0)^2 \approx k_\star^4(t - t_0)^2 \ , \tag{5.248}$$

where $k_\star^2 \equiv g|\dot{\phi}_0|$. The evolution equation for a Fourier mode of the ψ field is then

$$\ddot{\psi}_k + 3H\dot{\psi}_k + \underbrace{\left(\frac{k^2}{a^2} + k_\star^4(t - t_0)^2\right)}_{\equiv \omega_k^2(t)}\psi_k = 0 \ . \tag{5.249}$$

---

[86] This section is based mostly on [641, 686].

Particles are produced when the evolution becomes non-adiabatic,

$$|\dot{\omega}_k| > \omega_k^2 \ . \tag{5.250}$$

This occurs in the time interval $|t - t_0| < k_\star^{-1}$ and for momenta $k < k_\star$. Solving (5.249) gives the occupation number of the ψ particles [686, 829],[87]

$$n_k = e^{-\pi k^2 / k_\star^2} \ . \tag{5.251}$$

Shortly after $t = t_0$, the number density of ψ particles is

$$n_\psi(t_0) = \int \frac{d^3 k}{(2\pi)^3} \, n_k \approx \frac{k_\star^3}{(2\pi)^3} \ . \tag{5.252}$$

This depends on a combination of the coupling constant g and the inflaton speed $|\dot{\phi}_0|$. We assume that the ψ particles become sufficiently massive after the production event so that they can be treated as non-relativistic matter. The density of ψ particles then dilutes as a^{-3},

$$n_\psi(t) = \frac{k_\star^3}{(2\pi)^3} \frac{a^3(t_0)}{a^3(t)} \, \Theta(t - t_0) \ , \tag{5.253}$$

where Θ is the Heaviside function. The energy density of the ψ particles is $\rho_\psi(t) = m_\psi n_\psi(t)$.

Next, we determine how the density of ψ particles affects the evolution of the inflaton field ϕ. The effect can be estimated by using the mean-field equation [686, 829]

$$\ddot{\phi} + 3H\dot{\phi} + V' = -g^2(\phi - \phi_0)\langle \psi^2 \rangle \ , \tag{5.254}$$

where[88]

$$\langle \psi^2 \rangle \approx \frac{n_\psi(t)}{g|\phi - \phi_0|} \ . \tag{5.255}$$

Figure 5.15 shows a numerical solution of Eq. (5.254). We see that the inflaton velocity $\dot{\phi}$ decays after the production event, but then returns almost to its initial value as the effect of the particles gets diluted away. The cosmological evolution is affected only temporarily by the particle production.[89] In other words, a single particle production event does not lead to many e-folds of dissipative dynamics. Achieving inflation from dissipation requires the density of particles to be kept high by repeated particle production. The resulting inflationary model is called *trapped inflation*.[90] We first describe an effective field theory construction of trapped inflation and then present a string theory realization.

---

[87] This result assumes $k_\star > H$.

[88] For a derivation of (5.255) see [686, 829] .

[89] Particle production can be continuous if the inflaton is coupled to a gauge field (see Section 5.6.3).

[90] Trapped inflation was first proposed in [686, 831]. The inflationary mechanism, the spectrum and bispectrum, and possible microphysical embeddings were systematically analyzed in [641]. See also the related work [830].

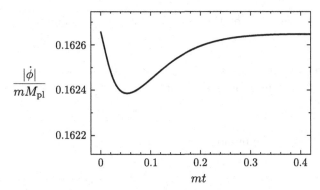

Fig. 5.15 Decay of the inflaton velocity due to particle production (figure adapted from [830]). The time $t = 0$ corresponds to the production event, the coupling is $g^2 = 0.1$, and $m^2 \equiv V''$.

Trapped inflation in effective field theory

To get repeated particle production, we replicate the coupling (5.246) at N points ϕ_i,

$$\mathcal{L}_{\text{int}} = -\frac{1}{2} g^2 \sum_{i=1}^{N} (\phi - \phi_i)^2 \psi_i^2 \ . \tag{5.256}$$

We take these points to be evenly spaced, with $\phi_{i+1} - \phi_i \equiv \Delta$, both to simplify the presentation and because uniform spacing is natural in microphysical models involving monodromy (see below). Particles are now produced periodically with densities given by (5.252),

$$n_{\psi_i}(t_i) \simeq \frac{\left(g\dot{\phi}(t_i)\right)^{3/2}}{(2\pi)^3} \equiv \frac{\dot{\varphi}^{3/2}(t_i)}{(2\pi)^3} \ . \tag{5.257}$$

We have ignored the effects of any finite pre-existing particle density on the particle production, and we have defined $\varphi \equiv g\phi$ for later convenience. Replacing ψ_i^2 by its expectation value $\langle \psi_i^2 \rangle$, we get an equation of motion for the inflaton,

$$\ddot{\phi} + 3H\dot{\phi} + V' + \sum_i \frac{g\dot{\varphi}^{3/2}(t_i)}{(2\pi)^3} \frac{a^3(t_i)}{a^3(t)} = 0 \ . \tag{5.258}$$

If the production events are spaced densely enough,[91] then we can replace the sum by an integral,

$$\sum_i \frac{g\dot{\varphi}^{3/2}(t_i)}{(2\pi)^3} \frac{a^3(t_i)}{a^3(t)} \approx \int^t \frac{dt'}{\Delta} \frac{\dot{\varphi}^{5/2}(t')}{(2\pi)^3} \frac{a^3(t')}{a^3(t)} \approx \frac{1}{3H\Delta} \frac{\dot{\varphi}^{5/2}(t)}{(2\pi)^3} \ , \tag{5.259}$$

---

[91] The necessary condition is $\Delta \ll \{ \dot{\phi}/H , \ \dot{\phi}^2/\ddot{\phi} \}$ [641].

and (5.258) becomes

$$\ddot{\phi} + 3H\dot{\phi} + V' + \frac{1}{24\pi^3} \frac{g^{5/2}}{H\Delta} \dot{\phi}^{5/2} = 0 . \tag{5.260}$$

Notice the extra friction term proportional to $\dot{\phi}^{5/2}$ provided by the finite density of ψ particles. Assuming slow-roll ($|\ddot{\phi}| \ll 3H|\dot{\phi}|$), and taking the damping to be dominated by particle production ($3H|\dot{\phi}| \ll V'$), we find

$$\dot{\varphi} = g\dot{\phi} \approx -\left(24\pi^3 H\Delta V'\right)^{2/5} . \tag{5.261}$$

Let us estimate the conditions for this solution to correspond to inflation. We assume that the Hubble parameter is dominated by the potential energy of the inflaton,

$$3M_{\rm pl}^2 H^2 = \rho_\phi + \rho_\psi \approx V(\phi) , \tag{5.262}$$

while its evolution is sourced by the ψ particles,

$$2M_{\rm pl}^2 \dot{H} \approx -\rho_\psi , \tag{5.263}$$

where we have used $\dot{\rho}_\psi \simeq -3H\rho_\psi \gg \dot{\rho}_\phi$. The Hubble slow-roll parameter is then

$$\varepsilon = -\frac{\dot{H}}{H^2} \approx \frac{3}{2}\frac{\rho_\psi}{V} , \tag{5.264}$$

where

$$\rho_\psi(t) = \sum_i g|\phi - \phi_i| \, n_{\psi_i}(t) \approx \int^t \frac{dt'}{\Delta} |\phi(t) - \phi(t')| \frac{\dot{\varphi}^{5/2}(t')}{(2\pi)^3} \frac{a^3(t')}{a^3(t)} . \tag{5.265}$$

Using $|\phi(t) - \phi(t')| \approx \dot{\phi}(t - t')$, we can approximate the integral in (5.265) in the same way as in (5.259),

$$\rho_\psi(t) \approx \frac{1}{(3H)^2} \frac{1}{g\Delta} \frac{\dot{\varphi}^{7/2}(t)}{(2\pi)^3} . \tag{5.266}$$

Using the solution (5.261) to replace $\dot{\varphi}$, we can write (5.264) as

$$\varepsilon \sim \frac{\epsilon^{7/10}}{g} \left(\frac{H}{M_{\rm pl}} \frac{\Delta^2}{M_{\rm pl}^2}\right)^{1/5} , \tag{5.267}$$

where ϵ is the potential slow-roll parameter (2.39) and we have dropped some unimportant numerical factors. We see that inflation can occur ($\varepsilon < 1$) even for a steep potential ($\epsilon > 1$). The parametric scaling of the answer in (5.267) is as expected: particle production is more efficient for larger coupling g and smaller spacing Δ; both of these effects correspond to smaller ε for fixed ϵ. Consistency conditions and further constraints on g and Δ were studied in [641].

Trapped inflation in string theory

The core requirement for trapped inflation is a closely spaced series of particle production events along the inflationary trajectory, as in the toy Lagrangian (5.256). This structure appears contrived in four-dimensional EFT, but readily arises in string theory as a consequence of monodromy (cf. Section 5.4.2). We will describe two approaches to a string theory embedding of trapped inflation [35, 641, 779].

Wrapped brane monodromy

We first examine a D4-brane in a nilmanifold compactification [779], where the replication responsible for serial particle production is most easily visualized. Consider the three-dimensional nilmanifold (or "twisted torus") $\mathcal{N}_3$ defined by coordinates u_1, u_2, x identified by

$$t_x : \quad (x, u_1, u_2) \mapsto (x + 1, u_1, u_2) \,, \tag{5.268}$$

$$t_{u_1} : \quad (x, u_1, u_2) \mapsto (x - Mu_2, u_1 + 1, u_2) \,, \tag{5.269}$$

$$t_{u_2} : \quad (x, u_1, u_2) \mapsto (x, u_1, u_2 + 1) \,, \tag{5.270}$$

with the line element

$$\frac{1}{\alpha'} ds^2 = L_{u_1}^2 du_1^2 + \underbrace{L_{u_2}^2 du_2^2 + L_x^2 \left(dx + Mu_1 du_2\right)^2}_{T^2} \,, \tag{5.271}$$

where L_{u_1}, L_{u_2}, and L_x are dimensionless constants. This geometry corresponds to a T^2 fibration over a circle parameterized by u_1, which we denote by $S_{u_1}^1$; for each value of u_1 there is a T^2 in u_2 and x,

$$\frac{1}{\alpha'} ds_{T^2}^2 (u_1) = L_{u_2}^2 du_2^2 + L_x^2 \left(dx + Mu_1 du_2\right)^2 \,. \tag{5.272}$$

The identification (5.269) shows that the fiber T^2 at $u_1 = 1$ is twisted by an $SL(2, \mathbb{Z})$ transformation before being glued to the fiber at $u_1 = 0$. More precisely, the complex structure of the torus shifts by M units, i.e. $\tau \mapsto \tau + M$ as $u_1 \mapsto u_1 + 1$. These equivalent tori are identified by the projection t_{u_1}. At M special locations around $S_{u_1}^1$, $Mu_1 = j \in \mathbb{Z}$, the tori are rectangular:

$$\frac{1}{\alpha'} ds_{T^2, \perp}^2 = L_x^2 dy_1^2 + L_{u_2}^2 dy_2^2 \,. \tag{5.273}$$

We have defined coordinates $y_1 \equiv x + ju_2$ and $y_2 \equiv u_2$ obtained from an $SL(2, \mathbb{Z})$ transformation of x and u_2.

The configuration of interest is type IIA string theory compactified on an orientifold of the product space $\mathcal{N}_3 \times \tilde{\mathcal{N}}_3$, with $\tilde{\mathcal{N}}_3$ a second nilmanifold. For the moment it suffices to consider a single $\mathcal{N}_3$. We consider a D4-brane wrapped on the one-cycle defined by $u_2 = \lambda$, or equivalently by $(y_1, y_2) = (j\lambda, \lambda)$. The role of the inflaton is played by the u_1 coordinate of the D4-brane. The key point is that if the D4-brane is transported in the u_1 direction, the fiber torus returns to an equivalent torus, but the one-cycle does *not*: e.g. at $u_1 = 0$, the brane wraps $(y_1, y_2) = (0, \lambda)$, while at $u_1 = 1$, the brane wraps $(y_1, y_2) = (M\lambda, \lambda)$. The D4-brane undergoes monodromy upon transport around $S^1_{u_1}$ [779].

To derive the dynamics of the wrapped D4-brane, we consult the DBI action,

$$S_{\text{D4}} = -\frac{1}{(2\pi)^4 g_\text{s}(\alpha')^2} \int d^4x \sqrt{-g} \sqrt{L^2_{u_2} + L^2_x M^2 u^2_1} \left(1 - \frac{1}{2}\alpha' L^2_{u_1} \dot{u}^2_1\right). \quad (5.274)$$

For $L_x M u_1 \gg L_{u_2}$, we get

$$S_{\text{D4}} = \int d^4x \sqrt{-g} \left(\frac{1}{2}\dot{\phi}^2 - \mu^{10/3}\phi^{2/3}\right), \quad (5.275)$$

where

$$\frac{\phi^2}{M^2_{\text{pl}}} \equiv \frac{2}{9}(2\pi)^3 g_\text{s} \frac{M}{L^3} \frac{L_{u_1}}{L_{u_2}} u^3_1, \quad (5.276)$$

$$\frac{\mu}{M_{\text{pl}}} \equiv \frac{M_\text{s}}{M_{\text{pl}}} \left(\frac{9}{4} \frac{M^2}{(2\pi)^8 g^2_\text{s}} \left(\frac{L_x}{L}\right)^3 \frac{L_{u_2}}{L_{u_1}}\right)^{1/10}, \quad (5.277)$$

with $L^3 \equiv L_{u_1} L_{u_2} L_x$. The field range can be super-Planckian if $L_{u_1} \gtrsim L_{u_2}$ and

$$\Delta u^3_1 \gg \frac{L^3}{M}. \quad (5.278)$$

This corresponds to moving around the S^1 many times (see Fig. 5.16).

Trapped inflation from wrapped branes

A small modification of the scenario of [779] allows for repeated particle production events, as in trapped inflation (see Fig. 5.17). In addition to the inflationary brane, we consider N D4-branes wrapping the $SL(2, \mathbb{Z})$ transforms of the one-cycle associated with the inflaton brane. The jth brane has a potential $(u_1 - j/M)^2 + \cdots$ and therefore minimizes its energy at $u_1 = j/M$. As the inflaton brane unwinds, it comes close to each of these spectator branes. The strings stretching between the mobile brane and the lattice of stationary branes

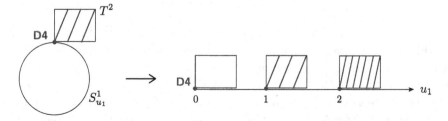

Fig. 5.16 Monodromy of a wrapped D4-brane on a nilmanifold.

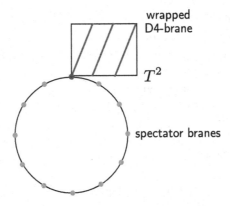

Fig. 5.17 Trapped inflation on a nilmanifold.

play the role of the extra fields ψ_j. Before accounting for moduli stabilization, the effective potential is of the form given in (5.256).

A serious obstacle to realizing trapped inflation in a nilmanifold compactification is that the presence of a large number of D4-branes tends to destabilize the moduli [641]. For the moduli stabilization constructions described in [302, 303], the D4-brane energy exceeds the scale of the moduli potential barriers when $N_{\mathrm{D4}} \gtrsim \mathcal{O}(10)$, which does not allow enough particle production events for trapped inflation.

Trapped inflation from axions

A closely related scenario in which instabilities are under better control [641] is the axion monodromy model [35] described in Section 5.4.2. Let us denote by ℓ^2 the volume of the two-cycle Σ_2 wrapped by the NS5-brane, cf. (5.160). A D3-brane wrapping Σ_2, with worldvolume flux

$$\int_{\Sigma_2} \mathcal{F}_2 = n \in \mathbb{Z} \,, \tag{5.279}$$

gives rise in four dimensions to a string with tension

$$T_{\mathrm{D3}/\Sigma_2} = T_3 \sqrt{\ell^2 + (c g_{\mathrm{s}} + n)^2} \,, \tag{5.280}$$

where $c \equiv \int_{\Sigma_2} C_2$ measures induced D1-brane charge in the D3-brane. For $c g_{\mathrm{s}} \in \mathbb{Z}$, this induced charge can be canceled by the quantized flux $\mathcal{F}_2$ in (5.279), if $n = -c g_{\mathrm{s}}$. We now notice that if $\ell \to 0$, this configuration gives rise to a *tensionless string* in four dimensions whenever integer values of $c g_{\mathrm{s}}$ are canceled by appropriate flux $\mathcal{F}_2$. These strings play the role of the ψ particles in the EFT discussion of Section 5.6.1; as c diminishes from a large initial vev, tensionless strings are produced at regularly spaced intervals.

A necessary condition for trapped inflation in this setting is a field range [641] $\Delta\phi/M_{\rm pl} \gtrsim \sqrt{60}/10$, which is a bit milder than the requirement for chaotic inflation in Section 5.4.2. On the other hand, a systematic study of moduli stabilization and backreaction would be necessary to determine sufficient conditions for trapped inflation in the context of axion monodromy, and to characterize corrections to the simple model of (5.256).

Weak coupling limit of DBI inflation

Trapped inflation is closely related to DBI inflation. Consider a D3-brane in the background of a stack of $(N-1)$ D3-branes, corresponding to a location on the Coulomb branch of $\mathcal{N} = 4$ super-Yang–Mills theory (see Section 5.3.3). At large $g_{\rm s}N$, the D3-brane is a probe of an $AdS_5 \times S^5$ geometry. Taking ϕ to be the canonical field representing the radial position of the D3-brane, one easily sees that strings stretching between the isolated D3-brane and the stack have mass proportional to ϕ. By analogy to the Higgs mechanism, the fields of mass $m \propto \phi$ are sometimes called "W-bosons" (though some of the relevant fields are fermions).

Now suppose that the D3-brane moves toward the stack, breaking supersymmetry by virtue of its kinetic energy. There are two important effects that can change its trajectory: *virtual* W-bosons induce quantum corrections to the action for ϕ, while pair production of *on-shell* W-bosons – caused by the time dependence of their mass – drains energy from the ϕ sector. The former effect leads to the DBI action (5.92); indeed, it is the fact that the W-bosons have $m \propto \phi$ that causes the non-renormalizable terms in (5.92) to be suppressed by ϕ, rather than by a fixed cutoff scale. The latter effect is precisely the particle production process described in Section 5.6.1. At large 't Hooft coupling, so that the gauge theory is strongly coupled but the supergravity background probed by the D3-brane is weakly curved, the dominant effect on the D3-brane dynamics comes from virtual W-bosons [39]; the evolution is governed by the DBI action, with negligible particle production. If instead the gauge theory is weakly coupled, particle production dominates. In a sense, trapped inflation is the weak-coupling analogue of DBI inflation, even though – as usual with strong–weak dualities – one rarely has control of both sides in the same setting. Indeed, we saw above that the most plausible string theory realizations of trapped inflation do not involve taking the weak coupling limit of the configurations studied in Section 5.3 (namely, D3-branes in Calabi–Yau cones); instead, compactifications involving monodromy are a more fruitful setting.

The phenomenology of trapped inflation will be discussed in Section 5.6.4.

5.6.2 Flux cascades

The *unwinding inflation* scenario [642, 643] combines bubble nucleation, dissipation/trapping, monodromy, the DBI effect, oscillations in the potential, and a hybrid exit through brane–antibrane annihilation. The basic setup is the following: a $(p+2)$-form flux F_{p+2} fills the noncompact spacetime and threads a $(p-2)$-cycle in the compact space. Initially there are $Q_0 \gg 1$ units of F_{p+2}, but the flux can be discharged by the nucleation of a p-brane/anti-p-brane pair, followed by $\mathcal{O}(Q_0)$ "unwindings," in which the brane and antibrane move in opposite directions around the compact cycle, reducing the flux and colliding with each other in every circuit.[92] In this section, we will give some of the details of unwinding inflation, and comment on the prospect of realizing this idea in string theory.[93]

The unwinding mechanism is applicable in a broad class of higher-dimensional gravity theories involving suitable fluxes, but, anticipating a UV completion in string theory, we will limit our discussion to string compactifications. Consider string theory in the ten-dimensional spacetime

$$\mathcal{M}_{10} = dS_4 \times X_6 \, , \tag{5.281}$$

for X_6 a compact manifold, and take $Q_0 \gg 1$ units of the R-R $(p+2)$-form flux F_{p+2} to fill the noncompact spacetime and thread a $(p-2)$-cycle Σ_{p-2} in X_6,

$$\int_{dS_4 \times \Sigma_{p-2}} F_{p+2} = Q_0 \gg 1 \, . \tag{5.282}$$

The flux induces an effective cosmological constant in four dimensions; this will play the role of the inflationary energy density. A Dp-brane carries electric charge under F_{p+2}, and nucleation of a bubble bounded by a Dp-brane creates a region (the bubble interior) in which the flux is reduced by one unit compared with the exterior, as in [836]; this is a higher-dimensional analogue of the Schwinger process in QED. The background flux creates a force on the bubble, driving it to expand in Σ_{p-2}.[94]

Eventually the Dp-brane bubble becomes so large that it "unwraps" Σ_{p-2} and collides with itself, dissipating energy into open string degrees of freedom. Because the portions of the bubble that collide have opposite orientation, this is locally a Dp-brane/anti-Dp-brane collision. Provided that the collision happens so rapidly that the brane–antibrane tachyon does not have time to condense, and provided that the dissipation is not strong enough to stall the unwinding process – see below for discussions of these important points – the brane and

---

[92] The first proposal to use self-collisions of a bubble in compact extra dimensions to drive inflation appears in [832]. Cascades following nucleation events were discussed in [833, 834]. A closely related scenario in which D-brane motion around the compact cycle discharges a flux is [835].

[93] This section is based on [642, 643].

[94] As a simple analogy [643], one can picture the flux as a rubber sheet that wraps repeatedly around Σ_{p-2}. The initial bubble nucleation corresponds to the appearance of an approximately spherical hole in one layer of the wrapped sheet. The tension of the rubber causes the hole to expand, unwinding layer after layer of the sheet.

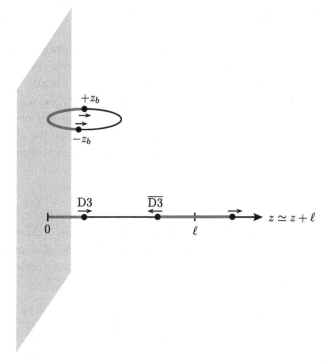

Fig. 5.18 A cascade of five-form flux on $dS_4 \times S^1$. The compact S^1 and its covering space are both shown.

the antibrane pass through each other and continue to unwind (see Fig. 5.18 for a five-dimensional example). Each subsequent collision reduces the flux by a further unit. In the four-dimensional effective theory, this appears as a slow reduction of the effective cosmological constant, mimicking the evolution during slow-roll inflation.[95] When the flux has dropped sufficiently, the branes stop moving relativistically. Tachyon condensation can then be efficient when the branes approach each other, and brane–antibrane annihilation provides a natural hybrid exit from inflation.

A number of important questions arise at this stage. What sort of bubble nucleation event leads to a flux discharge cascade? What is the four-dimensional effective action for the unwinding branes? Is dissipation a small correction to the background evolution, and to the scalar and tensor perturbations? What is the dynamics of the D-brane pair in the compact directions perpendicular to the flux? Are the requirements of unwinding inflation compatible with compactification and moduli stabilization? We will briefly address the first three points, following

---

[95] This is similar in spirit to *chain inflation* [837–843], although the microscopic details are quite different. It is also very similar to the unwinding of a wrapped D4-brane in monodromy inflation in nilmanifold compactifications [779], and to the reduction of induced D3-brane charge during axion monodromy inflation [35], cf. Section 5.4.2.

[642, 643], and then review some of the difficulties involved in embedding these ideas in string theory.

To discuss the effective action, it will be instructive to examine the simplified example of $dS_4 \times S^1$ with five-form flux F_5 [642]. In this case, bubble nucleation leads to the situation depicted in Fig. 5.18. The bubble is bounded by D3-branes at $+z_b$ and $-z_b$; because these branes have opposite charges, one can think of them as a brane–antibrane pair. The role of the inflaton is played by z_b, the radius of the bubble in the extra dimension.[96] To determine the effective action for z_b, [642] examined bubble nucleation in the Euclidean spacetime with metric

$$ds_E^2 = H^{-2}\left(d\xi^2 + \sin^2\xi \, d\Omega_3^2\right) + dz^2 \,, \tag{5.283}$$

where $d\Omega_3^2$ is the line element on S^3 and z is the coordinate for an S^1 of circumference ℓ. The bubbles of primary interest have initial size[97] $\Delta z \ll \ell$ and have the maximum possible symmetry (as this is characteristic of dominant instantons). Solving the Euclidean equations of motion and then continuing back to Lorentzian signature, [642] obtained the action

$$S = \int dz \int d\mathcal{H}_3 \, dt \, \frac{\sinh^3(Ht)}{H^3}\left(-2\sigma \, \delta(z - z_b)\sqrt{1 - (\partial z_b)^2} - \frac{F_5^2}{2 \cdot 5!}\right) \,, \tag{5.284}$$

where $d\mathcal{H}_3 \equiv \sinh^2(\rho)\,d\rho\,d\Omega_2$ is the integration measure on a three-hyperboloid, σ is the tension of the wall (a D3-brane), and

$$\frac{F_5^2}{5!} = \mu^5 Q^2 = \mu^5 \left(Q_0 + \sum_{j=-\infty}^{\infty}\left[\Theta(z - z_b + j\ell) - \Theta(z + z_b + j\ell)\right]\right)^2 \,. \tag{5.285}$$

Here, Q_0 stands for the initial flux before bubble nucleation, $\mu^{5/2}$ is the charge of the brane, and the sum is over all the image branes (cf. Fig. 5.18). Equation (5.285) is a consequence of Gauss's law, which requires that the flux changes by one unit of the brane charge across the bubble wall. Once the radius of curvature of the bubble becomes much larger than the Hubble radius, the four-dimensional spacetime inside the bubble can be approximated by flat de Sitter space. Integrating over the fifth dimension, the action (5.284) becomes

$$S = \int dt \, d^3x \, a^3(t)\left(-2\sigma\sqrt{1 - (\partial z_b)^2} - V(z_b)\right) \,, \tag{5.286}$$

where $a(t) \equiv e^{Ht}$.

---

[96] For Dp-branes in a compactification of critical string theory, there will be additional scalars describing the remaining coordinates of the D-branes (at least if $p < 8$, which includes all examples of interest). One should bear in mind that these fields could be crucial for the background evolution and the perturbations, so this five-dimensional toy model may not give a faithful representation of unwinding in a string compactification.

[97] A bubble of size $\Delta z \gtrsim \ell$ would correspond to an ordinary bubble of reduced flux in dS_4, and would expand in dS_4 without initiating a cascade.

Upon solving the equations of motion that follow from (5.286) – still neglecting dissipation during collisions – one finds that the D3-brane velocity $\dot{z}_b$ is relativistic, and approximately constant. The branes collide with image branes when $z_b = n\ell/2$, for $n \in \mathbb{Z}$. This leads to discrete jumps in the vacuum energy V perceived by an observer in the four-dimensional spacetime. On timescales that are large relative to the Kaluza–Klein scale, this reduction in the inflationary energy density appears continuous and can be approximated as

$$V(z_b) \sim \mu^5 \left(Q_0 - \frac{z_b}{\ell} \right)^2 . \tag{5.287}$$

Tachyon condensation ends unwinding inflation, so it is crucial that the brane and antibrane can collide $\mathcal{O}(Q_0)$ times without slowing down so much from the resulting dissipation that the tachyon condenses prematurely. Tachyon condensation is suppressed when the brane–antibrane collision is relativistic, and a priori one could construct a configuration in which the electric force from the flux accelerates the brane to an arbitrarily large γ, allowing a correspondingly large number of cycles. In a realistic cosmology, however, γ is bounded from above; the DBI kinetic term of the moving brane leads to equilateral non-Gaussianity, as described in Section 5.3, and the Planck upper limit requires that $\gamma \lesssim 24$, cf. Eq. (5.135). This limits the degree to which tachyon condensation can be deferred.

Computing the production of open and closed strings in a relativistic brane–antibrane collision – particularly in the most singular case of zero impact parameter – is highly nontrivial. If the branes are taken to be homogeneous, some aspects of the calculation can be performed in the four-dimensional EFT involving the lightest string modes, but there are subtleties in this approach. The naive EFT obtained by dimensional reduction of a compactification containing a stationary brane–antibrane pair does not correctly capture the spectrum of masses of stretched strings between a brane and an antibrane in *relativistic* relative motion [844]. To obtain the correct rate of open string production, one applies the optical theorem to the annulus amplitude for the moving branes [845], which reveals that the effective tension of the stretched string diminishes at large γ, increasing the pair production rate [742, 844]. Building on the results of [845], [642] argued that for the velocities allowed by (5.135), only the lowest few massive string modes are produced during the collision. One limitation of this approach is that the annulus amplitude provides information about pair production in a constant-velocity scattering process, while in practice the dissipation from a head-on collision may substantially (albeit temporarily) decelerate the brane–antibrane pair.[98] A direct calculation of the production of excited open

---

[98] This deceleration leads to bremsstrahlung, which is dramatically enhanced at large γ [742]. The values of γ allowed by limits on non-Gaussianity are not large enough for the results of [742], where an ultrarelativistic limit was assumed, to be directly applicable, but

strings in a series of relativistic scattering processes with varying velocity (and perhaps with inhomogeneities) would be a major technical challenge.

Even if the dissipation in each collision is a mild correction to the background evolution, dissipation could have a major impact on the perturbations. One possibility is that the periodic modulations of the Hubble constant will induce resonant contributions to the spectrum and bispectrum, as described for axion monodromy inflation in Section 5.4.2. More dramatically, the repeated production of open strings could source the dominant component of the scalar power spectrum, as for trapped inflation in Section 5.6.1.

The evolution described above assumes that only a single coordinate (the bubble radius) is relevant, and that production of particles and strings, as well as the eventual tachyon condensation, are not strongly inhomogeneous. These issues are linked; a fluctuation of the moving brane in a transverse direction changes the impact parameter of the collision, and both particle production and the tachyon mass depend sensitively on the impact parameter. Although a number of related consistency checks were performed in the toy models of [642], the geometries considered in [642] may be too simple to capture the dynamics of unwinding in a realistic compactification, and further investigation is warranted.

The essential mechanism of unwinding inflation is fairly simple, and appears to arise naturally in toy flux compactifications, with simple unwarped geometries and with the moduli stabilized by fiat. Above we have highlighted some limitations of the calculations of [642], as well as some ways in which the presence of extra compact dimensions – still stabilized by hand – could complicate, or prematurely terminate, the unwinding process. In closing, we will reemphasize the importance of complete and calculable moduli stabilization. For a proper perspective, one should recognize that *all* of the mechanisms for inflation in string theory that we have described thus far appear to succeed naturally in unstabilized toy compactifications, but (we would argue) *none* has been automatically successful after moduli stabilization and careful implementation of microphysical constraints. Because unwinding inflation tends to occur at a high scale [642], it faces the very general problem of achieving adequate barriers to destabilization and decompactification, which in various guises plagued the large-field axion models of Section 5.4. Addressing this issue by realizing unwinding inflation in a fully stabilized string compactification is an interesting problem for the future.

5.6.3 Magnetic drift

Another class of scenarios for dissipative inflation invokes couplings between the inflaton and one or more gauge fields. Suppose first [796] that the inflaton is

---

the losses to closed string radiation during unwinding inflation may nevertheless be significant, and deserve further study.

an axion ϕ that couples to N $U(1)$ gauge fields A_i via the standard axionic coupling

$$\mathcal{L} \supset - \sum_{i=1}^{N} \alpha_i \frac{\phi}{f} F_i \tilde{F}_i , \qquad (5.288)$$

where $F_i = \mathrm{d}A_i$ is the gauge field strength of the ith gauge group, and $\tilde{F}_i$ is its dual. A time-dependent axion vev, $\phi(t)$, breaks the conformal invariance of the action for the gauge field, leading to the production of quanta of the gauge field. It is natural to ask whether dissipation through production of gauge fields can slow ϕ sufficiently to give inflation. In [796], it was shown that a successful inflationary period with phenomenologically viable perturbations requires large couplings, $\alpha_i \sim \mathcal{O}(100)$, to a large number of gauge fields, $N \sim 10^5$. The top-down naturalness of this mechanism remains to be established, and in particular no complete string theory realization has been constructed.[99]

Another recent proposal, which we will now describe in some detail, is that inflation can be achieved by coupling an axion to non-Abelian gauge fields with suitable vevs [644, 828] (see [846–848] and the review article [849] for the related idea of *gauge-flation*). A large Chern–Simons coupling between the axion and the gauge fields transfers energy from the inflaton sector to the gauge fields – without dissipation – and allows slow-roll to occur even in the presence of a steep potential. The basic dynamics is similar to that of a charged particle in a magnetic field.[100]

As a concrete example, let us consider a stack of N D3-branes, with $SU(N)$ gauge theory on their worldvolume. The Chern–Simons coupling (3.30) includes the term

$$S_{\mathrm{CS}} = \frac{i}{2\pi} \int_{\mathcal{M}_4} C_0 \, \mathrm{Tr} \left[\mathcal{F}_2 \wedge \mathcal{F}_2 \right] . \qquad (5.289)$$

This is a topological term, and will not appear in the stress tensor. Combining this with the standard kinetic terms for the axion and the gauge field, and evaluating the action in a homogeneous FRW background, we find

$$S_{\mathrm{eff}} = \int \mathrm{d}^4 x \left\{ a^3 \left[\gamma_C \dot{C}_0^2 - V(C_0) + \gamma_A \frac{\mathrm{Tr}(\dot{A}^2)}{a^2} + \gamma_A \frac{\mathrm{Tr}([A,A]^2)}{a^4} \right] \right.$$
$$\left. + C_0 \, \mathrm{Tr}(\dot{A}[A,A]) \right\} , \qquad (5.290)$$

---

[99] One way to achieve large α is to fine-tune two axion decay constants to be nearly coincident [796].

[100] This section is based mostly on [644, 828].

where $\gamma_A \equiv 1/((2\pi)^2 g_s)$ and $\gamma_C \equiv g_s^2 M_{pl}^2$. We have left the axion potential, $V(C_0)$, unspecified. We introduce the canonically normalized inflaton

$$C_0(t) \equiv \frac{\phi(t)}{\sqrt{\gamma_C}} , \tag{5.291}$$

and choose an initial gauge field configuration with a rotationally invariant vacuum expectation value,

$$A_0 \equiv 0 , \qquad A_i(t) \equiv \frac{a(t)\psi(t)}{\sqrt{\gamma_A \nu}} J_i , \tag{5.292}$$

where J_i are the generators of $SU(2)$ in the N-dimensional representation and $\mathrm{Tr}[J_i J_j] = \frac{1}{12}N(N^2-1)\delta_{ij} \equiv \nu\delta_{ij}.$[101] Substituting (5.291) and (5.292) into (5.290), we find

$$S_{\text{eff}} = \int \mathrm{d}^4 x\, a^3 \left[\frac{1}{2}\dot{\phi}^2 - V(\phi) + \frac{3}{2}(\dot{\psi} + H\psi)^2 - \frac{3}{2}g^2\psi^4 \right.$$
$$\left. - \frac{3g\lambda}{f}\phi\,\psi^2(\dot{\psi} + H\psi) \right] , \tag{5.293}$$

where $f \equiv \sqrt{\gamma_C}$, $\lambda \equiv 1/\gamma_A$, and $g \equiv 1/\sqrt{\gamma_A \nu}$. The same action arises in phenomenological models of *chromo-natural inflation* [828].

The equation of motion for the inflaton is

$$\ddot{\phi} + 3H\dot{\phi} + V_{,\phi} = -3\frac{g\lambda}{f}\psi^2(\dot{\psi} + H\psi) . \tag{5.294}$$

In addition to the force from the bare axion potential, the field experiences the analogue of a *magnetic drift* force proportional to the coupling λ. For large λ, the two forces balance each other and hence allow a slow evolution of the inflaton. This is closely related to the magnetic drift phenomenon of a charged particle coupled to a magnetic field; see Fig. 5.19.

For sufficiently large λ, the effective action (5.293) leads to inflation. The maximum number of e-folds that can be achieved while the axion rolls to the minimum of its potential is found to be [828]

$$(N_e)_{\text{max}} \approx \frac{3}{5}\lambda . \tag{5.295}$$

Thus, successful chromo-natural inflation requires a large Chern–Simons coupling, $\lambda \gtrsim \mathcal{O}(100)$.

The crucial question is whether such a large coupling can be achieved in a controlled string compactification. In fact, it is easy to see that this cannot be achieved for a stack of D3-branes at weak coupling, as in that case the Chern–Simons coupling is fixed by the string coupling, $\lambda \sim g_s \ll 1$. A possible alternative is a stack of D7-branes wrapping a four-cycle Σ_4, with Euler number $\chi(\Sigma_4) =$

---

[101] Any $SU(N)$ group has an $SU(2)$ subgroup, and here we have identified the global part of this $SU(2)$ with the group $SO(3)$ of spatial rotations.

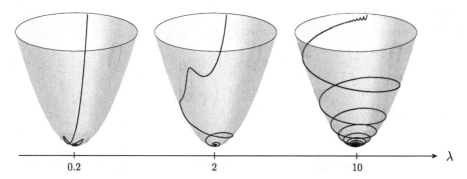

Fig. 5.19 Evolution of a charged particle in two dimensions with quadratic external potential and a coupling to a homogeneous magnetic field. For large enough coupling to the magnetic field, the particle experiences slow magnetic drift. (The numerics for these figures was kindly provided by Peter Adshead.)

$\int_{\Sigma_4} \hat{\chi}(R)$, and with worldvolume gauge field instanton number $c_2 = K$. The Chern–Simons interaction is then

$$S_{\mathrm{CS}} = \frac{i}{(2\pi)^3} \int_{\mathcal{M}_4 \times \Sigma_4} \frac{C_0}{24} \left(\mathrm{Tr}[\mathcal{F}_2 \wedge \mathcal{F}_2 \wedge \mathcal{F}_2 \wedge \mathcal{F}_2] + \frac{1}{2} \mathrm{Tr}[\mathcal{F}_2 \wedge \mathcal{F}_2] \, \hat{\chi}(R) \right) ,$$

$$(5.296)$$

where we have included the curvature coupling proportional to the Euler density $\hat{\chi}(R)$. The effective coupling in (5.293) is then found to be [644]

$$\lambda = \left[K + \frac{\chi(\Sigma_4)}{24} \right] \times g_{\mathrm{s}} \times \frac{\ell_{\mathrm{s}}^4}{\mathcal{V}_4} .$$

$$(5.297)$$

It appears difficult, but not impossible, to achieve $\lambda \gtrsim \mathcal{O}(100)$ in this setting while retaining control of the g_{s} and α' expansions. Further attempts to obtain large magnetic couplings in string theory are discussed in [644].

5.6.4 Phenomenology

The study of the primordial perturbations arising in dissipative models is comparatively new, and because realizations in string theory are also a work in progress, a definitive characterization of the phenomenology is not available at present. In this section, we will briefly describe some of the more robust signatures.

Trapped inflation

The perturbations in trapped inflation, while understood in some detail [641, 823], are not easily described analytically. We therefore present only the main results, referring the reader to the original literature for derivations [641, 823].

Power spectra

Particle production affects the spectrum of primordial perturbations. The inflaton fluctuations satisfy

$$\ddot{\delta\phi} + \left(M^2 + \frac{k^2}{a^2}\right)\delta\phi + \int^t dt'\, M^2 \left(\frac{5}{2}\dot{\delta\phi}(t') - 3H\delta\phi(t')\right)\frac{a^3(t')}{a^3(t)}$$

$$= -g\,\Delta n_\psi(k,t)\;, \tag{5.298}$$

where we have defined the time-dependent effective mass of inflaton fluctuations,

$$M^2 \equiv \frac{g^{5/2}}{(2\pi)^3}\frac{\dot{\phi}^{3/2}}{\Delta}\;, \tag{5.299}$$

and the variance in the number of produced ψ particles, $\Delta n_\psi \equiv g\sum_i \left(\psi_i^2 - \langle\psi_i^2\rangle\right)(\phi - \phi_i)$. Solving (5.298) leads to the power spectrum of curvature perturbations [641]

$$\Delta_{\mathcal{R}}^2 \approx g^{8/3}\frac{H}{M}\left(\frac{M}{\Delta}\right)^{2/3}\;. \tag{5.300}$$

Using (5.299), this can be written as

$$\Delta_{\mathcal{R}}^2 \approx g^{9/4}\left(\frac{H}{\Delta}\right)^{1/2}\left(\frac{H^2}{\dot{\phi}}\right)^{1/4}\;, \tag{5.301}$$

and the spectral tilt is

$$n_s - 1 = \frac{\dot{H}}{H^2} - \frac{1}{4}\frac{\ddot{\phi}}{H\dot{\phi}}\;. \tag{5.302}$$

For the specific example studied in [641], the tilt was found to be $n_s = 0.99$, but in general the tilt depends on the details of the model, such as the shape of the potential, the density of particle production events, and the properties of the particles that are produced.

If the power spectrum of tensors is dominated by vacuum fluctuations, cf. Eq. (1.34), then the tensor-to-scalar ratio is

$$r = g^{-8/2}\frac{HM}{M_{\rm pl}^2}\left(\frac{\Delta}{M}\right)^{2/3}\;. \tag{5.303}$$

For the regime of parameters that is consistent with constraints on the scalar fluctuations, [641] finds $r \ll 10^{-4}$.

Equilateral non-Gaussianity

The nonlinear couplings between the inflaton ϕ and the extra fields ψ_i lead to non-Gaussianity in the primordial curvature perturbations. The bispectrum peaks in the equilateral configuration and has amplitude

$$f_{\mathrm{NL}}^{\mathrm{equil}} \simeq \frac{M^2}{H^2} .$$
(5.304)

The bispectrum for trapped inflation still satisfies the single-field consistency condition [119, 120], as proved in [850].

Secondary gravitational waves

The produced ψ particles can also be a *classical* source of gravitational waves. References [270–272] studied EFT variations of trapped inflation in which this source of tensor fluctuations can sometimes be competitive with the quantum-mechanical result (1.34). Their examples go beyond the simplest models of trapped inflation and have not yet been realized in string theory. Moreover, it remains to be checked whether the regime that produces large tensors is consistent with existing constraints on non-Gaussianity.

Unwinding inflation

The phenomenology of unwinding inflation is just beginning to be explored. The signatures depend strongly on the D-brane dimension p; for $p = 3$, fluctuations in open string production provide the dominant source of perturbations, but the compactification must be highly anisotropic, while for $p = 4$ ($p = 5$) the scalar perturbations receive a 10% (1%) contribution from open strings.[102] Possible signatures include a modest level of tensor perturbations ($r \lesssim 10^{-2}$), equilateral non-Gaussianity, and (for $p = 4$) oscillations in the spectrum from modulations of the open string pair production rate.

Chromo-natural inflation

String theory realizations of chromo-natural inflation are not yet developed enough to make robust predictions for observables. Taken at face value, the original model (5.293) is in conflict with the CMB data [851]; it either predicts a spectral tilt that is too red, overproduces gravitational waves, or both. Nevertheless, it remains interesting to explore whether large Chern–Simons couplings can arise in consistent string compactifications, and whether models with viable phenomenology can be constructed.

---

[102] In the case of $p = 6$ the Lorentz factor exceeds the limit of Eq. (5.135), while $p = 7$ and $p = 8$ are clearly incompatible with metastable compactification [642].

6

Conclusions and outlook

Our mistake is not that we take our theories too seriously, but that we
do not take them seriously enough. It is always hard to realize that these
numbers and equations we play with at our desks have something to do with
the real world. Even worse, there often seems to be a general agreement that
certain phenomena are just not fit subjects for respectable theoretical and
observational effort.

Steven Weinberg, on the Big Bang model [852].

Consistent theories of quantum gravity do not grow on trees. After a search
spanning nearly a century, string theory is the only known example of such a
theory. Of course, it does not follow that string theory describes our universe;
mathematical consistency is a necessary requirement, but it is far from sufficient. To connect string theory to particle physics and cosmology, we must seek
guidance from terrestrial experiments and from observations of the cosmos. One
should not be surprised that experimental evidence is elusive, for quantum gravity is naturally relevant at scales many orders of magnitude beyond those accessed
on Earth. Running the theory to low energies and extracting predictions that are
sensitive to its high-scale origin have proved challenging. However, the early universe provides an arena where ideas about quantum gravity can be tested, and
the initial singularity of the Big Bang model is a prime example where a theory
of quantum gravity is compulsory. Quantum fluctuations of the metric during
inflation, imprinted in primordial B-mode perturbations of the CMB, are the
most vivid evidence conceivable for the reality of quantum gravity, and for the
significance of quantum gravity in the early history of our universe.

Inflation defers the singularity problem, allowing us to make predictions for the
initial conditions that emerge from the aftermath of the Big Bang. However, as
we have shown, the inflationary mechanism retains a subtle sensitivity to Planck-
scale interactions. This is both a challenge for microscopic theories of inflation, as
well as an opportunity for using the early universe as a window on Planck-scale

physics. To fulfill this promise, inflationary scenarios in string theory must be developed to an unprecedented level of completeness and sophistication.

The last decade of research on inflation in string theory has witnessed a number of significant advances. The development of methods of moduli stabilization has led to vastly improved technical capabilities, and in turn to a sharply improved understanding of metastable string compactifications and of the associated inflationary models. In special cases it has been possible to characterize the Planck-suppressed corrections to the inflaton action, leading to the first existence proofs of inflation in string theory. Furthermore, the symmetry structures required for large-field inflation are now better understood. In addition, techniques for studying the dynamics of theories with many moduli have recently emerged. Moreover, string inflation has expanded and refined our ideas for inflationary mechanisms in effective field theory. Consistency conditions in string theory have suggested that certain classes of models conceived as effective theories may not admit ultraviolet completions, and at the same time, confronting these restrictions has led to novel ideas for consistent low-energy field theories. Thus – as in many problems outside cosmology – string theory has frequently yielded solutions with unanticipated properties, and has served as a generating function for ideas that were hard to perceive from a purely low-energy point of view.

On the other hand, many critical challenges still remain. Our understanding of reheating and of the connection between the inflationary sector and the Standard Model degrees of freedom is tenuous. Non-supersymmetric solutions of string theory, particularly de Sitter solutions, remain much less controlled than their supersymmetric counterparts. This continues to be a zeroth-order challenge for deriving inflation from string theory, and has stymied many attempts to develop inflationary scenarios outside of type IIB string theory. Moreover, time-dependent solutions in string compactifications are barely understood beyond the adiabatic approximation. Furthermore, in most cases the Planck-suppressed corrections to the inflationary action are only partially characterized. Finally, and most importantly, there is not a single observation that gives direct evidence for a string-theoretic origin for inflation; although an unambiguous detection of gravitational waves produced by quantum fluctuations of the metric during inflation would directly prove the quantization of the gravitational field, discerning the character of the quantum gravity theory requires more refined observations. At present, we are led to inflation in string theory by a web of inference involving the success of inflation in effective field theory, the naturalness principle in particle physics, and the unique status of string theory as an ultraviolet completion of gravity.

We have largely avoided discussing deep issues involving the initial conditions for inflation, including the global view of eternal inflation [853–856], the associated *measure problem* [857–864], and the geodesic incompleteness of inflation in the past [865]. String theory has inspired several compelling approaches to these questions, but no complete solutions have been advanced. Many authors have noted that these unresolved problems threaten the predictivity of the inflationary paradigm. Of course, once an inflationary phase begins in a particular region of field space, clear and specific predictions do emerge. On the other hand, determining the relative probabilities of different inflationary models in a broader setting remains an important task. More generally, deriving specific predictions from the string landscape as a whole, rather than from individual models, is a distant goal that could require a new approach to the measure problem.

A striking feature of present observations is the extraordinary simplicity of the primordial curvature fluctuations, which are approximately Gaussian, adiabatic, and nearly scale-invariant. In contrast, the ultraviolet completions presented in this book are complex, involving many interacting fields and a landscape of quantized parameters. Should the "simple" observations be read as evidence against "complicated" models of inflation in string theory? We do not believe so; although the simplicity of the data motivates considering simple *effective* theories of inflation, it does not constrain the ultraviolet completions in the same way. As an analogy, the Fermi theory of beta decay is far simpler – in terms of a counting of parameters – than the Standard Model, but is merely a low-energy effective description. Indeed, the whole point in using effective theories is that they are simpler to use than their ultraviolet completions. Even so, it remains important to understand whether the simplicity of the data can emerge from the apparent complexity of the ultraviolet completion; one should determine which details of the short-distance physics decouple and which leave subtle traces in the data.

In closing, we would like to emphasize that the study of inflation in string theory has advanced to a stage where a properly constructed model can be falsified; indeed, many models have already been falsified by recent observations, while others are under observational pressure. Proper construction of models of string inflation, however, is a subtle art. We have railed against incorrect predictions rooted in oversimplified effective theories, and have catalogued the pitfalls in attempts to compute observational signatures. Our hope is that the reader will use the ideas and techniques presented in this book to derive predictions that illuminate the history of the universe and shed light on the nature of quantum gravity.

Appendix A

Mathematical preliminaries

In this appendix we summarize the key mathematical concepts that arise, directly or indirectly, in the research reported in this book. This is meant to refresh the memory rather than teach the subject; the reader is referred to books such as [352, 866–868] for more extensive background.

A.1 Manifolds and bundles

A key notion in geometry is that of a **manifold**.

> A **real manifold** of dimension n is a space M that is locally topologically isomorphic to $\mathbb{R}^n$.

To make this precise, we define a **coordinate chart** on M as a map,

$$\varphi_\alpha : U_\alpha \to \mathbb{R}^n , \qquad (A.1)$$

where $U_\alpha \subset M$ is an open set, or patch, in M. Given two intersecting charts U_α and U_β, the map $\varphi_\alpha \circ \varphi_\beta^{-1}$ defines a **transition function**, which is a homeomorphism between two open sets of $\mathbb{R}^n$. The transition function describes how M is assembled by gluing together patches that are homeomorphic to subsets of $\mathbb{R}^n$, and the set of patches and transition functions encodes all the structure of M. If the transition functions are differentiable, then M is said to be a **differentiable manifold**.

A **bundle** is a special kind of manifold that possesses a local product structure, consisting of a "base" and a "fiber."

A **fiber bundle** over a differentiable base manifold B, with fiber F, consists of a manifold E, known as the total space, and a map $\pi : E \rightarrow B$, known as the projection. For each $p \in B$, we require that there exists a neighborhood $U_p \subset B$ and a diffeomorphism

$$\phi : U_p \times F \rightarrow \pi^{-1}(U_p) \,, \tag{A.2}$$

such that on $\pi^{-1}(U_p) \subset E$, we have the projection

$$\pi \circ \phi : U_p \times F \rightarrow U_p \,. \tag{A.3}$$

The map ϕ is called the **local trivialization** of the bundle E over the neighborhood U_p.

The idea is that over each point of the base manifold B there is a fiber F, and B can be covered by charts U such that $\pi^{-1}(U)$ is diffeomorphic to $U \times F$; locally, the bundle can be trivialized to a product. At a point r in the intersection of two patches U_p and U_q, the map $(\phi_p^{-1}\phi_q)_r : F \rightarrow F$ defines the action on F of an element $g_{p,q}(r)$ of a group G, known as the **structure group** of E. That is, two overlapping local trivializations are related by the action of G on the fiber.

A **local section** s of E over $U \subset B$ is a map $s : U \rightarrow E$ that specifies a point in the fiber F over each point $p \in U$, such that $\pi \circ s$ is the identity map on U. When U can be taken to be all of B, the local section is called a **global section**.

A (real) **vector bundle** of rank k is a fiber bundle for which the fiber F is $\mathbb{R}^k$, and the structure group is a subgroup of $GL(k, \mathbb{R})$. An important example is the **tangent bundle** of a manifold B, for which the fiber at $p \in B$ is the tangent space at p, $T_p B$. A section of the tangent bundle is a vector field on B.

A.2 Holonomy

To characterize the geometry of a manifold equipped with a connection, it is useful to ask how tangent vectors evolve under parallel transport. For example, parallel transport of a horizontal stick along a closed curve on the Earth's surface that encloses a region of area $A = aR_E^2$ leads to rotation of the stick by an angle a. In this way a sequence of local operations leads to information about the curvature of the surface.

On a two-dimensional manifold a rotation of a tangent vector is characterized by a single angle, but on a manifold M of dimension $D > 2$ the tangent vector is free to rotate in more than one plane. Determining whether a tangent vector does in fact rotate in all possible directions upon parallel transport, or instead remains inside a subspace of the tangent space, provides a diagnostic of the degree of "symmetry" or "simplicity" of M.

The **holonomy group** $\text{Hol}_p(\Gamma)$, at a point p, of a connection Γ on the tangent bundle of a differentiable manifold M of dimension k is the set of transformations induced by parallel transport of tangent vectors around closed loops c in M that begin and end at p:

$$\text{Hol}_p(\Gamma) = \{G_c : T_pM \to T_pM\} \subset GL(k, \mathbb{R}) \,. \tag{A.4}$$

Given two points p and q in M that can be connected by a path, one easily shows that

$$\text{Hol}_p(\Gamma) = g \, \text{Hol}_q(\Gamma) g^{-1} \,, \tag{A.5}$$

for some $g \in GL(k, \mathbb{R})$. Because the holonomy groups at different points of a connected manifold are thus equal up to conjugation in $GL(k, \mathbb{R})$, the holonomy of a connected manifold does not depend on the base point.

If the connection Γ is compatible with a metric, then parallel transport leaves lengths invariant. In this case, for M an orientable manifold of real dimension n, G must be $SO(n)$ or a subgroup. We will be interested in the case where Γ is the Levi-Civita connection for a Riemannian metric on M; the corresponding holonomy group is known as the Riemannian holonomy group, and is denoted $\text{Hol}(M)$.

The possible holonomy groups of simply connected Riemannian manifolds have been classified. Let M be a simply connected n-dimensional Riemannian manifold. Then **Berger's classification** states that either M is a product of lower-dimensional manifolds; M is a symmetric space (i.e. a coset space G/H, for G a Lie group and $H \subset G$ a Lie subgroup); or one of the following cases holds:

1. $\text{Hol}(M) = SO(n)$.
2. $\text{Hol}(M) = U(k) \subset SO(2k)$, for $n = 2k$.
3. $\text{Hol}(M) = SU(k) \subset SO(2k)$, for $n = 2k$.
4. $\text{Hol}(M) = Sp(k) \subset SO(4k)$, for $n = 4k$.
5. $\text{Hol}(M) = Sp(k)Sp(1) \subset SO(4k)$, for $n = 4k$.
6. $\text{Hol}(M) = G_2 \subset SO(7)$, for $n = 7$.
7. $\text{Hol}(M) = Spin(7) \subset SO(8)$, for $n = 8$.

See [869] for a comprehensive discussion. Case (1), holonomy $SO(n)$, obtains in generic Riemannian manifolds, while the remaining cases (2)–(7) are known as *special* (or *reduced*) holonomy, because $\text{Hol}(M)$ is a proper subgroup of $SO(n)$. Reduced holonomy is of central importance in supersymmetric compactifications of string theory, because the holonomy group determines the number of ten-dimensional spinors η that are invariant under parallel transport. It is these invariant spinors that correspond to four-dimensional supercharges upon dimensional reduction. The larger the holonomy group, the more thoroughly spinors get transformed upon parallel transport, and for the generic case of holonomy $SO(n)$, there are no invariant spinors, and hence no supersymmetry in four dimensions. If instead $\mathrm{Hol}(M)$ is a proper subgroup of $SO(n)$, one or more components of a ten-dimensional spinor is covariantly constant, and the four-dimensional theory preserves some supersymmetry.

Each of the cases (2)–(7) is relevant in string compactifications:

- A manifold with holonomy $U(k)$, case (2), is a *Kähler manifold*. In this case the Christoffel symbols have only holomorphic (or only antiholomorphic) indices, and hence parallel transport does not transform holomorphic vectors into antiholomorphic vectors.
- A manifold with holonomy $SU(k)$, case (3), is a *Calabi–Yau manifold*. Such a manifold admits a Ricci-flat Kähler metric, i.e. a Kähler metric for which the Ricci tensor vanishes, $R_{i\bar{j}} = 0$. Taking $k = 3$, one has the celebrated case of Calabi–Yau threefolds.
- A manifold with holonomy $Sp(k) \subseteq SU(2k)$, case (4), is a *hyper-Kähler manifold*. Such a manifold admits a Ricci-flat Kähler metric, and is a special case of a Calabi–Yau manifold.
- A manifold with holonomy $Sp(k)Sp(1)$, case (5), is a *quaternionic Kähler manifold*. The corresponding metric is neither Kähler nor Ricci-flat.
- Manifolds with holonomy G_2 or $Spin(7)$, cases (6) and (7), are said to have *exceptional holonomy*. The corresponding metrics are Ricci-flat.

A.3 Differential forms and cohomology

Calculus on manifolds is naturally formulated in terms of **differential forms**.

A.3.1 Differential forms

An **r-form** ω is a totally antisymmetric tensor field of type $(0, r)$. That is, ω assigns to each point $p \in M$ a map

$$\omega_p \ : \ \underbrace{T_pM \times \cdots \times T_pM}_{r} \ \to \ \mathbb{R} \qquad\qquad \text{(A.6)}$$

that is totally antisymmetric and is linear in each variable.

We use $\Omega_p^r(M)$ to denote the vector space of r-forms at a point p, i.e. the space of tensors (not tensor fields) of type $(0, r)$, and $\Omega^r(M)$ to denote the vector space of r-forms on M. Note that a one-form at p is a cotangent vector, i.e. an element of the **cotangent space** T_p^*M dual to T_pM. Then, just as vector fields are sections of the tangent bundle TM, one-forms are sections of the cotangent bundle T^*M.

We construct a basis for $\Omega_p^r(M)$ using the **wedge product** $\wedge$ of r one-forms dx^μ:

$$dx^{\mu_1} \wedge \cdots \wedge dx^{\mu_r} = \sum_{P \in S_r} \text{sign}(P)\, dx^{\mu_{P(1)}} \otimes \cdots \otimes dx^{\mu_{P(r)}} \,, \tag{A.7}$$

where S_r is the permutation group, $\text{sign}(P) = +1\,(-1)$ for even (odd) permutations, and $\otimes$ is the tensor product. In local coordinates at p, an element $\omega \in \Omega_p^r(M)$ can be written as

$$\omega = \frac{1}{r!}\omega_{\mu_1 \ldots \mu_r}\, dx^{\mu_1} \wedge \cdots \wedge dx^{\mu_r} \,, \tag{A.8}$$

where the components $\omega_{\mu_1 \ldots \mu_r}$ are totally antisymmetric. We will sometimes define multi-indices $\boldsymbol{M} = (\mu_1 \ldots \mu_r)$ and write

$$\omega = \frac{1}{r!}\omega_{\boldsymbol{M}} dx^{\boldsymbol{M}} \,. \tag{A.9}$$

The wedge product of an r-form ω and an s-form η is

$$(\omega \wedge \eta)_{\mu_1 \ldots \mu_r \nu_1 \ldots \nu_s} = \frac{(r+s)!}{r!s!}\omega_{[\mu_1 \ldots \mu_r}\eta_{\nu_1 \ldots \nu_s]} \,, \tag{A.10}$$

where the square brackets denote antisymmetrization. The wedge product satisfies

$$\omega \wedge \eta = (-1)^{rs}\,\eta \wedge \omega \,. \tag{A.11}$$

Using the wedge product, we can build up r-forms at p as elements of the rth exterior power of T_p^*M, denoted $\Lambda^r(T_p^*M)$, so that r-forms $\omega \in \Omega^r(M)$ are sections of the vector bundle $\Lambda^r(T^*M)$.

The **exterior derivative** is a map $d : \Omega^r(M) \to \Omega^{r+1}(M)$, defined so that for an r-form ω given in local coordinates by (A.9), we have

$$d\omega = \frac{1}{r!}\frac{\partial\omega_{\boldsymbol{M}}}{\partial x^\nu}\, dx^\nu \wedge dx^{\boldsymbol{M}} \,. \tag{A.12}$$

The exterior derivative satisfies

$$d(\omega_r \wedge \eta_s) = d\omega_r \wedge \eta_s + (-1)^r \omega_r \wedge d\eta_s \,, \tag{A.13}$$

$$d^2\omega = 0 \,. \tag{A.14}$$

In fact, d is the unique linear operator that obeys (A.13) and (A.14), and that acts as the differential on smooth functions; this provides an alternative axiomatic definition of the exterior derivative.

Differential forms transform in a simple way under a change of coordinates:

$$\mathrm{d}x^1 \wedge \cdots \wedge \mathrm{d}x^D \mapsto \frac{\partial x^1}{\partial y^{\mu_1}} \mathrm{d}y^{\mu_1} \wedge \cdots \wedge \frac{\partial x^D}{\partial y^{\mu_D}} \mathrm{d}y^{\mu_D}$$

$$= \det \left(\frac{\partial x^\mu}{\partial y^\nu} \right) \mathrm{d}y^1 \wedge \cdots \wedge \mathrm{d}y^D . \tag{A.15}$$

We can therefore integrate an r-form over an r-dimensional manifold without specifying a metric or a choice of coordinates.

A.3.2 Cohomology and homology

Differential forms efficiently encode topological information about M.

An r-form ω is called **closed** if $\mathrm{d}\omega = 0$, and **exact** if there exists an $(r-1)$-form α such that $\omega = \mathrm{d}\alpha$.

Because $\mathrm{d}^2 = 0$, every exact form is closed. Moreover, the **Poincaré lemma** states that a closed r-form $(r > 0)$ on a contractible open subset of $\mathbb{R}^n$ is exact; that is, every closed form is locally exact. Whether a closed r-form is globally exact depends on the topology of M, as we shall see.

A cautionary remark is required concerning the definition of an exact form: the $(r-1)$-form α appearing in $\omega = \mathrm{d}\alpha$ must be smooth and defined everywhere, i.e. α must be a global section of $\Lambda^{r-1}(T^*M)$, in order for ω to be exact. As an example, consider Maxwell theory, with the vector potential $A \equiv A_\mu \mathrm{d}x^\mu$. The field strength F is defined locally by $F = \mathrm{d}A$. When the vector potential is a smooth one-form, F is exact. However, in many cases of interest it is *not* possible to define A everywhere: the vector potentials in different coordinate patches are related by nontrivial gauge transformations. In this case F is closed but not exact.

Let $Z^r(M)$ denote the set of closed r-forms and $B^r(M)$ the set of exact r-forms. The rth **de Rham cohomology group** is the vector space

$$H^r(M, \mathbb{R}) \equiv Z^r(M)/B^r(M) . \tag{A.16}$$

Two closed r-forms ω and ξ that differ by an exact r-form are said to be **cohomologous**, or to be in the same cohomology class. We write $[\omega] = [\xi]$.

The de Rham cohomology groups $H^r(M)$, which characterize nontrivial differential forms on M, are closely related to the **homology groups** $H_r(M)$, which characterize topologically nontrivial submanifolds of M. We leave a careful definition of homology to the references and present only a minimal account here.

An **r-simplex** σ is an r-dimensional generalization of a filled triangle (2-simplex) or solid tetrahedron (3-simplex). We define an **r-chain** C_r as a formal sum of finitely many r-simplices,

$$C_r = \sum_i k_i \sigma_i , \tag{A.17}$$

where the coefficients k_i are integers. Next, let the **boundary operator** ∂ be defined so that ∂C_r is the $(r-1)$-chain consisting of the boundaries of the r-simplices in C_r. This allows us to make the following definitions.

An r-chain C_r is called an **r-cycle** if C_r has no boundary, $\partial C_r = 0$. An r-chain C_r is called an **r-boundary** if C_r is the boundary of some $(r+1)$-chain, $C_r = \partial D_{r+1}$.

Heuristically, an r-cycle is a (sum of) r-dimensional submanifolds of M, without boundary. We define $Z_r(M)$ to be the space of all r-cycles, and $B_r(M)$ to be the space of all r-boundaries. We can now define homology groups.

The rth **homology group** of M is the vector space

$$H_r(M) \equiv Z_r(M)/B_r(M) . \tag{A.18}$$

Two r-cycles Σ and Σ' that differ by an r-boundary are said to be **homologous**, or to be in the same homology class. We write $[\Sigma] = [\Sigma']$.

A.3.3 De Rham's theorem and Poincaré duality

The homology group $H_r(M)$ and the cohomology group $H^r(M)$ are dual vector spaces. Partial intuition for this result comes from the fact that the integrals of r-forms over r-cycles depend only on the cohomology classes of the forms and the homology classes of the cycles. Suppose that ω is a closed r-form, and Σ is an r-cycle. We can define an inner product $(\Sigma, \omega) : H_r(M) \times H^r(M) \to \mathbb{R}$ via

$$(\Sigma, \omega) \equiv \int_\Sigma \omega . \tag{A.19}$$

To see that this inner product depends only on the cohomology class $[\omega]$ and the homology class $[\Sigma]$, but not on the particular choice of representatives of these classes, we use **Stokes' theorem**, which states that if $\omega \in \Omega^r(M)$ is an r-form and C_r is an r-chain, then

$$\int_{C_r} d\omega = \int_{\partial C_r} \omega \ , \qquad (A.20)$$

where ∂C_r is the boundary of C_r. Now suppose we add an exact r-form $\xi = d\lambda$ to ω, and an r-boundary $\Pi = \partial \Lambda$ to Σ. Then

$$\int_{\Sigma + \Pi} (\omega + \xi) = \int_{\Sigma} \omega + \int_{\partial \Sigma} \lambda + \int_{\Lambda} d\omega + \int_{\Lambda} d^2\lambda = \int_{\Sigma} \omega \ , \qquad (A.21)$$

establishing that the inner product is independent of representatives. Moreover, **de Rham's theorem** states that in fact $H_r(M)$ and $H^r(M)$ are dual vector spaces.

In a manifold of dimension D, a generic r-cycle and a generic $(D-r)$-cycle intersect at a finite number of points, so that one can define a pairing

$$H_r(M) \times H_{D-r}(M) \to \mathbb{R} \ . \qquad (A.22)$$

This pairing defines a vector space isomorphism (denoted by $\cong$), known as **Poincaré duality**:

$$H_r(M) \cong H_{D-r}(M) \ . \qquad (A.23)$$

Poincaré duality can also be seen in cohomology; given $\omega \in H^r(M)$ and $\eta \in H^{D-r}(M)$, we define a map $H^r(M) \times H^{D-r}(M) \to \mathbb{R}$ via

$$(\omega, \eta) \equiv \int_M \omega \wedge \eta \ , \qquad (A.24)$$

from which one finds that

$$H^r(M) \cong H^{D-r}(M) \ . \qquad (A.25)$$

Combined with de Rham's theorem relating cohomology and homology, we conclude that

$$H^{D-r}(M) \cong H^r(M) \cong H_r(M) \cong H_{D-r}(M) \ . \qquad (A.26)$$

The rth **Betti number**,

$$b^r \equiv \dim H^r(M, \mathbb{R}) \ , \qquad (A.27)$$

counts the nontrivial cohomology classes of r-forms on M, or equivalently the nontrivial homology classes of r-cycles. Poincaré duality implies that

$$b^r = b^{D-r} \ . \qquad (A.28)$$

The **Euler characteristic** of M is the alternating sum

$$\chi(M) \equiv \sum_{r=0}^{D} (-1)^r b^r \ . \qquad (A.29)$$

A.3.4 Hodge theory and harmonic forms

Thus far, we have used only topological characteristics of M. However, when a metric is available, more concrete statements are possible.

The **Hodge star** is an isomorphism $\star : \Omega^r(M) \to \Omega^{D-r}(M)$ that acts on the basis elements of $\Omega^r(M)$ as

$$\star (\mathrm{d}x^{\mu_1} \wedge \ldots \wedge \mathrm{d}x^{\mu_r})$$

$$= \frac{\sqrt{|g|}}{(D-r)!} \epsilon^{\mu_1 \cdots \mu_r}{}_{\nu_{r+1} \cdots \nu_D} \mathrm{d}x^{\nu_{r+1}} \wedge \ldots \wedge \mathrm{d}x^{\nu_D} , \qquad (A.30)$$

where $g \equiv \det(g_{\mu\nu})$, and the totally antisymmetric Levi-Civita symbol $\epsilon_{\mu_1 \cdots \mu_D}$ equals $+1$ if $\mu_1 \cdots \mu_D$ is an even permutation of $012 \cdots D$.

The Hodge star gives a natural (but metric-dependent) realization of the isomorphism between $H^r(M, \mathbb{R})$ and $H^{D-r}(M, \mathbb{R})$, which as we saw above was necessitated at the level of topology. Note that $\star 1$ is the invariant volume element

$$\star 1 = \sqrt{|g|}\, \mathrm{d}x^{\nu_1} \wedge \cdots \wedge \mathrm{d}x^{\nu_D} . \qquad (A.31)$$

The action of $\star$ on an r-form in D dimensions is

$$(\star \omega_r)_{M_1 \ldots M_{D-r}} = \frac{\sqrt{|g|}}{r!(D-r)!} \epsilon_{M_1 \ldots M_{D-r}}{}^{N_1 \ldots N_r} \omega_{N_1 \ldots N_r} . \qquad (A.32)$$

Applying $\star$ twice, we find

$$\star \star \omega_r = (-1)^{r(D-r)} \omega_r \qquad (A.33)$$

on a Riemannian manifold M. If M is Lorentzian, (A.33) has an additional minus sign. The Hodge star allows us to construct a D-form from two r-forms:

$$\omega \wedge \star \xi = \frac{1}{r!} \omega_{\mu_1 \ldots \mu_r} \xi^{\mu_1 \cdots \mu_r} \sqrt{|g|}\, \mathrm{d}x^1 \wedge \cdots \wedge \mathrm{d}x^D . \qquad (A.34)$$

The corresponding **inner product** of two r-forms is

$$\langle \omega, \xi \rangle \equiv \int \omega \wedge \star \xi = \int \mathrm{d}^D x \sqrt{|g|}\, \omega_{\mu_1 \ldots \mu_r} \xi^{\mu_1 \cdots \mu_r} . \qquad (A.35)$$

The **adjoint exterior derivative** is the map $\mathrm{d}^\dagger : \Omega^r(M) \to \Omega^{r-1}(M)$ given by

$$\mathrm{d}^\dagger = (-1)^{Dr+D+1} \star \mathrm{d} \star . \qquad (A.36)$$

For $\eta \in \Omega^{r-1}(M)$ and $\omega \in \Omega^r(M)$, we find

$$\langle \mathrm{d}\eta, \omega \rangle = \langle \eta, \mathrm{d}^\dagger \omega \rangle , \qquad (A.37)$$

so that $\mathrm{d}^\dagger$ is the adjoint of d.

An r-form ω is said to be **co-closed** if $\mathrm{d}^\dagger \omega = 0$, and **co-exact** if $\omega = \mathrm{d}^\dagger \beta$ for a globally defined $(r+1)$-form β. The **Laplacian** is the map $\Delta : \Omega^r(M) \to \Omega^r(M)$ given by

$$\Delta = \mathrm{d}\mathrm{d}^\dagger + \mathrm{d}^\dagger\mathrm{d} . \tag{A.38}$$

An r-form h with $\Delta h = 0$ is termed **harmonic**. A harmonic form is both closed and co-closed:

$$0 = \langle \Delta h, h \rangle = \langle \mathrm{d}h, \mathrm{d}h \rangle + \langle \mathrm{d}^\dagger h, \mathrm{d}^\dagger h \rangle \geq 0 . \tag{A.39}$$

We denote the vector space of harmonic r-forms on M by $\mathcal{H}^r(M)$.

The **Hodge decomposition** states that every form ω on a compact Riemannian manifold has a unique global decomposition as the sum of a harmonic form, an exact form, and a co-exact form:

$$\omega = h + \mathrm{d}\alpha + \mathrm{d}^\dagger\beta , \tag{A.40}$$

where $h \in \mathcal{H}^r(M)$ is harmonic. The terms appearing in (A.40) are orthogonal; we have $\langle \mathrm{d}\alpha, \mathrm{d}^\dagger\beta \rangle = \langle \mathrm{d}^2\alpha, \beta \rangle = 0$, $\langle \mathrm{d}\alpha, h \rangle = \langle \alpha, \mathrm{d}^\dagger h \rangle = 0$, and $\langle \mathrm{d}^\dagger\beta, h \rangle = \langle \beta, \mathrm{d}h \rangle = 0$, where we have used that h is closed and co-closed.

Harmonic forms are closely connected with de Rham cohomology groups. We have just seen that a harmonic r-form $h \in \mathcal{H}^r(M)$ is closed, and is orthogonal to an exact r-form, so $\mathcal{H}^r(M) \subset H^r(M, \mathbb{R})$. To show that $H^r(M, \mathbb{R}) \subset \mathcal{H}^r(M)$, we suppose that $\xi \in H^r(M, \mathbb{R})$ is a closed r-form, decomposed as in (A.40). Using $\mathrm{d}\xi = \mathrm{d}h = \mathrm{d}^2\alpha = 0$, we then find

$$0 = \langle \mathrm{d}\xi, \beta \rangle = \langle \mathrm{d}^\dagger\beta, \mathrm{d}^\dagger\beta \rangle . \tag{A.41}$$

Because the inner product is positive-definite, we conclude that $\mathrm{d}^\dagger\beta = 0$, and

$$\xi = h + \mathrm{d}\alpha . \tag{A.42}$$

To see that h is the *unique* harmonic representative of the cohomology class $[\xi]$, suppose that $\xi' = \xi + \mathrm{d}\eta$ is another representative of $[\xi]$. Because the vector spaces $\mathcal{H}^r(M)$ and $\mathrm{d}\Omega^{r-1}(M)$ are orthogonal subspaces of $\Omega^r(M)$, ξ' has the same harmonic representative h as does ξ; the addition of an exact form only affects directions perpendicular to the harmonic subspace. This establishes **Hodge's theorem**, which states that each de Rham cohomology class has a unique harmonic representative, and the vector spaces $H^r(M, \mathbb{R})$ and $\mathcal{H}^r(M)$ are isomorphic:

$$H^r(M, \mathbb{R}) \cong \mathcal{H}^r(M) . \tag{A.43}$$

A.3.5 Harmonic forms and moduli

To understand the role of Hodge's theorem in string theory, recall that the ten-dimensional effective action of each superstring theory contains terms of the form

$$S_{r+1} = \frac{c}{r+1!} \int d^{10}X \sqrt{|G|}\, F_{M_1\ldots M_{r+1}} F^{M_1\ldots M_{r+1}} = c\,\langle F_{r+1}, F_{r+1} \rangle \,, \quad (A.44)$$

where $F_{r+1} = dC_r$ is an $(r+1)$-form generalization of the Maxwell field strength, and c is a normalization constant. The action S_{r+1} is non-negative, and is minimized if $F_{r+1} = 0$. Varying (A.44), one finds the equation of motion

$$d^\dagger dC_r = 0 \,. \quad (A.45)$$

Making the gauge choice $d^\dagger C_r = 0$, analogous to the Lorentz gauge $\partial_\mu A^\mu = 0$ in electromagnetism, we can write (A.45) in the form

$$\Delta C_r = 0 \,. \quad (A.46)$$

Now suppose that string theory is compactified on the ten-dimensional manifold $\mathcal{M}_{10} = \mathcal{M}_4 \times M$, where $\mathcal{M}_4$ is a noncompact spacetime and M is a compact six-manifold. The Laplacian Δ decomposes as

$$\Delta = \Delta_4 + \Delta_M \,, \quad (A.47)$$

where $\Delta_4 \equiv \square$ is the d'Alembertian. We make the ansatz

$$C_r = A_p \wedge B_{r-p} \,, \quad (A.48)$$

where A_p is a p-form on $\mathcal{M}_4$ and B_{r-p} is a $(r-p)$-form on M. Now suppose that B_{r-p} is an eigenfunction of Δ_M with eigenvalue λ, i.e. $\Delta_M B_{r-p} = \lambda B_{r-p}$. The equation of motion (A.46) yields

$$(\square + \lambda)A_p = 0 \,. \quad (A.49)$$

Thus, as usual in Kaluza–Klein theories, the eigenvalue of the internal Laplacian Δ_M dictates the mass of the corresponding four-dimensional field. In particular, A_p is a massless p-form in four dimensions if and only if B_{r-p} is a **zero mode** of Δ_M: that is, if B_{r-p} is a harmonic $(r-p)$-form. We conclude that

> The dimensional reduction of an r-form potential C_r yields one independent massless p-form in four dimensions for each harmonic $(r-p)$-form on M. By (A.43), the number of such p-form fields is b_{r-p}.

The power of this result is that it relates solutions to the equations of motion, which are analytic objects, to topological invariants of M.

Because of the equivalence between harmonic forms and (co)homology classes, one can think of a four-dimensional field A_p as arising from integrating C_r over an $(r-p)$-cycle on M that represents a nontrivial homology class. For example, integrating an r-form potential over an r-cycle yields a massless scalar in four dimensions, while integrating over an $(r-1)$-cycle yields a massless vector.

A.4 Complex manifolds

Complex manifolds are even-dimensional real manifolds that are locally modeled on $\mathbb{C}^k$. Just as the specialization from real analysis on $\mathbb{R}^2$ to complex analysis on $\mathbb{C}$ reveals powerfully constraining structures, restricting from real manifolds to complex manifolds leads to dramatic refinements. Complex manifolds are significant in string theory for two major reasons. First, complex manifolds arise as moduli spaces in many supersymmetric theories. Second, complex geometry has been a key tool in the analysis of Ricci-flat compactification manifolds, i.e. vacuum solutions; compact Ricci-flat manifolds are not necessarily complex manifolds, but those that are complex are generally easier to construct and study. For further details on complex manifolds we recommend the lecture notes by Candelas [870].

A.4.1 Preliminaries

A **complex manifold** M is a real manifold of dimension $n = 2k$ for which the coordinate charts are maps $\varphi_\alpha : U_\alpha \to \mathbb{C}^k$, and the transition functions are holomorphic.

While the structure group of the tangent bundle of a generic Riemannian manifold M of real dimension $2k$ is $GL(2k, \mathbb{R})$, the structure group of the tangent bundle of a complex manifold of real dimension $2k$ is contained in $GL(k, \mathbb{C}) \subset GL(2k, \mathbb{R})$.

A useful equivalent definition of a complex manifold is as a real manifold of dimension $n = 2k$ equipped with a tensor field $\mathcal{J}$ of type (1,1) known as a complex structure.

An **almost complex structure** $\mathcal{J}$ on a real manifold M of dimension $n = 2k$ is a (1,1) tensor field satisfying $\mathcal{J}^2 = -1$. It is useful to think of $\mathcal{J}$ as a map $\mathcal{J} : TM \to TM$.

At any single point of M, we can always choose coordinates $z^{\mu_1}, \ldots, z^{\mu_k}$ and their complex conjugates $\bar{z}^{\bar{\mu}_1}, \ldots, \bar{z}^{\bar{\mu}_k}$ so that the components of $\mathcal{J}$ take the form

$$\mathcal{J}^{\mu}_{\nu} = i\delta^{\mu}_{\nu} \quad , \quad \mathcal{J}^{\bar{\mu}}_{\bar{\nu}} = -i\delta^{\bar{\mu}}_{\bar{\nu}} \ . \tag{A.50}$$

When all of M can be covered with a collection of patches U_{α}, in each of which $\mathcal{J}$ takes the form (A.50) for a suitable choice of local holomorphic coordinates, we say that $\mathcal{J}$ is a **complex structure**, and M is a complex manifold.

The **complexified tangent space** $T_p M^{\mathbb{C}}$ at a point p in a real manifold M of dimension n (with n not necessarily even) is defined as

$$T_p M^{\mathbb{C}} = \{X + iY | X, Y \in T_p M\} \ , \tag{A.51}$$

and complexifications of more general tensors are defined similarly. If M is a complex manifold, we can use the complex structure $\mathcal{J}$ to split $T_p M^{\mathbb{C}}$ into eigenspaces of $\mathcal{J}$. In terms of local holomorphic coordinates $z^{\mu} \equiv x^{\mu} + iy^{\mu}$, a basis for $T_p M^{\mathbb{C}}$ consists of the vectors $\partial/\partial z^{\mu}$ and $\partial/\partial \bar{z}^{\bar{\mu}}$. Defining the action of $\mathcal{J}$ on $T_p M^{\mathbb{C}}$ via $\mathcal{J}(X + iY) = \mathcal{J}(X) + i\mathcal{J}(Y)$, we find that at a point p,

$$\mathcal{J}(\partial/\partial z^{\mu}) = i\partial/\partial z^{\mu} \quad , \quad \mathcal{J}(\partial/\partial \bar{z}^{\bar{\mu}}) = -i\partial/\partial \bar{z}^{\bar{\mu}} \ . \tag{A.52}$$

It is therefore natural to divide $T_p M^{\mathbb{C}}$ into the positive eigenspace spanned by $\{\partial/\partial z^{\mu}\}$ and the negative eigenspace spanned by $\{\partial/\partial \bar{z}^{\bar{\mu}}\}$. The elements of these spaces are known as holomorphic vectors and antiholomorphic vectors, respectively.

A.4.2 Complex differential forms

To extend our results on differential forms to complex manifolds, we first consider formal sums of r-forms, with coefficients in $\mathbb{C}$.

If α_r and β_r are two real r-forms, the sum

$$\gamma_r \equiv \alpha_r + i\beta_r \tag{A.53}$$

is termed a **complex r-form**. We denote the vector space of complex r-forms by $\Omega^r_{\mathbb{C}}(M)$. The conjugate of γ_r is $\bar{\gamma}_r \equiv \alpha_r - i\beta_r$.

Just as we decomposed the complexified tangent space, we can decompose $\Omega^r_{\mathbb{C}}(M)$.

An (r, s)-**form** is a complex-valued differential form with r holomorphic indices and s anti-holomorphic indices. In local coordinates, a basis for (r, s)-forms is

$$\mathrm{d}z_{\mu_1} \wedge \ldots \wedge \mathrm{d}z_{\mu_r} \wedge \mathrm{d}\bar{z}_{\bar{\nu}_1} \wedge \ldots \wedge \mathrm{d}\bar{z}_{\bar{\nu}_s} \equiv \mathrm{d}z_{\boldsymbol{M}} \wedge \mathrm{d}\bar{z}_{\boldsymbol{N}} \ , \tag{A.54}$$

where we have defined the multi-indices $\boldsymbol{M} = (\mu_1 \ldots \mu_r)$ and $\boldsymbol{N} = (\nu_1 \ldots \nu_s)$. We denote the vector space of (r, s)-forms on a manifold M by $\Omega^{r,s}(M)$.

An element $\gamma_{r,s}$ of $\Omega^{r,s}(M)$ can be written as

$$\gamma_{r,s} = \frac{1}{r!s!}\gamma_{M\bar{N}}\,\mathrm{d}z^M \wedge \mathrm{d}\bar{z}^{\bar{N}}\,, \qquad (A.55)$$

where $\gamma_{M\bar{N}} = \gamma_{\mu_1\ldots\mu_r\bar{\nu}_1\ldots\bar{\nu}_s}$. Any complex k-form can be decomposed uniquely into a sum of (r,s)-forms

$$\gamma_k = \sum_{r+s=k} \gamma_{r,s}\,, \qquad (A.56)$$

so that

$$\Omega^k_{\mathbb{C}} = \bigoplus_{r+s=k} \Omega^{r,s}\,. \qquad (A.57)$$

The next step is to refine the definition of d to respect the complex structure.

The **Dolbeault operators** are maps $\partial : \Omega^{r,s} \to \Omega^{r+1,s}$ and $\bar{\partial} : \Omega^{r,s} \to \Omega^{r,s+1}$ whose actions on $\gamma_{r,s}$ are

$$\partial\gamma_{r,s} = \left(\frac{\partial}{\partial z^\kappa}\gamma_{M\bar{N}}\right)\mathrm{d}z^\kappa \wedge \mathrm{d}z^M \wedge \mathrm{d}\bar{z}^{\bar{N}}\,, \qquad (A.58)$$

$$\bar{\partial}\gamma_{r,s} = \left(\frac{\partial}{\partial\bar{z}^{\bar{\kappa}}}\gamma_{M\bar{N}}\right)\mathrm{d}\bar{z}^{\bar{\kappa}} \wedge \mathrm{d}z^M \wedge \mathrm{d}\bar{z}^{\bar{N}}\,. \qquad (A.59)$$

It is easy to show that

$$\mathrm{d} = \partial + \bar{\partial}\,, \qquad (A.60)$$

$$\partial^2 = \bar{\partial}^2 = \partial\bar{\partial} + \bar{\partial}\partial = 0\,. \qquad (A.61)$$

An $(r,0)$-form $\gamma_{r,0}$ is said to be **holomorphic** if and only if

$$\bar{\partial}\gamma_{r,0} = 0\,. \qquad (A.62)$$

A holomorphic zero-form is a holomorphic function.

A.4.3 Dolbeault cohomology

The Dolbeault operators can be used to define Dolbeault cohomology classes just as the exterior derivative d defined de Rham cohomology classes.

Let $Z^{r,s}_{\bar{\partial}}(M)$ denote the set of $\bar{\partial}$-closed (r,s)-forms and $B^{r,s}_{\bar{\partial}}(M)$ the set of $\bar{\partial}$-exact (r,s)-forms. The **Dolbeault cohomology group** is then defined as

$$H^{r,s}_{\bar{\partial}}(M,\mathbb{C}) \equiv Z^{r,s}_{\bar{\partial}}(M)/B^{r,s}_{\bar{\partial}}(M)\,. \qquad (A.63)$$

Furthermore, Hodge theory for complex manifolds can be defined in close analogy with the real case: by extending the Hodge star to act on the complexified tangent space, one obtains an inner product, and can correspondingly define adjoints $\partial^\dagger$ and $\bar{\partial}^\dagger$ of the operators ∂ and $\bar{\partial}$. One then defines two types of Laplacians,

$$\Delta_\partial = \partial\partial^\dagger + \partial^\dagger\partial , \quad \Delta_{\bar{\partial}} = \bar{\partial}\bar{\partial}^\dagger + \bar{\partial}^\dagger\bar{\partial} . \tag{A.64}$$

We denote by $\mathcal{H}_{\bar{\partial}}^{r,s}(M)$ the set of $\bar{\partial}$-harmonic (r,s)-forms, i.e. (r,s)-forms annihilated by $\Delta_{\bar{\partial}}$. Hodge's theorem then states that $H_{\bar{\partial}}^{r,s}(M,\mathbb{C}) \cong \mathcal{H}_{\bar{\partial}}^{r,s}(M)$. The **Hodge numbers** are the complex dimensions of the Dolbeault cohomology groups:

$$h^{r,s} = \dim H_{\bar{\partial}}^{r,s}(M,\mathbb{C}) . \tag{A.65}$$

By using the Hodge star, one finds that $h^{r,s} = h^{k-r,k-s}$ on a manifold of complex dimension k.

A.5 Kähler and Calabi–Yau geometry

Kähler manifolds, a class of complex manifolds enjoying many remarkable properties, play a distinguished role in string theory. The best-understood vacuum solutions are compactifications on Kähler manifolds, and more specifically on Calabi–Yau manifolds. Moreover, the moduli space of a theory with $\mathcal{N} = 1$ supersymmetry is a Kähler manifold, as are many moduli spaces of theories with $\mathcal{N} > 1$ supersymmetry.

A.5.1 Kähler manifolds

A **Hermitian metric** on a complex manifold is a Riemannian metric $g : TM \times TM \to \mathbb{R}$ that satisfies

$$g(\mathcal{J}X, \mathcal{J}X) = g(X,Y) . \tag{A.66}$$

In local coordinates, we have

$$g = g_{\mu\bar{\nu}}\, \mathrm{d}z^\mu \otimes \mathrm{d}\bar{z}^{\bar{\nu}} + g_{\bar{\mu}\nu}\, \mathrm{d}\bar{z}^{\bar{\mu}} \otimes \mathrm{d}z^\nu . \tag{A.67}$$

Reality of the metric implies $g_{\mu\bar{\nu}} = \overline{g_{\nu\bar{\mu}}}$. The condition that the metric is Hermitian is expressed locally as $g_{\mu\nu} = g_{\bar{\mu}\bar{\nu}} = 0$. A Hermitian metric can be defined on any complex manifold M: if g_0 is a Riemannian metric on M, one can define a Hermitian metric g via $g(X,Y) = g_0(X,Y) + g_0(\mathcal{J}X, \mathcal{J}X)$. A complex manifold equipped with a Hermitian metric is known as a **Hermitian manifold**.

Using the Hermitian metric, we can define the **Kähler form** of a Hermitian manifold as the two-form given in local coordinates by

$$J = ig_{\mu\bar{\nu}} \, dz^\mu \wedge d\bar{z}^{\bar{\nu}} \, . \tag{A.68}$$

We are now prepared for the key definition.

A **Kähler manifold** is a Hermitian manifold with closed Kähler form,

$$dJ = 0 \, . \tag{A.69}$$

Equivalently, a complex manifold M of complex dimension k is a Kähler manifold if $\mathrm{Hol}(M) = U(k)$.

A Kähler metric can be expressed locally, i.e. in a single coordinate patch U_α, in terms of a **Kähler potential** k:

$$J = i\partial\bar{\partial}k \quad \text{or} \quad g_{i\bar{j}} = \partial_i \partial_{\bar{j}} k \, . \tag{A.70}$$

Evidently, (A.70) is invariant under the **Kähler transformation**

$$k \mapsto k + f(z^i) + \bar{f}(\bar{z}^{\bar{i}}) \, . \tag{A.71}$$

If U_α and U_β are two intersecting coordinate patches, we define local holomorphic coordinates and Kähler potentials $k^{(\alpha)}$ and $k^{(\beta)}$ in each patch. The metric is then given in each patch by (A.70), and agrees on the overlap. However, we need not require that $k^{(\alpha)} = k^{(\beta)}$ on $U_\alpha \cap U_\beta$. Instead, we can allow equality up to a Kähler transformation (A.71); that is,

$$k^{(\alpha)} = k^{(\beta)} + f(z^i) + \bar{f}(\bar{z}^{\bar{i}}) \, . \tag{A.72}$$

On a nontrivial compact Kähler manifold the Kähler potential is not a globally defined function, but is instead stitched together via the Kähler transformations (A.72).

The only nonvanishing Christoffel symbols on a Kähler manifold are

$$\Gamma^\mu_{\nu\rho} = g^{\mu\bar{\sigma}} \partial_\nu g_{\rho\bar{\sigma}} \tag{A.73}$$

and its complex conjugate $\Gamma^{\bar{\mu}}_{\bar{\nu}\bar{\rho}}$. Up to terms related by symmetries and complex conjugation, the only nonvanishing components of the Riemann tensor are

$$R^\mu{}_{\nu\rho\bar{\sigma}} = -\partial_{\bar{\sigma}} \Gamma^\mu_{\nu\rho} \, . \tag{A.74}$$

The **Ricci form** is defined by

$$\mathcal{R} = iR_{\mu\bar{\nu}} \, dz^\mu \wedge d\bar{z}^{\bar{\nu}} \, , \tag{A.75}$$

where $R_{\mu\bar{\nu}} = R^\rho{}_{\rho\mu\bar{\nu}}$. On a Kähler manifold, we can use the relation

$$\Gamma^\mu_{\nu\mu} = \partial_\nu \ln \det\left(g_{\rho\bar{\sigma}}\right) \, , \tag{A.76}$$

so that after defining $\mathfrak{g} = \det(g_{\rho\bar{\sigma}})$, the Ricci tensor can be written in the form

$$R_{\nu\bar{\sigma}} = -\partial_{\bar{\sigma}}\Gamma^{\mu}_{\nu\mu} = -\partial_{\bar{\sigma}}\partial_{\nu}\ln\mathfrak{g} \ . \tag{A.77}$$

The Ricci form of a Kähler manifold is then given by

$$\mathcal{R} = i\partial\bar{\partial}\ln\mathfrak{g} \ . \tag{A.78}$$

Using $d(\bar{\partial} - \partial) = 2\partial\bar{\partial}$, it is easy to see that $\mathcal{R}$ is closed with respect to the ordinary exterior derivative d, i.e. $d\mathcal{R} = 0$. Because $\mathfrak{g}$ is not a scalar, (A.78) does not imply that $\mathcal{R}$ is exact. We conclude that

---

The Ricci form of a Kähler manifold defines a de Rham cohomology class known as the **first Chern class**:

$$c_1 \equiv \frac{1}{2\pi}[\mathcal{R}] \in H^2(M, \mathbb{R}) \ . \tag{A.79}$$

On a Kähler manifold, the Laplacians obtained from d, ∂, and $\bar{\partial}$ are related:

$$\Delta = 2\Delta_{\partial} = 2\Delta_{\bar{\partial}} \ . \tag{A.80}$$

Using (A.80), one finds that the Hodge numbers of a Kähler manifold obey

$$h^{r,s} = h^{s,r} \ . \tag{A.81}$$

Moreover, because of the decomposition (A.57) and the relationship (A.80), the Betti numbers $b_k = \dim H^k(M, \mathbb{R})$ are related to the Hodge numbers by

$$b_k = \sum_{r+s=k} h^{r,s} \ . \tag{A.82}$$

A useful consequence is that "on a Kähler manifold, the odd Betti numbers are even":

$$b_{2k-1} = 2 \times \sum_{\substack{r+s=2k-1, \\ r<s}} h^{r,s} \ , \tag{A.83}$$

where we have used (A.81). Because the Kähler form J is a closed but not exact two-form, we find that the even Betti numbers are nonvanishing, $b_{2k} \geq 1$. Finally, the Euler characteristic may be written

$$\chi(M) = \sum_{r,s}(-1)^{r+s}h^{r,s} \ . \tag{A.84}$$

A.5.2 Calabi–Yau manifolds

A Calabi–Yau manifold is, roughly speaking, a Ricci-flat Kähler manifold. More precisely,

A **Calabi–Yau k-fold** is a compact Kähler manifold M of complex dimension k that is simply connected and satisfies the following equivalent conditions:

1. M admits a Kähler metric with holonomy in $SU(k)$.
2. There exists a nowhere-vanishing $(k, 0)$-form Ω on M.
3. M admits a Kähler metric with vanishing Ricci curvature.
4. The first Chern class $c_1(M)$ vanishes.

Variation in definitions can be found in the literature; some authors do not require compactness, some require that $\mathrm{Hol}(M) = SU(k)$ and not a subgroup, and in many cases one allows certain physically understood singularities. The requirement that M is simply connected can be dispensed with, but then the above conditions are not all equivalent.

The existence of a Ricci-flat metric on a Calabi–Yau k-fold M implies that a compactification of string theory on M (for $k \leq 4$, which we assume implicitly for the remainder) satisfies the vacuum Einstein equations $R_{MN} = 0$. The overwhelming majority of known examples of compact manifolds satisfying the vacuum Einstein equations are Calabi–Yau manifolds.[1] Strictly speaking, each of the existing examples consists of a specification of the *topological* data of a Calabi–Yau k-fold, involving for example a demonstration that the first Chern class is a trivial cohomology class. Yau's theorem [872] then implies that a Ricci-flat metric exists, even though no explicit Ricci-flat metric on a Calabi–Yau manifold has been found to date.

Compactification on a Calabi–Yau k-fold M preserves some supersymmetry, because the holonomy $SU(k) \subset SO(2k)$ is small enough to leave invariant one or more spinors. Consider the case $k = 3$, where through the Lie algebra relation $\mathfrak{so}(6) \cong \mathfrak{su}(4)$, the spinors η and $\bar{\eta}$ on M are the $\mathbf{4}$ and $\bar{\mathbf{4}}$ of $SU(4)$, respectively. Decomposing η into irreducible representations of the holonomy group $SU(3)$, we have

$$\mathbf{4} = \mathbf{3} + \mathbf{1} \,. \tag{A.85}$$

The singlet $\mathbf{1}$ in this decomposition is the unique covariantly constant spinor (of positive chirality) on a Calabi–Yau threefold. We conclude that when a ten-dimensional string theory with $\mathcal{N} = 1$ supersymmetry is compactified on

---

[1] The only existing constructions of compact manifolds that admit Ricci-flat metrics, but are not Calabi–Yau k-folds [871], are Joyce's constructions [869] of seven-manifolds of holonomy G_2, and of eight-manifolds of holonomy $Spin(7)$. Note in particular that all known examples of compact Ricci-flat manifolds have reduced holonomy.

a Calabi–Yau threefold, the four-dimensional theory contains a single (four-real-component) spinor, and hence preserves $\mathcal{N} = 1$ supersymmetry in four dimensions.

The existence of a unique $(n, 0)$-form on a Calabi–Yau n-fold implies that $h^{0,n} = h^{n,0} = 1$. One can also show that $h^{0,m} = h^{m,0} = 0$ for $m < n$. Using $h^{r,s} = h^{n-r,n-s}$ and $h^{r,s} = h^{s,r}$, one finds that a Calabi–Yau threefold has the following nonvanishing Hodge numbers: $h^{0,0} = h^{3,3} = h^{3,0} = h^{0,3} = 1$; $h^{1,1}$; and $h^{2,1} = h^{1,2}$. The cohomology is therefore characterized by specifying two numbers, $h^{1,1}$ and $h^{2,1}$. Correspondingly, the Euler characteristic takes the very simple form

$$\chi(\mathrm{CY}_3) = 2(h^{1,1} - h^{2,1}) \, . \tag{A.86}$$

A.5.3 Moduli of Calabi–Yau manifolds

Given a Calabi–Yau threefold M, including the specification of a Ricci-flat Kähler metric $g_{i\bar{j}}$ on M, one can ask whether there exist infinitesimal deformations of the metric, $g \mapsto g + \delta g$, such that $g + \delta g$ is also a Ricci-flat Kähler metric. That is, we search for deformations such that

$$R_{i\bar{j}}(g) = R_{i\bar{j}}(g + \delta g) = 0 \, , \tag{A.87}$$

subject to the requirement that the new metric $g + \delta g$ is still Hermitian,

$$(g + \delta g)_{ij} = 0 \, . \tag{A.88}$$

A general deformation can be written

$$\delta g = \delta g_{i\bar{j}} \mathrm{d}z^i \mathrm{d}\bar{z}^{\bar{j}} + \delta g_{ij} \mathrm{d}z^i \mathrm{d}z^j + c.c. \tag{A.89}$$

Expanding (A.87) in the gauge

$$\nabla^m \delta g_{mn} - \frac{1}{2} \nabla_n \delta g^m{}_m = 0 \tag{A.90}$$

leads to the *Lichnerowicz equation*,

$$\nabla^m \nabla_m \delta g_{pq} + 2R_p{}^r{}_q{}^s \, \delta g_{rs} = 0 \, , \tag{A.91}$$

where the indices in (A.90) and (A.91) run over both holomorphic and anti-holomorphic directions. Using the Calabi–Yau condition, it can be shown that the equations governing $\delta g_{i\bar{j}}$ and δg_{ij} are decoupled, so that there are two independent classes of deformations that solve (A.91).

The components $\delta g_{i\bar{j}}$ lead to deformations of the Kähler form,

$$\delta J = i \, \delta g_{i\bar{j}} \, \mathrm{d}z^i \wedge \mathrm{d}\bar{z}^{\bar{j}} \, . \tag{A.92}$$

When (A.91) holds, δJ is a harmonic $(1, 1)$-form, and hence corresponds to a representative of a Dolbeault cohomology class: $\delta J \in H^{1,1}_{\bar{\partial}}(M, \mathbb{C})$.

> The **Kähler moduli** of M are deformations of the form (A.92), for $\delta J \in H_{\bar{\partial}}^{1,1}(M, \mathbb{C})$. The number of Kähler moduli is $h^{1,1}(M)$.

One might be tempted to set $\delta g_{ij} = 0$ in view of (A.88), but this would be incorrect. The original metric g is Hermitian with respect to an initial complex structure $\mathcal{J}$, and a deformation $\delta g_{ij} \neq 0$ certainly renders the new metric $g + \delta g$ non-Hermitian with respect to this original complex structure $\mathcal{J}$. But the requirement (A.88) is instead that $g + \delta g$ should be Hermitian with respect to *some* complex structure. So we must ask whether there exists an infinitesimal deformation

$$\mathcal{J} \mapsto \mathcal{J} + \delta \mathcal{J} \tag{A.93}$$

such that $g + \delta g$ is a Hermitian metric in the complex structure $\mathcal{J} + \delta \mathcal{J}$. Using the nonvanishing $(3,0)$-form Ω on M, we can construct the $(2,1)$-form

$$\chi = \Omega_{ijk} \delta \mathcal{J}^k{}_{\bar{l}} \, dz^i \wedge dz^j \wedge d\bar{z}^{\bar{l}} = \Omega_{ijk} \delta g_{\bar{h}\bar{l}} \, g^{k\bar{h}} dz^i \wedge dz^j \wedge d\bar{z}^{\bar{l}} . \tag{A.94}$$

Just as for the Kähler deformations, one can show that (A.91) implies that χ is harmonic, so that $\chi \in H_{\bar{\partial}}^{2,1}(M, \mathbb{C})$.

> The **complex structure moduli** of M are deformations of the form (A.94), for $\chi \in H_{\bar{\partial}}^{2,1}(M, \mathbb{C})$. The number of (complex) complex structure moduli is $h^{2,1}(M)$.

In total, the purely geometric deformations of a Calabi–Yau threefold M that preserve the Calabi–Yau condition can be parameterized by $h^{1,1} + 2h^{2,1}$ real parameters. In string theory, these moduli combine with additional non-geometric deformations parameterized by p-forms. Each modulus gives rise to a scalar field in four dimensions. At the classical level, at leading order in α', and in the absence of localized sources and of background fields beyond the metric, these fields are massless. This is the origin of the moduli problem in Calabi–Yau compactifications.

Appendix B

The effective theory of inflation

Physical systems with spontaneously broken symmetries are ubiquitous in nature. Key characteristics of these systems are captured by the low-energy effective theory of the Goldstone bosons associated with the symmetry breaking. Our goal in this appendix is to formulate inflation as an example of spontaneous symmetry breaking and discuss the associated Goldstone dynamics [51].[1] In this case, the broken symmetry is time translation invariance of the cosmological spacetime.

B.1 Spontaneous symmetry breaking

Before developing the effective theory of inflation, we review symmetry breaking in gauge theory [54].

B.1.1 Broken global symmetries

Consider a set of real scalar fields Φ_i, $i = 1, \ldots, N$, governed by an action that is invariant under the global symmetry transformation

$$\Phi_i \mapsto \Phi_i' = \left(e^{i\theta^a G^a} \right)_{ij} \Phi_j \, , \tag{B.1}$$

where the G^a are the generators of a group G, and the θ^a are spacetime-independent parameters. If the fields acquire vacuum expectation values $\langle \Phi_i \rangle \equiv v_i$, the symmetry G is said to be *spontaneously broken* to the subgroup H that leaves the v_i invariant. Transformations in the coset G/H act nontrivially on v_i: $(T^a)_{ij} v_j \neq 0$, where the T^a are the generators of the broken symmetry. However, because the symmetry is only spontaneously broken, spacetime-independent transformations along the directions of broken symmetry connect different vacua

---

[1] We will closely follow the original literature [50, 51] and the reviews [873–875].

with the same energy. As a result, for each generator of a broken symmetry there is one flat direction in the space of field configurations. Fluctuations along the flat directions are the *Goldstone bosons*. Goldstone's theorem [876] asserts the existence of one massless Goldstone boson π_a for each broken generator.

To introduce the Goldstone bosons, we act on the vacuum configuration $\langle \Phi_i \rangle \equiv v_i$ with the broken symmetry, but replace the constant transformation parameter θ^a with a *spacetime-dependent* parameter $\pi^a(x)$:

$$\Phi_i' = \left(e^{i\pi^a(x)T^a} \right)_{ij} v_j . \tag{B.2}$$

While the fields $\pi^a(x)$ parameterize massless excitations of the theory after spontaneous symmetry breaking, the remaining directions in field space are typically massive, and decouple from the dynamics of the Goldstone bosons. This makes the Goldstone modes the natural degrees of freedom in the low-energy effective theory.

The EFT of the Goldstone bosons is determined to a large degree by the symmetry-breaking pattern. In particular, the Goldstone effective action linearly realizes the unbroken symmetry H, but nonlinearly realizes the remaining symmetries G/H. To construct the effective action for the Goldstone bosons, we introduce the field

$$U(x) = e^{i\pi(x)\cdot T} , \tag{B.3}$$

where $\pi(x) \equiv \{\pi^a\}$ and $T \equiv \{T^a\}$.[2] At lowest order in a derivative expansion, the unique G-invariant Lagrangian is

$$\mathcal{L}_{\text{eff}}^{(0)} = -\frac{f_\pi^2}{4}\text{Tr}[\partial_\mu U \partial^\mu U^\dagger] , \tag{B.4}$$

where f_π is a parameter with dimensions of mass. There can be no terms without derivatives because $\text{Tr}[UU^\dagger] = 2$ is a constant. Expanding the Lagrangian (B.4) in powers of $\pi_c \equiv f_\pi \pi$, we get

$$\mathcal{L}_{\text{eff}}^{(0)} = -\frac{1}{2}\partial_\mu \pi_c \cdot \partial^\mu \pi_c - \frac{1}{6f_\pi^2}\left[(\pi_c \cdot \partial_\mu \pi_c)^2 - \pi_c^2(\partial_\mu \pi_c \cdot \partial^\mu \pi_c) \right] + \cdots , \tag{B.5}$$

where we have used $2\,\text{Tr}(T^a T^b) = \delta^{ab}$. Notice the appearance of an infinite series of non-renormalizable interactions. The broken symmetry dictates relations among these interactions, and the couplings are all determined by the single parameter f_π. These interactions are called *universal*.

At higher order in the derivative expansion we obtain additional, *non-universal* interactions. For example, at next-to-leading order, we find the following terms involving only single derivatives:[3]

---

[2] As a concrete example, we may think of π as the triplet of pion fields in QCD and of T as the triplet of Pauli matrices [877]. The effective Lagrangian for the pions is then the *chiral Lagrangian*.

[3] One could also include terms such as $\Box U \equiv \eta^{\mu\nu}\partial_\mu\partial_\nu U$, leading to a number of additional terms in the effective action.

$$\mathcal{L}_{\text{eff}} = -\frac{f_\pi^2}{4}\text{Tr}[\partial_\mu U \partial^\mu U^\dagger] + c_1 \left(\text{Tr}[\partial_\mu U \partial^\mu U^\dagger]\right)^2$$

$$+ c_2 \text{Tr}[\partial_\mu U \partial_\nu U^\dagger] \text{Tr}[\partial^\mu U \partial^\nu U^\dagger] + \cdots , \tag{B.6}$$

where c_1 and c_2 are model-dependent, dimensionless constants. Expanding (B.6) in terms of π_c, we find a further series of non-renormalizable interactions,

$$\mathcal{L}_{\text{eff}} \supset \frac{c_1}{4f_\pi^4}\left((\partial_\mu \pi_c)^4 - \frac{2}{3f_\pi^2}(\partial_\mu \pi_c)^2(\pi_c \cdot \partial_\mu \pi_c)^2 + \cdots\right) + \cdots . \tag{B.7}$$

Individual interactions are again related by the nonlinearly-realized symmetry.

A classic example is the *sigma model*,

$$\mathcal{L} = -\frac{1}{4}\text{Tr}[\partial_\mu \Phi \partial^\mu \Phi^\dagger] + \frac{\mu^2}{4}\text{Tr}[\Phi\Phi^\dagger] - \frac{\lambda}{16}\left(\text{Tr}[\Phi\Phi^\dagger]\right)^2 . \tag{B.8}$$

Writing $\Phi = [v + \rho(x)]U(x)$, with $U(x)$ defined in (B.3), we get

$$\mathcal{L} = -\frac{(v+\rho)^2}{4}\text{Tr}[\partial_\mu U \partial^\mu U^\dagger] - \frac{1}{2}(\partial_\mu \rho)^2 - \mu^2\rho^2 - \lambda v\rho^3 - \frac{\lambda}{4}\rho^4 , \tag{B.9}$$

where we have dropped a constant vacuum energy contribution. The angular directions π^a correspond to massless Goldstone bosons, while the radial direction ρ is a massive field. Integrating out ρ gives precisely the effective Lagrangian (B.6), with $f_\pi \equiv v$ and $c_1 \equiv v^2/8\mu^2$. Thus, the non-universal terms in (B.6) arise from integrating out the heavy fields of the UV completion. In cases where the UV completion of the theory is unknown, the effective action (B.6) provides a model-independent description of the low-energy dynamics.

B.1.2 Energy scales

In order to understand the dynamics of a theory, it is useful to identify the energy scales at which different phenomena become important.

We will first determine the *symmetry-breaking scale*. By definition, any theory with a continuous global symmetry has a conserved Noether current J^μ, even if the symmetry is spontaneously broken. Associated with the current is a conserved charge $Q = \int \mathrm{d}^3 x \, J^0(t, \boldsymbol{x})$. Since the charge is conserved it commutes with the Hamiltonian, $[Q, H] = i\partial_t Q = 0$. A signature of spontaneous symmetry breaking is that the vacuum is charged, $Q|0\rangle \neq 0$. At high energies, the symmetry is restored and $Q|0\rangle = 0$.

If E_0 is the energy of the vacuum state $|0\rangle$, then $Q|0\rangle$ is a state with the same energy,

$$HQ|0\rangle = [H, Q]|0\rangle + QH|0\rangle = E_0 Q|0\rangle . \tag{B.10}$$

Now, consider the state

$$|\pi_c(\boldsymbol{p})\rangle \equiv -\frac{2i}{f}\int \mathrm{d}^3 x \, e^{i\boldsymbol{p}\cdot\boldsymbol{x}} J^0(t, \boldsymbol{x})|0\rangle , \tag{B.11}$$

where f is a constant with dimensions of mass. It is easy to see that the state $|\pi_c(\boldsymbol{p})\rangle$ has energy $E_0 + E(\boldsymbol{p})$. (Note that we do not assume a Lorentz-invariant dispersion relation.) Moreover, since $|\pi_c(\mathbf{0})\rangle = -(2i/f)Q|0\rangle$ has energy E_0, the state (B.11) must be gapless, i.e. it must satisfy $E(\boldsymbol{p}) \to 0$ as $\boldsymbol{p} \to 0$. This is precisely what we expect from the Goldstone boson state. Multiplying (B.11) from the left by $\langle\pi_c(\boldsymbol{q})|$ and integrating over $\int \mathrm{d}^3 p\, e^{-i\boldsymbol{p}\cdot\boldsymbol{y}}$, we find

$$\langle\pi_c(\boldsymbol{q})|J^0(t,\boldsymbol{y})|0\rangle = iE(\boldsymbol{q})f e^{-i\boldsymbol{q}\cdot\boldsymbol{y}} , \tag{B.12}$$

where we have used $\langle\pi_c(\boldsymbol{q})|\pi_c(\boldsymbol{p})\rangle = 2E(\boldsymbol{q})(2\pi)^3\delta(\boldsymbol{q}-\boldsymbol{p})$. We see that the current interpolates between the vacuum state $|0\rangle$ and the Goldstone boson state $|\pi_c\rangle$, with a strength set by the scale f. This identifies f as the order parameter of the symmetry breaking. In particular, f determines the interactions of the Goldstone boson with itself and with other fields, such as the field ρ in the theory (B.9). At energies above f, these additional fields restore the symmetry. Thus, symmetry breaking occurs at f.

In order to identify the scale f in our previous discussion, we consider the current associated with the effective Lagrangian (B.5),

$$J^\mu = -f_\pi \partial^\mu \pi_c + \cdots . \tag{B.13}$$

If $|\pi_c\rangle$ is the state created by acting with the operator π_c on the vacuum state $|0\rangle$, then we find

$$\langle\pi_c(q)|J^\mu(y)|0\rangle = iq^\mu f(q^2)e^{iqy} , \tag{B.14}$$

with $f(q^2) = f_\pi + \cdots$. This is the Lorentz-invariant version of (B.12), with f_π playing the role of f. The symmetry is restored when the right-hand side of (B.14) vanishes. This occurs when higher-order corrections cancel the leading term in $f(q^2)$, i.e. at energies of order f_π.[4]

At energies below f_π, a description of the physics in terms of weakly coupled Goldstone bosons alone is appropriate, while above f_π degrees of freedom other than the Goldstone modes become relevant. For our later discussion, we just need to remember that the symmetry-breaking scale can be read off from the scale appearing in the Noether current for the canonically normalized field.

The regime of validity of an effective theory is not always obvious. Given a microscopic definition of the theory (i.e. a UV completion such as the sigma model (B.8) of Section B.1.1), the regime of validity is determined by the scales at which additional modes were integrated out. When only the effective description is known, these energy scales may not be transparent in the Lagrangian. Nevertheless, a fairly reliable method for identifying the cutoff of the effective

---

[4] An alternative way to determine the symmetry-breaking scale is to identify the energy scale below which the charge Q does not exist because its correlation functions are IR divergent. The conserved current and commutators of the conserved charge always remain well-defined.

theory is to determine the *strong coupling scale*, Λ. This is the energy scale at which the perturbative expansion of the effective action breaks down.

The higher-order interactions in the Goldstone action (B.5) are suppressed by powers of f_π, but become important at high energies. By dimensional analysis, the effective coupling is E/f_π, suggesting strong coupling for $E \sim f_\pi$. More formally, we can define the strong coupling scale as the energy scale at which the loop expansion breaks down or perturbative unitary of Goldstone boson scattering is violated. Through such an analysis one arrives at a refined definition of the strong coupling scale [54],

$$\Lambda = 4\pi f_\pi \ . \tag{B.15}$$

If the non-universal interactions in (B.7) have large coefficients, $c_n \gg 1$, the strong coupling scale can be lower than $4\pi f_\pi$.

B.1.3 Broken gauge symmetries

So far, we have studied the spontaneous breaking of global symmetries, but the generalization to gauge symmetries is straightforward. To gauge a global symmetry, we replace all partial derivatives acting on charged fields with covariant derivatives,

$$D_\mu \equiv \partial_\mu + ig A_\mu \cdot T \ , \tag{B.16}$$

where $A_\mu \equiv \{A_\mu^a\}$ are the gauge fields associated with the symmetry and g is the gauge coupling. Under a spacetime-dependent symmetry transformation, the gauge fields transform as

$$A_\mu(x) \mapsto U(x) \left(A_\mu(x) - \frac{i}{g}\partial_\mu \right) U^\dagger(x) \ . \tag{B.17}$$

The low-energy effective action is now

$$\mathcal{L}_{\text{eff}} = -\frac{1}{4}\text{Tr}[F_{\mu\nu}^2] - \frac{f_\pi^2}{4}\text{Tr}[D_\mu U (D^\mu U)^\dagger] + \cdots \ . \tag{B.18}$$

The quadratic Lagrangian for the Goldstone bosons and the gauge fields becomes

$$\mathcal{L}_{\text{eff}}^{(2)} = -\frac{1}{4}\text{Tr}[F_{\mu\nu}^2] - \frac{1}{2}(\partial_\mu \pi_c)^2 - \frac{1}{2}m^2 A_\mu^2 + m\,\partial_\mu \pi_c \cdot A^\mu \ , \tag{B.19}$$

where $m^2 \equiv f_\pi^2 g^2$. Observe that the gauge fields have a mass after spontaneous symmetry breaking. The action (B.19) still possesses a gauge symmetry that relates physically equivalent configurations of π and A_μ. This redundancy of description can be removed by fixing the gauge. For example, the gauge freedom can be used to set the Goldstone bosons to zero ($\pi \equiv 0$). This defines *unitary gauge*, and the resulting theory is described purely in terms of massive vector bosons A_μ. The reverse process of introducing the Goldstone boson and the associated gauge redundancy into the theory of massive vector bosons is the

Stückelberg trick. Describing the physics in terms of the Goldstone bosons has the advantage that it makes the high-energy behavior of the theory manifest. Specifically, at high energies the scattering of the longitudinal modes of the gauge fields is well described by the scattering of the Goldstone bosons. This is an example of the Goldstone boson equivalence theorem [878].

Because the mixing term $\partial_\mu \pi_c \cdot A^\mu$ in (B.19) has one fewer derivative than the pion kinetic term $-\frac{1}{2}(\partial_\mu \pi_c)^2$, one expects that the mixing will become irrelevant at sufficiently high energies. This is seen most easily by taking the *decoupling limit*

$$g \to 0 \ , \ m \to 0 \quad \text{for} \quad f_\pi = m/g = const. \tag{B.20}$$

In this limit, there is no mixing between π and A^μ, and the Goldstone boson part of the action (B.19) reduces to (B.4); the local symmetry has effectively become a global symmetry. For energies $E > E_{\text{mix}} = m$, the Goldstone bosons are the simplest way to describe the scattering of the massive vector fields. Restoring finite g and m gives corrections to the results from pure Goldstone boson scattering that are perturbative in m/E and in g^2.

Finally, we should comment on a possible source of confusion. Since a gauge symmetry is only a redundancy of description and not a real symmetry, what does it mean to break it? The set of gauge transformations G can be split into those that approach the identity at spatial infinity, $G_\star$, and those that do not, $G/G_\star$. We call the latter the *global part* of the gauge transformations. Only the global part of the gauge symmetry is broken. Even when we speak of the "spontaneous breaking of a gauge symmetry," the symmetry $G_\star$ is not broken: it represents the true gauge redundancy of the system, while $G/G_\star$ is a physical symmetry with an associated Noether current and a conserved charge.

B.2 Symmetry breaking in cosmology

B.2.1 Broken time translations

To apply methods from spontaneous symmetry breaking in field theory to cosmological dynamics, we think of the metric tensor $g_{\mu\nu}$ as a gauge field. The relevant gauge symmetry is the invariance of general relativity under spacetime diffeomorphisms, $x^\mu \mapsto x'^\mu(x^\nu)$. In cosmology, we will be primarily interested in the spontaneous breaking of *time translations*, the global part of the time diffeomorphisms $t \mapsto t'(x^\nu)$.

Time translations are broken in any spacetime that does not have at least a local timelike Killing vector. To understand the relevance of the local qualifier, consider de Sitter space, with the line element

$$ds^2 = -dt^2 + e^{2Ht}d\mathbf{x}^2 \ . \tag{B.21}$$

In spite of appearances, this spacetime does not break time translation symmetry, as the dilatation symmetry

$$t \mapsto t + \lambda, \qquad \boldsymbol{x} \mapsto e^{-H\lambda} \boldsymbol{x} , \tag{B.22}$$

implies the existence of a timelike Killing vector, defined locally in a patch of size H^{-1}. In a perfect de Sitter spacetime, there is hence no preferred time slicing, as all slices are related by gauge transformations.

The distinctive feature of inflation is the deviation from the perfect de Sitter limit. An order parameter, or "clock," measures how much inflationary expansion remains. The expansion rate H, which decreases during inflation, is a natural example of such a clock. For some purposes it is more convenient to consider the time-dependent expectation value(s) $\psi_m(t)$ of some matter field(s) ψ_m, which could correspond to the inflaton field ϕ or the inflationary vacuum energy density ρ. The expectation value $\psi_m(t)$ defines a preferred time slicing, with the different time slices defined and labeled by distinct values of ψ_m. Time translation invariance is therefore broken. Although the spacetime-independent transformation $t \mapsto t + \xi$, with ξ a constant parameter, remains a redundancy of the system even in the presence of a preferred time slicing, the spacetime-dependent transformation

$$t \mapsto t + \pi(t, \boldsymbol{x}) \tag{B.23}$$

does not leave the action invariant unless π is constant. An immediate consequence of the broken symmetry is the existence of a Goldstone excitation,[5] corresponding to a spacetime-dependent transformation along the broken generator: that is, $U(t, \boldsymbol{x}) \equiv t + \pi(t, \boldsymbol{x})$.

B.2.2 Adiabatic perturbations

A key feature of the Goldstone fluctuations π is that they are closely connected to cosmological observables. Recall from Chapter 1 the definition of _adiabatic fluctuations_ as the specific type of perturbations induced by a local, common shift in time:

$$\delta \psi_m(t, \boldsymbol{x}) \equiv \psi_m(t + \pi(t, \boldsymbol{x})) - \psi_m(t) . \tag{B.24}$$

At linear order, adiabatic fluctuations are therefore proportional to the Goldstone mode, $\delta \psi_m = \dot{\psi}_m \pi$. Since observations do not show any signs of deviation from purely adiabatic initial conditions, the Goldstone boson is the most efficient way to describe the data. In spatially flat gauge,

$$g_{ij} \equiv a^2(t) \delta_{ij} , \tag{B.25}$$

all metric perturbations are related to the Goldstone mode by the Einstein constraint equations – cf. Appendix C.

---

[5] In the case of spontaneously broken _spacetime_ symmetries, the number of Goldstone bosons does not have to match the number of broken symmetry generators [879]. Although we are breaking four symmetries during inflation [56, 880] (one dilatation and three special conformal transformations), there is only one Goldstone boson.

For purely adiabatic fluctuations, we can perform a time shift, $t \mapsto t - \pi(t, \boldsymbol{x})$ to remove all matter fluctuations, $\delta\psi_m \mapsto \delta\psi_m \equiv 0$. This is nothing but undoing (B.23). This transformation induces an isotropic perturbation to the spatial part of the metric,

$$\delta g_{ij} = a^2(t)e^{2\mathcal{R}(t,\boldsymbol{x})}\delta_{ij} \, , \quad \text{where} \quad \mathcal{R} = -H\pi \, . \tag{B.26}$$

Thus, the curvature perturbation $\mathcal{R}$ in comoving gauge is proportional to the Goldstone boson π in spatially flat gauge. For nearly constant H, we can therefore think of $\mathcal{R}$ and π interchangeably.

B.2.3 Effective action in unitary gauge

We could construct the action for the Goldstone boson by writing down the most general Lorentz-invariant action for the field $U \equiv t + \pi$ [275],

$$S = \int \mathrm{d}^4x\sqrt{-g}\,\mathcal{L}[U, (\partial_\mu U)^2, \Box U, \cdots] \, . \tag{B.27}$$

However, we will follow an alternative geometrical approach to obtain the effective action [50, 51]. Readers who do not wish to follow the details of the derivation can find a summary of the Goldstone action in Section B.3.4.

Systematics

The goal is to write the most general effective action for the metric perturbations around an FRW background. We have seen that in the case of purely adiabatic fluctuations, we can remove any matter perturbations $\delta\psi_m$ by a local time shift (see Fig. B.1). This takes us to unitary gauge ($\pi \equiv 0$), where the field π is "eaten" by the metric $g_{\mu\nu}$. After gauge fixing, the action does not have to respect the full diffeomorphism invariance, but only has to be invariant under time-dependent spatial diffeomorphisms, $x^i \mapsto x^i + \xi^i(t, x^j)$. Besides terms that are invariant under all diffeomorphisms (such as curvature invariants like R and $R_{\mu\nu\rho\sigma}R^{\mu\nu\rho\sigma}$), the reduced symmetry of the system now allows many new terms in the action.

In describing the metric perturbations on the constant time hypersurfaces Σ_t it is useful to introduce a unit four-vector n_μ orthogonal to Σ_t. In unitary gauge, this vector is

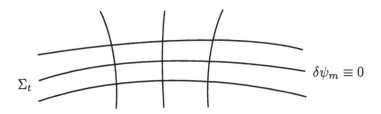

Fig. B.1 The time dependence of the background fields, $\bar{\psi}_m(t)$, foliates the spacetime into a series of spacelike hypersurfaces Σ_t. Metric perturbations on Σ_t describe adiabatic fluctuations.

$$n_\mu = -\frac{\delta_\mu^{\ 0}}{\sqrt{-g^{00}}} \, . \tag{B.28}$$

By contracting covariant tensors with n_μ, we produce objects with uncontracted upper 0 indices, such as g^{00} and R^{00}. It is easy to check that these are scalars under spatial diffeomorphisms. Functions of g^{00}, R^{00}, etc. are therefore allowed in the effective action. In general, products of any four-dimensional covariant tensors with free upper 0 indices, but with all spatial indices contracted, are allowed operators. Finally, we can have three-dimensional quantities describing the geometry of the hypersurfaces Σ_t, such as the *induced metric*, $h_{\mu\nu} = g_{\mu\nu} + n_\mu n_\nu$, the *extrinsic curvature tensor*, $K_{\mu\nu} = h_\mu^{\ \rho} \nabla_\rho n_\nu$, and the Riemann curvature $\hat{R}_{\alpha\beta\gamma\delta}$ of the induced metric.[6] A helpful connection between three-dimensional and four-dimensional quantities is given by the Gauss–Codazzi relation,

$$\hat{R}_{\alpha\beta\gamma\delta} = h_\alpha^{\ \mu} h_\beta^{\ \nu} h_\gamma^{\ \rho} h_\delta^{\ \sigma} R_{\mu\nu\rho\sigma} - K_{\alpha\gamma} K_{\beta\delta} + K_{\alpha\delta} K_{\beta\gamma} \, . \tag{B.29}$$

The most general action constructed with these ingredients is [51]

$$S = \int d^4x \sqrt{-g} \, \mathcal{L}[R_{\mu\nu\rho\sigma}, g^{00}, K_{\mu\nu}, \hat{R}_{\mu\nu}, \nabla_\mu; t] \, , \tag{B.30}$$

where the only free indices entering $\mathcal{L}$ are upper 0's. The spacetime indices in (B.30) are contracted with the four-dimensional metric $g^{\mu\nu}$. Terms involving the induced metric $h_{\mu\nu}$ explicitly do not lead to new operators.

Universal part of the action

Expanding (B.30) around a flat Robertson–Walker background with a given Hubble expansion rate $H(t)$, and with

$$\bar{g}^{00} = -1 \, , \quad \bar{R} \equiv \bar{R}^\mu_{\ \mu} = 12H^2 + 6\dot{H} \, , \quad \bar{K} \equiv \bar{K}^\mu_{\ \mu} = 3H \, , \tag{B.31}$$

we get

$$S = \int d^4x \sqrt{-g} \left[\Lambda_0(t) + c_1(t)[g^{00} + 1] + c_2(t)[K - 3H] + c_3(t)[R - \bar{R}(t)] + \cdots \right] . \tag{B.32}$$

For now we have only shown terms at linear order in fluctuations. Since time diffeomorphisms are broken, we have allowed for the coefficients in the action to depend explicitly on time. It is convenient to absorb $\bar{R}(t)$ into the zeroth-order term $\Lambda_0(t)$, and to remove the coefficient $c_3(t)$ by a conformal transformation of the metric. Moreover, the term linear in K can be traded for a function of g^{00} via integration by parts [51],

---

[6] All Greek indices in this section run over $0 \ldots 3$; the induced metric $h_\alpha^{\ \mu}$ serves to project tensors in the four-dimensional spacetime to tensors on Σ_t. See e.g. [881] for geometrical background.

$$\int d^4x\sqrt{-g}\,c_2(t)K = -\int d^4x\sqrt{-g}\,n^\mu\nabla_\mu c_2(t)$$

$$= +\int d^4x\sqrt{-g}\,(-g^{00})^{1/2}\,\dot{c}_2(t)\,. \qquad (B.33)$$

Equation (B.32) can therefore be written as

$$S = \int d^4x\sqrt{-g}\,\left[\frac{M_{\rm pl}^2}{2}R - \Lambda(t) - c(t)g^{00}\right] + \Delta S\,, \qquad (B.34)$$

where ΔS denotes terms of quadratic order and higher. The functions $\Lambda(t)$ and $c(t)$ are fixed by the FRW background about which we have expanded. Terms linear in fluctuations correspond to tadpoles and must vanish if we expand around a consistent background solution. Indeed, varying the linear terms in the action (B.34) with respect to the metric $g_{\mu\nu}$ gives the Friedmann equations

$$H^2 = \frac{1}{3M_{\rm pl}^2}\left[c(t) + \Lambda(t)\right]\,, \qquad (B.35)$$

$$\dot{H} + H^2 = -\frac{1}{3M_{\rm pl}^2}\left[2c(t) - \Lambda(t)\right]\,, \qquad (B.36)$$

which imply $\Lambda(t) = M_{\rm pl}^2(3H^2 + \dot{H})$ and $c(t) = -M_{\rm pl}^2\dot{H}$. Equation (B.34) therefore becomes

$$S = \int d^4x\sqrt{-g}\,\left[\frac{M_{\rm pl}^2}{2}R - M_{\rm pl}^2(3H^2 + \dot{H}) + M_{\rm pl}^2\dot{H}g^{00}\right] + \Delta S\,. \qquad (B.37)$$

This completes the derivation of the universal part of the action. Notice that this part of the action is fixed completely by specifying the FRW background, $H(t)$, while the differences among cosmological models are parameterized by the higher-order terms in ΔS.

Higher-order terms

We organize ΔS as an expansion in powers of fluctuations[7]

$$\Delta S = \int d^4x\sqrt{-g}\,\left[\frac{M_2^4(t)}{2}(\delta g^{00})^2 + \frac{M_3^4(t)}{3!}(\delta g^{00})^3 + \frac{M_4^4(t)}{4!}(\delta g^{00})^4 + \cdots\right.$$

$$- \frac{\bar{M}_1^3(t)}{2}\delta g^{00}\delta K - \frac{\bar{M}_2^2(t)}{2}\delta K^2 - \frac{\bar{M}_3^2(t)}{2}(\delta K^\mu{}_\nu)^2 + \cdots$$

$$\left.- \frac{\hat{M}_1^2(t)}{2}\delta g^{00}\hat{R} + \cdots\right]\,, \qquad (B.38)$$

---

[7] Instead of the four-dimensional curvature invariants, it is convenient to use three-dimensional quantities ($K_{\mu\nu}$, $\hat{R}_{\mu\nu}$, etc.), which do not contain higher time derivatives. Not shown in (B.38) are operators with derivatives, such as $g^{\mu\nu}\partial_\mu g^{00}\partial_\nu g^{00}$.

where $\delta g^{00} \equiv g^{00} + 1$ and $\delta K_{\mu\nu} \equiv K_{\mu\nu} - a^2 H h_{\mu\nu}$. Although not very explicit in this parameterization, the action (B.38) also contains a derivative expansion; δg^{00} is a scalar with zero derivatives acting on it, while $\delta K_{\mu\nu}$ and $\delta \hat{R}_{\mu\nu}$ are one- and two-derivative objects, respectively. In many situations the terms involving δg^{00} therefore dominate the dynamics, as we shall see.

B.2.4 Introducing the Goldstone boson

To make the dynamics of the theory defined by (B.38) more transparent, we introduce the Goldstone boson π associated with the spontaneous breaking of time-translation invariance. Through the Stückelberg trick, π also restores the gauge-invariance of the theory. Specifically, we perform a spacetime-dependent time reparameterization,

$$t \mapsto t' = t + \pi(t, x^i) \, , \tag{B.39}$$

$$x_i \mapsto x_i' = x_i \, . \tag{B.40}$$

The four-dimensional Ricci scalar and the volume element $\sqrt{-g} \, \mathrm{d}^4 x$ are invariant under general four-dimensional diffeomorphisms, so they do not transform under (B.39) and hence make no contribution to the Goldstone action. Let us see how the other terms in (B.38) transform under (B.39). First of all, any time-dependent coefficients transform as

$$f(t) \mapsto f(t + \pi) = f(t) + \dot{f}\pi + \frac{1}{2}\ddot{f}\pi^2 + \cdots \, . \tag{B.41}$$

The contravariant and covariant components of any tensor transform as

$$t^{\mu\nu} \mapsto \frac{\partial x^{\mu'}}{\partial x^\alpha} \frac{\partial x^{\nu'}}{\partial x^\beta} t^{\alpha\beta} = \left(\delta^\mu_\alpha + \delta^\mu_0 \partial_\alpha \pi\right)\left(\delta^\nu_\beta + \delta^\nu_0 \partial_\beta \pi\right) t^{\alpha\beta} \, , \tag{B.42}$$

$$t_{\mu\nu} \mapsto \frac{\partial x^\alpha}{\partial x^{\mu'}} \frac{\partial x^\beta}{\partial x^{\nu'}} t_{\alpha\beta} = \left(\delta^\alpha_\mu + \delta^\alpha_0 \partial_\mu \pi\right)^{-1}\left(\delta^\beta_\nu + \delta^\beta_0 \partial_\nu \pi\right)^{-1} t_{\alpha\beta} \, . \tag{B.43}$$

For the contravariant components of the metric we therefore find

$$g^{00} \mapsto g^{00} + 2\partial_\mu \pi g^{0\mu} + \partial_\mu \pi \partial_\nu \pi g^{\mu\nu} \, , \tag{B.44}$$

$$g^{0i} \mapsto g^{0i} + \partial_\mu \pi g^{\mu i} \, , \tag{B.45}$$

$$g^{ij} \mapsto g^{ij} \, , \tag{B.46}$$

while the covariant components can be written as an expansion in π. Care must be taken when transforming three-dimensional quantities on a fixed time slice Σ_t, such as the extrinsic curvature $K_{\mu\nu}$ or the intrinsic curvature $\hat{R}_{\mu\nu}$. A coordinate transformation changes the surface Σ_t relative to which $K_{\mu\nu}$ and $\hat{R}_{\mu\nu}$ are defined. We therefore should *not* transform $K_{\mu\nu}$ (and $\hat{R}_{\mu\nu}$) according to (B.43). Instead, we first write $K_{\mu\nu}$ in terms of the four-dimensional metric

$$K_{\mu\nu} = \frac{\delta^0_\nu \partial_\mu g^{00}}{2(-g^{00})^{3/2}} + \frac{\delta^0_\mu \delta^0_\nu g^{0\sigma} \partial_\sigma g^{00}}{2(-g^{00})^{5/2}} - \frac{g^{0\sigma}\left(\partial_\mu g_{\sigma\nu} + \partial_\nu g_{\sigma\mu} - \partial_\sigma g_{\mu\nu}\right)}{2(-g^{00})^{1/2}} \, , \tag{B.47}$$

and use the transformation of the metric to determine how $K_{\mu\nu}$ transforms. The result can be written as an expansion in π. Substituting these results into (B.37) and (B.38) gives the action for the Goldstone boson π.

B.2.5 Decoupling

Even just showing the terms coming from powers of δg^{00}, the Goldstone action appears rather complicated:

$$
S = \int \mathrm{d}^4 x \sqrt{-g} \left[\frac{M_{\mathrm{pl}}^2}{2} R - M_{\mathrm{pl}}^2 \left(3H^2(t+\pi) + \dot{H}(t+\pi) \right) \right.
$$
$$
+ M_{\mathrm{pl}}^2 \dot{H} \left(g^{00} + 2\partial_\mu \pi g^{0\mu} + \partial_\mu \pi \partial_\nu \pi g^{\mu\nu} \right)
$$
$$
\left. + \sum_{n=2}^{\infty} \frac{M_n^4(t+\pi)}{n!} \left(1 + g^{00} + 2\partial_\mu \pi g^{0\mu} + \partial_\mu \pi \partial_\nu \pi g^{\mu\nu} \right)^n \right] . \tag{B.48}
$$

The Goldstone mode π mixes with the metric perturbations in a nontrivial way. The same happened in the gauge theory example of Section B.1.3, and in that case a major simplification occurred at high energies, where the Goldstone mode decoupled from the gauge field. Exactly the same happens here: above a certain mixing scale ω_{mix}, the Goldstone mode decouples from the metric perturbations $\delta g^{\mu\nu}$. The analogue of the decoupling limit (B.20) is

$$
M_{\mathrm{pl}} \to \infty \ , \ \dot{H} \to 0 \quad \text{for} \quad M_{\mathrm{pl}}^2 \dot{H} = const. \tag{B.49}
$$

This limit suggests that gravitational perturbations decouple from the Goldstone mode for frequencies above $\omega_{\mathrm{mix}}^2 = |\dot{H}|$. As a result, for $\omega > \omega_{\mathrm{mix}}$ we can evaluate (B.48) in the *unperturbed* spacetime, $\bar{g}^{\mu\nu}$,

$$
S = \int \mathrm{d}^4 x \sqrt{-g} \left[\frac{M_{\mathrm{pl}}^2}{2} R - M_{\mathrm{pl}}^2 \left(3H^2(t+\pi) + \dot{H}(t+\pi) \right) \right.
$$
$$
+ M_{\mathrm{pl}}^2 \dot{H} \left(-1 - 2\dot{\pi} + (\partial_\mu \pi)^2 \right)
$$
$$
\left. + \sum_{n=2}^{\infty} \frac{M_n^4(t+\pi)}{n!} \left(-2\dot{\pi} + (\partial_\mu \pi)^2 \right)^n \right] , \tag{B.50}
$$

up to errors of order $(\omega_{\mathrm{mix}}/\omega)^2$.

B.3 The effective theory of inflation

Although the analysis of the previous section is valid for arbitrary FRW backgrounds, we will be interested in quasi-de Sitter spacetimes for which the fractional change in H and $\dot{H}$ per Hubble time is small. We will assume that the same holds for the time dependence of all coefficients in the effective action,[8] e.g. $|\dot{M}_2| \ll HM_2$.

B.3.1 Universal action

We first consider the universal part of the action, which in unitary gauge reads

$$S = \int \mathrm{d}^4 x \sqrt{-g} \left[\frac{M_{\mathrm{pl}}^2}{2} R - M_{\mathrm{pl}}^2 (3H^2 + \dot{H}) + M_{\mathrm{pl}}^2 \dot{H} g^{00} \right] . \tag{B.51}$$

For $-\dot{H} \ll H^2$, this is just single-field slow-roll inflation in disguise. To see this, consider the canonical inflaton Lagrangian

$$\mathcal{L}_{\mathrm{sr}} = -\frac{1}{2} g^{\mu\nu} \partial_\mu \phi \partial_\nu \phi - V(\phi) . \tag{B.52}$$

In unitary gauge, $\phi = \bar{\phi}(t)$, this becomes

$$\begin{aligned} \mathcal{L}_{\mathrm{sr}} &= -\frac{1}{2} \dot{\bar{\phi}}^2 g^{00} - V(\bar{\phi}) \\ &= M_{\mathrm{pl}}^2 \dot{H} g^{00} - M_{\mathrm{pl}}^2 (3H^2 + \dot{H}) , \end{aligned} \tag{B.53}$$

where in the second line we have used the Friedmann equations. The slow-roll action in unitary gauge therefore matches (B.51).

Quadratic action

We introduce the Goldstone boson into (B.51) as in (B.48):

$$\begin{aligned} \mathcal{L}_\pi = &- M_{\mathrm{pl}}^2 \left(3H^2(t+\pi) + \dot{H}(t+\pi) \right) \\ &+ M_{\mathrm{pl}}^2 \dot{H} \left(g^{00} + 2\partial_\mu \pi g^{0\mu} + \partial_\mu \pi \partial_\nu \pi g^{\mu\nu} \right) . \end{aligned} \tag{B.54}$$

The leading mixing of π with gravitational fluctuations[9] is $2M_{\mathrm{pl}}^2 \dot{H} \dot{\pi} \delta g^{00}$. This should be compared with the Goldstone kinetic term $-M_{\mathrm{pl}}^2 \dot{H} \dot{\pi}^2$. Using the Einstein constraint equation [882] (see Appendix C),

$$\delta g^{00} = 2\varepsilon H \pi , \tag{B.55}$$

the mixing term becomes $4M_{\mathrm{pl}}^2 \dot{H}(\varepsilon H \dot{\pi} \pi)$. Notice that $\dot{\pi}\pi$ is a total derivative in flat space, so we should integrate the mixing term by parts before we compare it with the kinetic term. This gives

$$4M_{\mathrm{pl}}^2 \dot{H}(\varepsilon H \dot{\pi}\pi) \rightarrow -6M_{\mathrm{pl}}^2 \dot{H}(\varepsilon H^2 \pi^2) + \cdots . \tag{B.56}$$

---

[8] Assuming that all coefficients vary slowly implies that the action for the *fluctuations* is approximately time translation invariant. This additional global symmetry should not be confused with the broken time translation symmetry of the *background*. See [791] for consequences of breaking the time translation symmetry of the perturbations.

[9] The contribution to the quadratic Goldstone action of the fluctuations δg^{0i} in (B.54) is canceled by a corresponding term from the Einstein–Hilbert action, and so can be neglected at this stage.

A similar term arises from the expansion

$$-3M_{\text{pl}}^2 H^2(t+\pi) = 3M_{\text{pl}}^2 \dot{H}(\varepsilon H^2 \pi^2) + \cdots . \tag{B.57}$$

Combining (B.56) and (B.57), the ratio of the mixing term and the kinetic term is therefore

$$\frac{\text{mixing}}{\text{kinetic}} = \frac{3\varepsilon H^2 \pi^2}{\dot{\pi}^2} = \frac{3\varepsilon H^2}{\omega^2} , \tag{B.58}$$

and the mixing is negligible in the decoupling limit (B.49):

$$\omega \gg \omega_{\text{mix}} \equiv \sqrt{\varepsilon} H . \tag{B.59}$$

It is a special feature of quasi-de Sitter backgrounds, where $\varepsilon \ll 1$, that decoupling happens at relatively low frequencies. In particular, horizon crossing, $\omega = H$, falls in the decoupling regime. Evaluating (B.54) in the unperturbed spacetime, we find

$$\mathcal{L}_\pi^{(2)} = M_{\text{pl}}^2 |\dot{H}| \left(\dot{\pi}^2 - \frac{(\partial_i \pi)^2}{a^2} \right) , \tag{B.60}$$

where $(\partial_i \pi)^2 \equiv \delta^{ij} \partial_i \pi \partial_j \pi$.

Following the logic of Section B.1.2, we can determine the symmetry-breaking scale by examining the Noether current for time translations, $J^\mu = T^{0\mu}$. Using (B.60), we get[10]

$$J^\mu = -(2M_{\text{pl}}^2 |\dot{H}|)^{1/2} \partial^\mu \pi_c + \mathcal{O}(\pi_c^2) , \tag{B.61}$$

where $\pi_c \equiv (2M_{\text{pl}}^2 |\dot{H}|)^{1/2} \pi$. We read off the symmetry-breaking scale from the normalization of the current:

$$f_\pi^4 \equiv 2M_{\text{pl}}^2 |\dot{H}| . \tag{B.62}$$

For slow-roll inflation driven by a scalar field, this reduces to $f_\pi^4 = \dot{\phi}^2$, which matches the intuition that the time variation of the inflaton is responsible for breaking the symmetry.

Had we included the mixing with gravity (B.55) in (B.54), we would have found a small mass term for the Goldstone boson,

$$\mathcal{L}_\pi^{(2)} = M_{\text{pl}}^2 |\dot{H}| \left(\dot{\pi}^2 - \frac{(\partial_i \pi)^2}{a^2} + 3\varepsilon H^2 \pi^2 \right) , \tag{B.63}$$

where we have kept terms at next-to-leading order in slow-roll parameters. Expressed in terms of the comoving curvature perturbation, $\mathcal{R} = -H\pi$, the action (B.63) takes the form

----

[10] The spacetime current J^μ in (B.61) has mass-dimension four, while the global current in (B.13) has mass-dimension three.

$$\mathcal{L}_{\mathcal{R}}^{(2)} = \frac{1}{2}\frac{f_\pi^4}{H^2}\left(\dot{\mathcal{R}}^2 - \frac{(\partial_i \mathcal{R})^2}{a^2}\right). \tag{B.64}$$

Although (B.63) involves a slow-roll approximation, (B.64) is actually an exact result. The action (B.64) has precisely the form that is required in order for $\mathcal{R}$ to be massless, and therefore conserved outside the horizon. In Appendix C, we will derive the power spectrum of vacuum fluctuations with the action (B.64) as a starting point, arriving at

$$\Delta_{\mathcal{R}}^2 = \frac{1}{4\pi^2}\left(\frac{H}{f_\pi}\right)^4. \tag{B.65}$$

The size of the fluctuations is determined by the ratio of the Hubble rate H to the symmetry-breaking scale f_π. This is a universal result that will continue to hold even for more complicated models of inflation [263].

Cubic action

The dominant interactions in slow-roll inflation come from the mixing with gravity, which is not visible in the decoupling limit. Deriving the cubic action in slow-roll inflation is therefore somewhat involved [119]. The size of the associated non-Gaussianity is small, of order the inflationary slow-roll parameters, and the bispectrum vanishes in the squeezed limit because of the single-field consistency relation [119, 120]. Larger non-Gaussianity can arise from deformations away from the slow-roll limit, as parameterized by the action (B.38). We discuss these theories next.

B.3.2 Single-derivative Goldstone action

We start by considering the following class of corrections:

$$\Delta\mathcal{L} = \sum_{n=2}^{\infty}\frac{M_n^4(t)}{n!}(\delta g^{00})^n. \tag{B.66}$$

This is equivalent to studying $P(X)$ theories [40, 124, 247] (see Section 2.2.3), in which the inflaton Lagrangian is a functional P of the inflaton kinetic term, $X \equiv -\frac{1}{2}(\partial\phi)^2$,

$$\mathcal{L}_{P(X)} = P(X, \phi). \tag{B.67}$$

Evaluating (B.67) in unitary gauge, $\phi = \bar{\phi}(t)$, we obtain

$$\mathcal{L}_{P(X)} = P\left(-\frac{1}{2}\dot{\bar{\phi}}^2 g^{00}, \bar{\phi}\right)$$
$$= M_{\rm pl}^2\dot{H}g^{00} - M_{\rm pl}^2\left(3H^2 + \dot{H}\right) + \sum_{n=2}^{\infty}\frac{M_n^4(t)}{n!}\left(\delta g^{00}\right)^n, \tag{B.68}$$

where

$$M_n^4(t) \equiv \left(-\tfrac{1}{2}\dot{\phi}^2 \right)^n \frac{\partial^n P}{\partial X^n} \ . \tag{B.69}$$

We will now analyze the dynamics of these theories in terms of the Goldstone fluctuations. We will build up the action in stages, starting with the operator $(\delta g^{00})^2$ and then adding higher-order terms.

Quadratic action

Introducing the Goldstone boson in the operator $(\delta g^{00})^2$, we get

$$\frac{M_2^4}{2}(\delta g^{00})^2 \ \mapsto \ \frac{M_2^4}{2}\left(1 + g^{00} + 2\partial_\mu \pi g^{0\mu} + \partial_\mu \pi \partial_\nu \pi g^{\mu\nu} \right)^2 \ . \tag{B.70}$$

For $M_2^4 \gg M_{\rm pl}^2|\dot{H}|$, the dominant mixing with gravity now comes from the interaction $M_2^4\dot{\pi}\delta g^{00}$, while the kinetic term becomes $M_2^4\dot{\pi}^2$. The Einstein constraint equation still implies [882]

$$\delta g^{00} = 2\varepsilon H\pi \ . \tag{B.71}$$

The condition for being able to ignore the coupling to the metric perturbations is therefore given, as before, by (B.59). Evaluating (B.70) in the unperturbed spacetime, we find

$$\frac{M_2^4}{2}(\delta g^{00})^2 \ \mapsto \ 2M_2^4\left(\dot{\pi}^2 + \dot{\pi}^3 - \frac{\dot{\pi}(\partial_i \pi)^2}{a^2} \right) + \mathcal{O}(\pi^4) \ . \tag{B.72}$$

We see that the operator $(\delta g^{00})^2$ changes the Goldstone kinetic term, $\dot{\pi}^2$, but not the spatial gradient term, $(\partial_i\pi)^2$, whose coefficient is fixed by the symmetries of the FRW background. Adding $(\delta g^{00})^2$ to the universal part of the Goldstone action therefore induces a nontrivial speed of sound for the fluctuations.[11] In particular, adding (B.72) to (B.60), we get the quadratic action

$$\mathcal{L}_\pi^{(2)} = \frac{M_{\rm pl}^2|\dot{H}|}{c_s^2}\left(\dot{\pi}^2 - c_s^2\frac{(\partial_i\pi)^2}{a^2} \right) \ , \tag{B.73}$$

where we have defined the *speed of sound*

$$c_s^2 \equiv \frac{M_{\rm pl}^2\dot{H}}{M_{\rm pl}^2\dot{H} - 2M_2^4} \ . \tag{B.74}$$

In order to avoid superluminal propagation for the fluctuations, we must impose $M_2^4 > 0$. For $M_2^4 \gg M_{\rm pl}^2|\dot{H}|$ we have $c_s \ll 1$.

---

[11] Further contributions to the sound speed can come from the operators $\delta g^{00}\delta K$, $\delta K^2 - (\delta K^\mu_{\ \nu})^2$, and $\delta g^{00}\hat{R}$ [50].

It is convenient to absorb the sound speed into a redefinition of the spatial coordinates, $x^i \mapsto \tilde{x}^i \equiv c_s^{-1} x^i$, so that fake Lorentz invariance is restored:

$$\tilde{\mathcal{L}}_\pi^{(2)} \equiv c_s^3 \mathcal{L}_\pi^{(2)} = M_{\rm pl}^2 |\dot{H}| c_s \left(\dot{\pi}^2 - \frac{(\tilde{\partial}_i \pi)^2}{a^2} \right) . \tag{B.75}$$

This allows us to read off the symmetry-breaking scale directly from the normalization of the kinetic term,

$$f_\pi^4 \equiv 2 M_{\rm pl}^2 |\dot{H}| c_s . \tag{B.76}$$

As required, this reduces to (B.62) in the limit $c_s \to 1$.

Had we included the mixing with gravity (B.71), instead of (B.75) we would have found

$$\tilde{\mathcal{L}}_\pi^{(2)} = \frac{1}{2} f_\pi^4 \left(\dot{\pi}^2 - \frac{(\tilde{\partial}_i \pi)^2}{a^2} + 3\varepsilon H^2 \pi^2 \right) , \tag{B.77}$$

or, in terms of $\mathcal{R} = -H\pi$,

$$\tilde{\mathcal{L}}_\mathcal{R}^{(2)} = \frac{1}{2} \frac{f_\pi^4}{H^2} \left(\dot{\mathcal{R}}^2 - \frac{(\tilde{\partial}_i \mathcal{R})^2}{a^2} \right) . \tag{B.78}$$

It is easy to see that (B.78) corresponds to (1.17) in the original coordinates x^i. Notice that the action (B.78) is identical to (B.64), but with rescaled spatial coordinates and with f_π defined by (B.76). The power spectrum of curvature perturbations is therefore still given by (B.65). Horizon crossing corresponds to $k = aH$ in the rescaled coordinates, and $c_s k = aH$ in the original coordinates.

Cubic action

The nonlinearly-realized time-translation symmetry relates a small sound speed, $c_s \ll 1$, to large interactions. This is manifest in (B.72), where several interaction terms are proportional to M_2^4, and therefore related to c_s. Consider the cubic part of the Goldstone action,

$$\mathcal{L}_\pi^{(3)} = \frac{M_{\rm pl}^2 |\dot{H}|}{c_s^2} (1 - c_s^2) \left(\dot{\pi}^3 - \frac{\dot{\pi}(\partial_i \pi)^2}{a^2} \right) , \tag{B.79}$$

or, in terms of the rescaled coordinates,

$$\tilde{\mathcal{L}}_\pi^{(3)} = \frac{1}{2} f_\pi^4 (1 - c_s^2) \left(\dot{\pi}^3 - \frac{1}{c_s^2} \frac{\dot{\pi}(\tilde{\partial}_i \pi)^2}{a^2} \right) . \tag{B.80}$$

For $c_s \ll 1$, the operator $\dot{\pi}(\tilde{\partial}_i \pi)^2$ has a boosted coefficient.

To complete the cubic Goldstone action at single-derivative level, we add the operator $(\delta g^{00})^3$, whose representation in terms of π is

$$\frac{M_3^4}{3!} (\delta g^{00})^3 \mapsto \frac{M_3^4}{3!} \left(-2\dot{\pi} + (\partial_\mu \pi)^2 \right)^3 = -\frac{4}{3} M_3^4 \dot{\pi}^3 + \cdots , \tag{B.81}$$

where the Goldstone interactions have been evaluated in the decoupling limit. Adding this to (B.79), we get

$$\mathcal{L}_\pi^{(3)} = \frac{M_{\rm pl}^2 \dot{H}}{c_s^2}(1 - c_s^2)\left[\frac{\dot{\pi}(\partial_i\pi)^2}{a^2} + \frac{A}{c_s^2}\dot{\pi}^3\right],\tag{B.82}$$

where we have defined

$$\frac{A}{c_s^2} \equiv -1 + \frac{2}{3}\frac{M_3^4}{M_2^4}.\tag{B.83}$$

The cubic Goldstone Lagrangian therefore has two independent operators, $\dot{\pi}^3$ and $\dot{\pi}(\partial_i\pi)^2$. In Appendix C, we will compute the two different shapes of non-Gaussianity associated with these operators. However, the expected amplitudes can be estimated without a detailed calculation.

Non-Gaussianity

As before, we rescale the spatial coordinates, $x^i \mapsto \tilde{x}^i \equiv c_s^{-1}x^i$, and define the canonically normalized field $\pi_c \equiv f_\pi^2\pi$. The cubic Goldstone Lagrangian then becomes

$$\tilde{\mathcal{L}}_\pi = -\frac{1}{2}(\tilde{\partial}_\mu\pi_c)^2 - \frac{1}{2\Lambda^2}\left[\frac{\dot{\pi}_c(\tilde{\partial}_i\pi_c)^2}{a^2} + A\dot{\pi}_c^3\right],\tag{B.84}$$

where we have defined the *strong coupling scale*

$$\Lambda^4 \equiv f_\pi^4 \frac{c_s^4}{(1 - c_s^2)^2} = 2M_{\rm pl}^2|\dot{H}|\frac{c_s^5}{(1 - c_s^2)^2}.\tag{B.85}$$

In order not to have an unnatural hierarchy between the scales associated with the two distinct operators $\dot{\pi}_c(\tilde{\partial}_i\pi_c)^2$ and $\dot{\pi}_c^3$, we require $A \sim \mathcal{O}(1)$ [884]. The two operators then produce non-Gaussian fluctuations of the same order of magnitude. A simple estimate for the amplitude of the non-Gaussianity is

$$f_{\rm NL}\mathcal{R} \equiv \left.\frac{\mathcal{L}_3}{\mathcal{L}_2}\right|_{\omega=H} \sim \frac{1}{2\Lambda^2}\frac{\dot{\pi}_c(\tilde{\partial}\pi_c)^2}{(\tilde{\partial}\pi_c)^2} \sim \left(\frac{f_\pi}{\Lambda}\right)^2 \mathcal{R} \sim \frac{1 - c_s^2}{c_s^2}\mathcal{R}.\tag{B.86}$$

The more precise analysis in Appendix C gives

$$f_{\rm NL}^{\dot{\pi}(\partial_i\pi)^2} = -\frac{85}{324}\left(\frac{f_\pi}{\Lambda}\right)^2,\tag{B.87}$$

$$f_{\rm NL}^{\dot{\pi}^3} = +\frac{5A}{81}\left(\frac{f_\pi}{\Lambda}\right)^2.\tag{B.88}$$

Projected onto the equilateral and orthogonal templates (1.67) and (1.68), one finds

$$f_{\rm NL}^{\rm equil} = \left(-0.27 + 0.08A\right)\left(\frac{f_\pi}{\Lambda}\right)^2,\tag{B.89}$$

$$f_{\rm NL}^{\rm ortho} = \left(+0.02 + 0.02A\right)\left(\frac{f_\pi}{\Lambda}\right)^2.\tag{B.90}$$

Measurements of primordial non-Gaussianity therefore constrain the ratio of the symmetry-breaking scale, f_π, and the cutoff scale of the effective theory, Λ. Current observations [10] give

$$f_{\mathrm{NL}}^{\dot\pi(\partial_i\pi)^2} = +8 \pm 73 , \qquad f_{\mathrm{NL}}^{\dot\pi^3} = +19 \pm 57 , \tag{B.91}$$

$$f_{\mathrm{NL}}^{\mathrm{equil}} = -42 \pm 75 , \qquad f_{\mathrm{NL}}^{\mathrm{ortho}} = -25 \pm 39 . \tag{B.92}$$

To understand the physical significance of these constraints we need to discuss the energy scales of the effective theory.

A theoretical threshold

Figure B.2 illustrates the relevant energy scales in the EFT of inflation. The relative sizes of these scales are determined by cosmological observables. First of all, the observed amplitude of curvature perturbations fixes the hierarchy between the symmetry-breaking scale and the freeze-out scale to be

$$f_\pi = 60H . \tag{B.93}$$

Similarly, an observation of primordial tensor modes would determine the ratio of the Hubble scale to the Planck scale,

$$H = 3 \times 10^{-5}\sqrt{r/0.1}\, M_{\mathrm{pl}} . \tag{B.94}$$

Finally, as we have seen above, the strong coupling scale Λ is determined by the non-Gaussianity of the fluctuations. The strong coupling scale is also closely related to the breakdown of perturbative unitarity of Goldstone boson scattering, which is found to happen when [275, 883]

$$\omega^4 > \frac{24\pi}{5}(1 - c_s^2)\Lambda^4 \equiv \Lambda_u^4 . \tag{B.95}$$

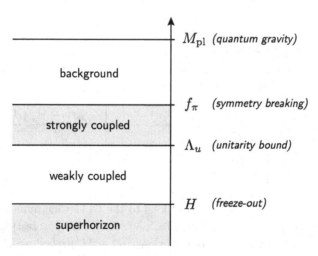

Fig. B.2 Illustration of the relevant energy scales in the EFT of inflation. Whether Λ_u is above or below f_π is an important qualitative distinction.

Whether the unitarity bound Λ_u is above or below f_π is physically relevant. If $\Lambda_u > f_\pi$, then the theory is close to a weakly coupled slow-roll background with perturbative higher-derivative corrections, while for $\Lambda_u < f_\pi$ the Goldstone action involves a completion below the symmetry-breaking scale, plausibly signaling strongly coupled dynamics. (This is like the distinction between the Higgs mechanism and technicolor in the Standard Model.) The critical value $\Lambda_u = f_\pi$ leads to an interesting threshold for the sound speed, $(c_s)_\star$, and consequently the amplitude of the non-Gaussian signal, $(f_{\rm NL})_\star$. Using (B.95), we find [883]

$$\frac{24\pi}{5}\left(\frac{2M_{\rm pl}^2|\dot{H}|(c_s)_\star^5}{1-(c_s)_\star^2}\right) = 2M_{\rm pl}^2|\dot{H}|(c_s)_\star \quad \Rightarrow \quad (c_s)_\star = 0.47 , \tag{B.96}$$

so that (B.87) gives

$$(f_{\rm NL}^{\dot{\pi}(\partial_i\pi)^2})_\star = -\frac{85}{324}\frac{1-(c_s)_\star^2}{(c_s)_\star^2} \simeq -0.93 . \tag{B.97}$$

Similar thresholds, $f_{\rm NL}^{\rm equil} \sim \mathcal{O}(1)$, can be derived for other operators in the EFT of inflation. Comparison with (B.91) and (B.92) shows that current observations are not sensitive enough to rule out non-Gaussianity above the threshold. Probing equilateral non-Gaussianity down to $f_{\rm NL}^{\rm equil} \sim \mathcal{O}(1)$ is an important target for future experiments.

B.3.3 Higher-derivative Goldstone action

At higher orders in the derivative expansion, we should also consider operators involving extrinsic and intrinsic curvature:

$$\Delta\mathcal{L} = -\frac{\bar{M}_1^3(t)}{2}\delta g^{00}\delta K - \frac{\bar{M}_2^2(t)}{2}\delta K^2 - \frac{\bar{M}_3^2(t)}{2}(\delta K^\mu{}_\nu)^2 + \cdots$$
$$- \frac{\hat{M}_1^2(t)}{2}\delta g^{00}\hat{R} + \cdots . \tag{B.98}$$

These operators are relevant in models of ghost inflation [115, 885] and Galileon inflation [276]. Introducing the Goldstone boson, we get

$$\Delta\mathcal{L}_\pi = -\frac{\bar{M}_1^3}{2}\left[H\frac{(\partial_i\pi)^2}{a^2} - \frac{\partial_i^2\pi(\partial_j\pi)^2}{a^4} + \cdots\right] - \frac{\bar{M}_2^2 + \bar{M}_3^2}{2}\frac{(\partial_i^2\pi)^2}{a^4} + \cdots$$
$$+ 2\hat{M}_1^2\left[H^2\frac{(\partial_i\pi)^2}{a^2} + \cdots\right] + \cdots . \tag{B.99}$$

The phenomenology of these contributions to the Goldstone dynamics was presented in detail in [115] (see also [276, 886, 887]). An interesting feature of (B.99) is the higher-derivative contribution $(\bar{M}_2^2 + \bar{M}_3^2)(\partial_i^2\pi)^2$ to the quadratic Lagrangian,

$$\mathcal{L}_\pi^{(2)} = \left(2M_2^4 - M_{\rm pl}^2 \dot{H}\right)\dot{\pi}^2 + M_{\rm pl}^2 \dot{H}\frac{(\partial_i \pi)^2}{a^2} - \frac{\bar{M}_2^2 + \bar{M}_3^2}{2}\frac{(\partial_i^2 \pi)^2}{a^4} + \cdots . \tag{B.100}$$

Let us assume that the operator $(\partial_i^2 \pi)^2$ dominates over $(\partial_i \pi)^2$ at horizon crossing. Formally, we can achieve this by taking the de Sitter limit, $\dot{H} \to 0$. The Goldstone boson then satisfies the non-relativistic dispersion relation

$$\omega = k^2/\rho , \tag{B.101}$$

where we have defined the scale

$$\rho^2 \equiv \frac{M_2^4}{\bar{M}_2^2 + \bar{M}_3^2} . \tag{B.102}$$

To avoid superluminal propagation, we require $\rho > H$. The dispersion relation (B.101) modifies the mode functions of the Goldstone boson, affecting the correlation functions of π (and $\mathcal{R} = -H\pi$) in a nontrivial way. The power spectrum of curvature perturbations takes the form

$$\Delta_\mathcal{R}^2 = \frac{\pi \Gamma[3/4]^2}{2}\frac{H^4}{M_2^4}\left(\frac{\rho}{H}\right)^{3/2} . \tag{B.103}$$

Because of the nonlinear dispersion relation (B.101), the cubic operators $\dot{\pi}(\partial_i \pi)^2$ and $\partial_i^2 \pi(\partial_j \pi)^2$ have the same scaling dimension, while the operator $\dot{\pi}^3$ is highly suppressed. The dominant operators in the cubic Lagrangian are therefore

$$\mathcal{L}_\pi^{(3)} = 2M_2^4 \frac{\dot{\pi}(\partial_i \pi)^2}{a^2} + \frac{\bar{M}_1^3}{2}\frac{\partial_i^2 \pi(\partial_j \pi)^2}{a^4} + \cdots . \tag{B.104}$$

The non-Gaussianity associated with the interactions in (B.104) is [115]

$$f_{\rm NL}^{\dot{\pi}(\partial_i \pi)^2} = 0.25\,\frac{\rho}{H} , \tag{B.105}$$

$$f_{\rm NL}^{\partial_i^2 \pi(\partial_j \pi)^2} = 0.13\,\frac{\bar{M}_1^3}{H(\bar{M}_2^2 + \bar{M}_3^2)} . \tag{B.106}$$

WMAP constraints on these types of non-Gaussianity can be found in [115].

B.3.4 Summary of the Goldstone action

Let us summarize the results of Section B.3. Up to cubic order, and taking the decoupling limit, the Lagrangian of the Goldstone boson is

$$\mathcal{L}_\pi = M_{\rm pl}^2 |\dot{H}| \left[\dot{\pi}^2 - \frac{(\partial_i \pi)^2}{a^2}\right] + 2M_2^4 \left[\dot{\pi}^2 - \frac{\dot{\pi}(\partial_i \pi)^2}{a^2}\right] + \left(2M_2^4 - \frac{4}{3}M_3^4\right)\dot{\pi}^3$$

$$- \frac{H\bar{M}_1^3}{2}\frac{(\partial_i \pi)^2}{a^2} - \frac{\bar{M}_2^2 + \bar{M}_3^2}{2}\frac{(\partial_i^2 \pi)^2}{a^4} + \frac{\bar{M}_1^3}{2}\frac{\partial_i^2 \pi(\partial_j \pi)^2}{a^4} + \cdots . \tag{B.107}$$

We have discussed various limits of the dynamics:

- Taking $M_n = \bar{M}_n = 0$, the action describes the fluctuations in single-field slow-roll inflation. In the decoupling limit, the mixing with metric perturbations can be ignored. As a consequence the Goldstone boson is a massless, free field and the induced curvature perturbations are nearly Gaussian.
- Turning on finite M_2 induces a nontrivial sound speed for the Goldstone modes,

$$c_s^2 \equiv \frac{M_{\mathrm{pl}}^2 \dot{H}}{M_{\mathrm{pl}}^2 \dot{H} - 2M_2^4} \, . \tag{B.108}$$

For small c_s, the dispersion relation $\omega = c_s k$ enhances spatial derivatives relative to time derivatives. The dominant cubic interaction is then $\dot{\pi}(\partial_i \pi)^2$. This leads to non-Gaussianity of equilateral shape, with amplitude,

$$f_{\mathrm{NL}}^{\dot{\pi}(\partial_i \pi)^2} \approx -0.25 \, c_s^{-2} \, . \tag{B.109}$$

A small sound speed also enhances the size of the curvature perturbations (at fixed H and ε) and therefore suppresses the tensor-to-scalar ratio,

$$r = 16\varepsilon c_s \, . \tag{B.110}$$

Detecting a large value of r would put a strong lower bound on the sound speed, $c_s \gtrsim 0.1$, if we take into account that ε is constrained by the scalar spectrum [883].

- It is unnatural to have small M_3 while considering large M_2. In fact, loop corrections induce $M_3^4 \sim M_2^4/c_s^2$. Such values of M_3 enhance the contribution of the operator $\dot{\pi}^3$ to be of the same order as that of the operator $\dot{\pi}(\partial_i \pi)^2$. The bispectra induced by the operators $\dot{\pi}^3$ and $\dot{\pi}(\partial_i \pi)^2$ have similar, but not identical, equilateral shapes. Certain linear combinations of the two bispectra (i.e. specific combinations of the parameters M_2 and M_3) produce a new shape known as the orthogonal shape. By definition, this shape has less than 70% correlation with both the equilateral and local shapes.

  While it is unnatural to have large M_2 and small M_3, the hierarchy $M_3 \gg M_2$ is technically natural [249]. Loops only generate $M_2^4 \sim M_{\mathrm{pl}}^2 \dot{H}$ (or small deviations from $c_s = 1$). This limit allows large equilateral non-Gaussianity even in the absence of a small sound speed.
- For large $\bar{M}_2$ and/or $\bar{M}_3$, the operator $(\partial_i^2 \pi)^2$ dominates over $(\partial_i \pi)^2$ at horizon crossing. The dispersion relation of the Goldstone is then $\omega = k^2/\rho$. The interactions $\dot{\pi}(\partial_i \pi)^2$ and $\partial_i^2 \pi (\partial_j \pi)^2$ have the same scaling dimensions and make the leading contributions to the bispectrum.

B.3.5 Additional degrees of freedom

In this appendix we have focused exclusively on the Goldstone boson π, but as explained in Chapter 4, ultraviolet completions of inflation typically involve additional fields, which we will collectively denote by ψ. If the fields ψ are sufficiently massive, with $m_\psi \gg H$, then they can be integrated out, and affect inflation only by determining the couplings of the effective theory for π. On the other hand, in the more general circumstance in which $m_\psi \lesssim H$, the fields ψ must be included in the effective theory of inflationary fluctuations; a description in terms of a single clock π is insufficient.

A general treatment of multi-field effective theories is challenging, because the interactions of ψ are typically much less constrained by symmetry than are those of π, and the number of possible interactions is correspondingly greatly increased. To reduce the complexity of the problem, one might assume the existence of additional symmetries, which can have the ancillary benefit of making small ψ masses technically natural. A systematic study of multi-field inflation with global symmetries acting on the fields ψ was presented in [888], while a supersymmetric effective theory of multi-field inflation was developed in [249].

The presence of additional degrees of freedom can also lead to alternative models of inflation involving strong dissipative effects. An EFT approach to the study of generic classes of dissipative models was put forward in [823, 850]. The idea is to systematically couple the EFT of inflation to a dissipative sector described by composite operators $(\mathcal{O}, \mathcal{O}_\mu, \mathcal{O}_{\mu\nu}, \ldots)$, whose correlation functions are constrained by symmetries. One can show that for strongly dissipative systems, the power spectrum depends on new parameters and is dominated by non-vacuum fluctuations.

B.4 A generalized Lyth bound

The EFT of inflation allows us to derive a generalization of the Lyth bound (2.103) that is valid away from the slow-roll regime [263]. Being a theory of inflationary fluctuations only, the EFT of inflation may seem ill-suited for deriving such a bound, which after all is a statement about the evolution of the background field. However, as we will show, the relevant information is still encoded in the couplings of the EFT Lagrangian.

B.4.1 Preliminaries

To derive the bound we only need the quadratic part of the Goldstone action. In fact, we only need the normalization of the kinetic term and the scaling dimension (or equivalently the dispersion relation) of π. Our starting point is therefore

$$\mathcal{L} = f^4 \dot{\pi}^2 + \cdots , \tag{B.111}$$

where $f^4 = M_{\rm pl}^2 |\dot{H}|$ for slow-roll inflation, $f^4 = M_{\rm pl}^2 |\dot{H}| c_s^{-2}$ for $P(X)$ theories, $f^4 = M^4$ for ghost inflation, and $f^4 = M_{\rm pl}^2 |\dot{H}| c_s^{-2}$ for Galileon inflation. We assume that, near horizon crossing, π has the dispersion relation,

$$\omega = k^n / \rho^{n-1} . \tag{B.112}$$

Often n will be an integer, but for the present argument it need not be. For $n = 1$, we write $\omega = c_s k$. Under the scaling $\omega \to \lambda \omega$, the field π transforms as $\pi = \lambda^\Delta \pi$, with

$$\Delta = \frac{3 - n}{2n} . \tag{B.113}$$

As before, the symmetry-breaking scale, f_π, can be read off from (B.111). While f^4 is an energy *density*, i.e. has units of $(\text{energy}) \times (\text{momentum})^3$, f_π^4 has units of $(\text{energy})^4$. Using the dispersion relation (B.112) to convert the units, we obtain the following relation between f_π and f:

$$\left(\frac{f_\pi}{H} \right)^{2(1+\Delta)} = c_p^3 \left(\frac{f}{H} \right)^4 , \tag{B.114}$$

where we have introduced the *phase velocity* at horizon crossing,

$$c_p \equiv \frac{\omega}{k} \bigg|_{\omega = H} = \left(\frac{H}{\rho} \right)^{\frac{2}{3}(1-\Delta)} . \tag{B.115}$$

To forbid superluminal propagation we must require $c_p \leq 1$.

B.4.2 A field range bound

The power spectrum of comoving curvature perturbations can be written in the elegant form [263]

$$\Delta_{\mathcal{R}}^2 = \frac{1}{4\pi^2} \left(\frac{H}{f_\pi} \right)^{2(1+\Delta)} . \tag{B.116}$$

All the model dependence of the scalar fluctuations is captured by the symmetry-breaking scale f_π and the scaling dimension Δ. Using (B.114), we can write (B.116) as

$$\Delta_{\mathcal{R}}^2 = \frac{c_p^{-3}}{4\pi^2} \left(\frac{H}{f} \right)^4 . \tag{B.117}$$

Tensor perturbations have no model dependence: their power spectrum is the same for all models of inflation,

$$\Delta_h^2 = \frac{2}{\pi^2} \frac{H^2}{M_{\text{pl}}^2} . \tag{B.118}$$

Using (B.117), the tensor-to-scalar ratio becomes

$$r = 8 c_p^3 \frac{f^4}{M_{\text{pl}}^2 H^2} . \tag{B.119}$$

Below we will argue that the natural size of Planck-suppressed corrections to the Goldstone action is determined by the scale f in (B.111). In particular, we will show that the physically relevant field range is $\Delta\varphi \equiv f^2 \Delta t$. Using (B.119), we find

$$\frac{\Delta\varphi}{M_{\rm pl}} = c_p^{-3/2} \sqrt{\frac{r}{8}} \, \Delta N \ . \qquad (B.120)$$

This reduces to the slow-roll version of the Lyth bound for $c_p = 1$. For $c_p < 1$, the bound (B.120) is stronger than the usual Lyth bound (2.103).

B.4.3 Interpretation

One way to see that f is a natural scale appearing in Planck-suppressed corrections to the Goldstone action is to consider the coupling of $\pi_c \equiv \sqrt{2} f^2 \pi$ to linearized metric perturbations $\gamma_{ij} \equiv M_{\rm pl} \delta g_{ij}$. The Lagrangian (B.111) implies a coupling of the form

$$\mathcal{L} = \frac{1}{8M_{\rm pl}^2} \gamma_{ij} \gamma^{ij} \dot{\pi}_c^2 \ . \qquad (B.121)$$

This becomes strongly coupled at $\omega \sim M_{\rm pl}$. New physics has to enter at or below $M_{\rm pl}$. In general, the new degrees of freedom at the Planck scale will couple to π_c. Moreover, since the Planck-scale physics breaks global symmetries, the couplings to π_c can be through non-derivative operators. Integrating out the new physics at the Planck scale leads to Planck-suppressed corrections of the form

$$3M_{\rm pl}^2 H^2 \ \rightarrow \ 3M_{\rm pl}^2 H^2 \left[1 + \sum_{n=1}^{\infty} c_n \left(\frac{f^4}{M_{\rm pl}^2} (t+\pi)^2 \right)^n \right] \ , \qquad (B.122)$$

where the c_n are Wilson coefficients. For slow-roll inflation, this is simply equivalent to

$$V(\phi) \ \rightarrow \ V(\phi) \left[1 + \sum_{n=1}^{\infty} c_n \frac{\phi^{2n}}{M_{\rm pl}^{2n}} \right] \ . \qquad (B.123)$$

From (B.122) we see that an infinite number of terms become important when $\Delta\varphi = f^2 \Delta t > M_{\rm pl}$. Of course, this is the characteristic problem of large-field inflation, so it seems reasonable to identify $\Delta\varphi$ as the physically relevant field range.

B.4.4 An explicit example

As an explicit example, let us consider $P(X)$ theories, with $X \equiv -\frac{1}{2}(\partial\phi)^2$. The naive generalization of (2.103) is [254]

$$\frac{\Delta\phi}{M_{\rm pl}} = (c_s P_{,X})^{-1/2} \sqrt{\frac{r}{8}} \, \Delta N \ . \qquad (B.124)$$

Taking $P_{,X} \gg 1$ for fixed c_s would seem to allow significant gravitational waves, $r \gtrsim 0.01$, even for $\Delta\phi \ll M_{\rm pl}$. While formally correct, this misses the point. The threshold $\Delta\phi \sim M_{\rm pl}$ is important in slow-roll models with canonical kinetic term because in such theories the effects of the ultraviolet completion of gravity include terms in the potential proportional to $(\phi/M_{\rm pl})^n$, with $n > 0$ (see Section 2.3.3 for details). But when $P_{,X} \gg 1$, the kinetic term for ϕ is far from canonical, and one should examine corrections to the entire $P(X)$ action, including Planck-suppressed corrections of the form

$$\Delta\mathcal{L} = P\left(X - V(\phi)\frac{\phi^2}{M_{\rm pl}^2},\phi\right) = P(X,\phi) - P_{,X}V(\phi)\frac{\phi^2}{M_{\rm pl}^2} + \cdots . \tag{B.125}$$

When $P_{,X} \gg 1$, these corrections are more important than Planck-suppressed corrections to $V(\phi)$. Making $P_{,X}$ large transforms, but does not remove, the problem of Planck-suppressed effects.

To connect this with our previous discussion, let us expand the inflaton as $\phi(t + \pi) = \bar{\phi} + \dot{\phi}\pi + \cdots$. We find

$$f^4 = c_s^{-2}XP_{,X} . \tag{B.126}$$

We can then write $\Delta\varphi = f^2\Delta t$ as

$$\Delta\varphi = c_s^{-3/2}(c_s P_{,X})^{1/2}\Delta\phi , \tag{B.127}$$

where $\Delta\phi \equiv \dot{\phi}\Delta t$. Substituting (B.124) into (B.127), we get

$$\frac{\Delta\varphi}{M_{\rm pl}} = c_s^{-3/2}\sqrt{\frac{r}{8}}\,\Delta N , \tag{B.128}$$

recovering (B.120), with $c_p = c_s$. Although $\Delta\phi$ can become small for $P_{,X} \gg 1$, the physically relevant field range $\Delta\varphi$ remains large for large tensors. We conclude that the UV sensitivity of observable tensors is universal and not a special feature of slow-roll inflation.

Appendix C

Primordial perturbations from inflation

This appendix gives a brief introduction to cosmological perturbation theory and derives the spectrum of primordial quantum fluctuations from inflation. This reproduces the classic results of [12, 14–17].

We will work in the context of the effective theory of inflation, as in Appendix B. We take the unitary gauge action for metric perturbations to be

$$S = \int d^4 x \sqrt{-g} \left[\frac{M_{\rm pl}^2}{2} R + \sum_{n=0}^{\infty} \frac{M_n^4}{n!} (\delta g^{00})^n \right] , \qquad (C.1)$$

with $M_0^4 \equiv -M_{\rm pl}^2 (3H^2 + 2\dot{H})$ and $M_1^4 \equiv M_{\rm pl}^2 \dot{H}$. In Section C.1, we study this action at quadratic order and derive the power spectrum of curvature perturbations $\mathcal{R}$. In Section C.2, we give the corresponding result for tensor fluctuations. In Section C.3, we explain how higher-order correlations are computed in the *in-in* formalism. As an example, we derive the two equilateral bispectra corresponding to the two cubic interactions arising from (C.1). Finally, in Section C.4, we describe some of the novel features of perturbations in multi-field inflation.

C.1 Free fields: scalar modes

For purely adiabatic fluctuations, there are five real scalar perturbations before accounting for gauge redundancy: one Goldstone mode π and four metric perturbations. As we will see, diffeomorphism invariance removes two perturbations, while the Einstein constraint equations eliminate two more. Hence, we are left with a single physical scalar degree of freedom whose dynamics we would like to describe. The physical interpretation of this degree of freedom will depend on the specific choice of gauge. Before proceeding, we will briefly describe the gauges used in this book.

C.1.1 Gauges

The starting point is a general first-order perturbation about a spatially flat FRW cosmology,

$$\mathrm{d}s^2 = -(1 + 2\Phi)\mathrm{d}t^2 + 2a(t)B_i\mathrm{d}x^i\,\mathrm{d}t + a^2(t)\Big[(1 - 2\Psi)\delta_{ij} + C_{ij}\Big]\mathrm{d}x^i\mathrm{d}x^j\ , \quad \text{(C.2)}$$

where C_{ij} is symmetric and traceless. For scalar perturbations, we can write $B_i = \partial_i B$ and $C_{ij} = (\partial_i\partial_j - \frac{1}{3}\delta_{ij}\nabla^2)C$. A convenient alternative parameterization of the same line element is

$$\mathrm{d}s^2 = -(1 + \delta g_{00})\mathrm{d}t^2 + 2\delta g_{i0}\mathrm{d}x^i\,\mathrm{d}t + a^2(t)g_{ij}\mathrm{d}x^i\mathrm{d}x^j\ . \quad \text{(C.3)}$$

In *spatially flat gauge*, the spatial part of the metric is taken to be unperturbed,

$$g_{ij} = \delta_{ij}\ . \quad \text{(C.4)}$$

The remaining metric perturbations δg_{00} and δg_{i0} are then related by the Einstein constraint equations to the Goldstone boson π, which encodes adiabatic perturbations. Spatially flat gauge is convenient for describing the quantization of inflationary perturbations.

In *comoving gauge* (also known as *unitary gauge*) the Goldstone boson is set to zero by a time reparameterization: $\pi \equiv 0$. This does not fix the gauge freedom completely, as one can still reparameterize the spatial coordinates. We will typically use that freedom to put the spatial metric into the isotropic form,

$$g_{ij} = e^{2\mathcal{R}}\delta_{ij}\ , \quad \text{(C.5)}$$

where $\mathcal{R}(t, \boldsymbol{x})$ is the comoving curvature perturbation. The remaining metric perturbations δg_{00} and δg_{i0} are related to the curvature perturbation $\mathcal{R}$ by the Einstein constraint equations. Comoving gauge is convenient for describing the evolution of adiabatic perturbations, because in the absence of entropy perturbations (see Section C.4), $\mathcal{R}$ is conserved on superhorizon scales [55].

Newtonian gauge is defined by $B_i = C_{ij} = 0$, with the line element

$$\mathrm{d}s^2 = -(1 + 2\Phi)\mathrm{d}t^2 + a^2(t)(1 - 2\Psi)\delta_{ij}\mathrm{d}x^i\mathrm{d}x^j\ . \quad \text{(C.6)}$$

In the absence of anisotropic stress, the Einstein equations enforce $\Psi = \Phi$. Notice the similarity of (C.6) to the weak-field limit of general relativity in Minkowski space. On superhorizon scales, the Newtonian potential Φ is related to the comoving curvature perturbation $\mathcal{R}$ via

$$\mathcal{R} = -\frac{5 + 3w}{3 + 3w}\Phi\ , \quad \text{(C.7)}$$

where w is the equation of state of the background fluid. During the matter-dominated era this implies that $\mathcal{R} = -\frac{5}{3}\Phi$. Newtonian gauge is convenient for studying the formation of large-scale structures and CMB anisotropies.

C.1.2 Quadratic action

After fixing the gauge, we consider the Einstein equations to impose constraints between the remaining perturbations. This is done most conveniently in the formalism of Arnowitt, Deser, and Misner (ADM) [889], in which the spacetime metric takes the form

$$ds^2 = -N^2 dt^2 + \hat{g}_{ij} \left(dx^i + N^i dt \right) \left(dx^j + N^j dt \right) , \tag{C.8}$$

where $N(t, \boldsymbol{x})$ is the *lapse* function, $N^i(t, \boldsymbol{x})$ is the *shift* vector, and $\hat{g}_{ij}(t, \boldsymbol{x})$ is the induced metric on three-dimensional hypersurfaces of constant time t. The geometry of these hypersurfaces is characterized by the intrinsic curvature $\hat{R}_{ij}$, i.e. the Ricci tensor of the induced metric, and by the extrinsic curvature

$$K_{ij} \equiv \frac{1}{2N} \left(\dot{\hat{g}}_{ij} - \nabla_i N_j - \nabla_j N_i \right) \equiv \frac{1}{N} E_{ij} , \tag{C.9}$$

where ∇_i is the covariant derivative associated with $\hat{g}_{ij}$. The four-dimensional Ricci scalar, R, can be written in terms of the three-dimensional Ricci scalar, $\hat{R}$, and the extrinsic curvature tensor as

$$R = \hat{R} + N^{-2} \left(E^{ij} E_{ij} - E^2 \right) , \tag{C.10}$$

where indices are raised with $\hat{g}^{ij}$, and $E \equiv \hat{g}^{ij} E_{ij}$. At quadratic order, the action (C.1) then becomes

$$S^{(2)} = \int d^4 x \sqrt{\hat{g}} \left\{ \frac{M_{\rm pl}^2}{2} \left[N\hat{R} + \frac{1}{N} \left(E^{ij} E_{ij} - E^2 \right) \right] + N M_0^4 \right.$$
$$\left. + N M_1^4 \left(1 - \frac{1}{N^2} \right) + \frac{N}{2} M_2^4 \left(1 - \frac{1}{N^2} \right)^2 \right\} . \tag{C.11}$$

Varying (C.11) with respect to N gives

$$M_{\rm pl}^2 \left[\hat{R} - \frac{1}{N^2} \left(E^{ij} E_{ij} - E^2 \right) \right] + M_0^4 + M_1^4 \left(1 + \frac{1}{N^2} \right)$$
$$+ \frac{M_2^4}{2} \left(1 + \frac{2}{N^2} - \frac{3}{N^4} \right) = 0 , \tag{C.12}$$

while varying it with respect to N^i gives

$$\nabla_j \left[\frac{1}{N} \left(E^j{}_i - \delta^j{}_i E \right) \right] = 0 . \tag{C.13}$$

Notice that these equations do not involve time derivatives of N and N^i, so they are constraint equations. We now expand the lapse and the shift in small perturbations: $N = 1 + \delta N$ and $N^i \equiv \partial^i \psi + N_T^i$, with $\partial_i N_T^i = 0$. At linear order, we have

$$\hat{R}_{ij} = -\partial_i \partial_j \mathcal{R} - \delta_{ij}\partial^2 \mathcal{R} \, , \tag{C.14}$$

$$E^i{}_j = (H + \dot{\mathcal{R}})\delta^i{}_j - \partial^i \partial_j \psi - \frac{1}{2}(\nabla^i N_j^T + \nabla_j N_T^i) \, , \tag{C.15}$$

so the momentum constraint (C.13) becomes

$$\delta N = \frac{\dot{\mathcal{R}}}{H} \, , \quad N_T^i = 0 \, , \tag{C.16}$$

while the Hamiltonian constraint (C.12) gives[1]

$$\partial^2 \psi = \left(\frac{2M_2^4}{M_{\rm pl}^2 H^2} - \frac{\dot{H}}{H^2} \right) \dot{\mathcal{R}} - \frac{\partial^2 \mathcal{R}}{Ha^2} \, ,$$

$$= \frac{\varepsilon}{c_s^2}\dot{\mathcal{R}} - \frac{\partial^2 \mathcal{R}}{Ha^2} \, , \tag{C.17}$$

We substitute (C.16) and (C.17) into the action (C.11), and use

$$\sqrt{\hat{g}} = a^3 e^{3\mathcal{R}} \, , \tag{C.18}$$

$$\hat{R} = \left(-4\partial^2 \mathcal{R} - 2(\partial_i \mathcal{R})^2 \right) a^{-2} e^{-2\mathcal{R}} \, , \tag{C.19}$$

$$E^{ij} E_{ij} - E^2 = -6(H + \dot{\mathcal{R}})^2 \, . \tag{C.20}$$

All terms that are linear in $\mathcal{R}$ vanish after using the background equations of motion. Remarkably, the mass term for $\mathcal{R}$ also vanishes and we are left with

$$S_{\mathcal{R}}^{(2)} = \frac{1}{2} \int {\rm d}\tau {\rm d}^3 x \, z^2 \left[(\mathcal{R}')^2 - c_s^2 (\partial_i \mathcal{R})^2 \right] \, , \quad \frac{z^2}{M_{\rm pl}^2} \equiv \frac{2a^2 \varepsilon}{c_s^2} \, , \tag{C.21}$$

where we have introduced conformal time τ and use a prime to denote the corresponding time derivative. The free field action (C.21) is the starting point for canonical quantization.

C.1.3 Canonical quantization

From the action (C.21) we can read off the conjugate momentum, $\Pi \equiv \partial \mathcal{L}/\partial \mathcal{R}' = z^2 \mathcal{R}'$. We promote the fields to operators, $\mathcal{R} \to \hat{\mathcal{R}}$ and $\Pi \to \hat{\Pi}$, and impose the equal-time canonical commutation relation,

$$[\hat{\mathcal{R}}(\tau, \boldsymbol{x}), \hat{\Pi}(\tau, \boldsymbol{x}')] = i\delta(\boldsymbol{x} - \boldsymbol{x}') \, , \quad \hbar \equiv 1 \, , \tag{C.22}$$

or, for each Fourier component of the operators,

$$[\hat{\mathcal{R}}_{\boldsymbol{k}}(\tau), \hat{\Pi}_{\boldsymbol{k}'}(\tau)] = i\delta(\boldsymbol{k} - \boldsymbol{k}') \, . \tag{C.23}$$

The equation of motion implies that the commutation relation (C.22) holds at all times if it is imposed at some initial time. Here we are working in the Heisenberg

---

[1] The solution for ψ only contributes boundary terms to the quadratic action, but makes nontrivial contributions to the cubic action.

picture, where operators vary in time and states are unchanging. The operator solution $\hat{\mathcal{R}}_k(\tau)$ is determined by the two initial conditions $\hat{\mathcal{R}}_k(0)$ and $\hat{\Pi}_k \propto \hat{\mathcal{R}}'_k(0)$. Since the evolution equation is linear, the solution is linear in these operators. It is convenient to trade $\hat{\mathcal{R}}_k(0)$ and $\hat{\Pi}_k(0)$ for a single time-independent non-Hermitian operator $\hat{a}_k$, in terms of which the solution can be written as

$$\hat{\mathcal{R}}_k(\tau) = \mathcal{R}_k(\tau)\hat{a}_k + \mathcal{R}_k^*(\tau)\hat{a}_{-k}^\dagger \ , \tag{C.24}$$

where $\hat{a}_k^\dagger$ is the Hermitian conjugate of $\hat{a}_k$. The mode function $\mathcal{R}_k(\tau)$ and its complex conjugate $\mathcal{R}_k^*(\tau)$ are two linearly independent solutions of the Mukhanov–Sasaki (MS) equation obtained from (C.21),

$$\frac{d^2\mathcal{R}_k}{d\tau^2} + \frac{2}{z}\frac{dz}{d\tau}\frac{d\mathcal{R}_k}{d\tau} + c_s^2 k^2 \mathcal{R}_k = 0 \ , \tag{C.25}$$

where

$$\frac{2}{z}\frac{dz}{d\tau} = 2(aH)\left(1 + \frac{\tilde{\eta} + \kappa}{2}\right) \ , \tag{C.26}$$

with

$$\varepsilon \equiv -\frac{\dot{H}}{H^2} \ , \quad \tilde{\eta} \equiv \frac{\dot{\varepsilon}}{H\varepsilon} \ , \quad \kappa \equiv \frac{\dot{c}_s}{Hc_s} \ . \tag{C.27}$$

Although the slow-roll parameters (C.27) appear in (C.25), we stress that (C.25) is exact: no slow-roll approximation has been used at this point. As indicated by dropping the vector notation $\boldsymbol{k}$ on the subscript, the mode functions, $\mathcal{R}_k(\tau)$ and $\mathcal{R}_k^*(\tau)$, are the same for all Fourier modes[2] with $|\boldsymbol{k}| = k$. If the mode functions satisfy the Wronskian normalization,

$$W[\mathcal{R}_k, \mathcal{R}_k^*] \equiv -iz^2(\mathcal{R}_k\partial_\tau\mathcal{R}_k^* - (\partial_\tau\mathcal{R}_k)\mathcal{R}_k^*) = 1 \ , \tag{C.28}$$

then (C.22) implies

$$[\hat{a}_k, \hat{a}_{k'}^\dagger] = \delta(\boldsymbol{k} - \boldsymbol{k}') \ . \tag{C.29}$$

For each Fourier mode $\boldsymbol{k}$, Eq. (C.29) is the commutation relation for the creation operator $\hat{a}_k^\dagger$ and the annihilation operator $\hat{a}_k$ of a simple harmonic oscillator. The vacuum state is defined by the condition

$$\hat{a}_k|0\rangle = 0 \ . \tag{C.30}$$

However, at this point, the vacuum is not yet completely specified. A change in $\hat{a}_k$ (and hence $|0\rangle$) can be compensated for by a change in the mode function $\mathcal{R}_k$, leaving the operator solution $\hat{\mathcal{R}}_k$ fixed. To fix the vacuum, we consider the limit of very early times, when all modes of cosmological interest were deep inside the

---

[2] Since the MS equation depends only on $k \equiv |\boldsymbol{k}|$, the evolution does not depend on direction. However, the constant operators $\hat{a}_k$ and $\hat{a}_k^\dagger$ define initial conditions that may depend on direction.

Hubble radius. By the equivalence principle, these modes should behave as in flat space. It is therefore conventional to impose the Bunch–Davies initial condition,

$$\lim_{k\tau \to -\infty} z(\tau)\mathcal{R}_k(\tau) = \frac{1}{\sqrt{2c_s k}} e^{-ic_s k\tau} \,, \tag{C.31}$$

which corresponds to the minimum energy state in the Minkowski limit.

Since our discussion has been exact up to this point, we could now find the exact Bunch–Davies mode functions by integrating the MS equation (C.25) numerically. However, further analytical progress can be made by working in the slow-roll approximation. Integrating the exact relation

$$\frac{d(aH)}{d\tau} = (aH)^2(1 - \varepsilon) \,, \tag{C.32}$$

under the assumption that ε is slowly varying, we find

$$aH = -\frac{1}{(1 - \varepsilon)\tau} \,. \tag{C.33}$$

Using (C.33) in (C.25), and assuming $\varepsilon \ll 1$, the MS equation takes the form

$$\frac{d^2\mathcal{R}_k}{d\tau^2} - \frac{2}{\tau}\left(1 + \varepsilon + \frac{\tilde{\eta} + \kappa}{2}\right)\frac{d\mathcal{R}_k}{d\tau} + c_s^2 k^2 \mathcal{R}_k = 0 \,, \tag{C.34}$$

to leading order in slow-roll. To solve (C.34) we assume that ε, $\tilde{\eta}$, and κ are small, and vary sufficiently slowly in time.[3] Imposing the Bunch–Davies initial condition (C.31) and the Wronskian normalization (C.28), we find

$$\mathcal{R}_k(\tau) = \left(\frac{\pi}{2}\right)^{1/2} \frac{(-c_s k\tau)^{-\nu + \frac{1}{2}}}{z(\tau)} \frac{(-c_s k\tau)^{\nu}}{\sqrt{2c_s k}} H_\nu^{(1)}(-c_s k\tau) \,, \tag{C.35}$$

where $H_\nu^{(1)}$ is a Hankel function of the first kind and

$$\nu \equiv \frac{3}{2} + \varepsilon + \frac{\tilde{\eta} + \kappa}{2} \,. \tag{C.36}$$

Since $z(\tau) \propto \tau^{-\nu + \frac{1}{2}}$, the second factor in (C.35) is time independent. In terms of the solution (C.35), the two-point function of curvature perturbations in the vacuum state can be written as

$$\langle 0|\hat{\mathcal{R}}(\tau, \boldsymbol{x})\hat{\mathcal{R}}(\tau, \boldsymbol{x}')|0\rangle = \int d\ln k \; \Delta_\mathcal{R}^2(\tau, k) \, J_0(k|\boldsymbol{x} - \boldsymbol{x}'|) \,, \tag{C.37}$$

where $J_0(y) \equiv \sin(y)/y$ and

$$\Delta_\mathcal{R}^2(\tau, k) \equiv \frac{k^3 |\mathcal{R}_k(\tau)|^2}{2\pi^2} \,. \tag{C.38}$$

The dimensionless power spectrum $\Delta_\mathcal{R}^2(\tau, k)$ is a key observable.

---

[3] The assumptions that ε and $\tilde{\eta}$ are small and slowly varying were already required to derive (C.34), but $\kappa \ll 1$ and $\dot{\kappa}/H\kappa \ll 1$ are needed only for the slow-roll solution of (C.34).

C.1.4 Power spectrum

We are interested in evaluating (C.38) in the superhorizon limit. Using

$$\lim_{y \to 0} H_\nu^{(1)}(y) \to -\frac{i\Gamma(\nu)}{\pi} \left(\frac{2}{y}\right)^\nu ,$$ (C.39)

we find

$$\mathcal{R}_k^{(o)} \equiv \lim_{k\tau \to 0} \mathcal{R}_k(\tau) = \frac{(-\tau)^{-\nu+\frac{1}{2}}}{z(\tau)} \frac{\Gamma(\nu)}{2\sqrt{\pi}} \frac{2^\nu}{(c_s k)^\nu} ,$$ (C.40)

where we have dropped an irrelevant phase, and the superscript denotes evaluation outside the horizon. The power spectrum after horizon crossing is therefore

$$\Delta_{\mathcal{R}}^2(k) \equiv \frac{k^3 |\mathcal{R}_k^{(o)}|^2}{2\pi^2} \equiv \Delta_{\mathcal{R}}^2(k_\star) \left(\frac{k}{k_\star}\right)^{n_s(k_\star)-1} ,$$ (C.41)

where

$$n_s(k_\star) - 1 = -2\varepsilon - \tilde{\eta} - \kappa .$$ (C.42)

The amplitude, $\Delta_{\mathcal{R}}(k_\star)$, and the spectral index, $n_s(k_\star)$, are defined with respect to a reference scale $k_\star$. To relate the amplitude $\Delta_{\mathcal{R}}(k_\star)$ to the inflationary background it is convenient to evaluate the mode function at the time $\tau_\star$ when the reference scale $k_\star$ crosses the *sound horizon*, i.e. for $\tau = \tau_\star$, defined so that $k_\star = a(\tau_\star)H(\tau_\star)/c_s(\tau_\star)$. Using the definition (C.21) of $z(\tau)$ and setting $2^\nu \Gamma(\nu) \approx 2^{3/2}\Gamma(3/2)$ in (C.40), we find

$$\Delta_{\mathcal{R}}^2(k_\star) \approx \frac{1}{8\pi^2} \frac{H^4}{M_{\text{pl}}^2 |\dot{H}| c_s} ,$$ (C.43)

where the right-hand side is evaluated at $\tau = \tau_\star$. In the slow-roll limit, $c_s \to 1$, the results can be written in terms of the shape of the inflaton potential,

$$\Delta_{\mathcal{R}}^2(k_\star) = \frac{1}{12\pi^2} \frac{V^3}{M_{\text{pl}}^6 (V')^2} ,$$ (C.44)

$$n_s(k_\star) - 1 = -3M_{\text{pl}}^2 \left(\frac{V'}{V}\right)^2 + 2M_{\text{pl}}^2 \frac{V''}{V} .$$ (C.45)

C.2 Free fields: tensor modes

C.2.1 Quadratic action

In close parallel with the scalar case just described, we now consider tensor perturbations to the spatial metric,

$$g_{ij} = a^2(\delta_{ij} + h_{ij}) ,$$ (C.46)

where we take the perturbations h_{ij} to be transverse and traceless, i.e. $\partial^i h_{ij} = \delta^{ij} h_{ij} = 0$. The quadratic action for tensor perturbations follows by substituting (C.46) into the Einstein–Hilbert action: one finds

$$S_h^{(2)} = \frac{1}{2} \int \mathrm{d}\tau \mathrm{d}^3 x \, z^2 \left[(h'_{ij})^2 - (\partial_k h_{ij})^2 \right] , \qquad z^2 \equiv \frac{a^2 M_{\text{pl}}^2}{4} . \tag{C.47}$$

Quantization proceeds analogously to the scalar case, so we will be brief.

C.2.2 Canonical quantization

We promote the field to an operator, $h_{ij} \to \hat{h}_{ij}$. The solution for $\hat{h}_{ij}(\tau, \boldsymbol{k})$ can be written as

$$\hat{h}_{ij}(\tau, \boldsymbol{k}) = \sum_{\lambda = +, \times} h_k(\tau) e_{ij}^\lambda(\hat{\boldsymbol{k}}) \hat{a}_{\boldsymbol{k}}^\lambda + \text{h.c.} , \tag{C.48}$$

where $e_{ij}^\lambda(\hat{\boldsymbol{k}})$ are the transverse and traceless polarization tensors for the two helicity modes of the graviton. The normalization of the polarization tensors is chosen to be

$$\delta^{ik} \delta^{jl} e_{ij}^\lambda(\hat{\boldsymbol{k}}) (e_{kl}^{\lambda'}(\hat{\boldsymbol{k}}))^* = \delta^{\lambda\lambda'} . \tag{C.49}$$

Both $\hat{a}_{\boldsymbol{k}}^+$ and $\hat{a}_{\boldsymbol{k}}^\times$ satisfy (C.29). The evolution of the mode function $h_k(\tau)$ is determined by

$$\frac{d^2 h_k}{d\tau^2} + \frac{2}{z} \frac{dz}{d\tau} \frac{dh_k}{d\tau} + k^2 h_k = 0 , \qquad \frac{2}{z} \frac{dz}{d\tau} = 2(aH) . \tag{C.50}$$

Using (C.33), the MS equation (C.50) becomes

$$\frac{d^2 h_k}{d\tau^2} + \frac{2(1 + \varepsilon)}{\tau} \frac{dh_k}{d\tau} + k^2 h_k = 0 . \tag{C.51}$$

Solving (C.51) subject to the initial condition

$$\lim_{k\tau \to -\infty} z(\tau) h_k(\tau) = \frac{1}{\sqrt{2k}} e^{-ik\tau} , \tag{C.52}$$

we find

$$h_k(\tau) = \left(\frac{\pi}{2} \right)^{1/2} \frac{(-k\tau)^{-\mu + \frac{1}{2}}}{z(\tau)} \frac{(-k\tau)^\mu}{\sqrt{2k}} H_\mu^{(1)}(-k\tau) , \tag{C.53}$$

where $\mu \equiv \frac{3}{2} + \varepsilon$.

C.2.3 Power spectrum

The superhorizon limit of the mode function (C.53) is

$$h_k^{(o)} \equiv \lim_{k\tau \to 0} h_k(\tau) = \frac{(-\tau)^{-\mu + \frac{1}{2}}}{z(\tau)} \frac{\Gamma(\mu)}{2\sqrt{\pi}} \frac{2^\mu}{k^\mu} . \tag{C.54}$$

The power spectrum of tensor modes after horizon crossing is therefore

$$\Delta_h^2(k) \equiv 2 \times \frac{k^3 |h_k^{(o)}|^2}{2\pi^2} \equiv \Delta_h^2(k_\star) \left(\frac{k}{k_\star}\right)^{n_t(k_\star)} , \qquad (C.55)$$

where we have added the two polarization modes (hence the factor of 2) and defined

$$n_t(k_\star) = -2\varepsilon . \qquad (C.56)$$

Evaluating the mode function at horizon crossing, $k_\star = a(\tau_\star)H(\tau_\star)$, gives

$$\Delta_h^2(k_\star) = \frac{2}{\pi^2} \frac{H^2}{M_{\rm pl}^2} . \qquad (C.57)$$

It is conventional to define the tensor-to-scalar ratio

$$r \equiv \frac{\Delta_h^2}{\Delta_{\mathcal{R}}^2} = 16\varepsilon c_s . \qquad (C.58)$$

In the slow-roll limit, $c_s \to 1$, we obtain a consistency condition between the tensor amplitude and its tilt,

$$n_t = -\frac{r}{8} . \qquad (C.59)$$

C.3 Interacting fields: non-Gaussianity

To study the non-universal part of the Goldstone action, we have to go beyond the free-field limit and study interactions. In this section, we will show how to compute the associated non-Gaussianities in the primordial fluctuations.

C.3.1 In-in formalism

To determine the correlation functions of a quantum field theory in an asymptotically Minkowski spacetime, one can compute the S-matrix describing the transition probability for a state $|in\rangle$ in the far past to become some state $|out\rangle$ in the far future,

$$\langle out| S|in\rangle = \langle out(+\infty)|in(-\infty)\rangle . \qquad (C.60)$$

The scattering particles are taken to be non-interacting at very early and very late times, when they are far from the interaction region, and the asymptotic states can be taken to be vacuum states of the free Hamiltonian H_0.

In cosmology, on the other hand, the task is to determine the expectation values of products of operators *at a fixed time*. For example, we will wish to compute n-point functions of the Goldstone boson, i.e. expectation values of operators such as $\hat{Q} = \hat{\pi}_{k_1} \hat{\pi}_{k_2} \cdots \hat{\pi}_{k_n}$,

$$\langle \hat{Q} \rangle \equiv \langle \Omega| \hat{Q}(\tau) |\Omega\rangle , \qquad (C.61)$$

where $|\Omega\rangle$ is the vacuum of the interacting theory at some moment τ_0 in the far past, and $\tau > \tau_0$ is some later time, such as horizon crossing or the end of inflation. Boundary conditions are only imposed at very early times, when their wavelengths are much smaller than the horizon; in this limit the interaction picture fields should have the same form as in Minkowski space. As we have seen in the previous section, this leads to the definition of the Bunch–Davies vacuum – the free vacuum in Minkowski space. In this section, we will describe the *in-in* formalism[4] to compute cosmological correlation functions as expectation values in two $|in\rangle$ states.

The time evolution of the operators in the Heisenberg picture is determined by[5]

$$\frac{d\hat{\pi}}{d\tau} = i[\hat{H}, \hat{\pi}] \ , \quad \frac{d\hat{p}_\pi}{d\tau} = i[\hat{H}, \hat{p}_\pi] \ , \tag{C.62}$$

where $\hat{H} = \hat{H}_0 + \hat{H}_{int}$ is the perturbed Hamiltonian. This time-evolution is complicated by the interactions inside $\hat{H}$, which lead to nonlinear equations of motion. We therefore introduce the *interaction picture* in which the leading time-dependence of the fields is determined by the quadratic Hamiltonian $\hat{H}_0$ (or equivalently, by the linear equations of motion)

$$\hat{\pi}_I' = i[\hat{H}_0, \hat{\pi}_I] \ , \quad \hat{p}_{\pi,I}' = i[\hat{H}_0, \hat{p}_{\pi,I}] \ . \tag{C.63}$$

The solution to these equations can be written as

$$\hat{\pi}_k^I(\tau) = \pi_k^I(\tau)\hat{a}_k + h.c. \ , \tag{C.64}$$

where $\pi_k^I(\tau)$ is the solution to the free-field Mukhanov–Sasaki equation (C.25) and the operators $\hat{a}_k$ define the free-field vacuum $|0\rangle$. Since we are interested in the limit of large non-Gaussianity, we will ignore slow-roll corrections to the mode functions and use the solution for de Sitter space,

$$\pi_k^I(\tau) = \pi_k^{(o)}\left(1 + ic_s k\tau\right)e^{-ic_s k\tau} \ , \quad \pi_k^{(o)} \equiv \frac{i}{2M_{\rm pl}\sqrt{\varepsilon c_s}}\frac{1}{k^{3/2}} \ . \tag{C.65}$$

Corrections to the evolution of the operators can then be treated perturbatively in $\hat{H}_{int}$. Relatively straightforward algebraic manipulations of (C.62) and (C.63) allow us to express an operator in the Heisenberg picture in terms of operators in the interaction picture [893],

$$\hat{Q}(\tau) = \hat{F}^{-1}(\tau, \tau_0)\,\hat{Q}^I(\tau)\,\hat{F}(\tau, \tau_0) \ , \tag{C.66}$$

where

$$\hat{F}(\tau, \tau_0) \equiv Te^{-i\int_{\tau_0}^{\tau} \hat{H}_{int}^I(\tau'')d\tau''} \ . \tag{C.67}$$

---

[4] This is also referred to as the Schwinger–Keldysh formalism [890]. The use of the *in-in* formalism in cosmology was pioneered in [119, 891–893] (see also [894, 895]) and is reviewed in [112, 113].

[5] The Goldstone boson π should not be confused with the corresponding conjugate momentum, which we denote by p_π.

The symbol T denotes time-ordering. We can think of $\hat{F}(\tau, \tau_0)$ as an operator evolving quantum states in the interaction picture,

$$|\Omega(\tau)\rangle = \hat{F}(\tau, \tau_0)|\Omega(\tau_0)\rangle , \tag{C.68}$$

where $|\Omega(\tau_0)\rangle \equiv |\Omega\rangle$. We would like to relate the vacuum of the interacting theory, $|\Omega\rangle$, to the vacuum of the free theory, $|0\rangle$. Inserting a complete set of energy eigenstates $\{|\Omega\rangle, |n\rangle\}$ of the full theory, where $|n\rangle$ are the excited states, we have

$$|0\rangle = |\Omega\rangle\langle\Omega|0\rangle + \sum_n |n\rangle\langle n|0\rangle , \tag{C.69}$$

and correspondingly

$$e^{-i\hat{H}(\tau-\tau_0)}|0\rangle = e^{-i\hat{H}(\tau-\tau_0)}|\Omega\rangle\langle\Omega|0\rangle + \sum_n e^{-iE_n(\tau-\tau_0)}|n\rangle\langle n|0\rangle . \tag{C.70}$$

Adding a small imaginary part to the initial time, $\tau_0 \to -\infty(1 - i\epsilon) \equiv -\infty^-$, will project out the excited states, $e^{-iE_n(\tau-\tau_0)} \to e^{-\infty\times\epsilon E_n}(\cdots) \to 0$. We are then left with

$$\hat{F}(\tau, -\infty^-)|\Omega\rangle = \frac{\hat{F}(\tau, -\infty^-)|0\rangle}{\langle\Omega|0\rangle} . \tag{C.71}$$

The $i\epsilon$ prescription has effectively turned off the interaction in the far past and projected the interacting vacuum $|\Omega\rangle$ onto the free vacuum $|0\rangle$. Setting $|\langle\Omega|0\rangle| \to 1$, we arrive at the *in-in* master formula,

$$\langle\hat{Q}(\tau)\rangle = \langle 0| \bar{T}e^{i\int_{-\infty^+}^{\tau} \hat{H}_{int}^I(\tau')d\tau'} \hat{Q}^I(\tau) \, Te^{-i\int_{-\infty^-}^{\tau} \hat{H}_{int}^I(\tau'')d\tau''} |0\rangle , \tag{C.72}$$

where $\bar{T}$ is the anti-time-ordering symbol and $\infty^\pm \equiv \infty(1\pm i\epsilon)$. The integration contour goes from $-\infty(1 - i\epsilon)$ to τ (where the correlation function is evaluated) and back to $-\infty(1+i\epsilon)$; see panel (a) of Fig. C.1. By expanding the exponentials in (C.72), we can compute the correlation function perturbatively in $\hat{H}_{int}$. For example, at leading order (tree level), we find

$$\langle\hat{Q}(\tau)\rangle = -i \int_{-\infty^-}^{\tau} d\tau' \, \langle 0| [\hat{Q}^I(\tau), \hat{H}_{int}^I(\tau')] |0\rangle . \tag{C.73}$$

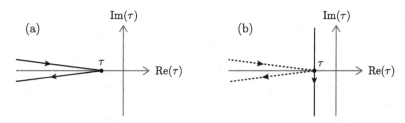

Fig. C.1 Illustration of the contours of integration in the *in-in* formalism.

This will be evaluated by using (C.64). We can use Feynman diagrams to organize higher orders in the power series, drawing interaction vertices for every power of H_{int}. Notice that in the *in-in* formalism there is *no time flow*: each vertex insertion is associated not just with momentum conservation, but also with a time integral.

C.3.2 Equilateral bispectra

As a concrete example, we will now use the *in-in* formalism to compute three-point correlation functions for the cubic Goldstone Lagrangian, which in the decoupling limit and at leading order in derivatives takes the form

$$\mathcal{L}_\pi = \frac{M_{\rm pl}^2 |\dot{H}|}{c_s^2} \left[\left(\dot{\pi}^2 - c_s^2 \frac{(\partial_i \pi)^2}{a^2} \right) - (1 - c_s^2) \left(\frac{\dot{\pi}(\partial_i \pi)^2}{a^2} + \frac{A}{c_s^2} \dot{\pi}^3 \right) \right]. \quad (C.74)$$

Our goal will be to compute the non-Gaussianities associated with the operators $\dot{\pi}^3$ and $\dot{\pi}(\partial_i \pi)^2$; specifically, we consider the bispectrum of Goldstone fluctuations after horizon crossing,

$$B_\pi(k_1, k_2, k_3) \equiv \lim_{\tau \to 0} \langle \hat{\pi}_{k_1}(\tau) \hat{\pi}_{k_2}(\tau) \hat{\pi}_{k_3}(\tau) \rangle', \quad (C.75)$$

where the prime indicates that an overall delta-function, $(2\pi)^3 \delta(k_1 + k_2 + k_3)$, has been omitted. Using (C.73) this becomes

$$B_\pi(k_1, k_2, k_3) = -i \int_{-\infty^-}^0 d\tau \langle [\hat{\pi}_{k_1}(0) \hat{\pi}_{k_2}(0) \hat{\pi}_{k_3}(0), \hat{H}_{int}(\tau)] \rangle', \quad (C.76)$$

where we have suppressed the index I on the interaction picture operators. At leading order, $H_{int} = -\int d^3x \, a^4 \mathcal{L}_\pi^{(3)}$, with $\mathcal{L}_\pi^{(3)}$ given by the cubic part of (C.74). Performing Wick contractions, we find

$$B_\pi(k_1, k_2, k_3) = M_{\rm pl}^2 |\dot{H}| \left(\frac{1}{c_s^2} - 1 \right) \mathrm{Re}\left[\pi_{k_1}^{(o)} \pi_{k_2}^{(o)} \pi_{k_3}^{(o)} \mathcal{I}(k_1, k_2, k_3) \right] + \text{perms.}, \quad (C.77)$$

where $\mathcal{I} \equiv \mathcal{I}_{\dot{\pi}(\partial_i \pi)^2} + \mathcal{I}_{\dot{\pi}^3}$, with

$$\mathcal{I}_{\dot{\pi}(\partial_i \pi)^2} \equiv \int_{-\infty^-}^0 \frac{d\tau}{H\tau} (\pi_{k_1}^*)' \pi_{k_2}^* \pi_{k_3}^* (k_1 \cdot k_2), \quad (C.78)$$

$$\mathcal{I}_{\dot{\pi}^3} \equiv \frac{A}{c_s^2} \int_{-\infty^-}^0 \frac{d\tau}{H\tau} (\pi_{k_1}^*)' (\pi_{k_2}^*)' (\pi_{k_3}^*)', \quad (C.79)$$

and primes now denote derivatives with respect to conformal time τ. We insert Eq. (C.65) for the mode functions. The resulting integrals converge due to the $i\epsilon$ in the integration limit $-\infty^- = -\infty(1 - i\epsilon)$. Since there are no poles in the complex τ plane, we can deform the contour of integration as shown in panel (b) of Fig. C.1. This contour corresponds to the Wick rotation $\tau \to i\tau$. The integrals in (C.78) and (C.79) can then be performed analytically (e.g. by using Mathematica).

The bispectrum for the curvature fluctuations $\mathcal{R}$ is obtained by a simple rescaling: $B_\mathcal{R}(k_1, k_2, k_3) = -H^3 B_\pi(k_1, k_2, k_3)$. We can write the bispectrum in terms of an amplitude f_{NL} and a shape function $\mathcal{S}(k_1, k_2, k_3)$:

$$\frac{(k_1 k_2 k_3)^2}{\Delta_\mathcal{R}^4} B_\mathcal{R}(k_1, k_2, k_3) = \frac{18}{5} f_{\mathrm{NL}} \, \mathcal{S}(k_1, k_2, k_3) \,, \tag{C.80}$$

where $\mathcal{S}(k, k, k) \equiv 1$. The shape functions associated with the two interactions in (C.74) are

$$\mathcal{S}_{\dot{\mathcal{R}}(\partial_i \mathcal{R})^2} = \frac{\hat{k}_1^2 - \hat{k}_2^2 - \hat{k}_3^2}{\hat{k}_1 \hat{k}_2 \hat{k}_3} \left(-1 + \sum_{i>j} \frac{\hat{k}_i \hat{k}_j}{9} + \frac{\hat{k}_1 \hat{k}_2 \hat{k}_3}{27} \right) + \text{perms.} \,, \tag{C.81}$$

$$\mathcal{S}_{\dot{\mathcal{R}}^3} = \hat{k}_1 \hat{k}_2 \hat{k}_3 \,, \tag{C.82}$$

where $\hat{k}_i \equiv k_i/K$ and $K \equiv \frac{1}{3}(k_1 + k_2 + k_3)$. The amplitudes associated with these two shapes are

$$f_{\mathrm{NL}}^{\dot{\mathcal{R}}(\partial_i \mathcal{R})^2} = -\frac{85}{324} \left(\frac{1}{c_s^2} - 1 \right) \,, \tag{C.83}$$

$$f_{\mathrm{NL}}^{\dot{\mathcal{R}}^3} = -\frac{5A}{81} \left(\frac{1}{c_s^2} - 1 \right) \,. \tag{C.84}$$

Both $\mathcal{S}_{\dot{\mathcal{R}}(\partial_i \mathcal{R})^2}$ and $\mathcal{S}_{\dot{\mathcal{R}}^3}$ peak in the equilateral momentum configuration, $k_1 = k_2 = k_3$, and the shapes are well-approximated by the equilateral template (1.67). However, the shapes are not identical. In fact, for $3 \lesssim A \lesssim 4$ the combined shape is nearly orthogonal to both the equilateral shape and the local shape [115], as measured by an inner product defined in [126]. The bispectrum in this parameter regime has been named the "orthogonal shape" and is highly correlated with the orthogonal template (1.68).

C.4 Multiple fields: entropy perturbations

As explained in Section C.1, every model of inflation, regardless of its microphysical origin, contains at least one physical scalar degree of freedom: this is the comoving curvature perturbation $\mathcal{R}(t, \boldsymbol{x})$ of (C.5) in comoving gauge, or the Goldstone boson $\pi(t, \boldsymbol{x})$ in spatially flat gauge. Inflationary scenarios involving more than one fluctuating scalar degree of freedom are known as *multi-field models*.

While single-field models appear simple and economical, there are several reasons to examine multi-field models. First, a priori notions of simplicity have had very limited success in the development of the Standard Model of particle physics, and it seems prudent to examine all self-consistent inflationary scenarios, rather than just those with the minimal field content. Second, as emphasized in Section 4.4, ultraviolet completions of inflationary effective theories very often involve scalar fields in addition to the inflaton candidate. Third, as we shall see, multi-field models make distinctive predictions that differentiate them from single-field scenarios.

The presence of fluctuations in addition to those of the Goldstone boson π makes the analysis of the primordial perturbations in multi-field inflation much more intricate than in the single-field case. In this section, we summarize the new features that arise for perturbations in multi-field inflation.[6]

C.4.1 Background evolution

At lowest order in derivatives, the most general Einstein-frame action describing the evolution of N real scalar fields ϕ^I is

$$S = \int \mathrm{d}^4 x \sqrt{-g} \left[\frac{M_{\mathrm{pl}}^2}{2} R - \frac{1}{2} G_{IJ}(\phi) \partial^\mu \phi^I \partial_\mu \phi^J - V(\phi^I) \right] , \qquad (C.85)$$

where G_{IJ} is the metric on field space.[7]

We first consider the evolution of spatially homogeneous fields $\phi^I(t)$ in a flat FRW background. The Friedmann equations are

$$3 M_{\mathrm{pl}}^2 H^2 = \frac{1}{2} G_{IJ} \dot\phi^I \dot\phi^J + V \quad \text{and} \quad M_{\mathrm{pl}}^2 \dot H = -\frac{1}{2} G_{IJ} \dot\phi^I \dot\phi^J , \qquad (C.86)$$

while the equations of motion for the scalar fields are

$$\mathcal{D}_t \dot\phi^I + 3 H \dot\phi^I + G^{IJ} V_{,J} = 0 , \qquad (C.87)$$

where $V_{,J} \equiv \partial_J V$, and we have defined the covariant derivative of a vector A^I as

$$\mathcal{D}_t A^I \equiv \dot A^I + \Gamma^I_{JK} \dot\phi^J A^K , \qquad (C.88)$$

with $\Gamma^I_{JK} \equiv \frac{1}{2} G^{IL}(\partial_J G_{KL} + \partial_K G_{JL} - \partial_L G_{JK})$. Just as in single-field inflation, we have

$$\frac{\ddot a}{a} = H^2 (1 - \varepsilon) , \qquad \varepsilon \equiv -\frac{\dot H}{H^2} = \frac{\frac{1}{2} G_{IJ} \dot\phi^I \dot\phi^J}{M_{\mathrm{pl}}^2 H^2} . \qquad (C.89)$$

A necessary condition for prolonged inflation is that the steepest unstable direction is sufficiently flat to allow $\varepsilon < 1$ over multiple e-folds. For definitions of slow-roll parameters and slow-roll hierarchies in multi-field inflation, see e.g. [896, 899, 900].

---

[6] A formalism for multi-field inflation in theories with two-derivative actions and arbitrary field space metrics was developed in [896, 897], while extensions to the $P(X)$ theories discussed in Section 2.2.3 were obtained in [753]. We will follow the treatment of [753, 898].

[7] Although the kinetic term in (C.85) cannot be put in the canonical form $G_{IJ} = \delta_{IJ}$ everywhere in field space (except in the trivial case where the Riemann curvature derived from G_{IJ} vanishes), in the context of inflation the phrase "nontrivial kinetic term" is generally reserved for kinetic terms involving more than two derivatives.

C.4.2 Quadratic action

To describe fluctuations around a background trajectory[8] $\bar{\phi}^I(t)$, it is convenient to work in the spatially flat gauge (B.25), in which the perturbations are parameterized by scalar field fluctuations on flat hypersurfaces,

$$\phi^I(t, \boldsymbol{x}) \equiv \bar{\phi}^I(t) + Q^I(t, \boldsymbol{x}) \,. \tag{C.90}$$

The basic dynamical objects are then the field perturbations $Q^I \equiv \delta\phi^I$. From now on, we will drop the bar over $\bar{\phi}^I(t)$.

We will restrict ourselves to linearized perturbations, whose evolution is captured by the quadratic action [753, 897, 901]

$$S_Q^{(2)} = \int \mathrm{d}t\, \mathrm{d}^3x\, a^3 \left[G_{IJ} \mathcal{D}_t Q^I \mathcal{D}_t Q^J - \frac{1}{a^2} G_{IJ} \partial_i Q^I \partial^i Q^J - M_{IJ} Q^I Q^J \right] \,, \tag{C.91}$$

where $\mathcal{D}_t$ was defined in (C.88), and the mass matrix is given by

$$M_{IJ} \equiv V_{;IJ} - R_{IKLJ} \dot{\phi}^K \dot{\phi}^L - \frac{1}{a^3} \mathcal{D}_t \left[\frac{a^3}{H} \dot{\phi}_I \dot{\phi}_J \right] \,, \tag{C.92}$$

where $V_{;IJ} \equiv \nabla_I \nabla_J V$ and R_{IKLJ} is the Riemann tensor computed from G_{IJ}. The equations of motion that follow from (C.91) are

$$\mathcal{D}_t \mathcal{D}_t Q^I + 3H \mathcal{D}_t Q^I + \left[\frac{k^2}{a^2} \delta^I{}_J + M^I{}_J \right] Q^J = 0 \,. \tag{C.93}$$

C.4.3 Entropic modes

It is illuminating to express the perturbations at each point along the trajectory $\phi^I(t)$ in terms of a natural basis at that point. Let us define [896, 897]

$$\phi^I_{(1)} \equiv \dot{\phi}^I \quad \text{and} \quad \phi^I_{(n)} \equiv \mathcal{D}_t^{(n-1)} \dot{\phi}^I \,, \quad \text{for} \quad 2 \leq n \leq N. \tag{C.94}$$

An orthonormal basis $\{e^I_{(1)}, \ldots, e^I_{(N)}\}$ can be obtained from (C.94) by the Gram–Schmidt procedure. The first vector $e^I_{(1)}$ points along the instantaneous trajectory, i.e. $e^I_{(1)}$ is parallel to the field velocity, while $e^I_{(2)}$ is parallel to the component of acceleration that is perpendicular to the velocity, which we term the "transverse acceleration." We now expand the perturbations in this orthonormal "kinematic" basis. The adiabatic perturbation is

$$Q_\sigma \equiv e_{\sigma I} Q^I \,, \tag{C.95}$$

with $e^I_\sigma \equiv e^I_{(1)} \equiv \dot{\phi}^I / \dot{\sigma}$, where $\dot{\sigma}^2 \equiv G_{IJ} \dot{\phi}^I \dot{\phi}^J$. The adiabatic perturbation Q_σ is a measure of the comoving curvature perturbation

$$\mathcal{R} \equiv \frac{H}{\dot{\sigma}} Q_\sigma \,, \tag{C.96}$$

---

[8] The index I runs over the N independent scalar fields in the theory, cf. (C.85), and should not be confused with the interaction picture superscript I of Section C.3.

and corresponds to a shift along the background trajectory; Q_σ is proportional to the field π in Appendix B. The $N-1$ entropic perturbations, in contrast, are fluctuations transverse to the background trajectory, i.e. in the directions spanned by $\{e^I_{(2)}, \ldots, e^I_{(N)}\}$. A preferred role is played by the "first entropic mode,"

$$Q_s \equiv e_{sI} Q^I , \qquad (C.97)$$

with $e^I_s \equiv e^I_{(2)}$. The first entropic mode measures the fluctuation along the direction of the component of the acceleration that is perpendicular to the velocity.

Let us now define a dimensionless parameter $\eta_\perp$ that measures the size of the coupling between the adiabatic mode and the first entropic mode:

$$\eta_\perp \equiv -\frac{V_{,s}}{H\dot\sigma} , \qquad (C.98)$$

where $V_{,s} \equiv e^I_s V_{,I}$ (and similarly we write $V_{,\sigma} \equiv e^I_\sigma V_{,I}$). The parameter $\eta_\perp$ measures the acceleration transverse to the instantaneous trajectory in field space. The equation of motion of the adiabatic mode then takes the form

$$\ddot{Q}_\sigma + 3H\dot{Q}_\sigma + \left(\frac{k^2}{a^2} + m_\sigma^2\right) Q_\sigma = \frac{d}{dt}(2H\eta_\perp Q_s) - \left(\frac{\dot{H}}{H} + \frac{V_{,\sigma}}{\dot\sigma}\right) 2H\eta_\perp Q_s , \quad (C.99)$$

where the mass of the adiabatic mode is

$$\frac{m_\sigma^2}{H^2} \equiv -\frac{3}{2}\tilde\eta - \frac{1}{4}\tilde\eta^2 - \frac{1}{2}\varepsilon\tilde\eta - \frac{1}{2}\frac{\dot{\tilde\eta}}{H} \approx -\frac{3}{2}\tilde\eta , \qquad (C.100)$$

with ε and $\tilde\eta$ the Hubble slow-roll parameters defined in (1.4) and (1.30), respectively. From (C.99) we learn that at the level of the quadratic action, the instantaneous adiabatic mode Q_σ couples only to the first entropic mode Q_s; the other $N-2$ entropic modes are decoupled from Q_σ.

To close the system of equations, we need the evolution equations for the entropic modes. We present only the equation of motion for the first entropic mode, Q_s:

$$\ddot{Q}_s + 3H\dot{Q}_s + \left(\frac{k^2}{a^2} + m_s^2\right) Q_s = 4\frac{H\eta_\perp}{\dot\sigma}\frac{k^2}{a^2}\Psi + \cdots , \qquad (C.101)$$

where Ψ is the metric perturbation in the Newtonian gauge (C.6), the ellipsis denotes couplings to the remaining entropic modes, and

$$\frac{m_s^2}{H^2} \equiv \frac{V_{;ss}}{H^2} - \eta_\perp^2 - \frac{\dot\sigma^2}{H^2}R_{s\sigma\sigma s} . \qquad (C.102)$$

The first entropic mode will be quantum-mechanically activated during inflation if $m_s^2 < \frac{3}{2}H^2$. For $m_s^2 > \frac{3}{2}H^2$, the vacuum fluctuations do not get amplified, and fluctuations on large scales are exponentially suppressed. In the superhorizon limit, we find

$$\ddot{Q}_s + 3H\dot{Q}_s + m_s^2 Q_s \approx 0 + \cdots , \tag{C.103}$$

implying that superhorizon evolution of the first entropic mode is not sourced by the adiabatic mode, but can be sourced by the higher entropic modes (represented by the ellipsis).

One might suspect that the higher entropic modes in the directions $e_{(n)}^I$, $n = 3, \cdots, N$, have subleading effects on the curvature perturbations; because they are instantaneously decoupled from the adiabatic mode, they can source curvature only through the bottleneck of the first entropic mode, Q_s. In fact, allowing modest violations of slow-roll, corresponding to $\eta_\perp \sim 1$, readily leads to trajectories in which the effect of Q_s on the curvature perturbations is negligible, while the effects of the higher entropic modes are large [265]; an example of this phenomenon is discussed in Section 5.1.6.

Two-field example

To develop intuition for the couplings of entropic modes, we consider a simple theory of two fields with field space $\mathbb{R}^2$. Figure C.2 shows the decomposition of an arbitrary field perturbation into an adiabatic component and an entropic component. Let θ be the angle between the direction of the instantaneous adiabatic field and the direction of one of the field perturbations, say $\delta\phi^1$. The parameter $\eta_\perp$ then measures the rate of change of this angle [902]:

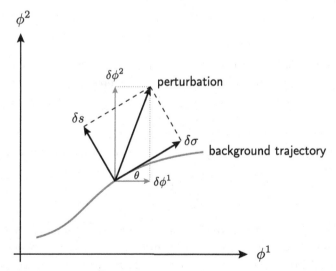

Fig. C.2 Decomposition of an arbitrary field perturbation into an adiabatic component ($\delta\sigma$) and an entropic component (δs).

$$\eta_\perp = \frac{\dot{\theta}}{H} \cdot \tag{C.104}$$

Hence, $\eta_\perp$ is large, and multi-field effects are significant, only when the background trajectory is strongly bent.

The time evolution of the curvature perturbation in this case is found to be [902]

$$\dot{\mathcal{R}} = \frac{H}{\dot{H}} \frac{k^2}{a^2} \Psi + 2 \frac{H^2}{\dot{\sigma}} \eta_\perp Q_s \,, \tag{C.105}$$

which in the superhorizon limit takes the simple form

$$\dot{\mathcal{R}} \approx 2 \frac{H^2}{\dot{\sigma}} \eta_\perp Q_s \,. \tag{C.106}$$

Entropic perturbations are, of course, absent in single-field inflation, $Q_s \equiv 0$, and $\mathcal{R}$ is correspondingly conserved on large scales. In multi-field inflation, however, the first entropic mode can source the superhorizon evolution of the curvature perturbation, via the coupling (C.105).

C.4.4 Transfer matrix

The rate at which the inflaton trajectory bends in field space, as captured by the parameter $\eta_\perp$, controls the couplings between adiabatic and entropic perturbations, and hence the possibility of deviations of the late-time curvature perturbation from the single-field result. A particularly simple class of multi-field models are those in which a *slow-turn approximation*,

$$\eta_\perp \ll 1 \,, \tag{C.107}$$

is valid at an appropriate stage.

Consider a two-field model in which the slow-turn approximation is applicable at the time of horizon crossing of some modes of interest – for example, the modes determining the observable multipoles of the CMB – but may be violated later in the evolution. In this case the adiabatic and entropic modes evolve independently around horizon crossing, and thereafter the entropic mode sources the adiabatic mode. To analyze this special case, it is convenient to define the rescaled entropy perturbation,

$$\mathcal{S} \equiv \frac{H}{\dot{\sigma}} Q_s \,, \tag{C.108}$$

so that the modes $\mathcal{R}$ and $\mathcal{S}$ have equal amplitudes at horizon crossing, $\mathcal{R}_\star = \mathcal{S}_\star$. The evolution after horizon exit is determined by the coupled mode equations (C.106) and (C.103), which can be written in the instructive schematic form [903]

$$\dot{\mathcal{R}} = \alpha H \mathcal{S} \,, \qquad \dot{\mathcal{S}} = \beta H \mathcal{S} \,, \tag{C.109}$$

where α and β are dimensionless time-dependent functions that depend on the background solution. Integrating (C.109) leads to a "transfer matrix" relationship between the large scale fluctuations at late times and the fluctuations at horizon crossing:

$$\begin{pmatrix} \mathcal{R} \\ \mathcal{S} \end{pmatrix} = \begin{pmatrix} 1 & T_{\mathcal{RS}} \\ 0 & T_{\mathcal{SS}} \end{pmatrix} \begin{pmatrix} \mathcal{R}_\star \\ \mathcal{S}_\star \end{pmatrix} , \tag{C.110}$$

where the model-dependent coefficients $T_{\mathcal{RS}}$ and $T_{\mathcal{SS}}$ are given by the formal integrals [903]

$$T_{\mathcal{SS}}(t_\star, t) = \exp\left(\int_{t_\star}^t \beta(t')\, H(t')\mathrm{d}t' \right) , \tag{C.111}$$

$$T_{\mathcal{RS}}(t_\star, t) = \int_{t_\star}^t \alpha(t')\, T_{\mathcal{SS}}(t_\star, t')\, H(t')\mathrm{d}t' . \tag{C.112}$$

The observed power spectra can then be written in terms of the initial power spectra as

$$\Delta_\mathcal{R}^2 = (1 + T_{\mathcal{RS}}^2)\Delta_{\mathcal{R}\star}^2 , \tag{C.113}$$

$$\Delta_\mathcal{S}^2 = T_{\mathcal{SS}}^2\Delta_{\mathcal{R}\star}^2 , \tag{C.114}$$

$$\Delta_{\mathcal{RS}}^2 = T_{\mathcal{RS}}T_{\mathcal{SS}}\Delta_{\mathcal{R}\star}^2 . \tag{C.115}$$

The cross-correlation is often written in terms of a correlation angle

$$\cos\delta \equiv \frac{\Delta_{\mathcal{RS}}^2}{\sqrt{\Delta_\mathcal{R}^2 \Delta_\mathcal{S}^2}} = \frac{T_{\mathcal{RS}}}{\sqrt{1 + T_{\mathcal{RS}}^2}} , \tag{C.116}$$

so that

$$\Delta_\mathcal{R}^2 = \frac{\Delta_{\mathcal{R}\star}^2}{\sin^2\delta} . \tag{C.117}$$

When the transfer matrix (C.110) is applicable, entropy perturbations lead to suppression of the tensor-to-scalar ratio,

$$\frac{r}{r_\star} = \sin^2\delta < 1 , \tag{C.118}$$

and to a modification of the single-field slow-roll consistency relation (C.59) [903, 904],

$$-\frac{r}{8n_t} = \sin^2\delta < 1 . \tag{C.119}$$

A cautionary remark about the transfer matrix approach is necessary. The assumption of a slow turn at horizon crossing underpins a key qualitative feature of (C.109), which is that the adiabatic mode does not source the entropic mode. This assumption is answerable for the specific conclusions (C.117), (C.118), and (C.119), namely that entropic contributions *increase* the scalar power and correspondingly diminish r. More general evolution in which $\eta_\perp \gtrsim 1$ during horizon crossing can violate each of these inequalities [265].

References

[1] **Supernova Cosmology Project** Collaboration, S. Perlmutter *et al.*, "Measurements of Omega and Lambda from 42 High Redshift Supernovae," *Astrophys.J.* **517** (1999) 565–586, arXiv:astro-ph/9812133 [astro-ph].

[2] **Supernova Search Team** Collaboration, A. Riess *et al.*, "Observational Evidence from Supernovae for an Accelerating Universe and a Cosmological Constant," *Astron.J.* **116** (1998) 1009–1038, arXiv:astro-ph/9805201 [astro-ph].

[3] **WMAP** Collaboration, D. Spergel *et al.*, "First-Year Wilkinson Microwave Anisotropy Probe (WMAP) Observations: Determination of Cosmological Parameters," *Astrophys.J.Suppl.* **148** (2003) 175–194, arXiv:astro-ph/0302209 [astro-ph].

[4] **WMAP** Collaboration, D. Spergel *et al.*, "Wilkinson Microwave Anisotropy Probe (WMAP) Three-Year Results: Implications for Cosmology," *Astrophys.J.Suppl.* **170** (2007) 377, arXiv:astro-ph/0603449 [astro-ph].

[5] **WMAP** Collaboration, E. Komatsu *et al.*, "Five-Year Wilkinson Microwave Anisotropy Probe (WMAP) Observations: Cosmological Interpretation," *Astrophys.J.Suppl.* **180** (2009) 330–376, arXiv:0803.0547 [astro-ph].

[6] **WMAP** Collaboration, E. Komatsu *et al.*, "Seven-Year Wilkinson Microwave Anisotropy Probe (WMAP) Observations: Cosmological Interpretation," *Astrophys.J.Suppl.* **192** (2011) 18, arXiv:1001.4538 [astro-ph.CO].

[7] **WMAP** Collaboration, G. Hinshaw *et al.*, "Nine-Year Wilkinson Microwave Anisotropy Probe (WMAP) Observations: Cosmological Parameter Results," arXiv:1212.5226 [astro-ph.CO].

[8] **Planck** Collaboration, P. Ade *et al.*, "Planck 2013 Results. XVI. Cosmological Parameters," arXiv:1303.5076 [astro-ph.CO].

[9] **Planck** Collaboration, P. Ade *et al.*, "Planck 2013 Results. XXII. Constraints on Inflation," arXiv:1303.5082 [astro-ph.CO].

[10] **Planck** Collaboration, P. Ade *et al.*, "Planck 2013 Results. XXIV. Constraints on Primordial Non-Gaussianity," arXiv:1303.5084 [astro-ph.CO].

[11] **SDSS** Collaboration, K. Abazajian *et al.*, "The Seventh Data Release of the Sloan Digital Sky Survey," *Astrophys.J.Suppl.* **182** (2009) 543–558, arXiv:0812.0649 [astro-ph].

[12] V. Mukhanov and G. Chibisov, "Quantum Fluctuation and Nonsingular Universe. (In Russian)," *JETP Lett.* **33** (1981) 532–535.

[13] G. Chibisov and V. Mukhanov, "Galaxy Formation and Phonons," *Mon.Not.Roy.Astron.Soc.* **200** (1982) 535–550.

[14] A. Guth and S. Pi, "Fluctuations in the New Inflationary Universe," *Phys.Rev.Lett.* **49** (1982) 1110–1113.

[15] S. Hawking, "The Development of Irregularities in a Single Bubble Inflationary Universe," *Phys.Lett.* **B115** (1982) 295.

[16] A. Starobinsky, "Dynamics of Phase Transition in the New Inflationary Universe Scenario and Generation of Perturbations," *Phys.Lett.* **B117** (1982) 175–178.

[17] J. Bardeen, P. Steinhardt, and M. Turner, "Spontaneous Creation of Almost Scale-Free Density Perturbations in an Inflationary Universe," *Phys.Rev.* **D28** (1983) 679.

[18] A. Guth, "The Inflationary Universe: A Possible Solution to the Horizon and Flatness Problems," *Phys.Rev.* **D23** (1981) 347–356.

[19] A. Linde, "A New Inflationary Universe Scenario: A Possible Solution of the Horizon, Flatness, Homogeneity, Isotropy and Primordial Monopole Problems," *Phys.Lett.* **B108** (1982) 389–393.

[20] A. Albrecht and P. Steinhardt, "Cosmology for Grand Unified Theories with Radiatively Induced Symmetry Breaking," *Phys.Rev.Lett.* **48** (1982) 1220–1223.

[21] Z. Bern, "Perturbative Quantum Gravity and its Relation to Gauge Theory," *Living Rev.Rel.* **5** (2002) 5, arXiv:gr-qc/0206071 [gr-qc].

[22] D. Baumann and L. McAllister, "Advances in Inflation in String Theory," *Ann.Rev.Nucl.Part.Sci.* **59** (2009) 67–94, arXiv:0901.0265 [hep-th].

[23] L. McAllister and E. Silverstein, "String Cosmology: A Review," *Gen.Rel.Grav.* **40** (2008) 565–605, arXiv:0710.2951 [hep-th].

[24] C. Burgess, "Lectures on Cosmic Inflation and its Potential Stringy Realizations," *Class.Quant.Grav.* **24** (2007) S795, arXiv:0708.2865 [hep-th].

[25] R. Kallosh, "On Inflation in String Theory," *Lect.Notes Phys.* **738** (2008) 119–156, arXiv:hep-th/0702059 [HEP-TH].

[26] **CMBPol Study Team** Collaboration, D. Baumann *et al.*, "CMBPol Mission Concept Study: Probing Inflation with CMB Polarization," *AIP Conf.Proc.* **1141** (2009) 10–120, arXiv:0811.3919 [astro-ph].

[27] D. Baumann, "TASI Lectures on Inflation," arXiv:0907.5424 [hep-th].

[28] A. Linde, "Inflation and String Cosmology," *Prog.Theor.Phys.Suppl.* **163** (2006) 295–322, arXiv:hep-th/0503195 [hep-th].

[29] F. Quevedo, "Lectures on String/Brane Cosmology," *Class.Quant.Grav.* **19** (2002) 5721–5779, arXiv:hep-th/0210292 [hep-th].

[30] J. Erdmenger, *String Cosmology: Modern String Theory Concepts from the Cosmic Structure.* Wiley, New York, NY, 2009.

[31] M. Cicoli and F. Quevedo, "String Moduli Inflation: An Overview," *Class.Quant.Grav.* **28** (2011) 204001, arXiv:1108.2659 [hep-th].

[32] C. Burgess, M. Cicoli, and F. Quevedo, "String Inflation After Planck 2013," arXiv:1306.3512 [hep-th].

[33] E. Silverstein, "Les Houches Lectures on Inflationary Observables and String Theory," arXiv:1311.2312 [hep-th].

[34] A. Linde, "Chaotic Inflation," *Phys.Lett.* **B129** (1983) 177–181.

[35] L. McAllister, E. Silverstein, and A. Westphal, "Gravity Waves and Linear Inflation from Axion Monodromy," *Phys.Rev.* **D82** (2010) 046003, arXiv:0808.0706 [hep-th].

[36] K. Freese, J. Frieman, and A. Olinto, "Natural Inflation with Pseudo-Nambu-Goldstone Bosons," *Phys.Rev.Lett.* **65** (1990) 3233–3236.

[37] R. Flauger, L. McAllister, E. Pajer, A. Westphal, and G. Xu, "Oscillations in the CMB from Axion Monodromy Inflation," *JCAP* **1006** (2010) 009, arXiv:0907.2916 [hep-th].

[38] R. Flauger and E. Pajer, "Resonant Non-Gaussianity," *JCAP* **1101** (2011) 017, arXiv:1002.0833 [hep-th].

[39] E. Silverstein and D. Tong, "Scalar Speed Limits and Cosmology: Acceleration from D-cceleration," *Phys.Rev.* **D70** (2004) 103505, arXiv:hep-th/0310221 [hep-th].

[40] C. Armendariz-Picon, T. Damour, and V. Mukhanov, "k-Inflation," *Phys.Lett.* **B458** (1999) 209–218, arXiv:hep-th/9904075 [hep-th].

[41] M. Alishahiha, E. Silverstein, and D. Tong, "DBI in the Sky," *Phys.Rev.* **D70** (2004) 123505, arXiv:hep-th/0404084 [hep-th].

[42] S. Kachru *et al.*, "Towards Inflation in String Theory," *JCAP* **0310** (2003) 013, arXiv:hep-th/0308055 [hep-th].

[43] D. Baumann, A. Dymarsky, S. Kachru, I. Klebanov, and L. McAllister, "D3-brane Potentials from Fluxes in AdS/CFT," *JHEP* **1006** (2010) 072, arXiv:1001.5028 [hep-th].

[44] M. Cicoli, C. Burgess, and F. Quevedo, "Fibre Inflation: Observable Gravity Waves from IIB String Compactifications," *JCAP* **0903** (2009) 013, arXiv:0808.0691 [hep-th].

[45] M. Douglas and S. Kachru, "Flux Compactification," *Rev.Mod.Phys.* **79** (2007) 733–796, arXiv:hep-th/0610102 [hep-th].

[46] F. Denef and M. Douglas, "Computational Complexity of the Landscape. I.," *Annals Phys.* **322** (2007) 1096–1142, arXiv:hep-th/0602072 [hep-th].

[47] T. Banks, M. Dine, P. Fox, and E. Gorbatov, "On the Possibility of Large Axion Decay Constants," *JCAP* **0306** (2003) 001, arXiv:hep-th/0303252 [hep-th].

[48] P. Svrcek and E. Witten, "Axions In String Theory," *JHEP* **0606** (2006) 051, arXiv:hep-th/0605206 [hep-th].

[49] D. Marsh, L. McAllister, and T. Wrase, "The Wasteland of Random Supergravities," *JHEP* **1203** (2012) 102, arXiv:1112.3034 [hep-th].

[50] P. Creminelli, M. Luty, A. Nicolis, and L. Senatore, "Starting the Universe: Stable Violation of the Null Energy Condition and Non-Standard Cosmologies," *JHEP* **0612** (2006) 080, arXiv:hep-th/0606090 [hep-th].

[51] C. Cheung, P. Creminelli, A. Fitzpatrick, J. Kaplan, and L. Senatore, "The Effective Field Theory of Inflation," *JHEP* **0803** (2008) 014, arXiv:0709.0293 [hep-th].

[52] C. Lineweaver, "Inflation and the Cosmic Microwave Background," arXiv:astro-ph/0305179 [astro-ph].

[53] P. Dirac, "The Relation between Mathematics and Physics," *Proc.R.Soc.Edinburgh* **59** (1939) .

[54] S. Weinberg, *The Quantum Theory of Fields. Vol. 2: Modern Applications.* Cambridge University Press, 2005.

[55] S. Weinberg, "Adiabatic Modes in Cosmology," *Phys.Rev.* **D67** (2003) 123504, arXiv:astro-ph/0302326 [astro-ph].

[56] K. Hinterbichler, L. Hui, and J. Khoury, "Conformal Symmetries of Adiabatic Modes in Cosmology," arXiv:1203.6351 [hep-th].

[57] S. Dodelson, *Modern Cosmology.* Academic Press, 2003.

[58] V. Mukhanov, *Physical Foundations of Cosmology.* Cambridge University Press, 2005.

[59] T. Jacobson, "Introduction to Quantum Fields in Curved Spacetime and the Hawking Effect," arXiv:gr-qc/0308048 [gr-qc].

[60] S. Hollands and R. Wald, "An Alternative to Inflation," *Gen.Rel.Grav.* **34** (2002) 2043–2055, arXiv:gr-qc/0205058 [gr-qc].

[61] A. Starobinsky, "Relict Gravitation Radiation Spectrum and Initial State of the Universe. (In Russian)," *JETP Lett.* **30** (1979) 682–685.

[62] W. Hu, "Lecture Notes on CMB Theory: From Nucleosynthesis to Recombination," arXiv:0802.3688 [astro-ph].

[63] A. Penzias and R. Wilson, "A Measurement of Excess Antenna Temperature at 4080-Mc/s," *Astrophys.J.* **142** (1965) 419–421.

[64] D. Fixsen, "The Temperature of the Cosmic Microwave Background," *Astrophys.J.* **707** (2009) 916–920, arXiv:0911.1955 [astro-ph.CO].

[65] A. Guth and S.-Y. Pi, "The Quantum Mechanics of the Scalar Field in the New Inflationary Universe," *Phys.Rev.* **D32** (1985) 1899–1920.

[66] D. Polarski and A. Starobinsky, "Semiclassicality and Decoherence of Cosmological Perturbations," *Class.Quant.Grav.* **13** (1996) 377–392, arXiv:gr-qc/9504030 [gr-qc].

[67] A. Perez, H. Sahlmann, and D. Sudarsky, "On the Quantum Origin of the Seeds of Cosmic Structure," *Class.Quant.Grav.* **23** (2006) 2317–2354, arXiv:gr-qc/0508100 [gr-qc].

[68] C. Burgess, R. Holman, and D. Hoover, "Decoherence of Inflationary Primordial Fluctuations," *Phys.Rev.* **D77** (2008) 063534, arXiv:astro-ph/0601646 [astro-ph].

[69] C. Kiefer and D. Polarski, "Why do cosmological perturbations look classical to us?," *Adv.Sci.Lett.* **2** (2009) 164–173, arXiv:0810.0087 [astro-ph].

[70] U. Seljak and M. Zaldarriaga, "A Line-of-Sight Integration Approach to Cosmic Microwave Background Anisotropies," *Astrophys.J.* **469** (1996) 437–444, arXiv:astro-ph/9603033 [astro-ph].

[71] A. Lewis, A. Challinor, and A. Lasenby, "Efficient Computation of CMB Anisotropies in Closed FRW Models," *Astrophys.J.* **538** (2000) 473–476, arXiv:astro-ph/9911177 [astro-ph].

[72] M. Rees, "Polarization and Spectrum of the Primeval Radiation in an Anisotropic Universe," *Astrophys.J.* **153** (July, 1968) L1.

[73] A. Polnarev, "Polarization and Anisotropy Induced in the Microwave Background by Cosmological Gravitational Waves," *Soviet Astronomy* **29** (Dec., 1985) 607–613.

[74] W. Hu and M. White, "A CMB Polarization Primer," *New Astron.* **2** (1997) 323, arXiv:astro-ph/9706147 [astro-ph].

[75] S. Chandrasekhar, *Radiative Transfer.* Courier Dover Publications, 2013.

[76] E. Newman and R. Penrose, "Note on the Bondi-Metzner-Sachs Group," *J.Math.Phys.* **7** (1966) 863–870.

[77] M. Kamionkowski, A. Kosowsky, and A. Stebbins, "Statistics of Cosmic Microwave Background Polarization," *Phys.Rev.* **D55** (1997) 7368–7388, arXiv:astro-ph/9611125 [astro-ph].

[78] M. Zaldarriaga and U. Seljak, "An All-Sky Analysis of Polarization in the Microwave Background," *Phys.Rev.* **D55** (1997) 1830–1840, arXiv:astro-ph/9609170 [astro-ph].

[79] G. Lemaitre, "The Expanding Universe," *Mon.Not.Roy.Astron.Soc.* **91** (1931) 490–501.

[80] F. Bernardeau, S. Colombi, E. Gaztanaga, and R. Scoccimarro, "Large-Scale Structure of the Universe and Cosmological Perturbation Theory," *Phys.Rept.* **367** (2002) 1–248, arXiv:astro-ph/0112551 [astro-ph].

[81] V. Springel, "The Cosmological Simulation Code GADGET-2," *Mon.Not. Roy.Astron.Soc.* **364** (2005) 1105–1134, arXiv:astro-ph/0505010[astro-ph].

[82] M. Bartelmann and P. Schneider, "Weak Gravitational Lensing," *Phys.Rept.* **340** (2001) 291–472, arXiv:astro-ph/9912508 [astro-ph].

[83] A. Refregier, "Weak Gravitational Lensing by Large-Scale Structure," *Ann.Rev.Astron.Astrophys.* **41** (2003) 645–668, arXiv:astro-ph/0307212 [astro-ph].

[84] A. Lewis and A. Challinor, "Weak Gravitational Lensing of the CMB," *Phys.Rept.* **429** (2006) 1–65, arXiv:astro-ph/0601594 [astro-ph].

[85] R. Hlozek *et al.*, "The Atacama Cosmology Telescope: a Measurement of the Primordial Power Spectrum," *Astrophys.J.* **749** (2012) 90, arXiv:1105.4887 [astro-ph.CO].

[86] A. Sanchez *et al.*, "The Clustering of Galaxies in the SDSS-III Baryon Oscillation Spectroscopic Survey: Cosmological Implications of the Large-Scale Two-Point Correlation Function," arXiv:1203.6616 [astro-ph.CO].

[87] **2dFGRS** Collaboration, S. Cole *et al.*, "The 2dF Galaxy Redshift Survey: Power Spectrum Analysis of the Final Dataset and Cosmological Implications," *Mon.Not.Roy.Astron.Soc.* **362** (2005) 505–534, arXiv:astro-ph/0501174 [astro-ph].

[88] **SDSS** Collaboration, D. Eisenstein *et al.*, "Detection of the Baryon Acoustic Peak in the Large-Scale Correlation Function of SDSS Luminous Red Galaxies," *Astrophys.J.* **633** (2005) 560–574, arXiv:astro-ph/0501171 [astro-ph].

[89] **SDSS** Collaboration, W. Percival *et al.*, "Baryon Acoustic Oscillations in the Sloan Digital Sky Survey Data Release 7 Galaxy Sample," *Mon.Not.Roy.Astron.Soc.* **401** (2010) 2148–2168, arXiv:0907.1660 [astro-ph.CO].

[90] N. Padmanabhan *et al.*, "A 2% Distance to $z = 0.35$ by Reconstructing Baryon Acoustic Oscillations: Methods and Application to the Sloan Digital Sky Survey," arXiv:1202.0090 [astro-ph.CO].

[91] L. Anderson *et al.*, "The Clustering of Galaxies in the SDSS-III Baryon Oscillation Spectroscopic Survey: Baryon Acoustic Oscillations in the Data Release 9 Spectroscopic Galaxy Sample," *Mon.Not.Roy.Astron.Soc.* **428** (2013) 1036–1054, arXiv:1203.6594 [astro-ph.CO].

[92] C. Blake *et al.*, "The WiggleZ Dark Energy Survey: Mapping the Distance-Redshift Relation with Baryon Acoustic Oscillations," *Mon.Not.Roy.Astron.Soc.* **418** (2011) 1707–1724, arXiv:1108.2635 [astro-ph.CO].

[93] F. Beutler *et al.*, "The 6dF Galaxy Survey: Baryon Acoustic Oscillations and the Local Hubble Constant," *Mon.Not.Roy.Astron.Soc.* **416** (2011) 3017–3032, arXiv:1106.3366 [astro-ph.CO].

[94] D. Weinberg *et al.*, "Observational Probes of Cosmic Acceleration," arXiv:1201.2434 [astro-ph.CO].

[95] G. Jungman, M. Kamionkowski, A. Kosowsky, and D. Spergel, "Cosmological Parameter Determination with Microwave Background Maps," *Phys.Rev.* **D54** (1996) 1332–1344, arXiv:astro-ph/9512139 [astro-ph].

[96] G. Jungman, M. Kamionkowski, A. Kosowsky, and D. Spergel, "Weighing the Universe with the Cosmic Microwave Background," *Phys.Rev.Lett.* **76** (1996) 1007–1010, arXiv:astro-ph/9507080 [astro-ph].

[97] J. Bond, G. Efstathiou, and M. Tegmark, "Forecasting Cosmic Parameter Errors from Microwave Background Anisotropy Experiments," *Mon.Not.Roy.Astron.Soc.* **291** (1997) L33–L41, arXiv:astro-ph/9702100 [astro-ph].

[98] M. Zaldarriaga, D. Spergel, and U. Seljak, "Microwave Background Constraints on Cosmological Parameters," *Astrophys.J.* **488** (1997) 1–13, arXiv:astro-ph/9702157 [astro-ph].

[99] S. Das *et al.*, "The Atacama Cosmology Telescope: A Measurement of the Cosmic Microwave Background Power Spectrum at 148 and 218 GHz from the 2008 Southern Survey," *Astrophys.J.* **729** (2011) 62, arXiv:1009.0847 [astro-ph.CO].

[100] J. Sievers *et al.*, "The Atacama Cosmology Telescope: Cosmological Parameters from Three Seasons of Data," arXiv:1301.0824 [astro-ph.CO].

[101] R. Keisler *et al.*, "A Measurement of the Damping Tail of the Cosmic Microwave Background Power Spectrum with the South Pole Telescope," *Astrophys.J.* **743** (2011) 28, arXiv:1105.3182 [astro-ph.CO].

[102] K. Story *et al.*, "A Measurement of the Cosmic Microwave Background Damping Tail from the 2500-Square-Degree SPT-SZ Survey," arXiv:1210.7231 [astro-ph.CO].

[103] **Planck** Collaboration, P. Ade *et al.*, "Planck 2013 Results. XVII. Gravitational Lensing by Large-Scale Structure," arXiv:1303.5077 [astro-ph.CO].

[104] **Planck** Collaboration, P. Ade *et al.*, "Planck 2013 Results. XXIII. Isotropy and Statistics of the CMB," arXiv:1303.5083 [astro-ph.CO].

[105] D. Spergel and M. Zaldarriaga, "CMB Polarization as a Direct Test of Inflation," *Phys.Rev.Lett.* **79** (1997) 2180–2183, arXiv:astro-ph/9705182 [astro-ph].

[106] S. Dodelson, "Coherent Phase Argument for Inflation," *AIP Conf.Proc.* **689** (2003) 184–196, arXiv:hep-ph/0309057 [hep-ph].

[107] A. Albrecht, D. Coulson, P. Ferreira, and J. Magueijo, "Causality and the Microwave Background," *Phys.Rev.Lett.* **76** (1996) 1413–1416, arXiv:astro-ph/9505030 [astro-ph].

[108] N. Turok, "A Causal Source Which Mimics Inflation," *Phys.Rev.Lett.* **77** (1996) 4138–4141, arXiv:astro-ph/9607109 [astro-ph].

[109] R. Durrer, M. Kunz, and A. Melchiorri, "Cosmic Structure Formation with Topological Defects," *Phys.Rept.* **364** (2002) 1–81, arXiv:astro-ph/0110348 [astro-ph].

[110] L. Knox, "Determination of Inflationary Observables by Cosmic Microwave Background Anisotropy Experiments," *Phys.Rev.* **D52** (1995) 4307–4318, arXiv:astro-ph/9504054 [astro-ph].

[111] E. Komatsu *et al.*, "Non-Gaussianity as a Probe of the Physics of the Primordial Universe and the Astrophysics of the Low Redshift Universe," arXiv:0902.4759 [astro-ph.CO].

[112] X. Chen, "Primordial Non-Gaussianities from Inflation Models," *Adv.Astron.* **2010** (2010) 638979, arXiv:1002.1416 [astro-ph.CO].

[113] Y. Wang, "Inflation, Cosmic Perturbations and Non-Gaussianities," arXiv:1303.1523 [hep-th].

[114] C. Bennett *et al.*, "Nine-Year Wilkinson Microwave Anisotropy Probe (WMAP) Observations: Final Maps and Results," arXiv:1212.5225 [astro-ph.CO].

[115] L. Senatore, K. Smith, and M. Zaldarriaga, "Non-Gaussianities in Single-Field Inflation and their Optimal Limits from the WMAP 5-Year Data," *JCAP* **1001** (2010) 028, arXiv:0905.3746 [astro-ph.CO].

[116] A. Gangui, F. Lucchin, S. Matarrese, and S. Mollerach, "The Three-Point Correlation Function of the Cosmic Microwave Background in Inflationary Models," *Astrophys.J.* **430** (1994) 447–457, arXiv:astro-ph/9312033 [astro-ph].

[117] L. Verde, L.-M. Wang, A. Heavens, and M. Kamionkowski, "Large-Scale Structure, the Cosmic Microwave Background, and Primordial Non-Gaussianity," *Mon.Not.Roy.Astron.Soc.* **313** (2000) L141–L147, arXiv:astro-ph/9906301 [astro-ph].

[118] E. Komatsu and D. Spergel, "Acoustic Signatures in the Primary Microwave Background Bispectrum," *Phys.Rev.* **D63** (2001) 063002, arXiv:astro-ph/0005036 [astro-ph].

[119] J. Maldacena, "Non-Gaussian Features of Primordial Fluctuations in Single Field Inflationary Models," *JHEP* **0305** (2003) 013, arXiv:astro-ph/0210603 [astro-ph].

[120] P. Creminelli and M. Zaldarriaga, "Single-Field Consistency Relation for the Three-Point Function," *JCAP* **0410** (2004) 006, arXiv:astro-ph/0407059 [astro-ph].

[121] V. Assassi, D. Baumann, and D. Green, "On Soft Limits of Inflationary Correlation Functions," *JCAP* **1211** (2012) 047, arXiv:1204.4207 [hep-th].

[122] K. Hinterbichler, L. Hui, and J. Khoury, "An Infinite Set of Ward Identities for Adiabatic Modes in Cosmology," arXiv:1304.5527 [hep-th].

[123] W. Goldberger, L. Hui, and A. Nicolis, "One-Particle-Irreducible Consistency Relations for Cosmological Perturbations," arXiv:1303.1193 [hep-th].

[124] X. Chen, M.-x. Huang, S. Kachru, and G. Shiu, "Observational Signatures and Non-Gaussianities of General Single-Field Inflation," *JCAP* **0701** (2007) 002, arXiv:hep-th/0605045 [hep-th].

[125] D. Seery and J. E. Lidsey, "Primordial Non-Gaussianities in Single-Field Inflation," *JCAP* **0506** (2005) 003, arXiv:astro-ph/0503692 [astro-ph].

[126] D. Babich, P. Creminelli, and M. Zaldarriaga, "The Shape of Non-Gaussianities," *JCAP* **0408** (2004) 009, arXiv:astro-ph/0405356 [astro-ph].

[127] P. Creminelli, R. Emami, M. Simonović, and G. Trevisan, "$ISO(4, 1)$ Symmetry in the EFT of Inflation," *JCAP* **1307** (2013) 037, arXiv:1304.4238 [hep-th].

[128] W. Hu and M. White, "Acoustic Signatures in the Cosmic Microwave Background," *Astrophys.J.* **471** (1996) 30–51, arXiv:astro-ph/9602019 [astro-ph].

[129] W. Hu, D. Spergel, and M. White, "Distinguishing Causal Seeds from Inflation," *Phys.Rev.* **D55** (1997) 3288–3302, arXiv:astro-ph/9605193 [astro-ph].

[130] **WMAP** Collaboration, H. Peiris *et al.*, "First-Year Wilkinson Microwave Anisotropy Probe (WMAP) Observations: Implications for Inflation," *Astrophys.J.Suppl.* **148** (2003) 213, arXiv:astro-ph/0302225 [astro-ph].

[131] D. Seckel and M. Turner, "Isothermal Density Perturbations in an Axion Dominated Inflationary Universe," *Phys.Rev.* **D32** (1985) 3178.

[132] A. Linde, "Axions in Inflationary Cosmology," *Physics Letters B* **259** (Apr., 1991) 38–47.

[133] M. Turner and F. Wilczek, "Inflationary Axion Cosmology," *Phys.Rev.Lett.* **66** (1991) 5–8.

[134] D. Lyth and D. Wands, "Generating the Curvature Perturbation without an Inflaton," *Phys.Lett.* **B524** (2002) 5–14, arXiv:hep-ph/0110002 [hep-ph].

[135] T. Moroi and T. Takahashi, "Effects of Cosmological Moduli Fields on Cosmic Microwave Background," *Phys.Lett.* **B522** (2001) 215–221, arXiv:hep-ph/0110096 [hep-ph].

[136] S. Weinberg, "Must cosmological perturbations remain non-adiabatic after multi-field inflation?," *Phys.Rev.* **D70** (2004) 083522, arXiv:astro-ph/0405397 [astro-ph].

[137] **BICEP2** Collaboration, P. Ade *et al.*, "BICEP2 I: Detection of B-mode Polarization at Degree Angular Scales," arXiv:1403.3985 [astro-ph.CO].

[138] D. Baumann and M. Zaldarriaga, "Causality and Primordial Tensor Modes," *JCAP* **0906** (2009) 013, arXiv:0901.0958 [astro-ph.CO].

[139] H. Chiang *et al.*, "Measurement of CMB Polarization Power Spectra from Two Years of BICEP Data," *Astrophys.J.* **711** (2010) 1123–1140, arXiv:0906.1181 [astro-ph.CO].

[140] K. Smith *et al.*, "On Quantifying and Resolving the BICEP2/Planck Tension Over Gravitational Waves," arXiv:1404.0373 [astro-ph.CO].

[141] C. Dvorkin, M. Wyman, D. Rudd, and W. Hu, "Neutrinos help reconcile Planck measurements with both early and local universe," arXiv:1403.8049 [astro-ph.CO].

[142] M. Mortonson and U. Seljak, "A Joint Analysis of Planck and BICEP2 B-modes Including Dust Polarization Uncertainty," arXiv:1405.5857 [astro-ph.CO].

[143] R. Flauger, J. Hill, and D. Spergel, "Toward an Understanding of Foreground Emission in the BICEP2 Region," arXiv:1405.7351 [astro-ph.CO].

[144] **BICEP2** Collaboration, P. Ade *et al.*, "BICEP2 II: Experiment and Three-Year Data Set," arXiv:1403.4302 [astro-ph.CO].

[145] B. Reichborn-Kjennerud *et al.*, "EBEX: A Balloon-Borne CMB Polarization Experiment," arXiv:1007.3672 [astro-ph.CO].

[146] A. Fraisse *et al.*, "SPIDER: Probing the Early Universe with a Suborbital Polarimeter," *JCAP* **1304** (2013) 047, arXiv:1106.3087 [astro-ph.CO].

[147] T. Essinger-Hileman *et al.*, "The Atacama B-Mode Search: CMB Polarimetry with Transition-Edge-Sensor Bolometers," arXiv:1008.3915 [astro-ph.IM].

[148] J. Eimer *et al.*, "The Cosmology Large Angular Scale Surveyor (CLASS): 40 GHz Optical Design," *Proc.SPIE Int.Soc.Opt.Eng.* **8452** (2012) 845220, arXiv:1211.0041 [astro-ph.IM].

[149] Z. Kermish *et al.*, "The POLARBEAR Experiment," arXiv:1210.7768 [astro-ph.IM].

[150] J. Austermann *et al.*, "SPTpol: An Instrument for CMB Polarization Measurements with the South Pole Telescope," *Proc.SPIE Int.Soc.Opt.Eng.* **8452** (2012) 84520E, arXiv:1210.4970 [astro-ph.IM].

[151] M. Niemack *et al.*, "ACTPol: A Polarization-Sensitive Receiver for the Atacama Cosmology Telescope," *Proc.SPIE Int.Soc.Opt.Eng.* **7741** (2010) 77411S, arXiv:1006.5049 [astro-ph.IM].

[152] N. Katayama and E. Komatsu, "Simple Foreground Cleaning Algorithm for Detecting Primordial B-mode Polarization of the Cosmic Microwave Background," *Astrophys.J.* **737** (2011) 78, arXiv:1101.5210 [astro-ph.CO].

[153] J. Caligiuri and A. Kosowsky, "Inflationary Tensor Perturbations After BICEP," arXiv:1403.5324 [astro-ph.CO].

[154] S. Dodelson, "How much can we learn about the physics of inflation?," arXiv:1403.6310 [astro-ph.CO].

[155] P. Creminelli, D. Nacir, M. Simonović, G. Trevisan, and M. Zaldarriaga, "ϕ^2 or not ϕ^2: Checking the Simplest Universe," arXiv:1404.1065 [astro-ph.CO].

[156] **DESI** Collaboration, M. Levi *et al.*, "The DESI Experiment, a Whitepaper for Snowmass 2013," arXiv:1308.0847 [astro-ph.CO].

[157] **PRISM** Collaboration, P. André *et al.*, "PRISM (Polarized Radiation Imaging and Spectroscopy Mission): An Extended White Paper," *JCAP* **1402** (2014) 006, arXiv:1310.1554 [astro-ph.CO].

[158] **COrE** Collaboration, F. Bouchet *et al.*, "COrE (Cosmic Origins Explorer) A White Paper," arXiv:1102.2181 [astro-ph.CO].

[159] A. Kogut *et al.*, "The Primordial Inflation Explorer (PIXIE): A Nulling Polarimeter for Cosmic Microwave Background Observations," *JCAP* **1107** (2011) 025, arXiv:1105.2044 [astro-ph.CO].

[160] T. Matsumura *et al.*, "Mission Design of LiteBIRD," arXiv:1311.2847 [astro-ph.IM].

[161] M. Takada, E. Komatsu, and T. Futamase, "Cosmology with High-Redshift Galaxy Survey: Neutrino Mass and Inflation," *Phys.Rev.* **D73** (2006) 083520, arXiv:astro-ph/0512374 [astro-ph].

[162] P. Adshead, R. Easther, J. Pritchard, and A. Loeb, "Inflation and the Scale-Dependent Spectral Index: Prospects and Strategies," *JCAP* **1102** (2011) 021, arXiv:1007.3748 [astro-ph.CO].

[163] C. Rimes and A. Hamilton, "Information Content of the Non-Linear Matter Power Spectrum," *Mon.Not.Roy.Astron.Soc.* **360** (2005) 82–86, arXiv:astro-ph/0502081 [astro-ph].

[164] M. Crocce and R. Scoccimarro, "Memory of Initial Conditions in Gravitational Clustering," *Phys.Rev.* **D73** (2006) 063520, arXiv:astro-ph/0509419 [astro-ph].

[165] **Euclid Theory Working Group** Collaboration, L. Amendola *et al.*, "Cosmology and Fundamental Physics with the Euclid Satellite," *Living Rev.Rel.* **16** (2013) 6, arXiv:1206.1225 [astro-ph.CO].

[166] D. Baumann, A. Nicolis, L. Senatore, and M. Zaldarriaga, "Cosmological Non-Linearities as an Effective Fluid," *JCAP* **1207** (2012) 051, arXiv:1004.2488 [astro-ph.CO].

[167] J. Carrasco, M. Hertzberg, and L. Senatore, "The Effective Field Theory of Cosmological Large Scale Structures," *JHEP* **1209** (2012) 082, arXiv:1206.2926 [astro-ph.CO].

[168] J. Carrasco, S. Foreman, D. Green, and L. Senatore, "The Effective Field Theory of Large Scale Structures at Two Loops," arXiv:1310.0464 [astro-ph.CO].

[169] R. Porto, L. Senatore, and M. Zaldarriaga, "The Lagrangian-Space Effective Field Theory of Large Scale Structures," arXiv:1311.2168 [astro-ph.CO].

[170] S. Carroll, S. Leichenauer, and J. Pollack, "A Consistent Effective Theory of Long-Wavelength Cosmological Perturbations," arXiv:1310.2920 [hep-th].

[171] L. Mercolli and E. Pajer, "On the Velocity in the Effective Field Theory of Large Scale Structures," arXiv:1307.3220 [astro-ph.CO].

[172] E. Pajer and M. Zaldarriaga, "On the Renormalization of the Effective Field Theory of Large Scale Structures," *JCAP* **1308** (2013) 037, arXiv:1301.7182 [astro-ph.CO].

[173] M. Hertzberg, "The Effective Field Theory of Dark Matter and Structure Formation: Semi-Analytical Results," arXiv:1208.0839 [astro-ph.CO].

[174] G. Dvali, A. Gruzinov, and M. Zaldarriaga, "A New Mechanism for Generating Density Perturbations from Inflation," *Phys.Rev.* **D69** (2004) 023505, arXiv:astro-ph/0303591 [astro-ph].

[175] M. Zaldarriaga, "Non-Gaussianities in Models with a Varying Inflaton Decay Rate," *Phys.Rev.* **D69** (2004) 043508, arXiv:astro-ph/0306006 [astro-ph].

[176] L. Kofman, "Probing String Theory with Modulated Cosmological Fluctuations," arXiv:astro-ph/0303614 [astro-ph].

[177] A. Linde and V. Mukhanov, "Non-Gaussian Isocurvature Perturbations from Inflation," *Phys.Rev.* **D56** (1997) 535–539, arXiv:astro-ph/9610219 [astro-ph].

[178] D. Lyth, C. Ungarelli, and D. Wands, "The Primordial Density Perturbation in the Curvaton Scenario," *Phys.Rev.* **D67** (2003) 023503, arXiv:astro-ph/0208055 [astro-ph].

[179] K. Abazajian *et al.*, "Inflation Physics from the Cosmic Microwave Background and Large Scale Structure," arXiv:1309.5381 [astro-ph.CO].

[180] T. Giannantonio, C. Porciani, J. Carron, A. Amara, and A. Pillepich, "Constraining Primordial Non-Gaussianity with Future Galaxy Surveys," *Mon.Not.Roy.Astron.Soc.* **422** (2012) 2854–2877, arXiv:1109.0958 [astro-ph.CO].

[181] C. Burgess, "Quantum Gravity in Everyday Life: General Relativity as an Effective Field Theory," *Living Rev.Rel.* **7** (2004) 5, arXiv:gr-qc/0311082 [gr-qc].

[182] D. B. Kaplan, "Five Lectures on Effective Field Theory," arXiv:nucl-th/0510023 [nucl-th].

[183] W. Skiba, "TASI Lectures on Effective Field Theory and Precision Electroweak Measurements," arXiv:1006.2142 [hep-ph].

[184] C. Burgess, "Introduction to Effective Field Theory," *Ann.Rev.Nucl.Part.Sci.* **57** (2007) 329–362, arXiv:hep-th/0701053 [hep-th].

[185] C. Burgess and D. London, "Uses and Abuses of Effective Lagrangians," *Phys.Rev.* **D48** (1993) 4337–4351, arXiv:hep-ph/9203216 [hep-ph].

[186] L. Susskind, "Dynamics of Spontaneous Symmetry Breaking in the Weinberg–Salam Theory," *Phys. Rev. D* **20** (Nov, 1979) 2619–2625.

[187] T. Appelquist and J. Carazzone, "Infrared Singularities and Massive Fields," *Phys.Rev.* **D11** (1975) 2856.

[188] H. Murayama, "Supersymmetry Phenomenology," arXiv:hep-ph/0002232 [hep-ph].

[189] G. Giudice, "Naturally Speaking: The Naturalness Criterion and Physics at the LHC," arXiv:0801.2562 [hep-ph].

[190] V. Weisskopf, "On the Self-Energy and the Electromagnetic Field of the Electron," *Phys.Rev.* **56** (1939) 72–85.

[191] M. Gaillard and B. Lee, "Rare Decay Modes of the K-Mesons in Gauge Theories," *Phys.Rev.* **D10** (1974) 897.

[192] D. Kaplan, M. Savage, and M. Wise, "Two Nucleon Systems from Effective Field Theory," *Nucl.Phys.* **B534** (1998) 329–355, arXiv:nucl-th/9802075 [nucl-th].

[193] P. Bedaque and U. van Kolck, "Effective Field Theory for Few Nucleon Systems," *Ann.Rev.Nucl.Part.Sci.* **52** (2002) 339–396, arXiv:nucl-th/0203055 [nucl-th].

[194] S. Weinberg, "Nuclear Forces from Chiral Lagrangians," *Phys.Lett.* **B251** (1990) 288–292.

[195] S. Weinberg, "Effective Chiral Lagrangians for Nucleon-Pion Interactions and Nuclear Forces," *Nucl.Phys.* **B363** (1991) 3–18.

[196] S. Weinberg, "Three-Body Interactions among Nucleons and Pions," *Phys.Lett.* **B295** (1992) 114–121, arXiv:hep-ph/9209257 [hep-ph].

[197] R. Bousso and J. Polchinski, "Quantization of Four-Form Fluxes and Dynamical Neutralization of the Cosmological Constant," *JHEP* **0006** (2000) 006, arXiv:hep-th/0004134 [hep-th].

[198] S. Weinberg, "Anthropic Bound on the Cosmological Constant," *Phys.Rev.Lett.* **59** (1987) 2607.

[199] G. 't Hooft, "Naturalness, Chiral Symmetry, and Spontaneous Chiral Symmetry Breaking," *NATO Adv.Study Inst.Ser.B Phys.* **59** (1980) 135.

[200] J. Donoghue, "Introduction to the Effective Field Theory Description of Gravity," arXiv:gr-qc/9512024 [gr-qc].

[201] S. Weinberg, "Understanding the Fundamental Constituents of Matter," *Proceedings: International School of Subnuclear Physics* (1978) .

[202] D. Gross and E. Witten, "Superstring Modifications of Einstein's Equations," *Nucl.Phys.* **B277** (1986) 1.

[203] T. Banks and N. Seiberg, "Symmetries and Strings in Field Theory and Gravity," *Phys.Rev.* **D83** (2011) 084019, arXiv:1011.5120 [hep-th].

[204] L. Abbott and M. Wise, "Wormholes and Global Symmetries," *Nucl.Phys.* **B325** (1989) 687.

[205] S. Coleman and K.-M. Lee, "Wormholes Made Without Massless Matter Fields," *Nucl.Phys.* **B329** (1990) 387.

[206] R. Kallosh, A. Linde, D. Linde, and L. Susskind, "Gravity and Global Symmetries," *Phys.Rev.* **D52** (1995) 912–935, arXiv:hep-th/9502069 [hep-th].

[207] J. Conlon, "Quantum Gravity Constraints on Inflation," *JCAP* **1209** (2012) 019, arXiv:1203.5476 [hep-th].

[208] T. Banks and L. Dixon, "Constraints on String Vacua with Space-Time Supersymmetry," *Nucl.Phys.* **B307** (1988) 93–108.

[209] H. Collins, R. Holman, and A. Ross, "Effective Field Theory in Time-Dependent Settings," arXiv:1208.3255 [hep-th].

[210] C. Burgess and M. Williams, "Who You Gonna Call? Runaway Ghosts, Higher Derivatives and Time-Dependence in EFTs," arXiv:1404.2236 [gr-qc].

[211] A. Avgoustidis *et al.*, "Decoupling Survives Inflation: A Critical Look at Effective Field Theory Violations During Inflation," *JCAP* **1206** (2012) 025, arXiv:1203.0016 [hep-th].

[212] C. Burgess, M. Horbatsch, and S. Patil, "Inflating in a Trough: Single-Field Effective Theory from Multiple-Field Curved Valleys," arXiv:1209.5701 [hep-th].

[213] P. Steinhardt and M. Turner, "A Prescription for Successful New Inflation," *Phys.Rev.* **D29** (1984) 2162–2171.

[214] A. Gefter, "What kind of bang was the big bang?," *New Scientist* (2012).

[215] D. Lyth and A. Riotto, "Particle Physics Models of Inflation and the Cosmological Density Perturbation," *Phys.Rept.* **314** (1999) 1–146, arXiv:hep-ph/9807278 [hep-ph].

[216] D. Lyth and A. Liddle, *The Primordial Density Perturbation: Cosmology, Inflation and the Origin of Structure.* Cambridge University Press, 2009.

[217] D. Lyth, "Particle Physics Models of Inflation," *Lect.Notes Phys.* **738** (2008) 81–118, arXiv:hep-th/0702128 [hep-th].

[218] J. Martin, C. Ringeval, and V. Vennin, "Encyclopaedia Inflationaris," arXiv:1303.3787 [astro-ph.CO].

[219] A. D. Linde, "Particle Physics and Inflationary Cosmology," *Contemp.Concepts Phys.* **5** (1990) 1–362, arXiv:hep-th/0503203 [hep-th].

[220] R. Kallosh, A. Linde, and A. Westphal, "Chaotic Inflation in Supergravity after Planck and BICEP2," arXiv:1405.0270 [hep-th].

[221] A. Linde, "Primordial Inflation without Primordial Monopoles," *Phys.Lett.* **B132** (1983) 317–320.

[222] L. Boubekeur and D. Lyth, "Hilltop Inflation," *JCAP* **0507** (2005) 010, arXiv:hep-ph/0502047 [hep-ph].

[223] L. Alabidi and D. Lyth, "Inflation Models and Observation," *JCAP* **0605** (2006) 016, arXiv:astro-ph/0510441 [astro-ph].

[224] D. Baumann, A. Dymarsky, I. Klebanov, L. McAllister, and P. Steinhardt, "A Delicate Universe," *Phys.Rev.Lett.* **99** (2007) 141601, arXiv:0705.3837 [hep-th].

[225] D. Baumann, A. Dymarsky, I. Klebanov, and L. McAllister, "Towards an Explicit Model of D-brane Inflation," *JCAP* **0801** (2008) 024, arXiv:0706.0360 [hep-th].

[226] R. Allahverdi, K. Enqvist, J. Garcia-Bellido, A. Jokinen, and A. Mazumdar, "MSSM Flat Direction Inflation: Slow Roll, Stability, Fine Tuning and Reheating," *JCAP* **0706** (2007) 019, arXiv:hep-ph/0610134 [hep-ph].

[227] B. Sanchez, K. Dimopoulos, and D. Lyth, "A-term Inflation and the MSSM," *JCAP* (2007) 015.

[228] A. Linde, "Hybrid Inflation," *Phys.Rev.* **D49** (1994) 748–754, arXiv:astro-ph/9307002 [astro-ph].

[229] D. E. Kaplan and N. Weiner, "Little Inflatons and Gauge Inflation," *JCAP* **0402** (2004) 005, arXiv:hep-ph/0302014 [hep-ph].

[230] N. Arkani-Hamed, H.-C. Cheng, P. Creminelli, and L. Randall, "Pseudonatural Inflation," *JCAP* **0307** (2003) 003, arXiv:hep-th/0302034 [hep-th].

[231] A. Starobinsky, "A New Type of Isotropic Cosmological Models Without Singularity," *Phys.Lett.* **B91** (1980) 99–102.

[232] R. Kallosh and A. Linde, "Superconformal Generalizations of the Starobinsky Model," *JCAP* **1306** (2013) 028, arXiv:1306.3214 [hep-th].

[233] J. Ellis, D. Nanopoulos, and K. Olive, "A No-Scale Supergravity Realization of the Starobinsky Model," arXiv:1305.1247 [hep-th].

[234] F. Farakos, A. Kehagias, and A. Riotto, "On the Starobinsky Model of Inflation from Supergravity," *Nucl.Phys.* **B876** (2013) 187–200, arXiv:1307.1137.

[235] J. Ellis, D. Nanopoulos, and K. Olive, "Starobinsky-like Inflationary Models as Avatars of No-Scale Supergravity," arXiv:1307.3537 [hep-th].

[236] S. Ferrara, A. Kehagias, and A. Riotto, "The Imaginary Starobinsky Model," arXiv:1403.5531 [hep-th].

[237] J. Dowker and R. Critchley, "Effective Lagrangian and Energy Momentum Tensor in de Sitter Space," *Phys.Rev.* **D13** (1976) 3224.

[238] C. Burgess, H. Lee, and M. Trott, "Power-Counting and the Validity of the Classical Approximation During Inflation," *JHEP* **0909** (2009) 103, arXiv:0902.4465 [hep-ph].

[239] M. Hertzberg, "On Inflation with Non-Minimal Coupling," *JHEP* **1011** (2010) 023, arXiv:1002.2995 [hep-ph].

[240] B. Spokoiny, "Inflation and Generation of Perturbations in Broken Symmetric Theory of Gravity ," *Phys.Lett.* **B147** (1984) 39–43.

[241] T. Futamase and K.-i. Maeda, "Chaotic Inflationary Scenario in Models Having Nonminimal Coupling With Curvature," *Phys.Rev.* **D39** (1989) 399–404.

[242] D. Salopek, J. Bond, and J. Bardeen, "Designing Density Fluctuation Spectra in Inflation," *Phys.Rev.* **D40** (1989) 1753.

[243] R. Fakir and W. Unruh, "Improvement on Cosmological Chaotic Inflation Through Non-Minimal Coupling," *Phys.Rev.* **D41** (1990) 1783–1791.

[244] D. Kaiser, "Primordial Spectral Indices from Generalized Einstein Theories," *Phys.Rev.* **D52** (1995) 4295–4306, arXiv:astro-ph/9408044 [astro-ph].

[245] E. Komatsu and T. Futamase, "Complete Constraints on a Non-Minimally Coupled Chaotic Inflationary Scenario from the Cosmic Microwave Background," *Phys.Rev.* **D59** (1999) 064029, arXiv:astro-ph/9901127 [astro-ph].

[246] F. Bezrukov and M. Shaposhnikov, "The Standard Model Higgs Boson as the Inflaton," *Phys.Lett.* **B659** (2008) 703–706, arXiv:0710.3755 [hep-th].

[247] J. Garriga and V. Mukhanov, "Perturbations in k-Inflation," *Phys.Lett.* **B458** (1999) 219–225, arXiv:hep-th/9904176 [hep-th].

[248] E. Copeland, A. Liddle, D. Lyth, E. Stewart, and D. Wands, "False Vacuum Inflation with Einstein Gravity," *Phys.Rev.* **D49** (1994) 6410–6433, arXiv:astro-ph/9401011 [astro-ph].

[249] D. Baumann and D. Green, "Signatures of Supersymmetry from the Early Universe," *Phys.Rev.* **D85** (2012) 103520, arXiv:1109.0292 [hep-th].

[250] N. Agarwal, R. Bean, L. McAllister, and G. Xu, "Universality in D-brane Inflation," *JCAP* **1109** (2011) 002, arXiv:1103.2775 [astro-ph.CO].

[251] D. Baumann and D. Green, "Inflating with Baryons," *JHEP* **1104** (2011) 071, arXiv:1009.3032 [hep-th].

[252] D. Lyth, "What would we learn by detecting a gravitational wave signal in the cosmic microwave background anisotropy?," *Phys.Rev.Lett.* **78** (1997) 1861–1863, arXiv:hep-ph/9606387 [hep-ph].

[253] R. Easther, W. Kinney, and B. Powell, "The Lyth Bound and the End of Inflation," *JCAP* **0608** (2006) 004, arXiv:astro-ph/0601276 [astro-ph].

[254] D. Baumann and L. McAllister, "A Microscopic Limit on Gravitational Waves from D-brane Inflation," *Phys.Rev.* **D75** (2007) 123508, arXiv:hep-th/0610285 [hep-th].

[255] M. Hertzberg, "Inflation, Symmetry, and B-Modes," arXiv:1403.5253 [hep-th].

[256] L. Smolin, "Gravitational Radiative Corrections as the Origin of Spontaneous Symmetry Breaking!," *Phys.Lett.* **B93** (1980) 95.

[257] N. Arkani-Hamed, L. Motl, A. Nicolis, and C. Vafa, "The String Landscape, Black Holes and Gravity as the Weakest Force," *JHEP* **0706** (2007) 060, arXiv:hep-th/0601001 [hep-th].

[258] I. Ben-Dayan and R. Brustein, "Cosmic Microwave Background Observables of Small Field Models of Inflation," *JCAP* **1009** (2010) 007, arXiv:0907.2384 [astro-ph.CO].

[259] Q. Shafi and J. Wickman, "Observable Gravity Waves From Supersymmetric Hybrid Inflation," *Phys.Lett.* **B696** (2011) 438–446, arXiv:1009.5340 [hep-ph].

[260] S. Hotchkiss, A. Mazumdar, and S. Nadathur, "Observable Gravitational Waves from Inflation with Small Field Excursions," *JCAP* **1202** (2012) 008, arXiv:1110.5389 [astro-ph.CO].

[261] A. Hebecker, S. Kraus, and A. Westphal, "Evading the Lyth Bound in Hybrid Natural Inflation," arXiv:1305.1947 [hep-th].

[262] S. Antusch and D. Nolde, "BICEP2 Implications for Single-Field Slow-Roll Inflation Revisited," arXiv:1404.1821 [hep-ph].

[263] D. Baumann and D. Green, "A Field Range Bound for General Single-Field Inflation," *JCAP* **1205** (2012) 017, arXiv:1111.3040 [hep-th].

[264] M. Berg, E. Pajer, and S. Sjors, "Dante's Inferno," *Phys.Rev.* **D81** (2010) 103535, arXiv:0912.1341 [hep-th].

[265] L. McAllister, S. Renaux-Petel, and G. Xu, "A Statistical Approach to Multi-Field Inflation: Many-Field Perturbations Beyond Slow-Roll," arXiv:1207.0317 [astro-ph.CO].

[266] G. Dvali, A. Gruzinov, and M. Zaldarriaga, "Cosmological Perturbations from Inhomogeneous Reheating, Freezeout, and Mass Domination," *Phys.Rev.* **D69** (2004) 083505, arXiv:astro-ph/0305548 [astro-ph].

[267] H. Collins, R. Holman, and T. Vardanyan, "Do Mixed States Save Effective Field Theory from BICEP?," arXiv:1403.4592 [hep-th].

[268] A. Ashoorioon, K. Dimopoulos, M. Sheikh-Jabbari, and G. Shiu, "Non-Bunch-Davies Initial State Reconciles Chaotic Models with BICEP and Planck," arXiv:1403.6099 [hep-th].

[269] A. Aravind, D. Lorshbough, and S. Paban, "Bogoliubov Excited States and the Lyth Bound," arXiv:1403.6216 [astro-ph.CO].

[270] J. Cook and L. Sorbo, "Particle Production During Inflation and Gravitational Waves Detectable by Ground-Based Interferometers," *Phys.Rev.* **D85** (2012) 023534, arXiv:1109.0022 [astro-ph.CO].

[271] L. Senatore, E. Silverstein, and M. Zaldarriaga, "New Sources of Gravitational Waves during Inflation," arXiv:1109.0542 [hep-th].

[272] N. Barnaby *et al.*, "Gravity Waves and Non-Gaussian Features from Particle Production in a Sector Gravitationally Coupled to the Inflaton," *Phys.Rev.* **D86** (2012) 103508, arXiv:1206.6117 [astro-ph.CO].

[273] D. Carney, W. Fischler, E. Kovetz, D. Lorshbough, and S. Paban, "Rapid Field Excursions and the Inflationary Tensor Spectrum," *JHEP* **1211** (2012) 042, arXiv:1209.3848 [hep-th].

[274] P. Creminelli, "On Non-Gaussianities in Single-Field Inflation," *JCAP* **0310** (2003) 003, arXiv:astro-ph/0306122 [astro-ph].

[275] D. Baumann and D. Green, "Equilateral Non-Gaussianity and New Physics on the Horizon," *JCAP* **1109** (2011) 014, arXiv:1102.5343 [hep-th].

[276] C. Burrage, C. de Rham, D. Seery, and A. J. Tolley, "Galileon Inflation," *JCAP* **1101** (2011) 014, arXiv:1009.2497 [hep-th].

[277] V. Assassi, D. Baumann, D. Green, and L. McAllister, "Planck-Suppressed Operators," arXiv:1304.5226 [hep-th].

[278] V. Kaplunovsky and J. Louis, "Model-Independent Analysis of Soft Terms in Effective Supergravity and in String Theory," *Phys.Lett.* **B306** (1993) 269–275, arXiv:hep-th/9303040 [hep-th].

[279] D. Green, M. Lewandowski, L. Senatore, E. Silverstein, and M. Zaldarriaga, "Anomalous Dimensions and Non-Gaussianity," arXiv:1301.2630 [hep-th].

[280] M. Green, J. Schwarz, and E. Witten, *Superstring Theory. Vol. 1: Introduction.* Cambridge University Press, 1987.

[281] M. Green, J. Schwarz, and E. Witten, *Superstring Theory. Vol. 2: Loop Amplitudes, Anomalies and Phenomenology.* Cambridge University Press, 1987.

[282] J. Polchinski, *String Theory. Vol. 1: An Introduction to the Bosonic String.* Cambridge University Press, 1998.

[283] J. Polchinski, *String Theory. Vol. 2: Superstring Theory and Beyond.* Cambridge University Press, 1998.

[284] D. Tong, "String Theory," arXiv:0908.0333 [hep-th].

[285] J. Polchinski, "What is String Theory?," arXiv:hep-th/9411028 [hep-th].

[286] E. Kiritsis, "Introduction to Superstring Theory," arXiv:hep-th/9709062 [hep-th].

[287] J. Bedford, "An Introduction to String Theory," arXiv:1107.3967 [hep-th].

[288] S. Forste, "Strings, Branes and Extra Dimensions," *Fortsch.Phys.* **50** (2002) 221–403, arXiv:hep-th/0110055 [hep-th].

[289] T. Mohaupt, "Introduction to String Theory," *Lect.Notes Phys.* **631** (2003) 173–251, arXiv:hep-th/0207249 [hep-th].

[290] A. Sen, "An Introduction to Nonperturbative String Theory," arXiv:hep-th/9802051 [hep-th].

[291] L. Ibanez and A. Uranga, *String Theory and Particle Physics: An Introduction to String Phenomenology.* Cambridge University Press, 2012.

[292] K. Becker, M. Becker, and J. Schwarz, *String Theory and M-theory: A Modern Introduction.* Cambridge University Press, 2007.

[293] C. Johnson, *D-branes.* Cambridge University Press, 2003.

[294] S. Deser and B. Zumino, "A Complete Action for the Spinning String," *Phys.Lett.* **B65** (1976) 369–373.

[295] L. Brink, P. Di Vecchia, and P. Howe, "A Locally Supersymmetric and Reparametrization Invariant Action for the Spinning String," *Phys.Lett.* **B65** (1976) 471–474.

[296] P. Goddard, J. Goldstone, C. Rebbi, and C. Thorn, "Quantum Dynamics of a Massless Relativistic String," *Nucl.Phys.* **B56** (1973) 109–135.

[297] D. Friedan, "Nonlinear Models in $2 + \epsilon$ Dimensions," *Annals Phys.* **163** (1985) 318.

[298] R. Myers, "Dielectric Branes," *JHEP* **9912** (1999) 022, arXiv:hep-th/9910053 [hep-th].

[299] G. Horowitz and A. Strominger, "Black Strings and P-branes," *Nucl.Phys.* **B360** (1991) 197–209.

[300] O. Aharony, S. Gubser, J. Maldacena, H. Ooguri, and Y. Oz, "Large N Field Theories, String Theory and Gravity," *Phys.Rept.* **323** (2000) 183–386, arXiv:hep-th/9905111 [hep-th].

[301] M. Dine, A. Morisse, A. Shomer, and Z. Sun, "IIA Moduli Stabilization With Badly Broken Supersymmetry," *JHEP* **0807** (2008) 070, arXiv:hep-th/0612189 [hep-th].

[302] E. Silverstein, "Simple de Sitter Solutions," *Phys.Rev.* **D77** (2008) 106006, arXiv:0712.1196 [hep-th].

[303] S. Haque, G. Shiu, B. Underwood, and T. Van Riet, "Minimal Simple de Sitter Solutions," *Phys.Rev.* **D79** (2009) 086005, arXiv:0810.5328 [hep-th].

[304] T. Grimm and J. Louis, "The Effective Action of $N = 1$ Calabi-Yau Orientifolds," *Nucl.Phys.* **B699** (2004) 387–426, arXiv:hep-th/0403067 [hep-th].

[305] S. Giddings, S. Kachru, and J. Polchinski, "Hierarchies from Fluxes in String Compactifications," *Phys.Rev.* **D66** (2002) 106006, arXiv:hep-th/0105097 [hep-th].

[306] A. Strominger, "Loop Corrections to the Universal Hypermultiplet," *Phys.Lett.* **B421** (1998) 139–148, arXiv:hep-th/9706195 [hep-th].

[307] K. Becker and M. Becker, "Instanton Action for Type II Hypermultiplets," *Nucl.Phys.* **B551** (1999) 102–116, arXiv:hep-th/9901126 [hep-th].

[308] I. Antoniadis, R. Minasian, S. Theisen, and P. Vanhove, "String Loop Corrections to the Universal Hypermultiplet," *Class.Quant.Grav.* **20** (2003) 5079–5102, arXiv:hep-th/0307268 [hep-th].

[309] X. Wen and E. Witten, "World Sheet Instantons and the Peccei-Quinn Symmetry," *Phys.Lett.* **B166** (1986) 397.

[310] E. Witten, "New Issues in Manifolds of $SU(3)$ Holonomy," *Nucl.Phys.* **B268** (1986) 79.

[311] M. Dine and N. Seiberg, "Non-Renormalization Theorems in Superstring Theory," *Phys.Rev.Lett.* **57** (1986) 2625.

[312] M. Dine, N. Seiberg, X. Wen, and E. Witten, "Nonperturbative Effects on the String World Sheet," *Nucl.Phys.* **B278** (1986) 769.

[313] T. Bachlechner, M. Dias, J. Frazer, and L. McAllister, "A New Angle on Chaotic Inflation," arXiv:1404.7496 [hep-th].

[314] M. Cicoli, J. Conlon, and F. Quevedo, "Dark Radiation in LARGE Volume Models," *Phys.Rev.* **D87** (2013) 043520, arXiv:1208.3562 [hep-ph].

[315] T. Higaki and F. Takahashi, "Dark Radiation and Dark Matter in Large Volume Compactifications," *JHEP* **1211** (2012) 125, arXiv:1208.3563 [hep-ph].

[316] J. Conlon and D. Marsh, "The Cosmo-Phenomenology of Axionic Dark Radiation," arXiv:1304.1804 [hep-ph].

[317] T. Higaki, K. Nakayama, and F. Takahashi, "Moduli-Induced Axion Problem," *JHEP* **1307** (2013) 005, arXiv:1304.7987 [hep-ph].

[318] J. Conlon and D. Marsh, "Searching for a 0.1-1 keV Cosmic Axion Background," arXiv:1305.3603 [astro-ph.CO].

[319] T. Higaki, K. Nakayama, and F. Takahashi, "Cosmological Constraints on Axionic Dark Radiation from Axion-Photon Conversion in the Early Universe," *JCAP* **1309** (2013) 030, arXiv:1306.6518 [hep-ph].

[320] M. Fairbairn, "Axionic Dark Radiation and the Milky Way's Magnetic Field," arXiv:1310.4464 [astro-ph.CO].

[321] S. Giddings and A. Maharana, "Dynamics of Warped Compactifications and the Shape of the Warped Landscape," *Phys.Rev.* **D73** (2006) 126003, arXiv:hep-th/0507158 [hep-th].

[322] G. Shiu, G. Torroba, B. Underwood, and M. Douglas, "Dynamics of Warped Flux Compactifications," *JHEP* **0806** (2008) 024, arXiv:0803.3068 [hep-th].

[323] M. Douglas and G. Torroba, "Kinetic Terms in Warped Compactifications," *JHEP* **0905** (2009) 013, arXiv:0805.3700 [hep-th].

[324] A. Frey, G. Torroba, B. Underwood, and M. Douglas, "The Universal Kähler Modulus in Warped Compactifications," *JHEP* **0901** (2009) 036, arXiv:0810.5768 [hep-th].

[325] F. Denef, M. Douglas, and S. Kachru, "Physics of String Flux Compactifications," *Ann.Rev.Nucl.Part.Sci.* **57** (2007) 119–144, arXiv:hep-th/0701050 [hep-th].

[326] M. Grana, "Flux Compactifications in String Theory: A Comprehensive Review," *Phys.Rept.* **423** (2006) 91–158, arXiv:hep-th/0509003 [hep-th].

[327] J. Maldacena and C. Nunez, "Supergravity Description of Field Theories on Curved Manifolds and a No-Go Theorem," *Int.J.Mod.Phys.* **A16** (2001) 822–855, arXiv:hep-th/0007018 [hep-th].

[328] D. Baumann, A. Dymarsky, S. Kachru, I. Klebanov, and L. McAllister, "Holographic Systematics of D-brane Inflation," *JHEP* **0903** (2009) 093, arXiv:0808.2811 [hep-th].

[329] S. Gukov, C. Vafa, and E. Witten, "CFT's from Calabi-Yau Four-Folds," *Nucl.Phys.* **B584** (2000) 69–108, arXiv:hep-th/9906070 [hep-th].

[330] J. Conlon, "Moduli Stabilisation and Applications in IIB String Theory," *Fortsch.Phys.* **55** (2007) 287–422, arXiv:hep-th/0611039 [hep-th].

[331] O. DeWolfe and S. Giddings, "Scales and Hierarchies in Warped Compactifications and Brane Worlds," *Phys.Rev.* **D67** (2003) 066008, arXiv:hep-th/0208123 [hep-th].

[332] K. Becker, M. Becker, M. Haack, and J. Louis, "Supersymmetry Breaking and Alpha-Prime Corrections to Flux-Induced Potentials," *JHEP* **0206** (2002) 060, arXiv:hep-th/0204254 [hep-th].

[333] M. Berg, M. Haack, and B. Kors, "String Loop Corrections to Kähler Potentials in Orientifolds," *JHEP* **0511** (2005) 030, arXiv:hep-th/0508043 [hep-th].

[334] M. Berg, M. Haack, and B. Kors, "On Volume Stabilization by Quantum Corrections," *Phys.Rev.Lett.* **96** (2006) 021601, arXiv:hep-th/0508171 [hep-th].

[335] M. Berg, M. Haack, and E. Pajer, "Jumping Through Loops: On Soft Terms from Large Volume Compactifications," *JHEP* **0709** (2007) 031, arXiv:0704.0737 [hep-th].

[336] G. von Gersdorff and A. Hebecker, "Kähler Corrections for the Volume Modulus of Flux Compactifications," *Phys.Lett.* **B624** (2005) 270–274, arXiv:hep-th/0507131 [hep-th].

[337] M. Grisaru, W. Siegel, and M. Rocek, "Improved Methods for Supergraphs," *Nucl.Phys.* **B159** (1979) 429.

[338] C. Burgess, C. Escoda, and F. Quevedo, "Non-Renormalization of Flux Superpotentials in String Theory," *JHEP* **0606** (2006) 044, arXiv:hep-th/0510213 [hep-th].

[339] V. Novikov, M. Shifman, A. Vainshtein, M. Voloshin, and V. Zakharov, "Supersymmetry Transformations of Instantons," *Nucl.Phys.* **B229** (1983) 394.

[340] V. Novikov, M. Shifman, A. Vainshtein, and V. Zakharov, "Instanton Effects in Supersymmetric Theories," *Nucl.Phys.* **B229** (1983) 407.

[341] S. Ferrara, L. Girardello, and H. Nilles, "Breakdown of Local Supersymmetry Through Gauge Fermion Condensates," *Phys.Lett.* **B125** (1983) 457.

[342] M. Dine, R. Rohm, N. Seiberg, and E. Witten, "Gluino Condensation in Superstring Models," *Phys.Lett.* **B156** (1985) 55.

[343] J. Derendinger, L. Ibanez, and H. Nilles, "On the Low-Energy $d = 4$, $N = 1$ Supergravity Theory Extracted from the $d = 10$, $N = 1$ Superstring," *Phys.Lett.* **B155** (1985) 65.

[344] M. Shifman and A. Vainshtein, "On Gluino Condensation in Supersymmetric Gauge Theories. $SU(N)$ and $O(N)$ Groups," *Nucl.Phys.* **B296** (1988) 445.

[345] L. Gorlich, S. Kachru, P. Tripathy, and S. Trivedi, "Gaugino Condensation and Non-perturbative Superpotentials in Flux Compactifications," *JHEP* **0412** (2004) 074, arXiv:hep-th/0407130 [hep-th].

[346] D. Baumann, A. Dymarsky, I. Klebanov, J. Maldacena, L. McAllister, and A. Murugan, "On D3-brane Potentials in Compactifications with Fluxes and Wrapped D-branes," *JHEP* **0611** (2006) 031, arXiv:hep-th/0607050 [hep-th].

[347] M. Berg, M. Haack, and B. Kors, "Loop Corrections to Volume Moduli and Inflation in String Theory," *Phys.Rev.* **D71** (2005) 026005, arXiv:hep-th/0404087 [hep-th].

[348] E. Witten, "Nonperturbative Superpotentials in String Theory," *Nucl.Phys.* **B474** (1996) 343–360, arXiv:hep-th/9604030 [hep-th].

[349] R. Blumenhagen, M. Cvetic, S. Kachru, and T. Weigand, "D-Brane Instantons in Type II Orientifolds," *Ann.Rev.Nucl.Part.Sci.* **59** (2009) 269–296, arXiv:0902.3251 [hep-th].

[350] K. Kodaira, "On The Structure Of Compact Complex Analytic Surfaces. I," *Amer. J. Math.* **86** (1964) 751–798.

[351] C. Vafa, "Evidence for F-theory," *Nucl.Phys.* **B469** (1996) 403–418, arXiv:hep-th/9602022 [hep-th].

[352] P. Griffiths and J. Harris, *Principles of Algebraic Geometry*. Wiley Classics Library. Wiley, 2011.

[353] F. Hirzebruch and M. Kreck, "On the Concept of Genus in Topology and Complex Analysis," *Notices of the AMS* **56** no. 6, (2009) .

[354] R. Kallosh and D. Sorokin, "Dirac Action on M5 and M2-branes with Bulk Fluxes," *JHEP* **0505** (2005) 005, arXiv:hep-th/0501081 [hep-th].

[355] N. Saulina, "Topological Constraints on Stabilized Flux Vacua," *Nucl.Phys.* **B720** (2005) 203–210, arXiv:hep-th/0503125 [hep-th].

[356] P. Tripathy and S. Trivedi, "D3-brane Action and Fermion Zero Modes in Presence of Background Flux," *JHEP* **0506** (2005) 066, arXiv:hep-th/0503072 [hep-th].

[357] R. Kallosh, A.-K. Kashani-Poor, and A. Tomasiello, "Counting Fermionic Zero Modes on M5 with Fluxes," *JHEP* **0506** (2005) 069, arXiv:hep-th/0503138 [hep-th].

[358] E. Bergshoeff, R. Kallosh, A.-K. Kashani-Poor, D. Sorokin, and A. Tomasiello, "An Index for the Dirac Operator on D3-branes with Background Fluxes," *JHEP* **0510** (2005) 102, arXiv:hep-th/0507069 [hep-th].

[359] L. Martucci, J. Rosseel, D. Van den Bleeken, and A. Van Proeyen, "Dirac Actions for D-branes on Backgrounds with Fluxes," *Class.Quant.Grav.* **22** (2005) 2745–2764, arXiv:hep-th/0504041 [hep-th].

[360] D. Robbins and S. Sethi, "A Barren Landscape?," *Phys.Rev.* **D71** (2005) 046008, arXiv:hep-th/0405011 [hep-th].

[361] R. Blumenhagen, S. Moster, and E. Plauschinn, "Moduli Stabilisation versus Chirality for MSSM like Type IIB Orientifolds," *JHEP* **0801** (2008) 058, arXiv:0711.3389 [hep-th].

[362] M. Bianchi, A. Collinucci, and L. Martucci, "Magnetized E3-brane Instantons in F-theory," *JHEP* **1112** (2011) 045, arXiv:1107.3732 [hep-th].

[363] M. Bianchi, G. Inverso, and L. Martucci, "Brane Instantons and Fluxes in F-theory," *JHEP* **1307** (2013) 037, arXiv:1212.0024 [hep-th].

[364] F. Denef, "Les Houches Lectures on Constructing String Vacua," arXiv:0803.1194 [hep-th].

[365] M. Dine and N. Seiberg, "Is the Superstring Weakly Coupled?," *Phys.Lett.* **B162** (1985) 299.

[366] E. Silverstein, "TASI / PiTP / ISS Lectures on Moduli and Microphysics," arXiv:hep-th/0405068 [hep-th].

[367] K. Bobkov, "Volume Stabilization via Alpha-Prime Corrections in Type IIB Theory with Fluxes," *JHEP* **0505** (2005) 010, arXiv:hep-th/0412239 [hep-th].

[368] S. Parameswaran and A. Westphal, "De Sitter String Vacua from Perturbative Kähler Corrections and Consistent D-terms," *JHEP* **0610** (2006) 079, arXiv:hep-th/0602253 [hep-th].

[369] S. Kachru, R. Kallosh, A. Linde, and S. Trivedi, "De Sitter Vacua in String Theory," *Phys.Rev.* **D68** (2003) 046005, arXiv:hep-th/0301240 [hep-th].

[370] V. Balasubramanian, P. Berglund, J. Conlon, and F. Quevedo, "Systematics of Moduli Stabilisation in Calabi-Yau Flux Compactifications," *JHEP* **0503** (2005) 007, arXiv:hep-th/0502058 [hep-th].

[371] F. Denef, M. Douglas, and B. Florea, "Building a Better Racetrack," *JHEP* **0406** (2004) 034, arXiv:hep-th/0404257 [hep-th].

[372] F. Denef, M. Douglas, B. Florea, A. Grassi, and S. Kachru, "Fixing all Moduli in a Simple F-theory Compactification," *Adv. Theor. Math. Phys.* **9** (2005) 861–929, arXiv:hep-th/0503124 [hep-th].

[373] K. Bobkov, V. Braun, P. Kumar, and S. Raby, "Stabilizing all Kähler Moduli in Type IIB Orientifolds," *JHEP* **1012** (2010) 056, arXiv:1003.1982 [hep-th].

[374] K. Choi, A. Falkowski, H. Nilles, M. Olechowski, and S. Pokorski, "Stability of Flux Compactifications and the Pattern of Supersymmetry Breaking," *JHEP* **0411** (2004) 076, arXiv:hep-th/0411066 [hep-th].

[375] S. de Alwis, "Effective Potentials for Light Moduli," *Phys.Lett.* **B626** (2005) 223–229, arXiv:hep-th/0506266 [hep-th].

[376] D. Lüst, S. Reffert, W. Schulgin, and S. Stieberger, "Moduli Stabilization in Type IIB Orientifolds (I): Orbifold Limits," *Nucl.Phys.* **B766** (2007) 68–149, arXiv:hep-th/0506090 [hep-th].

[377] G. Curio and V. Spillner, "On the Modified KKLT Procedure: A Case Study for the P(11169) [18] Model," *Int.J.Mod.Phys.* **A22** (2007) 3463–3492, arXiv:hep-th/0606047 [hep-th].

[378] H. Abe, T. Higaki, and T. Kobayashi, "Remark on Integrating Out Heavy Moduli in Flux Compactification," *Phys.Rev.* **D74** (2006) 045012, arXiv:hep-th/0606095 [hep-th].

[379] J. Conlon, F. Quevedo, and K. Suruliz, "Large Volume Flux Compactifications: Moduli Spectrum and D3/D7 Soft Supersymmetry Breaking," *JHEP* **0508** (2005) 007, arXiv:hep-th/0505076 [hep-th].

[380] M. Cicoli, J. Conlon, and F. Quevedo, "General Analysis of LARGE Volume Scenarios with String Loop Moduli Stabilisation," *JHEP* **0810** (2008) 105, arXiv:0805.1029 [hep-th].

[381] J. Gray *et al.*, "Calabi-Yau Manifolds with Large Volume Vacua," *Phys.Rev.* **D86** (2012) 101901, arXiv:1207.5801 [hep-th].

[382] M. Rummel and Y. Sumitomo, "Probability of Vacuum Stability in Type IIB Multi-Kähler Moduli Models," arXiv:1310.4202 [hep-th].

[383] S. Franco and A. Uranga, "Dynamical SUSY Breaking at Metastable Minima from D-branes at Obstructed Geometries," *JHEP* **0606** (2006) 031, arXiv:hep-th/0604136 [hep-th].

[384] R. Argurio, M. Bertolini, C. Closset, and S. Cremonesi, "On Stable Non-Supersymmetric Vacua at the Bottom of Cascading Theories," *JHEP* **0609** (2006) 030, arXiv:hep-th/0606175 [hep-th].

[385] B. Florea, S. Kachru, J. McGreevy, and N. Saulina, "Stringy Instantons and Quiver Gauge Theories," *JHEP* **0705** (2007) 024, arXiv:hep-th/0610003 [hep-th].

[386] R. Argurio, M. Bertolini, S. Franco, and S. Kachru, "Gauge/Gravity Duality and Metastable Dynamical Supersymmetry Breaking," *JHEP* **0701** (2007) 083, arXiv:hep-th/0610212 [hep-th].

[387] D.-E. Diaconescu, R. Donagi, and B. Florea, "Metastable Quivers in String Compactifications," *Nucl.Phys.* **B774** (2007) 102–126, arXiv:hep-th/0701104 [hep-th].

[388] I. Garcia-Etxebarria, F. Saad, and A. Uranga, "Supersymmetry Breaking Metastable Vacua in Runaway Quiver Gauge Theories," *JHEP* **0705** (2007) 047, arXiv:0704.0166 [hep-th].

[389] M. Buican, D. Malyshev, and H. Verlinde, "On the Geometry of Metastable Supersymmetry Breaking," *JHEP* **0806** (2008) 108, arXiv:0710.5519 [hep-th].

[390] D. Berenstein, C. Herzog, P. Ouyang, and S. Pinansky, "Supersymmetry Breaking from a Calabi-Yau Singularity," *JHEP* **0509** (2005) 084, arXiv:hep-th/0505029 [hep-th].

[391] S. Franco, A. Hanany, F. Saad, and A. Uranga, "Fractional Branes and Dynamical Supersymmetry Breaking," *JHEP* **0601** (2006) 011, arXiv:hep-th/0505040 [hep-th].

[392] M. Bertolini, F. Bigazzi, and A. Cotrone, "Supersymmetry Breaking at the End of a Cascade of Seiberg Dualities," *Phys.Rev.* **D72** (2005) 061902, arXiv:hep-th/0505055 [hep-th].

[393] K. Intriligator and N. Seiberg, "The Runaway Quiver," *JHEP* **0602** (2006) 031, arXiv:hep-th/0512347 [hep-th].

[394] A. Brini and D. Forcella, "Comments on the Non-Conformal Gauge Theories Dual to $Y^{p,q}$ Manifolds," *JHEP* **0606** (2006) 050, arXiv:hep-th/0603245 [hep-th].

[395] K. Intriligator, N. Seiberg, and D. Shih, "Dynamical SUSY Breaking in Metastable Vacua," *JHEP* **0604** (2006) 021, arXiv:hep-th/0602239 [hep-th].

[396] C. Burgess, R. Kallosh, and F. Quevedo, "De Sitter String Vacua from Supersymmetric D-terms," *JHEP* **0310** (2003) 056, arXiv:hep-th/0309187 [hep-th].

[397] C. Escoda, M. Gomez-Reino, and F. Quevedo, "Saltatory de Sitter String Vacua," *JHEP* **0311** (2003) 065, arXiv:hep-th/0307160 [hep-th].

[398] R. Brustein and S. de Alwis, "Moduli Potentials in String Compactifications with Fluxes: Mapping the Discretuum," *Phys.Rev.* **D69** (2004) 126006, arXiv:hep-th/0402088 [hep-th].

[399] A. Saltman and E. Silverstein, "The Scaling of the No-Scale Potential and de Sitter Model Building," *JHEP* **0411** (2004) 066, arXiv:hep-th/0402135 [hep-th].

[400] A. Saltman and E. Silverstein, "A New Handle on de Sitter Compactifications," *JHEP* **0601** (2006) 139, arXiv:hep-th/0411271 [hep-th].

[401] O. Lebedev, H. Nilles, and M. Ratz, "De Sitter Vacua from Matter Superpotentials," *Phys.Lett.* **B636** (2006) 126–131, arXiv:hep-th/0603047 [hep-th].

[402] O. Lebedev, V. Lowen, Y. Mambrini, H. Nilles, and M. Ratz, "Metastable Vacua in Flux Compactifications and Their Phenomenology," *JHEP* **0702** (2007) 063, arXiv:hep-ph/0612035 [hep-ph].

[403] A. Achucarro, B. de Carlos, J. Casas, and L. Doplicher, "De Sitter Vacua from Uplifting D-terms in Effective Supergravities from Realistic Strings," *JHEP* **0606** (2006) 014, arXiv:hep-th/0601190 [hep-th].

[404] E. Dudas, C. Papineau, and S. Pokorski, "Moduli Stabilization and Uplifting with Dynamically Generated F-terms," *JHEP* **0702** (2007) 028, arXiv:hep-th/0610297 [hep-th].

[405] A. Westphal, "De Sitter String Vacua from Kähler Uplifting," *JHEP* **0703** (2007) 102, arXiv:hep-th/0611332 [hep-th].

[406] D. Cremades, M.-P. Garcia del Moral, F. Quevedo, and K. Suruliz, "Moduli Stabilisation and de Sitter String Vacua from Magnetised D7-branes," *JHEP* **0705** (2007) 100, arXiv:hep-th/0701154 [hep-th].

[407] L. Covi, M. Gomez-Reino, C. Gross, G. Palma, and C. Scrucca, "Constructing de Sitter Vacua in No-Scale String Models without Uplifting," *JHEP* **0903** (2009) 146, arXiv:0812.3864 [hep-th].

[408] J. Polchinski and E. Silverstein, "Dual Purpose Landscaping Tools: Small Extra Dimensions in AdS/CFT," arXiv:0908.0756 [hep-th].

[409] S. Krippendorf and F. Quevedo, "Metastable SUSY Breaking, de Sitter Moduli Stabilisation and Kähler Moduli Inflation," *JHEP* **0911** (2009) 039, arXiv:0901.0683 [hep-th].

[410] X. Dong, B. Horn, E. Silverstein, and G. Torroba, "Micromanaging de Sitter Holography," *Class.Quant.Grav.* **27** (2010) 245020, arXiv:1005.5403 [hep-th].

[411] M. Rummel and A. Westphal, "A Sufficient Condition for de Sitter Vacua in Type IIB String Theory," *JHEP* **1201** (2012) 020, arXiv:1107.2115 [hep-th].

[412] J. Louis, M. Rummel, R. Valandro, and A. Westphal, "Building an Explicit de Sitter," *JHEP* **1210** (2012) 163, arXiv:1208.3208 [hep-th].

[413] M. Cicoli, A. Maharana, F. Quevedo, and C. Burgess, "De Sitter String Vacua from Dilaton-Dependent Non-perturbative Effects," *JHEP* **1206** (2012) 011, arXiv:1203.1750 [hep-th].

[414] M. Davidse, F. Saueressig, U. Theis, and S. Vandoren, "Membrane Instantons and de Sitter Vacua," *JHEP* **0509** (2005) 065, arXiv:hep-th/0506097 [hep-th].

[415] F. Saueressig, U. Theis, and S. Vandoren, "On de Sitter Vacua in Type IIA Orientifold Compactifications," *Phys.Lett.* **B633** (2006) 125–128, arXiv:hep-th/0506181 [hep-th].

[416] C. Caviezel *et al.*, "The Effective Theory of Type IIA *AdS*₄ Compactifications on Nilmanifolds and Cosets," *Class.Quant.Grav.* **26** (2009) 025014, arXiv:0806.3458 [hep-th].

[417] C. Caviezel *et al.*, "On the Cosmology of Type IIA Compactifications on SU(3)-structure Manifolds," *JHEP* **0904** (2009) 010, arXiv:0812.3551 [hep-th].

[418] R. Flauger, S. Paban, D. Robbins, and T. Wrase, "Searching for Slow-Roll Moduli Inflation in Massive Type IIA Supergravity with Metric Fluxes," *Phys.Rev.* **D79** (2009) 086011, arXiv:0812.3886 [hep-th].

[419] U. Danielsson, S. Haque, G. Shiu, and T. Van Riet, "Towards Classical de Sitter Solutions in String Theory," *JHEP* **0909** (2009) 114, arXiv:0907.2041 [hep-th].

[420] T. Wrase and M. Zagermann, "On Classical de Sitter Vacua in String Theory," *Fortsch.Phys.* **58** (2010) 906–910, arXiv:1003.0029 [hep-th].

[421] U. Danielsson, P. Koerber, and T. Van Riet, "Universal de Sitter Solutions at Tree-Level," *JHEP* **1005** (2010) 090, arXiv:1003.3590 [hep-th].

[422] U. Danielsson *et al.*, "De Sitter Hunting in a Classical Landscape," *Fortsch.Phys.* **59** (2011) 897–933, arXiv:1103.4858 [hep-th].

[423] G. Shiu and Y. Sumitomo, "Stability Constraints on Classical de Sitter Vacua," *JHEP* **1109** (2011) 052, arXiv:1107.2925 [hep-th].

[424] U. Danielsson and G. Dibitetto, "On the Distribution of Stable de Sitter Vacua," *JHEP* **1303** (2013) 018, arXiv:1212.4984 [hep-th].

[425] U. Danielsson, G. Shiu, T. Van Riet, and T. Wrase, "A Note on Obstinate Tachyons in Classical dS Solutions," *JHEP* **1303** (2013) 138, arXiv:1212.5178 [hep-th].

[426] E. Buchbinder and B. Ovrut, "Vacuum Stability in Heterotic M-theory," *Phys.Rev.* **D69** (2004) 086010, arXiv:hep-th/0310112 [hep-th].

[427] M. Becker, G. Curio, and A. Krause, "De Sitter Vacua from Heterotic M-theory," *Nucl.Phys.* **B693** (2004) 223–260, arXiv:hep-th/0403027 [hep-th].

[428] E. Buchbinder, "Raising Anti de Sitter Vacua to de Sitter Vacua in Heterotic M-theory," *Phys.Rev.* **D70** (2004) 066008, arXiv:hep-th/0406101 [hep-th].

[429] G. Curio and A. Krause, "S-track Stabilization of Heterotic de Sitter Vacua," *Phys.Rev.* **D75** (2007) 126003, arXiv:hep-th/0606243 [hep-th].

[430] S. Parameswaran, S. Ramos-Sanchez, and I. Zavala, "On Moduli Stabilisation and de Sitter Vacua in MSSM Heterotic Orbifolds," *JHEP* **1101** (2011) 071, arXiv:1009.3931 [hep-th].

[431] A. Maloney, E. Silverstein, and A. Strominger, "De Sitter Space in Noncritical String Theory," arXiv:hep-th/0205316 [hep-th].

[432] C. Hull, "De Sitter Space in Supergravity and M-theory," *JHEP* **0111** (2001) 012, arXiv:hep-th/0109213 [hep-th].

[433] R. Kallosh, A. Linde, S. Prokushkin, and M. Shmakova, "Gauged Supergravities, de Sitter Space and Cosmology," *Phys.Rev.* **D65** (2002) 105016, arXiv:hep-th/0110089 [hep-th].

[434] G. Gibbons and C. Hull, "De Sitter Space from Warped Supergravity Solutions," arXiv:hep-th/0111072 [hep-th].

[435] R. Kallosh, "$N = 2$ Supersymmetry and de Sitter Space," arXiv:hep-th/0109168 [hep-th].

[436] A. Chamblin and N. Lambert, "De Sitter Space from M-theory," *Phys.Lett.* **B508** (2001) 369–374, arXiv:hep-th/0102159 [hep-th].

[437] P. Fre, M. Trigiante, and A. Van Proeyen, "Stable de Sitter Vacua from $N = 2$ Supergravity," *Class.Quant.Grav.* **19** (2002) 4167–4194, arXiv:hep-th/0205119 [hep-th].

[438] M. Gomez-Reino, J. Louis, and C. A. Scrucca, "No Metastable de Sitter Vacua in $N = 2$ Supergravity with only Hypermultiplets," *JHEP* **0902** (2009) 003, arXiv:0812.0884 [hep-th].

[439] D. Roest and J. Rosseel, "De Sitter in Extended Supergravity," *Phys.Lett.* **B685** (2010) 201–207, arXiv:0912.4440 [hep-th].

[440] E. Dudas, N. Kitazawa, and A. Sagnotti, "On Climbing Scalars in String Theory," *Phys.Lett.* **B694** (2010) 80–88, arXiv:1009.0874 [hep-th].

[441] P. Fre, A. Sagnotti, and A. Sorin, "Integrable Scalar Cosmologies I. Foundations and Links with String Theory," *Nucl.Phys.* **B877** (2013) 1028–1106, arXiv:1307.1910 [hep-th].

[442] J. Russo, "Exact Solution of Scalar Tensor Cosmology with Exponential Potentials and Transient Acceleration," *Phys.Lett.* **B600** (2004) 185–190, arXiv:hep-th/0403010 [hep-th].

[443] S. Kachru, J. Pearson, and H. Verlinde, "Brane/Flux Annihilation and the String Dual of a Non-Supersymmetric Field Theory," *JHEP* **0206** (2002) 021, arXiv:hep-th/0112197 [hep-th].

[444] I. Klebanov and M. Strassler, "Supergravity and a Confining Gauge Theory: Duality Cascades and chi SB Resolution of Naked Singularities," *JHEP* **0008** (2000) 052, arXiv:hep-th/0007191 [hep-th].

[445] I. Bena, M. Grana, and N. Halmagyi, "On the Existence of Metastable Vacua in Klebanov-Strassler," *JHEP* **1009** (2010) 087, arXiv:0912.3519 [hep-th].

[446] I. Bena, G. Giecold, M. Grana, and N. Halmagyi, "On the Inflaton Potential from Antibranes in Warped Throats," *JHEP* **1207** (2012) 140, arXiv:1011.2626 [hep-th].

[447] A. Dymarsky, "On Gravity Dual of a Metastable Vacuum in Klebanov-Strassler Theory," *JHEP* **1105** (2011) 053, arXiv:1102.1734 [hep-th].

[448] I. Bena, G. Giecold, M. Grana, N. Halmagyi, and S. Massai, "On Metastable Vacua and the Warped Deformed Conifold: Analytic Results," *Class.Quant.Grav.* **30** (2013) 015003, arXiv:1102.2403 [hep-th].

[449] I. Bena, G. Giecold, M. Grana, N. Halmagyi, and S. Massai, "The Backreaction of Anti-D3-branes on the Klebanov-Strassler Geometry," arXiv:1106.6165 [hep-th].

[450] J. Blaback, U. Danielsson, and T. Van Riet, "Resolving Antibrane Singularities Through Time-Dependence," *JHEP* **1302** (2013) 061, arXiv:1202.1132 [hep-th].

[451] I. Bena *et al.*, "Persistent Anti-Brane Singularities," *JHEP* **1210** (2012) 078, arXiv:1205.1798 [hep-th].

[452] D. Kutasov and A. Wissanji, "IIA Perspective On Cascading Gauge Theory," *JHEP* **1209** (2012) 080, arXiv:1206.0747 [hep-th].

[453] I. Bena, M. Grana, S. Kuperstein, and S. Massai, "Anti-D3's – Singular to the Bitter End," arXiv:1206.6369 [hep-th].

[454] I. Bena, M. Grana, S. Kuperstein, and S. Massai, "Polchinski-Strassler does not uplift Klebanov-Strassler," arXiv:1212.4828 [hep-th].

[455] I. Bena, A. Buchel, and O. Dias, "Horizons Cannot Save the Landscape," arXiv:1212.5162 [hep-th].

[456] D. Junghans, "Backreaction of Localised Sources in String Compactifications," arXiv:1309.5990 [hep-th].

[457] A. Dymarsky and S. Massai, "Uplifting the Baryonic Branch: A Test for Backreacting Anti-D3-Branes," arXiv:1310.0015 [hep-th].

[458] D. Junghans, "Dynamics of Warped Flux Compactifications with Backreacting Anti-Branes," arXiv:1402.4571 [hep-th].

[459] K. Dasgupta, R. Gwyn, E. McDonough, M. Mia, and R. Tatar, "De Sitter Vacua in Type IIB String Theory: Classical Solutions and Quantum Corrections," arXiv:1402.5112 [hep-th].

[460] D. Junghans, D. Schmidt, and M. Zagermann, "Curvature-induced Resolution of Antibrane Singularities," arXiv:1402.6040 [hep-th].

[461] O. DeWolfe, S. Kachru, and M. Mulligan, "A Gravity Dual of Metastable Dynamical Supersymmetry Breaking," *Phys.Rev.* **D77** (2008) 065011, arXiv:0801.1520 [hep-th].

[462] P. McGuirk, G. Shiu, and Y. Sumitomo, "Non-supersymmetric Infrared Perturbations to the Warped Deformed Conifold," *Nucl.Phys.* **B842** (2011) 383–413, arXiv:0910.4581 [hep-th].

[463] J. Blaback *et al.*, "The Problematic Backreaction of SUSY-breaking Branes," *JHEP* **1108** (2011) 105, arXiv:1105.4879 [hep-th].

[464] J. Polchinski and M. Strassler, "The String Dual of a Confining Four-Dimensional Gauge Theory," arXiv:hep-th/0003136 [hep-th].

[465] N. Wolchover, "Is nature unnatural?," *Quanta Magazine* (2013).

[466] A. Strominger, "Superstrings with Torsion," *Nucl.Phys.* **B274** (1986) 253.

[467] W. Lerche, D. Lüst, and A. Schellekens, "Chiral Four-Dimensional Heterotic Strings from Selfdual Lattices," *Nucl.Phys.* **B287** (1987) 477.

[468] M. Douglas, "The Statistics of String / M-theory Vacua," *JHEP* **0305** (2003) 046, arXiv:hep-th/0303194 [hep-th].

[469]   J. Kumar, "A Review of Distributions on the String Landscape," *Int.J.Mod.Phys.* **A21** (2006) 3441–3472, arXiv:hep-th/0601053 [hep-th].

[470]   M. Dine, "Is There A String Theory Landscape: Some Cautionary Notes," arXiv:hep-th/0402101 [hep-th].

[471]   T. Banks, "Landskepticism or Why Effective Potentials Don't Count String Models," arXiv:hep-th/0412129 [hep-th].

[472]   T. Banks, "The Top 10^{500} Reasons Not to Believe in the Landscape," arXiv:1208.5715 [hep-th].

[473]   S. Ashok and M. Douglas, "Counting Flux Vacua," *JHEP* **0401** (2004) 060, arXiv:hep-th/0307049 [hep-th].

[474]   A. Giryavets, S. Kachru, and P. Tripathy, "On the Taxonomy of Flux Vacua," *JHEP* **0408** (2004) 002, arXiv:hep-th/0404243 [hep-th].

[475]   O. DeWolfe, A. Giryavets, S. Kachru, and W. Taylor, "Enumerating Flux Vacua with Enhanced Symmetries," *JHEP* **0502** (2005) 037, arXiv:hep-th/0411061 [hep-th].

[476]   A. Giryavets, S. Kachru, P. Tripathy, and S. Trivedi, "Flux Compactifications on Calabi-Yau Threefolds," *JHEP* **0404** (2004) 003, arXiv:hep-th/0312104 [hep-th].

[477]   F. Denef and M. Douglas, "Distributions of Nonsupersymmetric Flux Vacua," *JHEP* **0503** (2005) 061, arXiv:hep-th/0411183 [hep-th].

[478]   M. Mehta, *Random Matrices*. Pure and Applied Mathematics 142. Elsevier Science, 2004.

[479]   E. Wigner, "On the Statistical Distribution of the Widths and Spacings of Nuclear Resonance Levels," *Mathematical proceedings of the Cambridge Philosophical Society* **47** no. 4, (1951) 790–798.

[480]   L. Erdős, "Universality of Wigner Random Matrices: a Survey of Recent Results," arXiv:1004.0861v2 [math-ph].

[481]   A. Kuijlaars, "Universality," arXiv:1103.5922 [math-ph].

[482]   P. Deift, "Universality for Mathematical and Physical Systems," arXiv:math-ph/0603038 [math-ph].

[483]   J. H. Schenker and H. Schulz-Baldes, "Semicircle Law and Freeness for Random Matrices with Symmetries or Correlations," arXiv:math-ph/0505003.

[484]   T. Tao and V. Vu, "Random Covariance Matrices: Universality of Local Statistics of Eigenvalues," *Ann. Probab.* **40** no. 3, (Dec., 2012) 1285–1315, arXiv:0912.0966 [math.SP].

[485]   J. Distler and U. Varadarajan, "Random Polynomials and the Friendly Landscape," arXiv:hep-th/0507090 [hep-th].

[486]   D. Voiculescu, K. Dykema, and A. Nica, *Free Random Variables: A Noncommutative Probability Approach to Free Products with Applications to Random Matrices, Operator Algebras, and Harmonic Analysis on Free Groups*. CRM Monograph Series. American Mathematical Society, 1992.

[487]   C. Tracy and H. Widom, "Level-Spacing Distributions and the Airy Kernel," *Communications in Mathematical Physics* **159** no. 1, (1994) 151–174.

[488]   T. Bachlechner, D. Marsh, L. McAllister, and T. Wrase, *unpublished*.

[489]   A. Ben Arous, G. and Guionnet, "Large Deviations for Wigner's Law and Voiculescu's Non-Commutative Entropy," *Prob. Th. Rel. Fields* **108** (1997) 517–542.

[490]   D. Dean and S. Majumdar, "Large Deviations of Extreme Eigenvalues of Random Matrices," *Phys. Rev. Lett.* **97** (2006) 160201, arXiv:cond-mat/0609651 [cond-mat].

[491]   D. Dean and S. Majumdar, "Extreme Value Statistics of Eigenvalues of Gaussian Random Matrices," *Phys. Rev. E* **77** (2008) 041108, arXiv:0801.1730 [cond-mat.stat-mech].

[492]   A. Aazami and R. Easther, "Cosmology from Random Multifield Potentials," *JCAP* **0603** (2006) 013, arXiv:hep-th/0512050 [hep-th].

[493]   X. Chen, G. Shiu, Y. Sumitomo, and H. Tye, "A Global View on The Search for de-Sitter Vacua in (Type IIA) String Theory," *JHEP* **1204** (2012) 026, arXiv:1112.3338 [hep-th].

[494]   P. Breitenlohner and D. Freedman, "Positive Energy in Anti-De Sitter Backgrounds and Gauged Extended Supergravity," *Phys.Lett.* **B115** (1982) 197.

[495] T. Bachlechner, D. Marsh, L. McAllister, and T. Wrase, "Supersymmetric Vacua in Random Supergravity," *JHEP* **1301** (2013) 136, arXiv:1207.2763 [hep-th].

[496] A. Achucarro and K. Sousa, "F-term Uplifting and Moduli Stabilization Consistent with Kähler Invariance," *JHEP* **0803** (2008) 002, arXiv:0712.3460 [hep-th].

[497] L. Covi et al., "De Sitter Vacua in No-Scale Supergravities and Calabi-Yau String Models," *JHEP* **0806** (2008) 057, arXiv:0804.1073 [hep-th].

[498] L. Covi et al., "Constraints on Modular Inflation in Supergravity and String Theory," *JHEP* **0808** (2008) 055, arXiv:0805.3290 [hep-th].

[499] A. Borghese, D. Roest, and I. Zavala, "A Geometric Bound on F-term Inflation," *JHEP* **1209** (2012) 021, arXiv:1203.2909 [hep-th].

[500] J. Bausch, "On the Efficient Calculation of a Linear Combination of Chi-Square Random Variables with an Application in Counting String Vacua," arXiv:1208.2691 [math.PR].

[501] T. Bachlechner, "On Gaussian Random Supergravity," arXiv:1401.6187 [hep-th].

[502] R. Easther and L. McAllister, "Random Matrices and the Spectrum of N-flation," *JCAP* **0605** (2006) 018, arXiv:hep-th/0512102 [hep-th].

[503] D. Battefeld, T. Battefeld, and S. Schulz, "On the Unlikeliness of Multi-Field Inflation: Bounded Random Potentials and our Vacuum," *JCAP* **1206** (2012) 034, arXiv:1203.3941 [hep-th].

[504] A. Westphal, "Tensor Modes on the String Theory Landscape," *JHEP* **1304** (2013) 054, arXiv:1206.4034 [hep-th].

[505] F. Pedro and A. Westphal, "The Scale of Inflation in the Landscape," arXiv:1303.3224 [hep-th].

[506] D. Battefeld and T. Battefeld, "A Smooth Landscape: Ending Saddle Point Inflation Requires Features to be Shallow," arXiv:1304.0461 [hep-th].

[507] D. Marsh, L. McAllister, E. Pajer, and T. Wrase, "Charting an Inflationary Landscape with Random Matrix Theory," arXiv:1307.3559 [hep-th].

[508] C. Burgess, A. Maharana, and F. Quevedo, "Uber-naturalness: Unexpectedly Light Scalars from Supersymmetric Extra Dimensions," *JHEP* **1105** (2011) 010, arXiv:1005.1199 [hep-th].

[509] X. Chen and Y. Wang, "Quasi-Single Field Inflation and Non-Gaussianities," *JCAP* **1004** (2010) 027, arXiv:0911.3380 [hep-th].

[510] T. Noumi, M. Yamaguchi, and D. Yokoyama, "Effective Field Theory Approach to Quasi-Single-Field Inflation," *JHEP* **1306** (2013) 051, arXiv:1211.1624 [hep-th].

[511] E. Sefusatti, J. Fergusson, X. Chen, and E. Shellard, "Effects and Detectability of Quasi-Single-Field Inflation in the Large-Scale Structure and Cosmic Microwave Background," *JCAP* **1208** (2012) 033, arXiv:1204.6318 [astro-ph.CO].

[512] A. Achucarro, J.-O. Gong, S. Hardeman, G. Palma, and S. Patil, "Effective Theories of Single-Field Inflation when Heavy Fields Matter," *JHEP* **1205** (2012) 066, arXiv:1201.6342 [hep-th].

[513] S. Cespedes, V. Atal, and G. Palma, "On the Importance of Heavy Fields during Inflation," *JCAP* **1205** (2012) 008, arXiv:1201.4848 [hep-th].

[514] G. Shiu and J. Xu, "Effective Field Theory and Decoupling in Multi-Field Inflation: An Illustrative Case Study," *Phys.Rev.* **D84** (2011) 103509, arXiv:1108.0981 [hep-th].

[515] A. Achucarro, J.-O. Gong, S. Hardeman, G. Palma, and S. Patil, "Features of Heavy Physics in the CMB Power Spectrum," *JCAP* **1101** (2011) 030, arXiv:1010.3693 [hep-ph].

[516] S. Cremonini, Z. Lalak, and K. Turzynski, "Strongly Coupled Perturbations in Two-Field Inflationary Models," *JCAP* **1103** (2011) 016, arXiv:1010.3021 [hep-th].

[517] G. Villadoro and F. Zwirner, "$N = 1$ Effective Potential from Dual Type-IIA D6/O6 Orientifolds with General Fluxes," *JHEP* **0506** (2005) 047, arXiv:hep-th/0503169 [hep-th].

[518] O. DeWolfe, A. Giryavets, S. Kachru, and W. Taylor, "Type IIA Moduli Stabilization," *JHEP* **0507** (2005) 066, arXiv:hep-th/0505160 [hep-th].

[519] M. Ihl and T. Wrase, "Towards a Realistic Type IIA T^6/Z_4 Orientifold Model with Background Fluxes. Part 1. Moduli Stabilization," *JHEP* **0607** (2006) 027, arXiv:hep-th/0604087 [hep-th].

[520] M. Hertzberg, S. Kachru, W. Taylor, and M. Tegmark, "Inflationary Constraints on Type IIA String Theory," *JHEP* **0712** (2007) 095, arXiv:0711.2512 [hep-th].

[521] M. Hertzberg, M. Tegmark, S. Kachru, J. Shelton, and O. Ozcan, "Searching for Inflation in Simple String Theory Models: An Astrophysical Perspective," *Phys.Rev.* **D76** (2007) 103521, arXiv:0709.0002 [astro-ph].

[522] N. Ohta, "Accelerating Cosmologies and Inflation from M/Superstring Theories," *Int.J.Mod.Phys.* **A20** (2005) 1–40, arXiv:hep-th/0411230 [hep-th].

[523] J. M. Maldacena, "The Large N Limit of Superconformal Field Theories and Supergravity," *Adv.Theor.Math.Phys.* **2** (1998) 231–252, arXiv:hep-th/9711200 [hep-th].

[524] A. Achucarro, J.-O. Gong, S. Hardeman, G. Palma, and S. Patil, "Mass Hierarchies and Non-Decoupling in Multi-Scalar Field Dynamics," *Phys.Rev.* **D84** (2011) 043502, arXiv:1005.3848 [hep-th].

[525] A. Achucarro *et al.*, "Heavy Fields, Reduced Speeds of Sound and Decoupling During Inflation," *Phys.Rev.* **D86** (2012) 121301, arXiv:1205.0710 [hep-th].

[526] S. Hardeman, J. Oberreuter, G. Palma, K. Schalm, and T. van der Aalst, "The Ever-present Eta Problem: Knowledge of All Hidden Sectors Required," *JHEP* **1104** (2011) 009, arXiv:1012.5966 [hep-ph].

[527] L. Randall and R. Sundrum, "Out of this World Supersymmetry Breaking," *Nucl.Phys.* **B557** (1999) 79–118, arXiv:hep-th/9810155 [hep-th].

[528] A. Anisimov, M. Dine, M. Graesser, and S. Thomas, "Brane World SUSY Breaking," *Phys.Rev.* **D65** (2002) 105011, arXiv:hep-th/0111235 [hep-th].

[529] S. Kachru, J. McGreevy, and P. Svrcek, "Bounds on Masses of Bulk Fields in String Compactifications," *JHEP* **0604** (2006) 023, arXiv:hep-th/0601111 [hep-th].

[530] S. Kachru, L. McAllister, and R. Sundrum, "Sequestering in String Theory," *JHEP* **0710** (2007) 013, arXiv:hep-th/0703105 [hep-th].

[531] R. Blumenhagen, J. Conlon, S. Krippendorf, S. Moster, and F. Quevedo, "SUSY Breaking in Local String/F-Theory Models," *JHEP* **0909** (2009) 007, arXiv:0906.3297 [hep-th].

[532] M. Berg, D. Marsh, L. McAllister, and E. Pajer, "Sequestering in String Compactifications," *JHEP* **1106** (2011) 134, arXiv:1012.1858 [hep-th].

[533] M. Berg, J. Conlon, D. Marsh, and L. Witkowski, "Superpotential De-Sequestering in String Models," *JHEP* **1302** (2013) 018, arXiv:1207.1103 [hep-th].

[534] C. Burgess *et al.*, "The Inflationary Brane Anti-Brane Universe," *JHEP* **0107** (2001) 047, arXiv:hep-th/0105204 [hep-th].

[535] S. Shandera, B. Shlaer, H. Stoica, and H. Tye, "Interbrane Interactions in Compact Spaces and Brane Inflation," *JCAP* **0402** (2004) 013, arXiv:hep-th/0311207 [hep-th].

[536] P. Binetruy and G. Dvali, "D-term Inflation," *Phys.Lett.* **B388** (1996) 241–246, arXiv:hep-ph/9606342 [hep-ph].

[537] Z. Komargodski and N. Seiberg, "Comments on the Fayet-Iliopoulos Term in Field Theory and Supergravity," *JHEP* **0906** (2009) 007, arXiv:0904.1159 [hep-th].

[538] L. McAllister, "An Inflaton Mass Problem in String Inflation from Threshold Corrections to Volume Stabilization," *JCAP* **0602** (2006) 010, arXiv:hep-th/0502001 [hep-th].

[539] M. Kawasaki, M. Yamaguchi, and T. Yanagida, "Natural Chaotic Inflation in Supergravity," *Phys.Rev.Lett.* **85** (2000) 3572–3575, arXiv:hep-ph/0004243 [hep-ph].

[540] M. Gaillard, H. Murayama, and K. Olive, "Preserving Flat Directions During Inflation," *Phys.Lett.* **B355** (1995) 71–77, arXiv:hep-ph/9504307 [hep-ph].

[541] N. Arkani-Hamed, H.-C. Cheng, P. Creminelli, and L. Randall, "Extranatural Inflation," *Phys.Rev.Lett.* **90** (2003) 221302, arXiv:hep-th/0301218 [hep-th].

[542] D. Baumann and D. Green, "Desensitizing Inflation from the Planck Scale," *JHEP* **1009** (2010) 057, arXiv:1004.3801 [hep-th].

[543] D. Roest, M. Scalisi, and I. Zavala, "Kähler Potentials for Planck Inflation," arXiv:1307.4343 [hep-th].

[544] R. Kallosh, A. Linde, and T. Rube, "General Inflaton Potentials in Supergravity," *Phys.Rev.* **D83** (2011) 043507, arXiv:1011.5945 [hep-th].

[545] R. Kallosh and A. Linde, "New Models of Chaotic Inflation in Supergravity," *JCAP* **1011** (2010) 011, arXiv:1008.3375 [hep-th].

[546] S. Ferrara, R. Kallosh, A. Linde, A. Marrani, and A. Van Proeyen, "Superconformal Symmetry, NMSSM, and Inflation," *Phys.Rev.* **D83** (2011) 025008, arXiv:1008.2942 [hep-th].

[547] S. Ferrara, R. Kallosh, A. Linde, A. Marrani, and A. Van Proeyen, "Jordan Frame Supergravity and Inflation in NMSSM," *Phys.Rev.* **D82** (2010) 045003, arXiv:1004.0712 [hep-th].

[548] L. Alvarez-Gaume, C. Gomez, and R. Jimenez, "A Minimal Inflation Scenario," *JCAP* **1103** (2011) 027, arXiv:1101.4948 [hep-th].

[549] S. Dimopoulos, S. Kachru, J. McGreevy, and J. G. Wacker, "N-flation," *JCAP* **0808** (2008) 003, arXiv:hep-th/0507205 [hep-th].

[550] A. Liddle, A. Mazumdar, and F. Schunck, "Assisted Inflation," *Phys.Rev.* **D58** (1998) 061301, arXiv:astro-ph/9804177 [astro-ph].

[551] X. Dong, B. Horn, E. Silverstein, and A. Westphal, "Simple Exercises to Flatten Your Potential," *Phys.Rev.* **D84** (2011) 026011, arXiv:1011.4521 [hep-th].

[552] S. Panda, M. Sami, and S. Tsujikawa, "Prospects of Inflation in Delicate D-brane Cosmology," *Phys.Rev.* **D76** (2007) 103512, arXiv:0707.2848 [hep-th].

[553] S. Gandhi, L. McAllister, and S. Sjors, "A Toolkit for Perturbing Flux Compactifications," *JHEP* **1112** (2011) 053, arXiv:1106.0002 [hep-th].

[554] M. Dias, J. Frazer, and A. Liddle, "Multifield Consequences for D-brane Inflation," arXiv:1203.3792 [astro-ph.CO].

[555] M. Tegmark, "What does inflation really predict?," *JCAP* **0504** (2005) 001, arXiv:astro-ph/0410281 [astro-ph].

[556] H. Tye, J. Xu, and Y. Zhang, "Multi-Field Inflation with a Random Potential," *JCAP* **0904** (2009) 018, arXiv:0812.1944 [hep-th].

[557] H. Tye and J. Xu, "A Meandering Inflaton," *Phys.Lett.* **B683** (2010) 326–330, arXiv:0910.0849 [hep-th].

[558] D. Battefeld and T. Battefeld, "Multi-Field Inflation on the Landscape," *JCAP* **0903** (2009) 027, arXiv:0812.0367 [hep-th].

[559] J. Frazer and A. Liddle, "Exploring a String-like Landscape," *JCAP* **1102** (2011) 026, arXiv:1101.1619 [astro-ph.CO].

[560] J. Frazer and A. Liddle, "Multi-Field Inflation with Random Potentials: Field Dimension, Feature Scale and Non-Gaussianity," *JCAP* **1202** (2012) 039, arXiv:1111.6646 [astro-ph.CO].

[561] J. Frazer, "Predictions in Multi-Field Models of Inflation," arXiv:1303.3611 [astro-ph.CO].

[562] K. Enqvist and M. Sloth, "Adiabatic CMB Perturbations in Pre-Big Bang String Cosmology," *Nucl.Phys.* **B626** (2002) 395–409, arXiv:hep-ph/0109214 [hep-ph].

[563] R. Allahverdi, R. Brandenberger, F.-Y. Cyr-Racine, and A. Mazumdar, "Reheating in Inflationary Cosmology: Theory and Applications," *Ann.Rev.Nucl.Part.Sci.* **60** (2010) 27–51, arXiv:1001.2600 [hep-th].

[564] L. Kofman and P. Yi, "Reheating the Universe after String Theory Inflation," *Phys.Rev.* **D72** (2005) 106001, arXiv:hep-th/0507257 [hep-th].

[565] N. Barnaby, C. Burgess, and J. Cline, "Warped Reheating in Brane-Antibrane Inflation," *JCAP* **0504** (2005) 007, arXiv:hep-th/0412040 [hep-th].

[566] A. Frey, A. Mazumdar, and R. Myers, "Stringy Effects during Inflation and Reheating," *Phys.Rev.* **D73** (2006) 026003, arXiv:hep-th/0508139 [hep-th].

[567] D. Chialva, G. Shiu, and B. Underwood, "Warped Reheating in Multi-Throat Brane Inflation," *JHEP* **0601** (2006) 014, arXiv:hep-th/0508229 [hep-th].

[568] N. Barnaby and J. Cline, "Non-Gaussian and Non-Scale-Invariant Perturbations from Tachyonic Preheating in Hybrid Inflation," *Phys.Rev.* **D73** (2006) 106012, arXiv:astro-ph/0601481 [astro-ph].

[569] P. Langfelder, "On Tunnelling in Two-Throat Warped Reheating," *JHEP* **0606** (2006) 063, arXiv:hep-th/0602296 [hep-th].

[570] X. Chen and H. Tye, "Heating in Brane Inflation and Hidden Dark Matter," *JCAP* **0606** (2006) 011, arXiv:hep-th/0602136 [hep-th].

[571] R. Brandenberger, A. Frey, and L. Lorenz, "Entropy Fluctuations in Brane Inflation Models," *Int.J.Mod.Phys.* **A24** (2009) 4327–4354, arXiv:0712.2178 [hep-th].

[572] A. Berndsen, J. Cline, and H. Stoica, "Kaluza-Klein Relics from Warped Reheating," *Phys.Rev.* **D77** (2008) 123522, arXiv:0710.1299 [hep-th].

[573] J. Dufaux, L. Kofman, and M. Peloso, "Dangerous Angular KK/Glueball Relics in String Theory Cosmology," *Phys.Rev.* **D78** (2008) 023520, arXiv:0802.2958 [hep-th].

[574] S. Panda, M. Sami, and I. Thongkool, "Reheating the D-brane Universe via Instant Preheating," *Phys.Rev.* **D81** (2010) 103506, arXiv:0905.2284 [hep-th].

[575] A. Frey, R. Danos, and J. Cline, "Warped Kaluza–Klein Dark Matter," *JHEP* **0911** (2009) 102, arXiv:0908.1387 [hep-th].

[576] R. Brandenberger, A. Knauf, and L. Lorenz, "Reheating in a Brane Monodromy Inflation Model," *JHEP* **0810** (2008) 110, arXiv:0808.3936 [hep-th].

[577] R. Brandenberger, K. Dasgupta, and A.-C. Davis, "A Study of Structure Formation and Reheating in the D3/D7 Brane Inflation Model," *Phys.Rev.* **D78** (2008) 083502, arXiv:0801.3674 [hep-th].

[578] N. Bouatta, A.-C. Davis, R. Ribeiro, and D. Seery, "Preheating in Dirac–Born–Infeld Inflation," *JCAP* **1009** (2010) 011, arXiv:1005.2425 [astro-ph.CO].

[579] D. Green, "Reheating Closed String Inflation," *Phys.Rev.* **D76** (2007) 103504, arXiv:0707.3832 [hep-th].

[580] M. Cicoli and A. Mazumdar, "Reheating for Closed String Inflation," *JCAP* **1009** (2010) 025, arXiv:1005.5076 [hep-th].

[581] N. Barnaby, J. Bond, Z. Huang, and L. Kofman, "Preheating After Modular Inflation," *JCAP* **0912** (2009) 021, arXiv:0909.0503 [hep-th].

[582] C. Burgess *et al.*, "Non-Standard Primordial Fluctuations and Non-Gaussianity in String Inflation," *JHEP* **1008** (2010) 045, arXiv:1005.4840 [hep-th].

[583] J. Braden, L. Kofman, and N. Barnaby, "Reheating the Universe After Multi-Field Inflation," *JCAP* **1007** (2010) 016, arXiv:1005.2196 [hep-th].

[584] M. Cicoli and A. Mazumdar, "Inflation in String Theory: A Graceful Exit to the Real World," *Phys.Rev.* **D83** (2011) 063527, arXiv:1010.0941 [hep-th].

[585] M. Cicoli, G. Tasinato, I. Zavala, C. Burgess, and F. Quevedo, "Modulated Reheating and Large Non-Gaussianity in String Cosmology," *JCAP* **1205** (2012) 039, arXiv:1202.4580 [hep-th].

[586] J. Preskill, "Cosmological Production of Superheavy Magnetic Monopoles," *Phys.Rev.Lett.* **43** (1979) 1365.

[587] A. Vilenkin and E. Shellard, *Cosmic Strings and Other Topological Defects.* Cambridge Monographs on Mathematical Physics. Cambridge University Press, 2000.

[588] M. Hindmarsh and T. Kibble, "Cosmic Strings," *Rept.Prog.Phys.* **58** (1995) 477–562, arXiv:hep-ph/9411342 [hep-ph].

[589] J. Polchinski, "Introduction to Cosmic F- and D-Strings," arXiv:hep-th/0412244 [hep-th].

[590] T. Kibble, "Cosmic Strings Reborn?," arXiv:astro-ph/0410073 [astro-ph].

[591] E. Copeland and T. Kibble, "Cosmic Strings and Superstrings," *Proc.R.Soc.Lond.* **A466** (2010) 623–657, arXiv:0911.1345 [hep-th].

[592] T. Kibble, "Topology of Cosmic Domains and Strings," *J.Phys.* **A9** (1976) 1387–1398.

[593] J. Christiansen *et al.*, "Search for Cosmic Strings in the COSMOS Survey," *Phys.Rev.* **D83** (2011) 122004, arXiv:1008.0426 [astro-ph.CO].

[594] D. Chernoff and H. Tye, "Cosmic String Detection via Microlensing of Stars," arXiv:0709.1139 [astro-ph].

[595] K. Kuijken, X. Siemens, and T. Vachaspati, "Microlensing by Cosmic Strings," *Mon.Not.Roy.Astron.Soc.* (2007) , arXiv:0707.2971 [astro-ph].

[596] D. Chernoff, "Clustering of Superstring Loops," arXiv:0908.4077 [astro-ph.CO].

[597] N. Kaiser and A. Stebbins, "Microwave Anisotropy Due to Cosmic Strings," *Nature* **310** (1984) 391–393.

[598] R. Brandenberger, R. Danos, O. Hernandez, and G. Holder, "The 21 cm Signature of Cosmic String Wakes," *JCAP* **1012** (2010) 028, arXiv:1006.2514 [astro-ph.CO].

[599] F. Bouchet, P. Peter, A. Riazuelo, and M. Sakellariadou, "Is there evidence for topological defects in the BOOMERANG data?," *Phys.Rev.* **D65** (2002) 021301, arXiv:astro-ph/0005022 [astro-ph].

[600] E. Jeong and G. Smoot, "Search for Cosmic Strings in CMB Anisotropies," *Astrophys.J.* **624** (2005) 21–27, arXiv:astro-ph/0406432 [astro-ph].

[601] L. Pogosian, M. Wyman, and I. Wasserman, "Observational Constraints on Cosmic Strings: Bayesian Analysis in a Three Dimensional Parameter Space," *JCAP* **0409** (2004) 008, arXiv:astro-ph/0403268 [astro-ph].

[602] M. Wyman, L. Pogosian, and I. Wasserman, "Bounds on Cosmic Strings from WMAP and SDSS," *Phys.Rev.* **D72** (2005) 023513, arXiv:astro-ph/0503364 [astro-ph].

[603] N. Bevis, M. Hindmarsh, M. Kunz, and J. Urrestilla, "Fitting CMB Data with Cosmic Strings and Inflation," *Phys.Rev.Lett.* **100** (2008) 021301, arXiv:astro-ph/0702223 [astro-ph].

[604] **LIGO and Virgo** Collaboration, J. Abadie *et al.*, "Upper Limits on a Stochastic Gravitational-Wave Background using LIGO and Virgo Interferometers at 600-1000 Hz," *Phys.Rev.* **D85** (2012) 122001, arXiv:1112.5004 [gr-qc].

[605] X. Siemens, V. Mandic, and J. Creighton, "Gravitational Wave Stochastic Background from Cosmic (Super)strings," *Phys.Rev.Lett.* **98** (2007) 111101, arXiv:astro-ph/0610920 [astro-ph].

[606] T. Damour and A. Vilenkin, "Gravitational Wave Bursts from Cosmic Strings," *Phys.Rev.Lett.* **85** (2000) 3761–3764, arXiv:gr-qc/0004075 [gr-qc].

[607] T. Damour and A. Vilenkin, "Gravitational Wave Bursts from Cusps and Kinks on Cosmic Strings," *Phys.Rev.* **D64** (2001) 064008, arXiv:gr-qc/0104026 [gr-qc].

[608] T. Damour and A. Vilenkin, "Gravitational Radiation from Cosmic (Super)strings: Bursts, Stochastic Background, and Observational Windows," *Phys.Rev.* **D71** (2005) 063510, arXiv:hep-th/0410222 [hep-th].

[609] H. Motohashi and T. Suyama, "Detecting Cosmic String Passage through the Earth by Consequent Global Earthquake," arXiv:1305.6676 [astro-ph.CO].

[610] X. Martin and A. Vilenkin, "Gravitational Radiation from Monopoles Connected by Strings," *Phys.Rev.* **D55** (1997) 6054–6060, arXiv:gr-qc/9612008 [gr-qc].

[611] L. Leblond, B. Shlaer, and X. Siemens, "Gravitational Waves from Broken Cosmic Strings: The Bursts and the Beads," *Phys.Rev.* **D79** (2009) 123519, arXiv:0903.4686 [astro-ph.CO].

[612] H. Nielsen and P. Olesen, "Vortex Line Models for Dual Strings," *Nucl.Phys.* **B61** (1973) 45–61.

[613] E. Witten, "Cosmic Superstrings," *Phys.Lett.* **B153** (1985) 243.

[614] N. Jones, H. Stoica, and H. Tye, "Brane Interaction as the Origin of Inflation," *JHEP* **0207** (2002) 051, arXiv:hep-th/0203163 [hep-th].

[615] S. Sarangi and H. Tye, "Cosmic String Production towards the End of Brane Inflation," *Phys.Lett.* **B536** (2002) 185–192, arXiv:hep-th/0204074 [hep-th].

[616] N. Jones, H. Stoica, and H. Tye, "The Production, Spectrum and Evolution of Cosmic Strings in Brane Inflation," *Phys.Lett.* **B563** (2003) 6–14, arXiv:hep-th/0303269 [hep-th].

[617] K. Becker, M. Becker, and A. Krause, "Heterotic Cosmic Strings," *Phys.Rev.* **D74** (2006) 045023, arXiv:hep-th/0510066 [hep-th].

[618] A. Sen, "$SO(32)$ Spinors of Type I and Other Solitons on Brane/Anti-brane Pair," *JHEP* **9809** (1998) 023, arXiv:hep-th/9808141 [hep-th].

[619] G. Dvali and A. Vilenkin, "Formation and Evolution of Cosmic D-Strings," *JCAP* **0403** (2004) 010, arXiv:hep-th/0312007 [hep-th].

[620] E. Copeland, R. C. Myers, and J. Polchinski, "Cosmic F and D-Strings," *JHEP* **0406** (2004) 013, arXiv:hep-th/0312067 [hep-th].

[621] M. Jackson, N. Jones, and J. Polchinski, "Collisions of Cosmic F and D-Strings," *JHEP* **0510** (2005) 013, arXiv:hep-th/0405229 [hep-th].

[622] J. Harvey and A. Strominger, "The Heterotic String is a Soliton," *Nucl.Phys.* **B449** (1995) 535–552, arXiv:hep-th/9504047 [hep-th].

[623] J. Schwarz, "An $SL(2,Z)$ Multiplet of Type IIB Superstrings," *Phys.Lett.* **B360** (1995) 13–18, arXiv:hep-th/9508143 [hep-th].

[624] H. Tye, I. Wasserman, and M. Wyman, "Scaling of Multi-Tension Cosmic Superstring Networks," *Phys.Rev.* **D71** (2005) 103508, arXiv:astro-ph/0503506 [astro-ph].

[625] L. Leblond and H. Tye, "Stability of D1-Strings Inside a D3-brane," *JHEP* **0403** (2004) 055, arXiv:hep-th/0402072 [hep-th].

[626] H. Firouzjahi and H. Tye, "Brane Inflation and Cosmic String Tension in Superstring Theory," *JCAP* **0503** (2005) 009, arXiv:hep-th/0501099 [hep-th].

[627] G. Dvali and H. Tye, "Brane Inflation," *Phys.Lett.* **B450** (1999) 72–82, arXiv:hep-ph/9812483 [hep-ph].

[628] K. Dasgupta, C. Herdeiro, S. Hirano, and R. Kallosh, "D3/D7 Inflationary Model and M-theory," *Phys.Rev.* **D65** (2002) 126002, arXiv:hep-th/0203019 [hep-th].

[629] J. Hsu, R. Kallosh, and S. Prokushkin, "On Brane Inflation with Volume Stabilization," *JCAP* **0312** (2003) 009, arXiv:hep-th/0311077 [hep-th].

[630] K. Dasgupta, J. Hsu, R. Kallosh, A. Linde, and M. Zagermann, "D3/D7 Brane Inflation and Semilocal Strings," *JHEP* **0408** (2004) 030, arXiv:hep-th/0405247 [hep-th].

[631] J. Hsu and R. Kallosh, "Volume Stabilization and the Origin of the Inflaton Shift Symmetry in String Theory," *JHEP* **0404** (2004) 042, arXiv:hep-th/0402047 [hep-th].

[632] M. Haack *et al.*, "Update of D3/D7-Brane Inflation on $K3 \times T^2/Z_2$," *Nucl.Phys.* **B806** (2009) 103–177, arXiv:0804.3961 [hep-th].

[633] A. Hebecker, S. Kraus, D. Lüst, S. Steinfurt, and T. Weigand, "Fluxbrane Inflation," *Nucl.Phys.* **B854** (2012) 509–551, arXiv:1104.5016 [hep-th].

[634] A. Hebecker, S. Kraus, M. Kuntzler, D. Lüst, and T. Weigand, "Fluxbranes: Moduli Stabilisation and Inflation," *JHEP* **1301** (2013) 095, arXiv:1207.2766 [hep-th].

[635] E. Buchbinder, "Five-brane Dynamics and Inflation in Heterotic M-theory," *Nucl.Phys.* **B711** (2005) 314–344, arXiv:hep-th/0411062 [hep-th].

[636] K. Becker, M. Becker, and A. Krause, "M-theory Inflation from Multi-M5-brane Dynamics," *Nucl.Phys.* **B715** (2005) 349–371, arXiv:hep-th/0501130 [hep-th].

[637] J. Blanco-Pillado *et al.*, "Racetrack Inflation," *JHEP* **0411** (2004) 063, arXiv:hep-th/0406230 [hep-th].

[638] J. Blanco-Pillado *et al.*, "Inflating in a Better Racetrack," *JHEP* **0609** (2006) 002, arXiv:hep-th/0603129 [hep-th].

[639] J. Conlon and F. Quevedo, "Kähler Moduli Inflation," *JHEP* **0601** (2006) 146, arXiv:hep-th/0509012 [hep-th].

[640] M. Cicoli, F. Pedro, and G. Tasinato, "Poly-instanton Inflation," *JCAP* **1112** (2011) 022, arXiv:1110.6182 [hep-th].

[641] D. Green, B. Horn, L. Senatore, and E. Silverstein, "Trapped Inflation," *Phys.Rev.* **D80** (2009) 063533, arXiv:0902.1006 [hep-th].

[642] G. D'Amico, R. Gobbetti, M. Schillo, and M. Kleban, "Inflation from Flux Cascades," arXiv:1211.3416 [hep-th].

[643] G. D'Amico, R. Gobbetti, M. Kleban, and M. Schillo, "Unwinding Inflation," *JCAP* **1303** (2013) 004, arXiv:1211.4589 [hep-th].

[644] E. Martinec, P. Adshead, and M. Wyman, "Chern-Simons EM-flation," *JHEP* **1302** (2013) 027, arXiv:1206.2889 [hep-th].

[645] G. Dvali, Q. Shafi, and S. Solganik, "D-brane Inflation," arXiv:hep-th/0105203 [hep-th].

[646] S. Alexander, "Inflation from D-anti-D-brane Annihilation," *Phys.Rev.* **D65** (2002) 023507, arXiv:hep-th/0105032 [hep-th].

[647] C. Chan, P. Paul, and H. Verlinde, "A Note on Warped String Compactification," *Nucl.Phys.* **B581** (2000) 156–164, arXiv:hep-th/0003236 [hep-th].

[648] I. Klebanov and A. Tseytlin, "Gravity Duals of Supersymmetric $SU(N) \times SU(N + M)$ Gauge Theories," *Nucl.Phys.* **B578** (2000) 123–138, arXiv:hep-th/0002159 [hep-th].

[649] A. Kehagias and E. Kiritsis, "Mirage Cosmology," *JHEP* **9911** (1999) 022, arXiv:hep-th/9910174 [hep-th].

[650] S. Kachru and L. McAllister, "Bouncing Brane Cosmologies from Warped String Compactifications," *JHEP* **0303** (2003) 018, arXiv:hep-th/0205209 [hep-th].

[651] C. Germani, N. Grandi, and A. Kehagias, "A Stringy Alternative to Inflation: The Cosmological Slingshot Scenario," *Class.Quant.Grav.* **25** (2008) 135004, arXiv:hep-th/0611246 [hep-th].

[652] D. Easson, R. Gregory, G. Tasinato, and I. Zavala, "Cycling in the Throat," *JHEP* **0704** (2007) 026, arXiv:hep-th/0701252 [HEP-TH].

[653] P. Candelas and X. de la Ossa, "Comments on Conifolds," *Nucl.Phys.* **B342** (1990) 246–268.

[654] I. Klebanov and E. Witten, "Superconformal Field Theory on Three-Branes at a Calabi-Yau Singularity," *Nucl.Phys.* **B536** (1998) 199–218, arXiv:hep-th/9807080 [hep-th].

[655] C. Vafa, "Superstrings and Topological Strings at Large N," *J.Math.Phys.* **42** (2001) 2798–2817, arXiv:hep-th/0008142 [hep-th].

[656] C. Burgess, J. Cline, K. Dasgupta, and H. Firouzjahi, "Uplifting and Inflation with D3-branes," *JHEP* **0703** (2007) 027, arXiv:hep-th/0610320 [hep-th].

[657] A. Krause and E. Pajer, "Chasing Brane Inflation in String Theory," *JCAP* **0807** (2008) 023, arXiv:0705.4682 [hep-th].

[658] D. Arean, D. Crooks, and A. Ramallo, "Supersymmetric Probes on the Conifold," *JHEP* **0411** (2004) 035, arXiv:hep-th/0408210 [hep-th].

[659] S. Kuperstein, "Meson Spectroscopy from Holomorphic Probes on the Warped Deformed Conifold," *JHEP* **0503** (2005) 014, arXiv:hep-th/0411097 [hep-th].

[660] D. Baumann, A. Dymarsky, S. Kachru, I. Klebanov, and L. McAllister, "Compactification Effects in D-brane Inflation," *Phys.Rev.Lett.* **104** (2010) 251602, arXiv:0912.4268 [hep-th].

[661] A. Ceresole, G. Dall'Agata, R. D'Auria, and S. Ferrara, "Spectrum of Type IIB Supergravity on $AdS_5 \times T^{1,1}$: Predictions on $N = 1$ SCFT's," *Phys.Rev.* **D61** (2000) 066001, arXiv:hep-th/9905226 [hep-th].

[662] P. Koerber and L. Martucci, "From Ten to Four and Back Again: How to Generalize the Geometry," *JHEP* **0708** (2007) 059, arXiv:0707.1038 [hep-th].

[663] A. Ali, R. Chingangbam, S. Panda, and M. Sami, "Prospects of Inflation with Perturbed Throat Geometry," *Phys.Lett.* **B674** (2009) 131–136, arXiv:0809.4941 [hep-th].

[664] A. Ali, A. Deshamukhya, S. Panda, and M. Sami, "Inflation with Improved D3-brane Potential and the Fine-Tunings Associated with the Model," *Eur.Phys.J.* **C71** (2011) 1672, arXiv:1010.1407 [hep-th].

[665] R. Kallosh and A. Linde, "Landscape, the Scale of SUSY Breaking, and Inflation," *JHEP* **0412** (2004) 004, arXiv:hep-th/0411011 [hep-th].

[666] A. Sen, "Rolling Tachyon," *JHEP* **0204** (2002) 048, arXiv:hep-th/0203211 [hep-th].

[667] A. Sen, "Tachyon Matter," *JHEP* **0207** (2002) 065, arXiv:hep-th/0203265 [hep-th].

[668] N. Lambert, H. Liu, and J. Maldacena, "Closed Strings from Decaying D-branes," *JHEP* **0703** (2007) 014, arXiv:hep-th/0303139 [hep-th].

[669] S. Mukohyama, "Reheating a Multi-Throat Universe by Brane Motion," *Gen.Rel.Grav.* **41** (2009) 1151–1163, arXiv:0706.3214 [hep-th].

[670] S. Dimopoulos, S. Kachru, N. Kaloper, A. Lawrence, and E. Silverstein, "Small Numbers from Tunneling between Brane Throats," *Phys.Rev.* **D64** (2001) 121702, arXiv:hep-th/0104239 [hep-th].

[671] O. Aharony, Y. Antebi, and M. Berkooz, "Open String Moduli in KKLT Compactifications," *Phys.Rev.* **D72** (2005) 106009, arXiv:hep-th/0508080 [hep-th].

[672] H. Firouzjahi and H. Tye, "Closer towards Inflation in String Theory," *Phys.Lett.* **B584** (2004) 147–154, arXiv:hep-th/0312020 [hep-th].

[673] N. Iizuka and S. Trivedi, "An Inflationary Model in String Theory," *Phys.Rev.* **D70** (2004) 043519, arXiv:hep-th/0403203 [hep-th].

[674] J. Cline and H. Stoica, "Multibrane Inflation and Dynamical Flattening of the Inflaton Potential," *Phys.Rev.* **D72** (2005) 126004, arXiv:hep-th/0508029 [hep-th].

[675] L. Hoi and J. Cline, "How Delicate is Brane-Antibrane Inflation?," *Phys.Rev.* **D79** (2009) 083537, arXiv:0810.1303 [hep-th].

[676] J. Cline, L. Hoi, and B. Underwood, "Dynamical Fine Tuning in Brane Inflation," *JHEP* **0906** (2009) 078, arXiv:0902.0339 [hep-th].

[677] D. Easson and R. Gregory, "Circumventing the Eta Problem," *Phys.Rev.* **D80** (2009) 083518, arXiv:0902.1798 [hep-th].

[678] M. Bastero-Gil, A. Berera, and J. Rosa, "Warming up Brane-Antibrane Inflation," *Phys.Rev.* **D84** (2011) 103503, arXiv:1103.5623 [hep-th].

[679] S. Kachru, D. Simic, and S. Trivedi, "Stable Non-Supersymmetric Throats in String Theory," *JHEP* **1005** (2010) 067, arXiv:0905.2970 [hep-th].

[680] R. Brustein and P. Steinhardt, "Challenges for Superstring Cosmology," *Phys.Lett.* **B302** (1993) 196–201, arXiv:hep-th/9212049 [hep-th].

[681] B. Underwood, "Brane Inflation is Attractive," *Phys.Rev.* **D78** (2008) 023509, arXiv:0802.2117 [hep-th].

[682] S. Bird, H. Peiris, and D. Baumann, "Brane Inflation and the Overshoot Problem," *Phys.Rev.* **D80** (2009) 023534, arXiv:0905.2412 [hep-th].

[683] B. Freivogel, M. Kleban, M. Rodriguez Martinez, and L. Susskind, "Observational Consequences of a Landscape," *JHEP* **0603** (2006) 039, arXiv:hep-th/0505232 [hep-th].

[684] K. Dutta, P. Vaudrevange, and A. Westphal, "The Overshoot Problem in Inflation after Tunneling," *JCAP* **1201** (2012) 026, arXiv:1109.5182 [hep-th].

[685] N. Itzhaki and E. Kovetz, "Inflection Point Inflation and Time Dependent Potentials in String Theory," *JHEP* **0710** (2007) 054, arXiv:0708.2798 [hep-th].

[686] L. Kofman *et al.*, "Beauty is Attractive: Moduli Trapping at Enhanced Symmetry Points," *JHEP* **0405** (2004) 030, arXiv:hep-th/0403001 [hep-th].

[687] S. Watson, "Moduli Stabilization with the String Higgs Effect," *Phys.Rev.* **D70** (2004) 066005, arXiv:hep-th/0404177 [hep-th].

[688] B. Greene, S. Judes, J. Levin, S. Watson, and A. Weltman, "Cosmological Moduli Dynamics," *JHEP* **0707** (2007) 060, arXiv:hep-th/0702220 [hep-th].

[689] D. Battefeld and T. Battefeld, "A Terminal Velocity on the Landscape: Particle Production near Extra Species Loci in Higher Dimensions," *JHEP* **1007** (2010) 063, arXiv:1004.3551 [hep-th].

[690] S. Shandera and H. Tye, "Observing Brane Inflation," *JCAP* **0605** (2006) 007, arXiv:hep-th/0601099 [hep-th].

[691] H. Peiris, D. Baumann, B. Friedman, and A. Cooray, "Phenomenology of D-Brane Inflation with General Speed of Sound," *Phys.Rev.* **D76** (2007) 103517, arXiv:0706.1240 [astro-ph].

[692] H.-Y. Chen and J.-O. Gong, "Towards a Warped Inflationary Brane Scanning," *Phys.Rev.* **D80** (2009) 063507, arXiv:0812.4649 [hep-th].

[693] O. DeWolfe, S. Kachru, and H. Verlinde, "The Giant Inflaton," *JHEP* **0405** (2004) 017, arXiv:hep-th/0403123 [hep-th].

[694] E. Pajer, "Inflation at the Tip," *JCAP* **0804** (2008) 031, arXiv:0802.2916 [hep-th].

[695] H.-Y. Chen, J.-O. Gong, and G. Shiu, "Systematics of Multi-Field Effects at the End of Warped Brane Inflation," *JHEP* **0809** (2008) 011, arXiv:0807.1927 [hep-th].

[696] H.-Y. Chen, J.-O. Gong, K. Koyama, and G. Tasinato, "Towards Multi-Field D-brane Inflation in a Warped Throat," *JCAP* **1011** (2010) 034, arXiv:1007.2068 [hep-th].

[697] X. Chen, S. Sarangi, H. Tye, and J. Xu, "Is Brane Inflation Eternal?," *JCAP* **0611** (2006) 015, arXiv:hep-th/0608082 [hep-th].

[698] T. Kobayashi, S. Mukohyama, and S. Kinoshita, "Constraints on Wrapped DBI Inflation in a Warped Throat," *JCAP* **0801** (2008) 028, arXiv:0708.4285 [hep-th].

[699] M. Becker, L. Leblond, and S. Shandera, "Inflation from Wrapped Branes," *Phys.Rev.* **D76** (2007) 123516, arXiv:0709.1170 [hep-th].

[700] S. Chang, M. Kleban, and T. Levi, "When Worlds Collide," *JCAP* **0804** (2008) 034, arXiv:0712.2261 [hep-th].

[701] S. Chang, M. Kleban, and T. Levi, "Watching Worlds Collide: Effects on the CMB from Cosmological Bubble Collisions," *JCAP* **0904** (2009) 025, arXiv:0810.5128 [hep-th].

[702] S. Feeney, M. Johnson, D. Mortlock, and H. Peiris, "First Observational Tests of Eternal Inflation: Analysis Methods and WMAP 7-Year Results," *Phys.Rev.* **D84** (2011) 043507, arXiv:1012.3667 [astro-ph.CO].

[703] S. Feeney, M. Johnson, D. Mortlock, and H. Peiris, "First Observational Tests of Eternal Inflation," *Phys.Rev.Lett.* **107** (2011) 071301, arXiv:1012.1995 [astro-ph.CO].

[704] S. Osborne, L. Senatore, and K. Smith, "Collisions with other Universes: the Optimal Analysis of the WMAP Data," arXiv:1305.1964 [astro-ph.CO].

[705] J. Elliston, D. Mulryne, D. Seery, and R. Tavakol, "Evolution of f_{NL} to the Adiabatic Limit," *JCAP* **1111** (2011) 005, arXiv:1106.2153 [astro-ph.CO].

[706] R. Bean, X. Chen, G. Hailu, H. Tye, and J. Xu, "Duality Cascade in Brane Inflation," *JCAP* **0803** (2008) 026, arXiv:0802.0491 [hep-th].

[707] C. Burgess, J. Cline, H. Stoica, and F. Quevedo, "Inflation in Realistic D-brane Models," *JHEP* **0409** (2004) 033, arXiv:hep-th/0403119 [hep-th].

[708] H. Firouzjahi, L. Leblond, and H. Tye, "The (p, q) String Tension in a Warped Deformed Conifold," *JHEP* **0605** (2006) 047, arXiv:hep-th/0603161 [hep-th].

[709] C. Burgess, J. Cline, and M. Postma, "Axionic D3-D7 Inflation," *JHEP* **0903** (2009) 058, arXiv:0811.1503 [hep-th].

[710] P. Aspinwall and R. Kallosh, "Fixing all Moduli for M-theory on $K3 \times K3$," *JHEP* **0510** (2005) 001, arXiv:hep-th/0506014 [hep-th].

[711] P. Binetruy, G. Dvali, R. Kallosh, and A. Van Proeyen, "Fayet–Iliopoulos Terms in Supergravity and Cosmology," *Class.Quant.Grav.* **21** (2004) 3137–3170, arXiv:hep-th/0402046 [hep-th].

[712] Z. Komargodski and N. Seiberg, "Comments on the Fayet–Iliopoulos Term in Field Theory and Supergravity," *JHEP* **0906** (2009) 007, arXiv:0904.1159 [hep-th].

[713] K. Dienes and B. Thomas, "On the Inconsistency of Fayet–Iliopoulos Terms in Supergravity Theories," *Phys.Rev.* **D81** (2010) 065023, arXiv:0911.0677 [hep-th].

[714] R. Gwyn, M. Sakellariadou, and S. Sypsas, "Theoretical Constraints on Brane Inflation and Cosmic Superstring Radiation," *JHEP* **1109** (2011) 075, arXiv:1105.1784 [hep-th].

[715] Z. Komargodski and N. Seiberg, "Comments on Supercurrent Multiplets, Supersymmetric Field Theories and Supergravity," *JHEP* **1007** (2010) 017, arXiv:1002.2228 [hep-th].

[716] M. Berg, M. Haack, and B. Kors, "On the Moduli Dependence of Nonperturbative Superpotentials in Brane Inflation," arXiv:hep-th/0409282 [hep-th].

[717] J. Garcia-Bellido, R. Rabadan, and F. Zamora, "Inflationary Scenarios from Branes at Angles," *JHEP* **0201** (2002) 036, arXiv:hep-th/0112147 [hep-th].

[718] R. Blumenhagen, B. Kors, D. Lüst, and T. Ott, "Hybrid Inflation in Intersecting Brane Worlds," *Nucl.Phys.* **B641** (2002) 235–255, arXiv:hep-th/0202124 [hep-th].

[719] M. Gomez-Reino and I. Zavala, "Recombination of Intersecting D-branes and Cosmological Inflation," *JHEP* **0209** (2002) 020, arXiv:hep-th/0207278 [hep-th].

[720] A. Avgoustidis, D. Cremades, and F. Quevedo, "Wilson Line Inflation," *Gen.Rel.Grav.* **39** (2007) 1203–1234, arXiv:hep-th/0606031 [hep-th].

[721] A. Avgoustidis and I. Zavala, "Warped Wilson Line DBI Inflation," *JCAP* **0901** (2009) 045, arXiv:0810.5001 [hep-th].

[722] T. Grimm, M. Kerstan, E. Palti, and T. Weigand, "On Fluxed Instantons and Moduli Stabilisation in IIB Orientifolds and F-theory," *Phys.Rev.* **D84** (2011) 066001, arXiv:1105.3193 [hep-th].

[723] A. Ashoorioon and A. Krause, "Power Spectrum and Signatures for Cascade Inflation," arXiv:hep-th/0607001 [hep-th].

[724] A. Krause, "Large Gravitational Waves and Lyth Bound in Multi-Brane Inflation," *JCAP* **0807** (2008) 001, arXiv:0708.4414 [hep-th].

[725] L. Anderson, J. Gray, A. Lukas, and B. Ovrut, "Stabilizing All Geometric Moduli in Heterotic Calabi-Yau Vacua," *Phys.Rev.* **D83** (2011) 106011, arXiv:1102.0011 [hep-th].

[726] M. Cicoli, S. de Alwis, and A. Westphal, "Heterotic Moduli Stabilization," arXiv:1304.1809 [hep-th].

[727] S. Gukov, S. Kachru, X. Liu, and L. McAllister, "Heterotic Moduli Stabilization with Fractional Chern-Simons Invariants," *Phys.Rev.* **D69** (2004) 086008, arXiv:hep-th/0310159 [hep-th].

[728] P. Brax, S. Davis, and M. Postma, "The Robustness of $n_s \lesssim 0.95$ in Racetrack Inflation," *JCAP* **0802** (2008) 020, arXiv:0712.0535 [hep-th].

[729] J. Urrestilla, A. Achucarro, and A. Davis, "D-term Inflation without Cosmic Strings," *Phys.Rev.Lett.* **92** (2004) 251302, arXiv:hep-th/0402032 [hep-th].

[730] L. Randall and R. Sundrum, "A Large Mass Hierarchy from a Small Extra Dimension," *Phys.Rev.Lett.* **83** (1999) 3370–3373, arXiv:hep-ph/9905221 [hep-ph].

[731] R. Bean, S. Shandera, H. Tye, and J. Xu, "Comparing Brane Inflation to WMAP," *JCAP* **0705** (2007) 004, arXiv:hep-th/0702107 [hep-th].

[732] M. Spalinski, "On Power Law Inflation in DBI Models," *JCAP* **0705** (2007) 017, arXiv:hep-th/0702196 [hep-th].

[733] M. Spalinski, "A Consistency Relation for Power Law Inflation in DBI Models," *Phys.Lett.* **B650** (2007) 313–316, arXiv:hep-th/0703248 [HEP-TH].

[734] X. Chen, "Multi-Throat Brane Inflation," *Phys.Rev.* **D71** (2005) 063506, arXiv:hep-th/0408084 [hep-th].

[735] X. Chen, "Inflation from Warped Space," *JHEP* **0508** (2005) 045, arXiv:hep-th/0501184 [hep-th].

[736] X. Chen, "Running Non-Gaussianities in DBI inflation," *Phys.Rev.* **D72** (2005) 123518, arXiv:astro-ph/0507053 [astro-ph].

[737] C. de Rham and A. Tolley, "DBI and the Galileon Reunited," *JCAP* **1005** (2010) 015, arXiv:1003.5917 [hep-th].

[738] G. Goon, K. Hinterbichler, and M. Trodden, "Symmetries for Galileons and DBI Scalars on Curved Space," *JCAP* **1107** (2011) 017, arXiv:1103.5745 [hep-th].

[739] S. Gubser, I. Klebanov, and A. Polyakov, "Gauge Theory Correlators from Non-Critical String Theory," *Phys.Lett.* **B428** (1998) 105–114, arXiv:hep-th/9802109 [hep-th].

[740] E. Witten, "Anti-de Sitter space and Holography," *Adv.Theor.Math.Phys.* **2** (1998) 253–291, arXiv:hep-th/9802150 [hep-th].

[741] S. Gandhi, L. McAllister, and S. Sjors, *to appear*.

[742] T. Bachlechner and L. McAllister, "D-brane Bremsstrahlung," arXiv:1306.0003 [hep-th].

[743] R. Bean, X. Chen, H. Peiris, and J. Xu, "Comparing Infrared Dirac-Born-Infeld Brane Inflation to Observations," *Phys.Rev.* **D77** (2008) 023527, arXiv:0710.1812 [hep-th].

[744] E. Sefusatti, M. Liguori, A. Yadav, M. Jackson, and E. Pajer, "Constraining Running Non-Gaussianity," *JCAP* **0912** (2009) 022, arXiv:0906.0232 [astro-ph.CO].

[745] J. Lidsey and I. Huston, "Gravitational Wave Constraints on Dirac-Born-Infeld Inflation," *JCAP* **0707** (2007) 002, arXiv:0705.0240 [hep-th].

[746] J. Ward, "DBI N-flation," *JHEP* **0712** (2007) 045, arXiv:0711.0760 [hep-th].

[747] D. Easson, R. Gregory, D. Mota, G. Tasinato, and I. Zavala, "Spinflation," *JCAP* **0802** (2008) 010, arXiv:0709.2666 [hep-th].

[748] M.-x. Huang, G. Shiu, and B. Underwood, "Multifield DBI Inflation and Non-Gaussianities," *Phys.Rev.* **D77** (2008) 023511, arXiv:0709.3299 [hep-th].

[749] D. Langlois, S. Renaux-Petel, D. Steer, and T. Tanaka, "Primordial Perturbations and Non-Gaussianities in DBI and General Multi-Field Inflation," *Phys.Rev.* **D78** (2008) 063523, arXiv:0806.0336 [hep-th].

[750] D. Langlois, S. Renaux-Petel, D. Steer, and T. Tanaka, "Primordial Fluctuations and Non-Gaussianities in Multi-Field DBI Inflation," *Phys.Rev.Lett.* **101** (2008) 061301, arXiv:0804.3139 [hep-th].

[751] D. Langlois, S. Renaux-Petel, and D. Steer, "Multi-Field DBI Inflation: Introducing Bulk Forms and Revisiting the Gravitational Wave Constraints," *JCAP* **0904** (2009) 021, arXiv:0902.2941 [hep-th].

[752] S. Renaux-Petel, "Combined Local and Equilateral Non-Gaussianities from Multi-Field DBI Inflation," *JCAP* **0910** (2009) 012, arXiv:0907.2476 [hep-th].

[753] D. Langlois and S. Renaux-Petel, "Perturbations in Generalized Multi-Field Inflation," *JCAP* **0804** (2008) 017, arXiv:0801.1085 [hep-th].

[754] R. Gregory and D. Kaviani, "Spinflation with Angular Potentials," *JHEP* **1201** (2012) 037, arXiv:1107.5522 [hep-th].

[755] D. Kaviani, "Spinflation with Backreaction," *Int.J.Mod.Phys.* **D22** (2013) 1350062, arXiv:1212.5831 [hep-th].

[756] L. Leblond and S. Shandera, "Cosmology of the Tachyon in Brane Inflation," *JCAP* **0701** (2007) 009, arXiv:hep-th/0610321 [hep-th].

[757] J. Zhang, Y. Cai, and Y.-S. Piao, "Preheating in a DBI Inflation Model," arXiv:1307.6529 [hep-th].

[758] F. Adams, J. Bond, K. Freese, J. Frieman, and A. Olinto, "Natural Inflation: Particle Physics Models, Power Law Spectra for Large-Scale Structure, and Constraints from COBE," *Phys.Rev.* **D47** (1993) 426–455, arXiv:hep-ph/9207245 [hep-ph].

[759] J. Kim, H. Nilles, and M. Peloso, "Completing Natural Inflation," *JCAP* **0501** (2005) 005, arXiv:hep-ph/0409138 [hep-ph].

[760] T. Grimm, "Axion Inflation in Type II String Theory," *Phys.Rev.* **D77** (2008) 126007, arXiv:0710.3883 [hep-th].

[761] E. Copeland, A. Mazumdar, and N. Nunes, "Generalized Assisted Inflation," *Phys.Rev.* **D60** (1999) 083506, arXiv:astro-ph/9904309 [astro-ph].

[762] A. Jokinen and A. Mazumdar, "Inflation in Large N Limit of Supersymmetric Gauge Theories," *Phys.Lett.* **B597** (2004) 222, arXiv:hep-th/0406074 [hep-th].

[763] A. Ashoorioon, H. Firouzjahi, and M. Sheikh-Jabbari, "M-flation: Inflation from Matrix Valued Scalar Fields," *JCAP* **0906** (2009) 018, arXiv:0903.1481 [hep-th].

[764] A. Ashoorioon, H. Firouzjahi, and M. Sheikh-Jabbari, "Matrix Inflation and the Landscape of its Potential," *JCAP* **1005** (2010) 002, arXiv:0911.4284 [hep-th].

[765] A. Ashoorioon and M. Sheikh-Jabbari, "Gauged M-flation, its UV sensitivity and Spectator Species," *JCAP* **1106** (2011) 014, arXiv:1101.0048 [hep-th].

[766] A. Ashoorioon, U. Danielsson, and M. Sheikh-Jabbari, "$1/N$ Resolution to Inflationary η-Problem," *Phys.Lett.* **B713** (2012) 353–357, arXiv:1112.2272 [hep-th].

[767] R. Kallosh, N. Sivanandam, and M. Soroush, "Axion Inflation and Gravity Waves in String Theory," *Phys.Rev.* **D77** (2008) 043501, arXiv:0710.3429 [hep-th].

[768] M. Cicoli, K. Dutta, and A. Maharana, "N-flation with Hierarchically Light Axions in String Compactifications," arXiv:1401.2579 [hep-th].

[769] S. Kim and A. Liddle, "N-flation: Multi-Field Inflationary Dynamics and Perturbations," *Phys.Rev.* **D74** (2006) 023513, arXiv:astro-ph/0605604 [astro-ph].

[770] Y.-S. Piao, "On Perturbation Spectra of N-flation," *Phys.Rev.* **D74** (2006) 047302, arXiv:gr-qc/0606034 [gr-qc].

[771] S. Kim and A. Liddle, "N-flation: Non-Gaussianity in the Horizon-Crossing Approximation," *Phys.Rev.* **D74** (2006) 063522, arXiv:astro-ph/0608186 [astro-ph].

[772] T. Battefeld and R. Easther, "Non-Gaussianities in Multi-Field Inflation," *JCAP* **0703** (2007) 020, arXiv:astro-ph/0610296 [astro-ph].

[773] S. Kim and A. Liddle, "N-flation: Observable Predictions from the Random Matrix Mass Spectrum," *Phys.Rev.* **D76** (2007) 063515, arXiv:0707.1982 [astro-ph].

[774] D. Battefeld and T. Battefeld, "Non-Gaussianities in N-flation," *JCAP* **0705** (2007) 012, arXiv:hep-th/0703012 [hep-th].

[775] D. Battefeld and S. Kawai, "Preheating after N-flation," *Phys.Rev.* **D77** (2008) 123507, arXiv:0803.0321 [astro-ph].

[776] P. Adshead, R. Easther, and E. Lim, "Cosmology With Many Light Scalar Fields: Stochastic Inflation and Loop Corrections," *Phys.Rev.* **D79** (2009) 063504, arXiv:0809.4008 [hep-th].

[777] S. Kim, A. Liddle, and D. Seery, "Non-Gaussianity in Axion N-flation Models," *Phys.Rev.Lett.* **105** (2010) 181302, arXiv:1005.4410 [astro-ph.CO].

[778] S. Kim, A. Liddle, and D. Seery, "Non-Gaussianity in Axion N-flation Models: Detailed Predictions and Mass Spectra," *Phys.Rev.* **D85** (2012) 023532, arXiv:1108.2944 [astro-ph.CO].

[779] E. Silverstein and A. Westphal, "Monodromy in the CMB: Gravity Waves and String Inflation," *Phys.Rev.* **D78** (2008) 106003, arXiv:0803.3085 [hep-th].

[780] N. Kaloper and L. Sorbo, "A Natural Framework for Chaotic Inflation," *Phys.Rev.Lett.* **102** (2009) 121301, arXiv:0811.1989 [hep-th].

[781] N. Kaloper, A. Lawrence, and L. Sorbo, "An Ignoble Approach to Large-Field Inflation," *JCAP* **1103** (2011) 023, arXiv:1101.0026 [hep-th].

[782] S. Dubovsky, A. Lawrence, and M. Roberts, "Axion Monodromy in a Model of Holographic Gluodynamics," *JHEP* **1202** (2012) 053, arXiv:1105.3740 [hep-th].

[783] E. Palti and T. Weigand, "Towards Large r from $[p, q]$-Inflation," arXiv:1403.7507 [hep-th].

[784] L. McAllister, E. Silverstein, A. Westphal, and T. Wrase, "The Powers of Monodromy," arXiv:1405.3652 [hep-th].

[785] M. Aganagic, C. Beem, J. Seo, and C. Vafa, "Geometrically Induced Metastability and Holography," *Nucl.Phys.* **B789** (2008) 382–412, arXiv:hep-th/0610249 [hep-th].

[786] J. Conlon, "Brane–Antibrane Backreaction in Axion Monodromy Inflation," *JCAP* **1201** (2012) 033, arXiv:1110.6454 [hep-th].

[787] E. Pajer and M. Peloso, "A Review of Axion Inflation in the Era of Planck," arXiv:1305.3557 [hep-th].

[788] N. Barnaby, E. Pajer, and M. Peloso, "Gauge Field Production in Axion Inflation: Consequences for Monodromy, Non-Gaussianity in the CMB, and Gravitational Waves at Interferometers," *Phys.Rev.* **D85** (2012) 023525, arXiv:1110.3327 [astro-ph.CO].

[789] N. Barnaby and M. Peloso, "Large Non-Gaussianity in Axion Inflation," *Phys.Rev.Lett.* **106** (2011) 181301, arXiv:1011.1500 [hep-ph].

[790] A. Linde, S. Mooij, and E. Pajer, "Gauge Field Production in SUGRA Inflation: Local Non-Gaussianity and Primordial Black Holes," arXiv:1212.1693 [hep-th].

[791] S. Behbahani, A. Dymarsky, M. Mirbabayi, and L. Senatore, "(Small) Resonant Non-Gaussianities: Signatures of a Discrete Shift Symmetry in the Effective Field Theory of Inflation," *JCAP* **1212** (2012) 036, arXiv:1111.3373 [hep-th].

[792] X. Chen, R. Easther, and E. Lim, "Generation and Characterization of Large Non-Gaussianities in Single Field Inflation," *JCAP* **0804** (2008) 010, arXiv:0801.3295 [astro-ph].

[793] H. Peiris, R. Easther, and R. Flauger, "Constraining Monodromy Inflation," arXiv:1303.2616 [astro-ph.CO].

[794] R. Flauger, *private communication*.

[795] R. Easther and R. Flauger, "Planck Constraints on Monodromy Inflation," arXiv:1308.3736 [astro-ph.CO].

[796] M. Anber and L. Sorbo, "Naturally Inflating on Steep Potentials through Electromagnetic Dissipation," *Phys.Rev.* **D81** (2010) 043534, arXiv:0908.4089 [hep-th].

[797] P. Meerburg and E. Pajer, "Observational Constraints on Gauge Field Production in Axion Inflation," arXiv:1203.6076 [astro-ph.CO].

[798] L. Sorbo, "Parity Violation in the Cosmic Microwave Background from a Pseudoscalar Inflaton," *JCAP* **1106** (2011) 003, arXiv:1101.1525 [astro-ph.CO].

[799] A. Lue, L.-M. Wang, and M. Kamionkowski, "Cosmological Signature of New Parity Violating Interactions," *Phys.Rev.Lett.* **83** (1999) 1506–1509, arXiv:astro-ph/9812088 [astro-ph].

[800] S. Crowder, R. Namba, V. Mandic, S. Mukohyama, and M. Peloso, "Measurement of Parity Violation in the Early Universe using Gravitational Wave Detectors," arXiv:1212.4165 [astro-ph.CO].

[801] P. Binetruy and M. Gaillard, "Candidates for the Inflaton Field in Superstring Models," *Phys.Rev.* **D34** (1986) 3069–3083.

[802] T. Banks, M. Berkooz, S. Shenker, G. Moore, and P. Steinhardt, "Modular Cosmology," *Phys.Rev.* **D52** (1995) 3548–3562, arXiv:hep-th/9503114 [hep-th].

[803] M. Badziak and M. Olechowski, "Volume Modulus Inflection Point Inflation and the Gravitino Mass Problem," *JCAP* **0902** (2009) 010, arXiv:0810.4251 [hep-th].

[804] A. Linde and A. Westphal, "Accidental Inflation in String Theory," *JCAP* **0803** (2008) 005, arXiv:0712.1610 [hep-th].

[805] B. Greene and A. Weltman, "An Effect of α' Corrections on Racetrack Inflation," *JHEP* **0603** (2006) 035, arXiv:hep-th/0512135 [hep-th].

[806] J. Conlon, R. Kallosh, A. Linde, and F. Quevedo, "Volume Modulus Inflation and the Gravitino Mass Problem," *JCAP* **0809** (2008) 011, arXiv:0806.0809 [hep-th].

[807] W. Buchmuller, K. Hamaguchi, O. Lebedev, and M. Ratz, "Maximal Temperature in Flux Compactifications," *JCAP* **0501** (2005) 004, arXiv:hep-th/0411109 [hep-th].

[808] M. Graesser and M. Salem, "The Scale of Gravity and the Cosmological Constant Within a Landscape," *Phys.Rev.* **D76** (2007) 043506, arXiv:astro-ph/0611694 [astro-ph].

[809] M. Badziak and M. Olechowski, "Volume Modulus Inflation and a Low Scale of SUSY Breaking," *JCAP* **0807** (2008) 021, arXiv:0802.1014 [hep-th].

[810] T. He, S. Kachru, and A. Westphal, "Gravity Waves and the LHC: Towards High-Scale Inflation with Low-Energy SUSY," *JHEP* **1006** (2010) 065, arXiv:1003.4265 [hep-th].

[811] A. Linde, Y. Mambrini, and K. Olive, "Supersymmetry Breaking due to Moduli Stabilization in String Theory," *Phys.Rev.* **D85** (2012) 066005, arXiv:1111.1465 [hep-th].

[812] M. Cicoli, J. Conlon, and F. Quevedo, "Systematics of String Loop Corrections in Type IIB Calabi-Yau Flux Compactifications," *JHEP* **0801** (2008) 052, arXiv:0708.1873 [hep-th].

[813] R. Blumenhagen, X. Gao, T. Rahn, and P. Shukla, "Moduli Stabilization and Inflationary Cosmology with Poly-Instantons in Type IIB Orientifolds," *JHEP* **1211** (2012) 101, arXiv:1208.1160 [hep-th].

[814] J. Bond, L. Kofman, S. Prokushkin, and P. Vaudrevange, "Roulette Inflation with Kähler Moduli and their Axions," *Phys.Rev.* **D75** (2007) 123511, arXiv:hep-th/0612197 [hep-th].

[815] J. Blanco-Pillado, D. Buck, E. Copeland, M. Gomez-Reino, and N. Nunes, "Kähler Moduli Inflation Revisited," *JHEP* **1001** (2010) 081, arXiv:0906.3711 [hep-th].

[816] P. Candelas, X. De La Ossa, A. Font, S. Katz, and D. Morrison, "Mirror Symmetry for Two-Parameter Models. 1.," *Nucl.Phys.* **B416** (1994) 481–538, arXiv:hep-th/9308083 [hep-th].

[817] R. Blumenhagen and M. Schmidt-Sommerfeld, "Power Towers of String Instantons for $N = 1$ Vacua," *JHEP* **0807** (2008) 027, arXiv:0803.1562 [hep-th].

[818] I. Garcia-Etxebarria and A. Uranga, "Nonperturbative Superpotentials Across Lines of Marginal Stability," *JHEP* **0801** (2008) 033, arXiv:0711.1430 [hep-th].

[819] R. Blumenhagen, X. Gao, T. Rahn, and P. Shukla, "A Note on Poly-Instanton Effects in Type IIB Orientifolds on Calabi-Yau Threefolds," *JHEP* **1206** (2012) 162, arXiv:1205.2485 [hep-th].

[820] D. Lüst and X. Zhang, "Four Kähler Moduli Stabilisation in Type IIB Orientifolds with K3-fibred Calabi-Yau Threefold Compactification," *JHEP* **1305** (2013) 051, arXiv:1301.7280 [hep-th].

[821] M. Cicoli, S. Downes, and B. Dutta, "Power Suppression at Large Scales in String Inflation," arXiv:1309.3412 [hep-th].

[822] F. Pedro and A. Westphal, "Low-l CMB Power Loss in String Inflation," arXiv:1309.3413 [hep-th].

[823] D. Lopez Nacir, R. Porto, L. Senatore, and M. Zaldarriaga, "Dissipative Effects in the Effective Field Theory of Inflation," *JHEP* **1201** (2012) 075, arXiv:1109.4192 [hep-th].

[824] A. Berera, "Warm Inflation," *Phys.Rev.Lett.* **75** (1995) 3218–3221, arXiv:astro-ph/9509049 [astro-ph].

[825] A. Berera and T. Kephart, "The Ubiquitous Inflaton in String-Inspired Models," *Phys.Rev.Lett.* **83** (1999) 1084–1087, arXiv:hep-ph/9904410 [hep-ph].

[826] A. Berera, I. Moss, and R. Ramos, "Warm Inflation and its Microphysical Basis," *Reports on Progress in Physics* **72** no. 2, (2009) 026901.

[827] D. Battefeld, T. Battefeld, C. Byrnes, and D. Langlois, "Beauty is Distractive: Particle Production during Multifield Inflation," *JCAP* **1108** (2011) 025, arXiv:1106.1891 [astro-ph.CO].

[828] P. Adshead and M. Wyman, "Chromo-Natural Inflation: Natural Inflation on a Steep Potential with Classical Non-Abelian Gauge Fields," *Phys.Rev.Lett.* **108** (2012) 261302, arXiv:1202.2366 [hep-th].

[829] L. Kofman, A. Linde, and A. Starobinsky, "Towards the Theory of Reheating after Inflation," *Phys.Rev.* **D56** (1997) 3258–3295, arXiv:hep-ph/9704452 [hep-ph].

[830] N. Barnaby, Z. Huang, L. Kofman, and D. Pogosyan, "Cosmological Fluctuations from Infra-Red Cascading During Inflation," *Phys.Rev.* **D80** (2009) 043501, arXiv:0902.0615 [hep-th].

[831] L. Kofman, *unpublished.*

[832] A. Brown, "Boom and Bust Inflation: a Graceful Exit via Compact Extra Dimensions," *Phys.Rev.Lett.* **101** (2008) 221302, arXiv:0807.0457 [hep-th].

[833] R. Easther, J. Giblin, L. Hui, and E. Lim, "A New Mechanism for Bubble Nucleation: Classical Transitions," *Phys.Rev.* **D80** (2009) 123519, arXiv:0907.3234 [hep-th].

[834] M. Kleban, K. Krishnaiyengar, and M. Porrati, "Flux Discharge Cascades in Various Dimensions," *JHEP* **1111** (2011) 096, arXiv:1108.6102 [hep-th].

[835] B. Shlaer, "Chaotic Brane Inflation," arXiv:1211.4024 [hep-th].

[836] J. Brown and C. Teitelboim, "Neutralization of the Cosmological Constant by Membrane Creation," *Nucl.Phys.* **B297** (1988) 787–836.

[837] K. Freese and D. Spolyar, "Chain Inflation: 'Bubble Bubble Toil and Trouble'," *JCAP* **0507** (2005) 007, arXiv:hep-ph/0412145 [hep-ph].

[838] B. Feldstein and B. Tweedie, "Density Perturbations in Chain Inflation," *JCAP* **0704** (2007) 020, arXiv:hep-ph/0611286 [hep-ph].

[839] Q.-G. Huang, "Simplified Chain Inflation," *JCAP* **0705** (2007) 009, arXiv:0704.2835 [hep-th].

[840] D. Chialva and U. Danielsson, "Chain Inflation and the Imprint of Fundamental Physics in the CMBR," *JCAP* **0903** (2009) 007, arXiv:0809.2707 [hep-th].

[841] A. Ashoorioon, K. Freese, and J. Liu, "Slow Nucleation Rates in Chain Inflation with QCD Axions or Monodromy," *Phys.Rev.* **D79** (2009) 067302, arXiv:0810.0228 [hep-ph].

[842] A. Ashoorioon and K. Freese, "Gravity Waves from Chain Inflation," arXiv:0811.2401 [hep-th].

[843] J. Cline, G. Moore, and Y. Wang, "Chain Inflation Reconsidered," *JCAP* **1108** (2011) 032, arXiv:1106.2188 [hep-th].

[844] L. McAllister and I. Mitra, "Relativistic D-brane Scattering is Extremely Inelastic," *JHEP* **0502** (2005) 019, arXiv:hep-th/0408085 [hep-th].

[845] C. Bachas, "D-brane Dynamics," *Phys.Lett.* **B374** (1996) 37–42, arXiv:hep-th/9511043 [hep-th].

[846] A. Maleknejad and M. Sheikh-Jabbari, "Non-Abelian Gauge Field Inflation," *Phys.Rev.* **D84** (2011) 043515, arXiv:1102.1932 [hep-ph].

[847] A. Maleknejad and M. Sheikh-Jabbari, "Gauge-flation: Inflation From Non-Abelian Gauge Fields," *Phys.Lett.* **B723** (2013) 224–228, arXiv:1102.1513 [hep-ph].

[848] M. Sheikh-Jabbari, "Gauge-flation vs Chromo-Natural Inflation," *Phys.Lett.* **B717** (2012) 6–9, arXiv:1203.2265 [hep-th].

[849] A. Maleknejad, M. Sheikh-Jabbari, and J. Soda, "Gauge Fields and Inflation," arXiv:1212.2921 [hep-th].

[850] D. Lopez Nacir, R. Porto, and M. Zaldarriaga, "The Consistency Condition for the Three-Point Function in Dissipative Single-Clock Inflation," *JCAP* **1209** (2012) 004, arXiv:1206.7083 [hep-th].

[851] P. Adshead, E. Martinec, and M. Wyman, "Perturbations in Chromo-Natural Inflation," arXiv:1305.2930 [hep-th].

[852] S. Weinberg, *The First Three Minutes. A Modern View of the Origin of the Universe.* Basic Books, 1977.

[853] A. Vilenkin, "The Birth of Inflationary Universes," *Phys.Rev.* **D27** (1983) 2848.

[854] A. Linde, "Eternally Existing Selfreproducing Chaotic Inflationary Universe," *Phys.Lett.* **B175** (1986) 395–400.

[855] A. Linde, D. Linde, and A. Mezhlumian, "From the Big Bang Theory to the Theory of a Stationary Universe," *Phys.Rev.* **D49** (1994) 1783–1826, arXiv:gr-qc/9306035 [gr-qc].

[856] A. Linde and A. Mezhlumian, "Stationary Universe," *Phys.Lett.* **B307** (1993) 25–33, arXiv:gr-qc/9304015 [gr-qc].

[857] A. Vilenkin, "Predictions from Quantum Cosmology," *Phys.Rev.Lett.* **74** (1995) 846–849, arXiv:gr-qc/9406010 [gr-qc].

[858] J. Garriga, D. Schwartz-Perlov, A. Vilenkin, and S. Winitzki, "Probabilities in the Inflationary Multiverse," *JCAP* **0601** (2006) 017, arXiv:hep-th/0509184 [hep-th].

[859] J. Garriga and A. Vilenkin, "Holographic Multiverse," *JCAP* **0901** (2009) 021, arXiv:0809.4257 [hep-th].

[860] R. Bousso, "Holographic Probabilities in Eternal Inflation," *Phys.Rev.Lett.* **97** (2006) 191302, arXiv:hep-th/0605263 [hep-th].

[861] R. Bousso, B. Freivogel, and I.-S. Yang, "Eternal Inflation: The Inside Story," *Phys.Rev.* **D74** (2006) 103516, arXiv:hep-th/0606114 [hep-th].

[862] R. Bousso, B. Freivogel, and I.-S. Yang, "Properties of the Scale Factor Measure," *Phys.Rev.* **D79** (2009) 063513, arXiv:0808.3770 [hep-th].

[863] A. De Simone *et al.*, "Boltzmann Brains and the Scale-Factor Cutoff Measure of the Multiverse," *Phys.Rev.* **D82** (2010) 063520, arXiv:0808.3778 [hep-th].

[864] G. Gibbons and N. Turok, "The Measure Problem in Cosmology," *Phys.Rev.* **D77** (2008) 063516, arXiv:hep-th/0609095 [hep-th].

[865] A. Borde, A. Guth, and A. Vilenkin, "Inflationary Spacetimes are Incomplete in Past Directions," *Phys.Rev.Lett.* **90** (2003) 151301, arXiv:gr-qc/0110012 [gr-qc].

[866] G. Bredon, *Topology and Geometry.* Graduate Texts in Mathematics. Springer, 1993.

[867] M. Nakahara, *Geometry, Topology and Physics.* Taylor & Francis, 1990.

[868] K. Hori, S. Katz, A. Klemm, R. Pandharipande, R. Thomas *et al.*, *Mirror Symmetry.* Clay Mathematics Monographs, 2003.

[869] D. Joyce, *Compact Manifolds with Special Holonomy.* Oxford University Press, 2000.

[870] P. Candelas, "Lectures on Complex Manifolds." Unpublished.

[871] A. Moroianu, *Lectures on Kähler Geometry.* Cambridge University Press, 2007.

[872] S.-T. Yau, "On the Ricci Curvature of a Compact Kähler Manifold and the Complex Monge-Ampére Equation, I," *Communications on Pure and Applied Mathematics* **31** no. 3, (1978) 339–411.

[873] F. Piazza and F. Vernizzi, "Effective Field Theory of Cosmological Perturbations," arXiv:1307.4350 [hep-th].

[874] L. Senatore, "TASI Lectures on Inflation." Unpublished.

[875] P. Creminelli, "Les Houches Lectures on Non-Gaussianities." Unpublished.

[876] J. Goldstone, A. Salam, and S. Weinberg, "Broken Symmetries," *Phys.Rev.* **127** (1962) 965–970.

[877] J. Donoghue, E. Golowich, and B. R. Holstein, "Dynamics of the Standard Model," *Camb.Monogr.Part.Phys.Nucl.Phys.Cosmol.* **2** (1992) 1–540.

[878] J. Cornwall, D. Levin, and G. Tiktopoulos, "Derivation of Gauge Invariance from High-Energy Unitarity Bounds on the S-Matrix," *Phys.Rev.* **D10** (1974) 1145.

[879] I. Low and A. Manohar, "Spontaneously Broken Spacetime Symmetries and Goldstone's Theorem," *Phys.Rev.Lett.* **88** (2002) 101602, arXiv:hep-th/0110285 [hep-th].

[880] V. Assassi, D. Baumann, and D. Green, "Symmetries and Loops in Inflation," *JHEP* **1302** (2013) 151, arXiv:1210.7792 [hep-th].

[881] R. Wald, *General Relativity.* University of Chicago Press, 2010.

[882] C. Cheung, A. L. Fitzpatrick, J. Kaplan, and L. Senatore, "On the Consistency Relation of the Three-Point Function in Single-Field Inflation," *JCAP* **0802** (2008) 021, arXiv:0709.0295 [hep-th].

[883] D. Baumann, D. Green, and R. Porto, "B-modes and the Nature of Inflation," arXiv:1407.2621 [hep-th].

[884] L. Senatore and M. Zaldarriaga, "A Naturally Large Four-Point Function in Single-Field Inflation," *JCAP* **1101** (2011) 003, arXiv:1004.1201 [hep-th].

[885] N. Arkani-Hamed, P. Creminelli, S. Mukohyama, and M. Zaldarriaga, "Ghost Inflation," *JCAP* **0404** (2004) 001, arXiv:hep-th/0312100 [hep-th].

[886] N. Bartolo, M. Fasiello, S. Matarrese, and A. Riotto, "Large Non-Gaussianities in the Effective Field Theory Approach to Single-Field Inflation: the Bispectrum," *JCAP* **1008** (2010) 008, arXiv:1004.0893 [astro-ph.CO].

[887] P. Creminelli, G. D'Amico, M. Musso, J. Norena, and E. Trincherini, "Galilean Symmetry in the Effective Theory of Inflation: New Shapes of Non-Gaussianity," *JCAP* **1102** (2011) 006, arXiv:1011.3004 [hep-th].

[888] L. Senatore and M. Zaldarriaga, "The Effective Field Theory of Multifield Inflation," *JHEP* **1204** (2012) 024, arXiv:1009.2093 [hep-th].

[889] R. Arnowitt, S. Deser, and C. Misner, "The Dynamics of General Relativity," arXiv:gr-qc/0405109 [gr-qc].

[890] J. Schwinger, "Brownian Motion of a Quantum Oscillator," *J.Math.Phys.* **2** (1961) 407–432.

[891] E. Calzetta and B. Hu, "Closed Time Path Functional Formalism in Curved Space-Time: Application to Cosmological Back Reaction Problems," *Phys.Rev.* **D35** (1987) 495.

[892] R. Jordan, "Effective Field Equations for Expectation Values," *Phys.Rev.* **D33** (1986) 444–454.

[893] S. Weinberg, "Quantum Contributions to Cosmological Correlations," *Phys.Rev.* **D72** (2005) 043514, arXiv:hep-th/0506236 [hep-th].

[894] D. Seery, "One-Loop Corrections to a Scalar Field during Inflation," *JCAP* **0711** (2007) 025, arXiv:0707.3377 [astro-ph].

[895] P. Adshead, R. Easther, and E. Lim, "The 'in-in' Formalism and Cosmological Perturbations," *Phys.Rev.* **D80** (2009) 083521, arXiv:0904.4207 [hep-th].

[896] S. Groot Nibbelink and B. van Tent, "Density Perturbations Arising From Multiple Field Slow-Roll Inflation," arXiv:hep-ph/0011325 [hep-ph].

[897] S. Groot Nibbelink and B. van Tent, "Scalar Perturbations During Multiple Field Slow-Roll Inflation," *Class.Quant.Grav.* **19** (2002) 613–640, arXiv:hep-ph/0107272 [hep-ph].

[898] D. Langlois, "Lectures on Inflation and Cosmological Perturbations," *Lect.Notes Phys.* **800** (2010) 1–57, arXiv:1001.5259 [astro-ph.CO].

[899] B. van Tent, "Cosmological Inflation with Multiple Fields and the Theory of Density Fluctuations," PhD thesis, Utrecht University, 2002.

[900] R. Easther and J. Giblin, "The Hubble Slow Roll Expansion for Multi Field Inflation," *Phys.Rev.* **D72** (2005) 103505, arXiv:astro-ph/0505033 [astro-ph].

[901] M. Sasaki and E. Stewart, "A General Analytic Formula for the Spectral Index of the Density Perturbations Produced During Inflation," *Prog.Theor.Phys.* **95** (1996) 71–78, arXiv:astro-ph/9507001 [astro-ph].

[902] C. Gordon, D. Wands, B. Bassett, and R. Maartens, "Adiabatic and Entropy Perturbations from Inflation," *Phys.Rev.* **D63** (2001) 023506, arXiv:astro-ph/0009131 [astro-ph].

[903] D. Wands, N. Bartolo, S. Matarrese, and A. Riotto, "An Observational Test of Two-Field Inflation," *Phys.Rev.* **D66** (2002) 043520, arXiv:astro-ph/0205253 [astro-ph].

[904] N. Bartolo, S. Matarrese, and A. Riotto, "Adiabatic and Isocurvature Perturbations from Inflation: Power Spectra and Consistency Relations," *Phys.Rev.* **D64** (2001) 123504, arXiv:astro-ph/0107502 [astro-ph].

Index

Printed in the United States
By Bookmasters